Fractional Programming

Mathematics and Its Applications

Volume 409

Fractional Programming

Theory, Methods and Applications

by

I.M. Stancu-Minasian

Centre for Mathematical Statistics,
The Romanian Academy,
Bucharest, Romania

KLUWER ACADEMIC PUBLISHERS

DORDRECHT / BOSTON / LONDON

A C.I.P. Catalogue record for this book is available from the Library of Congress.

ISBN-13: 978-94-010-6504-7 e-ISBN-13: 978-94-009-0035-6

DOI: 10:1007/ 978-94-009-0035-6

Published by Kluwer Academic Publishers,
P.O. Box, 3300 AA Dordrecht, The Netherlands.

Sold and distributed in the U.S.A. and Canada
by Kluwer Academic Publishers,
101 Philip Drive, Norwell, MA 02061, U.S.A.

In all other countries, sold and distributed
by Kluwer Academic Publishers,
P.O. Box 322, 3300 AH Dordrecht, The Netherlands.

This is a complete revised and updated translation of the original work
Metode de Rezolvare a Problemelor de Programare Fracţionară
Editura Academiei Române, Bucharest ©1992.
Translated from the Romanian by Victor Giurgiutiu.

TABLE OF CONTENTS

INTRODUCTION

Mathematical programming has know a spectacular diversification in the last few decades. This process has happened both at the level of mathematical research and at the level of the applications generated by the solution methods that were created. To write a monograph dedicated to a certain domain of mathematical programming is, under such circumstances,especially difficult. In the present monograph we opt for the domain of fractional programming.

Interest of this subject was generated by the fact that various optimization problems from engineering and economics consider the minimization of a ratio between physical and/or economical functions, for example cost/time, cost/volume,cost/profit, or other quantities that measure the efficiency of a system. For example, the productivity of industrial systems, defined as the ratio between the realized services in a system within a given period of time and the utilized resources, is used as one of the best indicators of the quality of their operation. Such problems, where the objective function appears as a ratio of functions, constitute *fractional programming* problem.

Due to its importance in modeling various decision processes in management science, operational research, and economics, and also due to its frequent appearance in other problems that are not necessarily economical, such as information theory, numerical analysis, stochastic programming, decomposition algorithms for large linear systems, etc., the fractional programming method has received particular attention in the last three decades.

Starting from the optimization problem with linear constraints and with the objective function ratio of linear affine functions, this domain has known constant enrichment as new mathematical programming problems are being posed and solved: quadratic programming, nonlinear programming, integer programming, multi-objective programming, complex programming, inexact programming with set inclusive-constraints, etc. Actually, all the areas of mathematical programming have been reflected, in one way or another, in the posing and resolving of fractional programming problems.

In this book, we intend to present the state of the art in fractional programming. Using a vast bibliography, we set forth to present in a systematic way, the theory and the applications of fractional programming. The book has ten chapters. Each chapter ends with a review of historical dates and some commentaries. We are aware of the fact that, due to the restricted length of the book, we have not been able to cover several aspects such as: the extreme point fractional programming problem, the generalized fractional programming problem, the sensitivity analysis and parametric programming. Besides, the problems that are presented in the book could have been described more extensively. We tried to compensate for these shortcomings through the historical notes at the end of the chapters.

Chapter 1, "Fractional programming applications", sets forth to present several opportunities for using fractional programming in the modeling of economic processes. Applications to blending problems, to optimum cutting of raw materials, to information theory, to minimum-risk problems in stochastic programming, to macro-economic planning, to production efficiency, to mathematical taxonomy, to optimal graph paths, to repartition problems, to game theory, etc. Several applications to other areas, such as Physics (Optics) as well as to mathematical programming are also presented.

The importance of convex functions in the general mathematical programming is very clear. But the functions intervening in fractional programming are neither convex nor concave. To study functions that appear as a ratio of functions, the class of convex functions had to be extended. Thus quasiconvex, pseudoconvex, logarithmic-convex, αm-convex, invex functions, etc. had to be defined. Chapter 2, "Convex, quasiconvex, pseudoconvex, logarithmic-convex, αm-convex, and invex functions", discusses certain properties of such functions. A special emphasis is placed on the extremes of the generalized convex functions. Since, in fractional programming, certain types of functions were especially studied, Section 2.5 of this chapter discussed the convexity of such classes of functions.

Chapter 3, "Methods of solving linear fractional programming problems", is dedicated to the presentation of solution methods for linear fractional programming problems. The Primal Simplex algorithm is presented in Section 3.1. In Section 3.2, we develop the necessary and sufficient optimality conditions using the classical Lagrange method of the differential calculus. The optimality conditions for linear fractional programming problems with nonlinear and differentiable constraints are given in Section 3.3. Several solution methods are given in the subsequent sections: the change of variable method (Section 3.4), approximate and iterative methods (Section 3.5), parametric methods (Section 3.6), the convex simplex method (Section 3.7), and the multiplex method (Section 3.8). Section 3.9 presents a solution method for the complementary programming problems, *i.e.* those problems in which, besides classical constraints, one also has restrictions requiring that the product of two variables is zero. In Section 3.10, we study the linear fractional programming problem in which, to the usual constraints given as equalities or inequalities, one also adds supplementary constraints.

Chapter 4, "Nonlinear fractional programming", gives a number of methods for the solution of the nonlinear fractional programming problems in which the objective function appears as the ratio of nonlinear functions. Section 4.1 gives the optimality conditions. Section 4.2 presents the change of variable method, which is subsequently applied to homogeneous programming (Section 4.3). The linearization method is presented in Section 4.4, and the parametric method in Section 4.5. The fractional optimal control problem is presented in Section 4.6.

Chapter 5, "Duality in fractional programming", presents several aspects of duality in fractional programming. It is shown that, for fractional programming problems, the dual constructed according to Wolfe definition is not of practical interest. A natural way of constructing the dual is to transform the fractional problem into a non-fractional problem and to construct the dual of the latter in the classical sense. This is illustrated in Section 5.1 for linear fractional programming, and in Section 5.3 for the nonlinear fractional programming. Section 5.2 presents the dual for a particular problem of linear fractional programming which has the property that the objective function is convex, whereas Section 5.4 presents duality results for nonlinear and non-differentiable fractional programming problems. Dual problems for linear and nonlinear fractional programming *via* the duality principle in geometric programming are presented in Section 5.5.

Chapter 6, "Fractional programming with multiple objective functions", treat the case in which several objective functions appear simultaneously in the problem. As it is well known, when several objective functions are present, the optimal solution for a function is not necessarily optimal solution for the other functions. The notion of "best compromise" solutions is introduced (Section 6.1). Among them, the most general is the efficient solution. Section 6.2 gives the Kuhn-Tucker conditions for the linear fractional problem with multiple objective functions. Section 6.3 shows that, for the linear fractional problem with multiple objective functions, any efficient solution is properly efficient, and that the set of weak-efficient solutions is closed. Section 6.4 presents several method for determining efficient solutions for linear fractional programming problem with multiple objective functions, methods that are based on the linearization of the objective functions, or on the association of a non-fractional problem to the initial problem. Section 6.5 gives the optimality conditions for the linear fractional problem with multiple objective functions, and Section 6.6 gives methods for determining efficient solutions for nonlinear fractional vector programming problems.

Chapter 7, "Fractional programming in the complex space", uses the change of variable method to solve the linear fractional programming problem in the complex space (Section 7.2), for solving the nonlinear fractional programming problem in the complex space with nonlinear constraints (Section 7.4) and for the solution of the nonlinear and non-differentiable fractional programming problem in the complex space (Section 7.5). Section 7.3 considers the linear fractional programming problem in the complex space with nonlinear constraints, and gives a necessary and sufficient existence condition for the optimum. The duality for the linear fractional programming problem in the complex space is given in Section 7.6. The characterization of efficiency for multi-objective linear fractional programming problem in the complex space is given in Section 7.7. Using a change of variable, the problem is reduced to a multiple objective linear programming problem in the complex space. A theorem of the alternative for complex non-homogeneous linear inequality is also given.

Chapter 8, "Special linear fractional programming problems", presents certain special problems. Section 8.1 considers a linear fractional programming problem in which the variables in the objective functions appear in absolute-value, whereas the constraints do not have the non-negativity condition. It is shown that, under certain conditions, it is possible to apply the Simplex algorithm used in the solution conventional linear fractional programming problems. Section 8.2 considers the solution to the separable linear fractional programming problem, which is reduced to solution of a sequence of quadratic problems. Section 8.3 studies the linear fractional programming problem with disjunctive constraints, presents a solution method and constructs the dual problem. In Section 8.4, several approaches to fractional stochastic programming with one objective function are presented. In Section 8.5, the vectorial minimum-risk problem with linear and linear fractional objective function is considered. These problems are shown to be equivalent with deterministic linear fractional programming problems with multiple objective functions. In Section 8.6, Kataoka's problem in the case of multiple criteria is considered. It is shows that the problem is equivalent to some deterministic multi-objective nonlinear programming problem with fractional objective functions and further, it is equivalent to another deterministic multi-objective nonlinear programming problem with nonfractional objective functions. In Section 8.7, goal programming problems with random objective functions or target values are considered. Probabilistic target values are replaced by their deterministic equivalents and thus the initial problem is reduced to a linear fractional min-max problem with linear constraints. If the objective functions are random normal then the problem is equivalent to a deterministic linear fractional min-max problems with nonlinear constraints. The simply randomized objective functions is also considered. In Section 8.8, the stochastic linear and bilinear fractional max-min problem is considered. The minimum-risk approach is used in order to solve these problems. By an appropriate transformation of the variables, the linear (bilinear) - fractional stochastic max-min problem is transformed into the linear stochastic max-min problem and deterministic min-max bilinear problem. An algorithm is presented for solving the deterministic bilinear fractional problem. Kataoka's problem for bilinear fractional max-min problem is considered.Section 8.9 considers the linear fractional programming problem with inexact data and with set-inclusive constraints. Inexact multi-objective programs, goal programming with inexact data, and generalized pseudo-fractional max-min problems are presented. In Section 8.10, the multi-objective fuzzy linear fractional programming problem is analyzed, and an interactive solution method is presented.

Chapter 9, "Integer and mixed integer linear fractional programming", gives methods for the solution of the linear fractional programming problem when some (or all) of the problem variables are either 0-1, or integers. Section 9.1 studies the bivalent programming (with or without constraints). In the case with constraints, references are also made to the parametric solution method. Several solution methods are presented for the integer and mixed integer linear fractional programming problem, such as the cutting planes method

(Section 9.2) and the branch-and-bound method (Section 9.3).

Chapter 10, "The fractional transportation problem", studies transport problems having a linear fractional objective function. Section 10.1 presents the method of double potentials for the solution of the fractional transportation problem. Section 10.2 considers the transportation problem with limited capacities, and solves it by transformation to a three-dimensional transportation problem (without the capacity constraint on the transported quantity). Section 10.3 considers a special case of the fractional transportation problem and establishes a condition for local minimum. Section 10.4 studies the duality problem.

The book contains some original results developed by the author in the area of fractional programming and a vast bibliography on fractional programming. The original results are the outcome of intensive research and have been previously published in national and international journals. The backbone of the bibliography was formed by the exhaustive bibliographies on fractional programming, previously published by the author, of which three in the Indian journal of international circulation "Pure and Applied Mathematika Sciences", and one in the journal "Optimization". This extensive bibliography will be an useful guide to the readers working in fractional programming problems, and/or wishing to obtain more detailed information on fractional programming. The book will be useful to mathematicians, engineers, economists, students, and to everybody engaged in the application of optimization theory to various areas of social and economic activity.

Acknowledgments

I wish to address my warmest thanks to Academician Marius Iosifescu, Head of the Center for Mathematical Statistics of the Romanian Academy, for his sincere encouragement and full support from the beginning to the end of this work. I thank my collaborators Prof. D. I. Duca (Cluj-Napoca University, Romania), Dr. V. N. Patkar (B.M.R.D.A., Bombay, India), and Prof. St. Tigan (University of Medicine and Pharmacy, Cluj-Napoca, Romania) for accepting to have the results of our common research included in this book. I thank to Professor Victor Giurgiutiu (Virginia Polytechnic Institute and State University, USA) for this collaboration and for the fact that, in spite of geographic distance between us, has managed to bring this common endeavor to a successful conclusion. I am deeply grateful to Kluwer Academic Publishers for accepting to publish this book.

I thank my wife, Mihaela, and my daughter, Andreea-Madalina, for their patience and support. The book is dedicated to them.

I. M. Stancu-Minasian

Bucharest, September 1996

CHAPTER 1

FRACTIONAL PROGRAMMING APPLICATIONS

The applications of linear programming to various branches of human activity, and especially to economics, are well known. The applications of fractional programming are less known, and, until now, less numerous. Of course, the linearity of a problem makes it easier to tackle, and hence contributes to its wide recognition. However, not all real-life economic problems can be described by linear models, and hence are not likely applications of linear programming. Fractional programming is a nonlinear programming method that has known increasing exposure recently. Its importance in solving concrete problems is steadily increasing, and its nonlinear optimization models have been found to describe practical problems much better than the linear optimization models.

The present chapter will illustrate how fractional programming can be used in the modeling of economic processes. Besides the economic applications, that make the bulk of our examples, we will also present some applications in other domains, such as Physics (Optics), or even those that appear in the solution of some mathematical programming problems. Isbell and Marlow [504] have identifies fractional programming problems in the strategic problem of fire power distribution on enemy targets.

The fractional programming problems are particularly useful in the solution of economic problems in which various activities use certain resources in various proportions, while the objective is to optimize a certain indicator, usually the most favorable return-on-allocation ratio subject to the constraint imposed on the availability of goods. Examples of such situations are financial and corporate planning (debt/equity ratio), production planning (inventory/sales, output/employee), health care and hospital planning (cost/patient, nurse/patient ratio) etc. Charnes, Cooper and Rhodes [291] introduced a fractional programming method to measure the efficiency of a decision-making unit. Patkar [713] proposed fractional programming models for sharing the urban development responsibilities. Fractional resource allocation with S-shaped return functions were presented by Mjelde [665] (see also [659], [661], [662], [663], [667], [668]). A number of examples are detailed next following Stancu-Minasian [903].

1.1. BLENDING PROBLEMS

Munteanu and Radó [686] considered the problem of the determination, through fractional programming, of the batches for cast iron furnaces in order to obtain the liquid cast iron of a required composition with a minimal expenditure, while using some scrap iron at no cost. This work, that appeared in 1960, is among the first works in fractional programming, although the specialized literature seems to accept the idea that the first work in fractional programming was that of Charnes and Cooper [281].

Assume that there are n raw materials $M_1, \ldots, M_p$ (unrefined cast iron of normal quality), $M_{p+1}, \ldots, M_q$ (unrefined cast iron of special quality), $M_{q+1}, \ldots, M_{n-2}$ (scrap cast iron), M_{n-1} (scrap steel), M_n (own scrap) to obtain liquid cast iron of a certain composition. The data are presented in Table 1.1; percentage limits (p_i, p_i') were proposed for each element.

Table 1.1

Raw material	Contents of 1 kg raw material				Value of 1 kg of raw material
	Si (kg)	Mn (kg)	P (kg)	S (kg)	
$M_j (j = 1, \ldots, n)$	a_{1j}	a_{2j}	a_{3j}	a_{4j}	c_j
	Contents of 1 kg of liquid cast iron				
Liquid cast iron, %	p_1, p_1'	p_2, p_2'	p_3, p_3'	p_4, p_4'	

Due to a number of factors (the furnace and fuel types, melting time, air flow, melting temperature, etc.), the Si, Mn, P, S content of the batch was different from that of the liquid cast iron. Thus, if $b_1, \ldots, b_4$ and $b_1', \ldots, b_4'$, respectively, are the quantities of Si, Mn, P, S contained in a kilogram of the batch load, the following empiric formulae will apply:

$$p_i = \alpha_i + \beta_i b_i, \quad p_i' = \alpha_i + \beta_i b_i', \quad i = 1,2,3,4,$$

with constant α_i and β_i.

Let x_j be the quantity of raw material M_j in 1 kg of liquid cast iron. The following fractional programming model arises:

$$\min_{x \in D} F = \frac{\sum_{j=1}^{n-1} c_j x_j}{\sum_{j=1}^{n-1} x_j} = \frac{\sum_{j=1}^{n-1} c_j x_j}{1 - x_n},$$

where

$$D = \{x \mid \sum_{j=1}^{n} x_j = 1,\ b_i \le \sum_{j=1}^{n} a_{ij} x_i \le b_i',\ i = 1,2,3,4,\ x_j \ge 0,\ j = 1, \ldots, n\}.$$

Other constraints can also appear. For example, it may be required to introduce at least $p\%$ special cast iron in order to obtain a sufficiently fine granular structure $\sum_{j=p+1}^{q} x_j \ge p/100$, or to introduce old cast iron (or own rejects, respectively) at most $p'\%$ ($p''\%$, respectively) $\sum_{j=q+1}^{n-2} x_j \le p'/100,\ x_n \le p''/100$.

Note: Nakayama [688] presents a similar problem, namely material blending in cement production of an industrial company in Japan. Three linear fractional criteria are considered.

1.2. OPTIMAL CUTTING STOCK PROBLEMS

Cutting of materials is of great importance in the furniture manufacture, electric transformers production, civil engineering, ship building, consumer goods industry, etc. We refer to the cutting stock problem for linear materials (steel rods, pipes, timber, paper rolls, etc.). Consider a linear mass-production that comes in standard length and must be cut, according to a production order, in various lengths, such that one gets at least a quantity n_i of each length l_i ($i = 1, \ldots, m$). Obviously, the problem has solution if $l_i \le l$ ($i=1, \ldots, m$), where l is the length of the bars.

The problem is to realize a given order with the minimum number of cut bars (which is equivalent with the minimization of the total quantity of scrap material). For a given bar of length l, there are more than one possible ways to cut bars of length l_i. We call a cutting pattern (or a cutting solution) a particular way of cutting the bar, for example a_1 bars of length l_1, a_2 bars of length l_2, etc. Assume that, through some procedure, all the n distinct cutting solutions have been found. Denote by x_j the number of times the pattern j have been used, and denote by a_{ij} the number of bars of length l_i that are cut according to the pattern j. The cutting stock problem is formulated as:

$$\min \{ F = \sum_{j=1}^{n} x_j \mid \sum_{j=1}^{n} a_{ij} x_j \geq n_i, \quad i=1, ..., m, \quad x_j \geq 0, \quad x_j \text{ integer} \}$$

The problem is particularly difficult since, in real cases, n is very large, and since x_j must be integer. Gilmore and Gomory [409] showed that, in practical cases, a variation domain for the demand n_i is accepted without asking that x_i are integers and, in this case, the objective is to minimize the percentage of scrap material. The scrap resulting from using the pattern j is

$$w_j = l - \sum_{i=1}^{m} l_i a_{ij} .$$

Utilizing the pattern j for x_j times, one gets a total percentage of scrap equal to $100 \, (\sum_{j=1}^{n} w_j x_j \, / \, l \sum_{j=1}^{n} x_j)$. Thus, the cutting stock problem can be stated as a fractional programming problem

$$\min F = \sum_{j=1}^{n} w_j x_j \, / \, \sum_{j=1}^{n} x_j$$

subject to

$$n_i' \leq \sum_{j=1}^{n} a_{ij} x_j \leq n_i'', \quad i=1, ..., m,$$
$$x_j \geq 0, \quad j = 1, ..., n,$$

where n_i' and n_i'' are the limits introduced for n_i.

1.3. INFORMATION THEORY

Meister and Oettli [651] and Aggarwal and Sharma [26] applied fractional programming to calculate maximum transmission rate in an information channel. Consider a constant and discrete transmission channel with m input symbols and n output symbols, characterized by a transition matrix $P = ((p_{ij}))$, $i=1, ..., n$; $j=1, ..., m$, $p_{ij} \geq 0$, $\sum_i p_{ij} = 1$ where p_{ij} represents the probability of getting the symbol i at the output subject to the constraint that the input symbol was j. Assume that the matrix P does not have zero lines, *i. e.*, there are no output symbols that are never received. A cost $t_j > 0$ (energy, duration) is associated with the transmission of the input symbol j. Denote by $x = (x_j)$ the probability distribution function of the inputs. Obviously, $x_j \geq 0$ and $\sum_j x_j = 1$.

Define the transmission rate of the channel as[1]:

$$T(x) = \frac{\sum_i \sum_j x_j p_{ij} \log \left(p_{ij} \,/\, \sum_k x_k p_{ik} \right)}{\sum_j t_j x_j}$$

The relative capacity, C, of the channel is defined as the maximum of $T(x)$. We get the following nonlinear fractional programming problem:

$$G = \max_{x \in X} \{ T(x) | \; x_j \geq 0, \; \sum_j x_j = 1 \}. \qquad (1.1)$$

With the notations $c_j = \sum_i p_{ij} \log p_{ij}$; $c = (c_j)$; $y_i = \sum_j p_{ij} x_j$; $y = (y_j)$; $z = (x_j, y_i)$ (z being a vector with $m+n$ components), problem (1.1) becomes

$$\max \{ T(z) = \frac{c'x - y' \log y}{t'x} \; | \; x_j \geq 0, \; \sum_j x_j = 1, \; y_i = \sum_j p_{ij} x_j \}.$$

Closely connected with this case is also the following example.

1.4. OPTICAL PROCESSING OF INFORMATION

Fractional programming can be also applied in some Physics problems [386]. In spectral filters for the detection of quadratic law for infrared radiation, the problem of maximizing the signal-to-noise ratio appears. Mathematically, this means to maximize the filter-function.

$$\Phi(x) = \frac{(a'x)^2}{x'Bx + \beta}$$

on the domain $S = \{ x \in R^n, \; 0 \leq x_i \leq 1, \; i = 1, ..., n \}$ in which a and β are strict positive vector, and constant, respectively, B is a symmetric and positive definite matrix, $a'x$ represents the input signal, and $x'Bx + \beta$ represents the variance in the background signal. The domain of the feasible solutions S illustrates the fact that the filter cannot transmit more than 100% and less than 0% of the total energy. The optical filtering problems are very important in today's information technology, especially in coherent light applications, and optically based computers have already been built.

---

[1]F. M. Reza, *An Introduction to Information Theory*, McGraw-Hill, New York, 1961.

1.5. MINIMUM-RISK PROBLEMS IN STOCHASTIC PROGRAMMING

The minimum-risk problems in stochastic programming were first considered by Bereanu [171], [172] and by A. Charnes and W. W. Cooper[2]. Consider the usual linear programming problem

$$\max_{x} \{ c'x \mid Ax = b, \ x \geq 0 \} \tag{1.2}$$

and let D be the feasible domain. Such a problem appear frequently in the planning of a company when the various indicators need to be maximized. Such indicators, for example, the total cash outlay, can be considered to be the maximum of a linear programming problem. If the linear programming problem also contains random variables, it is natural that, among the feasible plans, to select that which has the maximum probability that the realized indicator is at least equal to the planned value.

Considering again problem (1.2), if the vector c is random, *i.e.*, its elements are random variables on the probability space $\{\Omega, K, P\}$, then, one fixes a level k of the resources and of the expenditure and then maximizes the probability that the objective function takes at least the value k

$$\max_{x \in D} P \{ \omega \in \Omega \mid c'(\omega)x \geq k \}.$$

If the costs c_j are independent random variables and if c has its mean value vector $\bar{c}$ whereas V is its covariance matrix, then

$$\max_{x \in D} P \{ \omega \mid c'(\omega)x \geq k \} = \max_{x \in D} P \{ \omega \mid \frac{c'(\omega)x - \bar{c}'x}{\sqrt{x'Vx}} \geq \frac{k - \bar{c}'x}{\sqrt{x'Vx}} \}$$

$$= 1 - \Phi \left(\max \frac{\bar{c}'x - k}{\sqrt{x'Vx}} \right),$$

where $\Phi(t) = \frac{1}{\sqrt{2\pi}} \int_{-\infty}^{t} e^{-\frac{s^2}{2}} ds$ is the Laplace function. Hence, the determination of the minimum-risk solution results in solving the fractional programming problem

$$\max \{ \frac{\bar{c}'x - k}{\sqrt{x'Vx}} \mid Ax = b \ ; \ x \geq 0 \}.$$

---

[2] *Deterministic equivalents for optimizing and satisficing under chance constraints*, Oper. Res. **11**, 18 - 39, 1963.

Extensions of the minimum-risk criterion to nonlinear programming, such as the linear Tchebysheff problem [919], [1007], the bottleneck transportation problem [923], the linear max-min problem [1008] and the bilinear max-min problem [920], the linear plus linear fractional programming problem [919] also yield fractional programming problems [928]. Consider, for example, the Tchebysheff problem [919] [1007]

$$\min_{x \in D} \max_{1 \le i \le r} \left\{ Z_i(x) = c^i x + \alpha^i \right\}$$

where c^i and α^i $(i = 1, ..., r)$ are constant vectors and scalar constants, respectively. Let now c^i and α^i assume random values $c^i(\omega) = c_1^i + t(\omega) c_2^i$; $\alpha^i(\omega) = \alpha_1^i + t(\omega)\alpha_2^i$, where c_1^i, c_2^i $(i = 1, ..., r)$ are constants vectors, α_1^i, α_2^i $(i = 1, ..., r)$ are scalar constants and $t(\omega)$ is a random variable. We assume that

$$z_i''(x) = c_2^i x + \alpha_2^i \neq 0 \text{ for all } x \in D \text{ and for all } i \in M = \{1, ..., r\} \tag{1.3}$$

If assumption (1.3) holds and the distribution function $T(z)$ of $t(\omega)$ is continuous and strictly increasing, then the minimum-risk solution of the Tchebysheff problem does not depend on $T(z)$ and can be obtained by solving the deterministic piecewise linear fractional programming problems

$$\max_{x} \min_{i \in M} \frac{k - c_1^i x - \alpha_1^i}{c_2^i x + \alpha_2^i} \quad \text{if } z_i''(x) > 0 \tag{1.4}$$

or

$$\min_{x} \max_{i \in M} \frac{k - c_1^i x - \alpha_1^i}{c_2^i x + \alpha_2^i} \quad \text{if } z_i''(x) < 0 \tag{1.5}$$

1.6. MACROECONOMIC PLANNING

One of the most significant applications of fractional programming, is that of dynamic modeling of macroeconomic planning using the input-output method. Let $Y(t)$ be the national income created in year t and let $Y_i(t)$ be the national income created in the branch i in the year t. Obviously, $Y(t) = \sum_i Y_i(t)$. If we denote by $C_{ik}(t)$ the consumption, in branch k, of goods of type i (that were created in branch i) and by $I_{ik}(t)$ the part of the national income created in branch i and allocated to investment in branch k, then the following repartition equation applies to the national income created in branch i :

$$Y_i(t) = \sum_k \left(C_{ik}(t) + I_{ik}(t)\right).$$

The increase of the national income in branch k is function of the investment made in this branch

$$\Delta Y_k(t) = Y_k(t+1) - Y_k(t) = b_k I_k(t) + b_{k0},$$

where $I_k = \sum_i I_{ik}$. In these conditions, the macroeconomic planning consists in the maximization of the increase rate of the national income

$$\frac{\Delta Y}{Y} = \frac{\sum_k \left(b_k I_k(t) + b_{k0}\right)}{\sum_i \sum_k \left(C_{ik}(t) + I_{ik}(t)\right)}$$

subject to the constraints $C_k(t) \geq \max\,(\bar{C}_k,\ C_k(0))$, where $C_k(t) = \sum_i C_{ik}(t)$, $I_k(0) \leq I_k(t) \leq I_k(\max)$, and $\bar{C}_k$ represents minimum consumption attributed to branch k whereas $I_k(\max)$ is the maximum level of investments for branch k.

1.7. THE PRODUCTION EFFICIENCY PROBLEM

Bradley and Frey, Jr. [201] considered the example of maximizing the profit/cost ratio, with the profit being a Cobb-Douglas production function. Through production function one understands the relation between the input variable specifying the production factors and the output variables specifying the results of production activity. Consider a set of projects $i = 1, \ldots, m$, and the total benefit value assigned to these projects given by a Cobb-Douglas production function

$$B = \sum_{j=1}^{n} a_0 x_j^{a_j};$$

where $\sum_{j=1}^{n} a_j = 1$. The variables x_j designate the production factors. Let the total cost, C, be a linear function of the allocations to the various projects

$$C = \sum_{j=1}^{n} c_j x_j + c_0.$$

It is required to maximize the profit/cost ratio with constraints on the expenditure.

The profit/cost ratio is an efficiency indicator, *i.e.*, the ration between what is obtained and the expenditure. The following production efficiency model is obtained

$$\max \left(\prod_{j=1}^{n} a_0 x_j^{a_j} \,/\, \sum_{j=1}^{n} a_j x_j + c_0 \right)$$

subject to

$$\sum_{j=1}^{n} a_{ij} x_j \geq b_i, \quad i = 1, \ldots, m \;;$$
$$x_j \geq 0, \quad j = 1, \ldots, n,$$

where b_i represents the profit that can be obtained at project i and a_{ij} represents the contribution of the production factor j to project i to realize the profit b_i.

1.8. CLUSTER ANALYSIS

Some problems, regarding classification of sets of elements in a given number of classes lead to fractional programming models [746]. Let A be a set of N elements and let M be the number of classes in which we want to divide its elements. Assume that, for each pair $(i, j) \in A \times A$, the distance d_{ij} between the elements i and j, was defined. It is required to divide into M classes the elements of the set A such that the total of the averages of the sum of the square of the distances in each class to be minimum. Define the 0 - 1 variables x_{ij} in the following way:

$$x_{ij} = \begin{cases} 1 & \text{if element i belongs to class } C_j \\ 0 & \text{otherwise} \end{cases}$$

The determination of an optimal division in classes of the elements of the set A becomes the solution of the following fractional non-linear 0 - 1 programming problem:

$$\min \sum_{k=1}^{M} \left[\left(\sum_{i=1}^{N-1} \sum_{j=i+1}^{N} d_{ij}^2 x_{ik} x_{jk} \right) / \sum_{i=1}^{N} x_{ik} \right],$$

subject to

$$\sum_{k=1}^{M} x_{ik} = 1, \quad i = 1, 2, \ldots, N,$$
$$x_{ik} \geq 0, \; i = 1, \ldots, N, \; k = 1, \ldots, M,$$
$$x_{ik} = 0 \text{ or } 1 \text{ for any } i \text{ and } k.$$

Let's analyze the case when the number of groups is equal to two i.e. M = 2. Define the 0 - 1 variable x_i as follows:

$$x_i = \begin{cases} 1 \text{ if entity } i \text{ belongs to class } C_1 \\ 0 \quad -//- \quad C_2 \end{cases}$$

The problem becomes

$$\min \frac{\sum_{i=1}^{N-1} \sum_{j=i+1}^{N} d_{ij}^2 x_i x_j}{\sum_{i=1}^{N} x_i} + \frac{\sum_{i=1}^{N-1} \sum_{j=i+1}^{N} d_{ij}^2 (1 - x_i)(1 - x_j)}{N - \sum_{i=1}^{N} x_i}$$

$x_i = 0$ or 1 for all i . This is a fractional non-linear 0 - 1 programming problem with no additional constraints.

1.9. OPTIMAL PATHS IN GRAPH THEORY

Typical optimization problems in graph theory are: optimal path problems, minimal ratio spanning trees, and max-min problems. Peteanu and Ţigan [725], [726], [727] (see also [639]) arrived at fractional programming problems when studying the optimum path problem in double weighted graphs. The fractional objective function was defined on the set of graph paths. Let $G = (X, U)$ be a connected graph without cycles in which $X = \{ x_1, ..., x_n \}$ represents the nodes (vertices) set, and $U \subseteq X \times X$ the arcs set. Let a_{ij} and b_{ij} be two positive weights associated to each arc $(x_i, x_j) \in U$, and let $\mathcal{L}(G)$ be the set of paths from node x_1 to node x_n. Determine a path $L^* \in \mathcal{L}(G)$ such that

$$\min \left\{ \frac{\sum_{(x_i, x_j) \in L} a_{ij} + \alpha}{\sum_{(x_i, x_j) \in L} b_{ij} + \beta} \mid L \in \mathcal{L}(G) \right\} = \frac{\sum_{(x_i, x_j) \in L^*} a_{ij} + \alpha}{\sum_{(x_i, x_j) \in L^*} b_{ij} + \beta} ,$$

where α and β are given real numbers. The above problem constitutes a fractional programming problem on a finite set (the paths set). This problem was previously treated by Dantzig, Blattner and Rao [343] and Fox [398] with the only difference that L constitute a circuit and that a_{ij} and b_{ij}, respectively, signified a cost, and a time, respectively, associated to arc (x_i, x_j).

Another type of fractional problem is mentioned in ref. [925], *viz.*, the determination of a path $L^* \in \mathcal{L}(G)$ such that:

$$\min \left\{ \frac{\max\limits_{(x_i, x_j) \in L} \{a_{ij}\}}{\max\limits_{(x_i, x_j) \in L} \{b_{ij}\}} \mid L \in \mathcal{L}(G) \right\} = \frac{\max\limits_{(x_i, x_j) \in L^*} \{a_{ij}\}}{\max\limits_{(x_i, x_j) \in L^*} \{b_{ij}\}}$$

Let $\mathcal{E}$ and $\mathcal{F}$ be two non-empty subgraphs sets of the graph G. In particular, these sets can be paths sets between two nodes or the set spanning trees of graph G. On set U, define the functions $p_i : U \to R (i = 1, 2)$ and $q_i : U \to R^+ (i = 1, 2)$. In ref. [727], the following fractional max-min problem on graph G was considered: Determine $E' \in \mathcal{E}$ and $F' \in \mathcal{F}$ such that (E', F') is optimal solution for the problem

$$\max_{E \in \mathcal{E}} \min_{F \in \mathcal{F}} \frac{\sum\limits_{u \in E} p_1(u) + \sum\limits_{u \in F} q_1(u)}{\sum\limits_{u \in E} p_2(u) + \sum\limits_{u \in F} q_2(u)}$$

Now, we consider an undirected connected graph $G = (N, E)$ with node set N, edge set E and a pair of numbers (a_e, b_e) assigned to each edge $e \in E$. Assume that $\sum\limits_{e \in T} b_e$ has the same sign for all spanning trees T. Chandrasekaran [277] (see also [278]) considered the problem of the determination of a spanning tree T that minimizes the ratio $\left(\sum\limits_{e \in T} a_e / \sum\limits_{e \in T} b_e \right)$.

1.10. REPARTITION PROBLEMS

In the case of the problem of dynamic repartition of resources, the maximization of the global income realized through complete utilization of resources is considered. Another optimal repartition mode can be considered, *viz.*, that in which one maximizes the efficiency of utilization of resources (the average income on unit time). Ţigan [993] arrived at the following separable fractional programming model.

$$\max F(x) = \frac{\sum\limits_{i=1}^{n} f_i(x_i) + \alpha}{\sum\limits_{i=1}^{n} g_i(x_i) + \beta},$$

subject to

$$\sum_{i=1}^{n} w_i x_i \leq b, \quad \alpha_i \geq x_i \geq \beta_i \geq 0, \quad x_i \text{ integers}$$

where w_i, α_i, β_i ($i = 1, ..., n$), α, β and b are given real numbers. Jagannathan [510] considered the general allocation problem in multiple character studies which he modeled as a separable fractional programming:

$$\min \left\{ F(x_1, ..., x_n) = \sum_{j=1}^{n} \frac{c_j}{x_j + s_j} \mid Ax \leq b \; ; \; x \geq 0 \right\} \text{ with } c_j \geq 0 \text{ and } s_j > 0 \; .$$

1.11. TRANSPORTATION PROBLEMS WITH SEVERAL OBJECTIVE FUNCTIONS

Fractional programming can be used to solve mathematical programming problems with several objective functions. In many of such problems, the concordance of the criteria is not obvious, some criteria being even contradictory. Among these, we mention the transportation problem with three efficiency functions: the total transportation cost, which must be minimum, the total transportation time, which also must be minimum, and the transported quantity which must be maximum. The domain of the feasible solutions is

$$D = \{ X = (x_{ij}) \mid \sum_{j=1}^{n} x_{ij} \leq a_i, \quad i = 1, ..., m \; ;$$

$$\sum_{i=1}^{m} x_{ij} \geq b_j, \quad j = 1, ..., n; \quad 0 \leq x_{ij} \leq g_{ij}, \quad i = 1, ..., m; \quad j = 1, ..., n \}$$

In ref. [901], a solution for this problem is considered through the construction of a synthesis function of the form

$$F = \frac{F_1 F_2}{F_3^2} = \frac{\left[\sum_{i=1}^{m} \sum_{j=1}^{n} c_{ij} x_{ij} \right] \left[\sum_{i=1}^{m} \sum_{j=1}^{n} t_{ij} x_{ij} \right]}{\left[\sum_{i=1}^{m} \sum_{j=1}^{n} x_{ij} \right]^2} \tag{1.6}$$

which must minimized on domain D. It is shown that, through the solution of the transportation problem with the objective function (1.6) one gets efficient solutions. In Chapter 10, this problem will be considered again, in a more general case.

1.12. INVESTMENTS ALLOCATION PROBLEMS

Ziemba, Brooks-Hill and Parkan [1081] arrived to a fractional programming model by considering the optimal selection problem for an investment portfolio, under certain economic assumptions. An investor wishes to invest in n production activities a certain amount of money. Denote by x_i the percentage of the fund which is going to be invested in activity i. The vector $x = (x_1, ..., x_n)$ satisfies the constraints

$$\sum_{i=1}^{n} x_i = 1, \quad x_i \geq 0,$$

to which one can add other constraints based on economic considerations. If ξ_i constitutes the income obtained when investing the full amount in activity i, then the income corresponding to the investment strategy x is

$$V(x) = \sum_{i=1}^{n} \xi_i x_i$$

The problem becomes more complicated since ξ_i are not constants, but rather random variables. If ξ_i are normal with mean values $\bar{\xi}_i$ and the covariance matrix $\Sigma = (\sigma_{ij})$ (positive definite matrix, *i.e.*, $x'\Sigma x > 0$ for every $x \neq \mathbf{0}$), then the average value of the total income is

$$E(V(x)) = \sum_{i=1}^{n} \bar{\xi}_i x_i$$

and the standard deviation is

$$D(V(x)) = \sqrt{x'\Sigma x}\ .$$

By considering the two functions together (*i.e.*, the maximization of the average value of the total income and the minimization of its standard deviation) one arrives [1081] to the following fractional programming problem

$$\max F(x) = \frac{\sum_{i=1}^{n} \bar{\xi}_i x_i}{(x'\Sigma x)^{1/2}} = \frac{\bar{\xi} x}{(x'\Sigma x)^{1/2}}$$

with constraints

$$\sum_{i=1}^{n} x_i = 1, \ x_i \geq 0, \ x \in S,$$

where S results from other economic constraints.

1.13. GEOGRAPHICAL REPARTITION OF INVESTMENTS

Another example, which can be modeled through fractional programming, was considered by Trattner and Zidăroiu [981], and it referred to the social-economic efficiency of investments as function of the cost of creating new jobs. Assume that, in a certain area, m investments are considered with the objective of providing jobs for a number of unemployed workers. Let s_i be the number of modules where the investment i can be done $(1 \leq i \leq m)$, let I_{ij} be the cost of the investment of type i which is realized at module j $(1 \leq j \leq s_j)$, let f'_{ij} (f''_{ij}, respectively) be the minimum work force (maximum, respectively) needed per unit time i, in the case in which is realized at module j, let F be the work force available in the area, and let α_1 (α_2, respectively) be the minimum (maximum) coefficient of the available work force that can be attracted by the new investments. The concept of module can be understood as placement options for the investments in the area.

If we denote by x_i the work force needed to be employed in the industrial units i $(i = 1, \ldots, m)$, and by y_{ij}, a 0 - 1 variable which takes the value 1 if the investment i is realized at module j, and 0 otherwise, the mathematical model takes the following form:

$$\min \frac{\sum_{i=1}^{m}\sum_{j=1}^{s_i} I_{ij}}{\sum_{i=1}^{m} x_i}\ ; \tag{1.7}$$

subject to

$$\alpha_1 F \leq \sum_{i=1}^{m} x_i \leq \alpha_2 F\ , \tag{1.8}$$

$$\sum_{j=1}^{s_i} f'_{ij} y_{ij} \leq x_i \leq \sum_{j=1}^{s_i} f''_{ij} y_{ij},\quad 1 \leq i \leq m, \tag{1.9}$$

$$\sum_{j=1}^{s_i} y_{ij} \leq 1,\quad 1 \leq i \leq m, \tag{1.10}$$

$$x_i \geq 0\ ;\ y_{ij} = 0 \text{ or } 1,\quad 1 \leq i \leq m\ ;\ 1 \leq j \leq s_i\ . \tag{1.11}$$

The efficiency function (1.7) represents the average cost of creating, through investment, of a new job in a given area, and this cost must be minimized. The constraint (1.8) refers to the attraction, in satisfactory numbers, for the respective period, of the work force from the surrounding area. The constraint (1.9) expresses the fact that the work force to be used in the newly created unit i must be found between two given limits. The constraints (1.10) expresses the fact that each unit i can be constructed, at most, in one capacity module. Hence, the problem of optimal selection of investments in terms of the cost of creating new jobs has been reduced to a mixed model, with the linear constraints and with linear fractional objective function.

1.14. GAME THEORY PROBLEMS

Fractional functions appear in the solution of certain games. Schroeder [823] considered the determination of the minimax solutions for a game with the payoff matrix of the form

$$F(X, Y) = X'AY \,/\, X'BY \tag{1.12}$$

in which A and B are real matrices ($B > 0$), and $X = (x_1, ..., x_n)'$ and $Y = (y_1, ..., y_n)'$ are the vectors of mixed strategies. At each move, the first player makes a move i with probability x_i while the second player makes a move j with probability y_j. These situations appear for example thus : Consider a stochastic game where the matrix A contains the payments for each move and B is a matrix of the probabilities that the game will end before the next move. At each move, the players must consider, besides the immediate payments, also the probability of payments in future moves. Then, $F(X, Y)$ will represent the expectation of total payoff for all the moves:

$$F(X, Y) = \sum_{n=0}^{\infty} X'AY(1 - X'BY)^n = X'AY \,/\, X'BY \,.$$

Since $X'BY$ represents a probability, we have $|X'BY| < 1$ and, hence, the above geometric series is convergent. Functions of form (1.12) appear in games with nonzero sum. If A is the wins matrix for the first player, and B for the second player, then a player tries to maximize the ratio of the wins, while the opponent will minimize this ratio. One gets the fractional minimax problem

$$\min_{Y} \ \max_{X} \ F(X, Y)$$

subject to

$$x_i \geq 0, \ \sum_{i=1}^{m} x_i = 1, \qquad i = 1, ..., m$$

$$y_j \geq 0, \ \sum_{j=1}^{n} y_j = 1, \qquad j = 1, ..., n$$

We also mention the fractional programming model which appears in game theory when the strategy spaces of the players are not independent. Such a model was presented by Cook, Kirby and Mehndiratta [323] in the following form

$$\max_{x} \ \min_{y} \ \left\{ \frac{cx + dy + \alpha}{fx + gy + \beta} \ \middle| \ Ax + By \leq b \ ; \ x, y \geq 0 \right\}$$

where $x, c \in R^p$; $y, d \in R^q$; $b \in R^n$, $A \in R^{m+p}$, $B \in R^{m+q}$.

1.15. INVENTORY PROBLEMS

Although inventory problems appear in a larger number of cases, most often they are met in the selling and production of consumer good. Considered the simple case of just one product and one level. In a given time period T, the demand R with constant intensity $r = R/T$ must be satisfied. The production rate (k) is assumed constant ($k > r$). The initial (time zero) inventory is zero, and it increases continually for a period t_1, until it reaches the maximum value S, then decreases continually down to zero for a period t_2. A period t_3 follows in which unsatisfied orders accumulate until the maximum deficit s is reached. At the end of the period t_3 the production starts again, and the back-log diminishes until it is completely eliminated after the period t_4. The cycle is restarted. The average expenditure on unit time $F(t_1, t_2, t_3, t_4)$ is obtained by summing up the storage expenditure, the outlay, and the fixed expenditure, and by dividing the results by the length of the cycle:

$$F(t_1, t_2, t_3, t_4) = \frac{\frac{1}{2}[c_1 S(t_1 + t_2) + c_2 s(t_3 + t_4)] + c_3}{(t_1 + t_2 + t_3 + t_4)},$$

where c_1 = the storage cost of a unit of the product for a unit time; c_2 = the cost of having a lack of products for unit time; c_3 = the fixed costs (for a production cycle). The above function takes the form:

$$F(t_2, t_3) = \frac{\frac{1}{2}kr\left(c_1 t_2^2 + c_2 t_3^2\right) + c_3(k - r)}{k\left(t_2 + t_3\right)}$$

The problem is to determine t_2 and t_3 such that the fractional function $F(t_2, t_3)$ is minimum, subject to some constraints on the t_i.

1.16. SENSITIVITY ANALYSIS FOR A LINEAR PROGRAMMING PROBLEM

This problem constitutes a very important domain in mathematical programming. The problem consists in the study of the general variation of the optimal solution of a mathematical programming problem to variations in the input data. In what follows, we consider the repricing theorems considered by Charnes and Cooper [283] for testing the sensitivity of a linear programming model, problem which leads to a model of fractional programming. Consider the usual transportation problem:

$$\min \{ Z = \sum_{i=1}^{m}\sum_{j=1}^{n} c_{ij}x_{ij} \mid \sum_{j=1}^{n} x_{ij} = a_i , \ \sum_{i=1}^{m} x_{ij} = b_j , \ x_{ij} \geq 0 , \quad i = 1, ..., m; j = 1, ..., n \}$$

Let $\{x_{ij}^*\}$ be its optimal solution and let Z^* be the minimum transportation cost. The possibility exists that by changing the prices c_{ij} to change the value of Z^*. Reference [283] shows that Z^* is a concave function of c_{ij} and it is piecewise linear. For simplification, consider a given price value, say, c_{pq}. We want to determine how big can this price be (the interval in which it can vary) such that the transportation cost Z^* to diminish with at least kZ^* (k is a proportionality factor $0 < k < 1$). The problem becomes

$$\max c_{pq}$$

with the constraint

$$\sum_{\substack{i=1 \\ i \neq p}}^{m}\sum_{\substack{j=1 \\ j \neq q}}^{n} c_{ij}x_{ij} + c_{pq}x_{pq} \leq Z^*(1 - k)$$

plus the constraints of the transportation problems. In this model, the unknowns are x_{ij} and c_{pq}. This problem is nonlinear since the constraints contain the product of two variables. Through the change of variables $y_{pq} = c_{pq}x_{pq}$, the nonlinearity is transferred to the objective function. The above model becomes the fractional programming model

$$\max y_{pq}/x_{pq}$$

with the constraint

$$\sum_{i=1}^{m}\sum_{j=1}^{n} c_{ij}x_{ij} + y_{pq} \leq Z^*(1 - k)$$

plus the constraints of the classical transportation problem.

1.17. MARITIME TRANSPORTATION PROBLEMS [192]

Assume that in a port A, we have to load a ship (of limited capacity) with a load made up of goods of various types regarding their weights and values, goods which must be transported in another port B. It is required to determine how much of each type of goods to load such that the loading of the ship to yield maximum profit. Let N be the number of types of available goods, Q_i the maximum available quantity of goods of type i, V

maximum load the ship can carry, f_i the profit obtained through loading of a unit of the i-th goods type (monetary unit/ton), g_i the weight of the i-th type of goods. Denoting by x_i the number of goods of type i which are loaded, the mathematical model of the loading problem is:

$$\max \{ \sum_{i=1}^{N} f_i x_i \mid x_i \le Q_i, \ \sum_{i=1}^{N} g_i x_i \le V, \ x_i \ge 0 \} .$$

Seen in this way, the problem, most of the times, is not correctly modeled through the above model. Since the loading time differs from a ship to another ship, we need also to consider the loading time since it will also influence the transportation cost. Let y_i be the loading rate (tons/day) of the goods i, let c_1 be the docking cost of the ship, let c_2 be the daily cost of the ship on sea, let s be the ship's speed, and let d be the total distance. The total time for which the ship must be used is the sum of the docking time and the time on sea:

$$T = d/s + \sum_{i=1}^{N} (x_i/y_i)$$

and the profit will be:

$$P = \sum_{i=1}^{N} f_i x_i - \sum_{i=1}^{N} [(c_1 x_i/y_i) + c_2(d/s)] .$$

More realistically will be to consider the average profit on unit time (day, month, year) *i.e.*, to optimize a function of the form:

$$F = \frac{P}{T} = \frac{\sum_{i=1}^{N} f_i x_i - \sum_{i=1}^{N} [(c_1 x_i/y_i) + c_2(d/s)]}{d/s + \sum_{i=1}^{N} (x_i/y_i)}$$

1.18. OPTIMUM TRANSPORTATION PLAN [52].

This example also refers to a maritime transportation problem. Find an optimal loading-unloading plan for a ship which goes on a given itinerary formed of N ordered ports. The ship makes $M (M \le N)$ stops (representing disjunctive sets of ports and which maintains the order of the sets) for loading-unloading. In each port, the loadings are available to be transported to the next ports. The loads available at each stop, are independent random variables defined on a finite set of discrete values. At each stop m ($m = 1, ..., M$), we know the decisions taken at the preceding stops, as well as the realization of the available loading vector (the loadings available at all the preceding stops). The problem is to determine the quantities and the destination of the loads to be loaded in the ports of the current stage such that to maximize the average of the sum of the incomes per unit time at coming stages (stops) of the itinerary. The objective function appears as the sum of some linear fractional and separable functions. One makes the assumption that the loading/unloading times and costs depend linearly on the loads, and that only one load type is transported from one port to another. Let Q_{ij} be the quantity of load (percentage of ship's loading capacity) shipped from port i to port j

$$0 \le Q_{ij} \le 1, \quad i = 1, ..., N \, ; \quad j = i + 1, ..., N + 1$$

and let S_{ij} be the quantity of load (percentage of ship's loading capacity) available from i to j, $0 \le S_{ij} \le 1$ for all i and j. Let T_{ij} be the time needed to load a ship completely in port i and to unload it in port j, calculated in percentage of the total time (the time on sea and the time in ports) used to cover the subroute representing the set to which the port i belongs, $T_{ij} > 0$ for all i and j. Let R_{ij} be the net income which results from the transportation of a unit load from port i to port j, divided by the time needed to load and unload this unit, $R_{ij} > 0$ for all i and j. Let C_m be the difference between the costs of a unit time on sea and in port, multiplied by the ratio between the time on sea and the total time on subroute representing the set m, $C_m > 0$, $m = 1, ..., M$.

Let $P_m = \{ \, i \mid i \text{ is in the } m\text{-th set of indices} \, \}$, $m = 1, ..., M$,

$$\bar{Q}_i = (Q_{i,i+1}, ..., Q_{i,N+1}) \quad \bar{Q} = (\bar{Q}_1, ..., \bar{Q}_N) \, ,$$
$$\bar{Q}^m = \{\bar{Q}_i \mid i \in P_m\}, \quad \bar{S}^m = \{\bar{S}_i \mid i \in P_m\},$$
$$\bar{S}_i = (S_{i,i+1}, ..., S_{i,N+1}), \quad \bar{S} = (\bar{S}_1, ..., \bar{S}_N).$$

The problem can be seen as a decision process in M steps. The mathematical model for step m is

$$\max f_m(\bar{Q}^m) = \frac{\sum_{i \in P_m} \sum_{j > i} T_{ij} R_{ij} Q_{ij} - C_m}{\sum_{i \in P_m} \sum_{j > i} T_{ij} Q_{ij} + 1}$$

with the constraints $\sum_{j>i} Q_{ij} + \sum_{p<i} \sum_{q>i} Q_{pq} \le 1$, $\forall \, i$, and $0 \le Q_{ij} \le S_{ij}$, $\forall \, i, j$.

Similar problems are described in references [51] and [54].

1.19. INDEFINITE PROGRAMMING PROBLEMS

Fractional programming problems appear in the solution of certain mathematical programming problems, for example, the indefinite programming or product programming. Swarup [945] considered the quadratic indefinite programming problem

$$\max \left\{ (c'x + c_0)(d'x + d_0) \mid Ax \leq b \,;\, x \geq 0 \right\}$$

which was solved using another problem, which was equivalent but convex. Through the change of variable $y = tx$ with $t \geq 0$ chosen such that $d'y + d_0 t = \delta$ (δ = const $\neq 0$) the problem becomes

$$\max \frac{(c'y + c_0 t)}{t^2} = \frac{1}{\min \dfrac{t^2}{c'y + c_0 t}}$$

on the domain

$$D = \left\{ (y, t) \mid Ay - bt \leq 0, \;\; d'y + d_0 t = \delta, \;\; y \geq 0, \;\; t \geq 0 \right\}$$

Since the function $t^2/(c'y + c_0 t)$ is a convex function, it follows that we have obtained a convex programming problem.

1.20. DECOMPOSITION PROBLEMS

In the solution of programming problems with many constraints, fractional programming problems appear. Consider the problem

$$\min \left\{ c'x \mid A_1 x = b_1, \;\; A_2 x = b_2, \;\; x \geq 0 \right\}, \tag{1.13}$$

where A_1 has m_1 lines and A_2 has m_2 lines. If the convex polyhedron

$$S = \{ x \mid A_2 x = b_2, \;\; x \geq 0 \}$$

is bounded, then, any of its elements, x, can be expressed in the form

$$x = \sum_j \lambda_j x^j, \;\; \sum_j \lambda_j = 1, \;\; \lambda_j \geq 0$$

(x^j are extreme points of S). Hence, problem (1.13) becomes

$$\min \{ \sum_j (cx^j)\lambda_j \mid \sum_j (A_1x^j)\lambda_i = b_1, \ \sum_j \lambda_j = 1, \ \lambda_j \geq 0 \} \tag{1.14}$$

If $\{\lambda_j^0\}$ is the optimal solution of problem (1.14), then $x^0 = \sum_j \lambda_j^0 x^j$ is a solution of problem (1.13). Different from problem (1.13) which has $m_1 + m_2$ lines, problem (1.14) has only $m_1 + 1$ lines, but it has more columns. Bell [164] applied the primal-dual method to find the solution of problem (1.14). He started with the finding of an initial basic solution and of complementary slackness of the dual problem which is

$$\max \{(\pi b_1 + \pi_0) \mid \pi A_1 x^j + \pi_0 \leq cx^j \text{ for all } j\}.$$

The dual has $m_1 + 1$ variables and many constraints. Choose the pivot column. Assume that the pivot element is in line r. Let B be the current (infeasible) base for problem (1.14) and let b_r be the line r of B^{-1}. The dual simplex method requires the determination of the minimum value:

$$\min_{\bar{a}_{rj} < 0} (-\bar{c}_1 / \bar{a}_{rj}) \tag{1.15}$$

If b_r is partitioned such that $b_r = (\hat{b}_r, b_r^0)$, where $\hat{b}_r$ is a m_1 - vector and b_r^0 is a scalar, then, for problem (1.14), we have:

$$-\bar{c}_j = (\pi A_1 - c)x^j + \pi_0, \quad \bar{a}_{rj} = (\hat{b}_r A_1)x^j + b_r^0$$

and hence (1.15) becomes:

$$\min \frac{(\pi A_1 - c)x^j + \pi_0}{(\hat{b}_r A_1)x^j + b_r^0}$$

where the minimum is considered with respect to all the extreme points x^j of S, for which the denominator is negative. Hence, the determination of column to enter the base is made by solving a fractional program.

1.21. MARKOV CHAIN DECISION PROBLEMS [348], [397]

Certain optimal control problems for dynamic systems can be modeled through fractional programming. Consider a controlled stochastic dynamic system which can exist at various time moments $t = 0, 1, \ldots$ in one of the states $0, 1, \ldots, L$. From a certain state, the system is directed towards another state through one of the decisions $d_1 \ldots d_k$ which will determine the probability law for the change of the system. To regulate the decision, the sequence of states of the system at various time moments $\{X_t\}$, $t = 0, 1, \ldots$ constitutes a Markov chain with stationary transition probabilities

$$p_{ij} = \sum_{k=1}^{K} q_{ij}(k) D_{ik} \qquad i, j = 0, ..., L$$

where the transition matrix $q_{ij}(k)$ depends on the decision variable $k = 1, 2, ..., K$, and D_{ik} is the probability of taking the decision k when the system is in state i. Let $w_{ik}(t) > 0,\ i = 0, ..., L;\ k = 1, ..., K;\ t = 0, 1 ...$ be finite values denoting the expected cost ascribed to time t, given that the system is observed in state i at that time and that decision d_k is made. We assume that $w_{ik}(t) = w_{ik}$, independent of t. A decision rule is required, such that the ratio of mean cost per unit time is minimized, comparing two sets of costs, w'_{ik} and w''_{ik}. This leads to minimizing the function

$$\frac{\sum_{i=0}^{L} \sum_{k=1}^{K} \pi_i D_{ik} w'_{ik}}{\sum_{i=1}^{L} \sum_{k=1}^{K} \pi_i D_{ik} w''_{ik}}$$

with respect to the decision variables D_{ik}, and where the stationary probabilities π_j satisfy $\sum_{i \in I} \pi_i\, p_{ij} = \pi_j,\ \sum_{j \in I} \pi_j = 1,\ \pi_j \geq 0$ (I the set of states)

This method was applied in ref. [556] to the determination of the optimal inspection, maintenance and replacement policy, for systems that have a failure mechanism that can be modeled with a finite Markov chain of discrete time parameter. Consider a machinery which can have a failure which is detected during an inspection. After inspection the machinery either is replaced, or is kept. In the latter case, repairs may be made, and the next inspection period is establish. Assume that the failure law of the system (machinery) is Markovian *i.e.*, the sequence of successive states forms a discrete Markov chain with stationary transition probabilities q_{ij} $(i, j = 0, ..., L)$. This assumption is plausible for real cases, for example, in electronic systems with many components, each component having (approximately) an exponential deterioration, in which case it is expected that the whole system will have a Markov description.

Consider now a new Markov chain with the states $i = 0,1, ..., L$, the decisions $k = 1, ..., L$ and the equilibrium probabilities π_j, in which, different from the first system, the transitions from one state to the other take place after the inspection and not after each time unit. The problem is to choose the inspection maintenance and replacement policy, which minimizes the average cost per unit time as expresses through the ratio between the average inspection cost and the time between inspections.

$$\min \left(\sum_i \sum_{s,k} x_{isk} h_{isk} \right) / \left(\sum_i \sum_{s,k} k x_{isk} \right)$$

subject to

$$\sum_{s,k} x_{jsk} = \sum_i \sum_{s,k} x_{isk} v_{skj} \quad (j = 0,1, \ldots, L(K-1))$$

$$x_{isk} \geq 0 \,, \quad \sum_i \sum_{s,k} x_{isk} = 1 \quad (k = 1, \ldots, K)$$

where h_{isk} are cost coefficients, $x_{isk} = \pi_i D_{isk}$, with D_{isk} the probability of deciding the transition from state i to state s and of inspecting after k periods, and v_{skj} are certain functions of q_{ij}.

1.22. LOADING PROBLEM [463].

Assume that it is required to fabricate n products $P_1, P_2, \ldots, P_n$ in the minimum quantities $p_1, p_2, \ldots, p_n$ using m types of machine tools, $M_1, \ldots, M_m$. Available are q_1 machine tools of type $M_1, \ldots, q_m$ machine tools of type M_m. Assume that the machine tools of type M_i have the same time period in which they are available for the period in which the programming is being done, and let that period be T_i. Denote by t_{ik} the necessary time for the execution of a unit of product P_i on the machine of type M_k.
Define the utilization index of a machine as the ratio between the effective time in which the machine is being actually used and the maximum time for which the machine could be used. Similarly, define the utilization index of an assembly of machines. The problem to be solved consists in the determination of the quantities to be produced of each item such that the utilization index of the assembly of machines is maximized. To do this, let x_i be the number of units produced from P_i and u_k the number of machines of type M_k used. In these conditions, the mathematical model is the following:

$$\max \frac{\sum_{k=1}^{m} \sum_{i=1}^{n} t_{ik} x_i}{\sum_{k=1}^{m} T_k u_k}$$

on the domain

$$D = \{x_i | x_i \geq p_i, \ 0 \leq u_k \leq q_k, \ \sum_{i=1}^{n} t_{ik} x_i \leq T_k u_k, \ k=1, \ldots, m, \ x_i, u_k - \textit{integers}\}$$

1.23. SET COVERING PROBLEM [76]

An airline company must perform m flights and has n crews. Denote by $c_j > 0$ the cost paid by the company when the j -th crew is used on a flight, and by d_j the commission received by the company, and let $\beta > 0$ be the amount of money received by the company for each crew that is being used. The company is interested to distribute the crews to the m flights such that $\sum c_j \,/\, (\sum d_j + \beta)$ is minimum. Let $I = \{1, ..., m\}$ be the set of flights, $J = \{1, ..., n\}$ the set of crews, and $P_j \subseteq I$ the set of flights which can be performed by the j -th crew. It is required to determine a set J^* of crews to cover the m flights and which minimizes the ratio $\sum_{j \in J^*} c_j \,/\, (\sum_{j \in J^*} d_j + \beta)$ where $J^* \subseteq J$ and $\bigcup_{j \in J^*} P_j = I$.

Define the variables

$$x_j = \begin{cases} 1 & \text{if the crew } j \text{ is asigned} \\ 0 & \text{otherwise} \end{cases}$$

and the matrix $A = (a_{ij})$

$$a_{ij} = \begin{cases} 1 & \text{if the flight } i \text{ is covered by the crew } j \\ 0 & \text{otherwise} \end{cases}$$

The set covering problem with linear fractional function becomes

$$\min \frac{\sum_{j=1}^{n} c_j x_j}{\sum_{j=1}^{n} d_j x_j + \beta}$$

subject to

$$x \in D = \{ x \in R^n \mid \sum_{j=1}^{n} a_{ij} x_j \geq 1, \; i = 1, ..., m; \; x_j = 0 \text{ or } 1 \}$$

1.24. OTHER FRACTIONAL PROGRAMMING EXAMPLES

24.1. A particular case of fractional programming, having only one constraint [379], is

$$\max \{ z(x) = \frac{cx + c_0}{dx + d_0} \mid ax = b, \ 0 \leq x \leq u \}$$

Such a problem appears when an amount of money S must divided to a set of n activities while maximizing a certain indicator (for example, the return index). In particular, assume that we wish to optimize the investment into a project. More precisely, to determine an optimal allocation in GERT network in which nodes are of the "exclusive-or" type. Initially, one maximizes the probability that the project is performed before a given date. The determinist equivalent represents a fractional problem.

24.2. Let $\vec{v}_i(b_i, a_i)$ $(i = 1, ..., T)$ be vectors in the real half-plane $R_+^2 = \{ (x, y) \mid x > 0, -\infty < y < \infty \}$. The extremity of vector $\vec{v}_i$ defines the point $P_i(b_i, a_i)$. Consider the sum vector $\vec{v}_i + \vec{v}_j = (b_i + b_j, \ a_i + a_j)$, with the position of it extremity obtained using the parallelogram rule. The slope of the line which unites the origin with the point $P_i(b_i, a_i)$ is a_i/b_i and that of the line which unites the origin with the point $(b_i + b_j, \ a_i + a_j)$ is $(a_i + a_j)/(b_i + b_j)$. Among the T vectors, find that subset made up of n $(n < T)$ vectors with the property that their sum has maximum slope. Introduce the variables

$$x_i = \begin{cases} 1 & \text{if the vector } \vec{v}_i \text{ was selected} \\ 0 & \text{otherwise} \end{cases}$$

and obtain the following fractional programming problem in variables 0,1:

$$\max \{ \sum_{j=1}^{T} a_j x_j \,/\, \sum_{j=1}^{T} b_j x_j \mid \sum_{j=1}^{T} x_j = n, \ x_j = 0, 1 \ \forall \ j = 1, ..., T \}$$

24.3. For a whole class of economic problems, the objective function appears as a combination of a linear function and a linear fractional function [483], [488], [643], [801]. For example, this type of problem is obtained when one optimizes the salary fund and the profitability of a company:

$$\max \{ F(x) = c'x + \frac{d_1'x}{d_2'x} \mid Ax \leq b \ ; \ x \geq 0 \}$$

24.4. The problems of selection of product variants [110] lead to fractional models. Assume that we know m constructive variants of a given product made up from n functional subassemblies. For the subassembly $i (i = 1, \ldots, n)$, let $A_{i1}, \ldots, A_{im}$ be the constructive variants. For each A_{ij}, we know a performance coefficient a_{ij} and a cost coefficient b_{ij}. The problem is to find, for each subassembly i, the proper constructive variant k_i^* such that the combination $(k_1^*, k_2^*, \ldots, k_n^*)$ to maximize the expression

$$V(A, B) = \sum_{i=1}^{n} a_{ik_i} \Big/ \sum_{i=1}^{n} b_{ik_i}, \quad 1 \leq k_i \leq m$$

If we introduce the 0 - 1 variable

$$x_{ij} = \begin{cases} 1 & \text{if the element on the line } i \text{ and column } j \text{ was selected} \\ 0 & \text{otherwise} \end{cases}$$

we obtain the following fractional programming model in the variables 0 - 1:

$$\max \left\{ \left(\sum_{i=1}^{n}\sum_{j=1}^{m} a_{ij}x_{ij}\right) / \left(\sum_{i=1}^{n}\sum_{j=1}^{m} b_{ij}x_{ij}\right) \mid \sum_{j=1}^{n} x_{ij} = 1,\ i=1, \ldots, m,\ x_{ij} = 0 \ \ or \ \ 1 \right\}$$

where the numbers $a_{ij} \geq 0$, $b_{ij} > 0$ are given.

24.5. Okabe [699] considered the spatial distributions maximizing or minimizing Geary's contiguity ratio

$$\max\ (\min)\ G = \frac{n - 1}{2\sum_{i=1}^{n} L_i} \; \frac{\sum_{i=1}^{n}\sum_{j=1}^{n} \delta_{ij}\,(x_i - x_j)^2}{\sum_{i=1}^{n} x_i^2}$$

subject to

$$\sum_{i=1}^{n} x_i = 0 \tag{1.16}$$

$$\sum_{i=1}^{n} x_i^2 = 1 \tag{1.17}$$

where n is the number of counties covering a region, L_i is the number of counties being adjacent to county i, x_i is an attribute value of county i satisfying (1.16) and

$$\delta_{ij} = \begin{cases} 1 \text{ if county } i \text{ is joined to county } j, \ (i \neq j), \\ 0 \qquad\qquad \text{if otherwise or } i=j \end{cases}$$

The constraint (1.17) is added because Geary's ratio is unique up to a scale transformation.

24.6. An example of application to probability is given next [822]. Let A_1 and A_2 be two events in a probability space, and let $p_i = P(A_i)(i=1, 2)$ be the probabilities of realization of these events. We want to find the lower bound to the probability of simultaneous realization of events A_1 and A_2 such that at least one of the events A_i has been realized, *i.e.*,, for $P(A_1 \cap A_2 \mid A_1 \cup A_2) = P(A_1 \cap A_2)/P(A_1 \cup A_2)$.

Let $E_1 = A_1 \cap A_2$; $E_2 = A_1 \cap \mathbf{C}A_2$; $E_3 = \mathbf{C}A_1 \cap A_2$; $E_4 = \mathbf{C}A_1 \cap \mathbf{C}A_2$ ($\mathbf{C}A$ is the event complementary to A) be a complete system of events, *i.e.*, an incompatible system of events and their reunion is equal to unitary event, $A_1 \cup A_2 = E_1 + E_2 + E_3$, $A_1 = E_1 \cup E_2$, $A_2 = E_1 \cup E_3$, $A_1 \cap A_2 = E_1$.

Let $x_i = P(E_i)$. Finding a lower margin for $P(A_1 \cap A_2 \mid A_1 \cup A_2)$ is equivalent to finding the solution to the fractional programming problem

$$\min \{x_1/(x_1+x_2+x_3) \mid x_1+x_2+x_3+x_4=1, \ x_1+x_2+x_3>0, \ x_1+x_2=p_1, \ x_1+x_3 = p_2, \ x_i \geq 0 \}$$

24.7. A few applications of disjunctive linear fractional programming in machine flow-shop scheduling (where the objective is of finding minimum processing-time of a product in a machine flow-shop), technology selection, regional planning etc. were identified in [719] and it seems possible to treat some of the problems in economic planning through this development.

Recently, Golovanov and Sarkisyan [417] considered a problem of forecasting of vehicle system for future where the use of fractional criterion, namely minimization of ratio of expenses to effect, is discussed. Here, besides the general constraints on resources, those associated with each particular vehicle system are also to be included. Thus this is a situation where disjunctive fractional programming would be applicable because generally the choice of one system excludes the others. This is particularly true in choosing the defence equipment for future where the choice of one technology rules out the other.

In regional planning, especially in deciding the relocation of certain activities like industries, markets etc., the objective of decongestion is to be achieved through maximization of the benefit to the cost. However, the linkages between various activities often rule out shifting of certain activities which could be expressed by disjunctive constraints.

Similar situation arising in portfolio management, media selection in advertising, selection of appropriate technology could also be treated by the development presented in this book.

CHAPTER 2

CONVEX, QUASICONVEX, PSEUDOCONVEX, LOGARITHMIC CONVEX, αm-CONVEX, AND INVEX FUNCTIONS

The importance of convex functions is well known in optimization problems. Convex functions come up in many mathematical models used in economics, engineering, etc. More often than not, convexity appears as a natural property of the various functions and domains encountered in such models. Furthermore, the property of convexity is invariant with respect to certain operations and transformations. Numerous efficient theoretical and practical methods are available in the literature for determining the minimum of a convex function. But for many problems encountered in economics and engineering the notion of convexity does no longer suffice. Hence it was necessary to extend the notion of convexity into the notions of pseudoconvexity, quasiconvexity, etc. Such functions will be introduced and characterized in this chapter.

Let R^n be the n-dimensional Euclidean space. An element $\boldsymbol{x} \in R^n$ of this space will be denoted with bold lower case letters (e.g. $\boldsymbol{x}_0$, $\boldsymbol{x}$, $\boldsymbol{y}$, $\boldsymbol{x}^1$, $\boldsymbol{y}^1$) and will be considered to be a column vector. For $\boldsymbol{x}, \boldsymbol{y} \in R^n$ we denote by $\boldsymbol{x}'\boldsymbol{y} = x_1y_1 + \ldots + x_ny_n$ the scalar product between $\boldsymbol{x}$ and $\boldsymbol{y}$, where $\boldsymbol{x}'$signifies the transpose of a vector or matrix. The vector norm of $\boldsymbol{x} \in R^n$ is defined as $\|\boldsymbol{x}\| = \sqrt{\boldsymbol{x}'\boldsymbol{x}} = (\sum_{i=1}^{n} x_i^2)^{1/2}$. We assume the reader to be familiar with the topology of R^n space, and that the notions of sphere, neighborhood of a point, open (closed) set, bounded set, compact set, and convex set are known and need not be defined. Let's consider the convex and non-empty set $X \subset R^n$ and the function $f : X \rightarrow R$. We denote by $C^m(X)$ the class of continuous functions with partial derivatives up to order m continuous on an open set $Y \supseteq X$. If f is differentiable in $\boldsymbol{x}_0 \in X$, then

$$\nabla f(\boldsymbol{x}_0) = \left[\frac{\partial f(\boldsymbol{x}_0)}{\partial x_1}, \ldots, \frac{\partial f(\boldsymbol{x}_0)}{\partial x_n}\right]'$$

is the gradient of f in $\boldsymbol{x}_0$; and if $f \in C^2(X)$, then

$$Hf(x_0) = \nabla^2 f(x_0) = \left[\frac{\partial^2 f(x_0)}{\partial x_i \partial x_j} \right], \quad i, j = 1, \ldots, n,$$

is the Hessian matrix of function f in x_0. If $f : R^n \to R^m$ is a differentiable vector function of vector variable, then

$$\nabla f(x_0) = \left[\frac{\partial f_i(x_0)}{\partial x_j} \right], \quad i = 1, \ldots, m; \quad j = 1, \ldots, n,$$

If $g : R^m \times R^k \to R$ is a real and differentiable function, then

$$\nabla_1 g(y, z) = \left(\frac{\partial g(y, z)}{\partial y_1}, \ldots, \frac{\partial g(y, z)}{\partial y_m} \right)',$$

and

$$\nabla_2 g(y, z) = \left(\frac{\partial g(y, z)}{\partial z_1}, \ldots, \frac{\partial g(y, z)}{\partial z_k} \right)'.$$

2.1. CONVEX SETS

Although we have assumed that convex set theory in R^n is known to the reader, it is useful to briefly review some of its main features.

Definition 2.1.1. A non-empty set $X \subset R^n$ is *convex* if $\forall\, x, y \in X$ and $\forall\, t \in [0, 1]$ one has

$$[x, y] = \{ x' = tx + (1 - t)y \mid t \in [0, 1]\} \subset X$$

We denoted by $[x, y]$ the closed segment between x and y. Examples of convex sets are:

a) The empty set $\varnothing$, and the whole space R^n;

b) The hyperplane
$$H(a, \alpha) = \{ x \in R^n \mid a'x = \alpha \}, \quad a \in R^n, \quad a \neq 0, \quad \alpha \in R;$$

c) The linear variety $V(a, b) = \{ x \in R^n \mid Ax = b \}$ where A is an $m \times n$ matrix, and $b \in R^m$;

d) The closed half-space
$$S(a, \alpha) = \{ x \in R^n \mid a'x \leq \alpha \}, \quad a \in R^n, \quad \alpha \in R;$$

e) The polyhedral convex set $T(A, b) = \{ x \in R^n \mid Ax \leq b \}$

f) The open sphere of center a and radius r,
$$S(a, r) = \{ x \in R^n \mid \|x - a\| < r \}, \quad a \in R^n, \quad r > 0;$$

g) The ellipsoid $S = \{ x \in R^n \mid x'Cx \leq \gamma \}$, where C is an $n \times n$ positive definite matrix, and $\gamma > 0$.

It can be easily proved that the intersection of any family of convex sets is a convex set, whereas the reunion generally is not convex. The set of convex sets is closed towards summation, and multiplication by scalar values, i.e. if $X, Y \subset R^n$ are convex, and $\alpha \in R$, the sets $X + Y = \{ x + y \mid x \in X, \ y \in Y \}$ and $\alpha X = \{ \alpha x \mid x \in X \}$ are convex.

Definition 2.1.2. A convex linear combination of points $x^1, ..., x^m \in R^n$ is a point $x \in R^n$ of the form $x = \sum_{i=1}^{m} t_i x^i$, $t_i \geq 0$ $(i = 1, ..., m)$ and $\sum_{i=1}^{m} t_i = 1$.
Equivalent definitions for convex sets are given by the following theorem:

Theorem 2.1.1. *Let $X \subset R^n$, non-empty. Then the following statements are equivalent: a) X is convex; b) X contains all the convex linear combinations of points from X; c) for any $x \in X$ and any $y \in R^n$, the set $T(x, y) = \{ t \in R \mid x + ty \in X \}$ is convex.*

Proof. a) ⇒ b) The proof is obtained through induction with respect to m. For $m = 1$ the property b) is obvious. Let's assume it is true for $m - 1$, and then let's show that it also holds for m. Let $t_i \geq 0$ $(i = 1, ..., m)$, $\sum_{i=1}^{m} t_i = 1$ and $x = \sum_{i=1}^{m} t_i x^i$ with $x^i \in X$, $(i = 1, ..., m)$. There is at least one $t_i \neq 1$, since if we had $t_i = 1$ for any $i = 1, ..., m$, then we would get $\sum_{i=1}^{m} t_i = m$. Assume that $t_1 \neq 1$. Then

$$x = t_1 x^1 + \sum_{i=2}^{m} t_i x^i = t_1 x^1 + (1 - t_1) \sum_{i=2}^{m} \frac{t_i}{1 - t_1} x^i \in X$$

due to $\sum_{i=2}^{m} \frac{t_i}{1 - t_1} = 1$, and to the induction initial assumption. The implication b) ⇒ a) is trivial.
a) ⇒ c) Let X be convex, $t_1, t_2 \in T(x, y)$, and $\lambda \in [0, 1]$. Let's prove that $\lambda t_1 + (1 - \lambda) t_2 \in T(x, y)$. Since $t_1, t_2 \in T(x, y)$, it follows that $\bar{x} = x + t_1 y \in X$, $\bar{y} = x + t_2 y \in X$; and since X is convex, it results that $\lambda \bar{x} + (1 - \lambda)\bar{y} \in X$. But

$$\lambda \bar{x} + (1 - \lambda)\bar{y} = x + [\lambda t_1 + (1 - \lambda) t_2] y \in X \tag{2.1.1}$$

implying $\lambda t_1 + (1 - \lambda) t_2 \in T(x, y)$.
c) ⇒ a) If $\bar{x}, \bar{y} \in X$ and $T(x, y)$ is convex, then relation (2.1.1) implies $\lambda \bar{x} + (1 - \lambda)\bar{y} \in X$, i.e. X is convex.

Definition 2.1.3. The *convex hull* of a set $X \subset R^n$ (denoted *conv* X), is the intersection of all convex sets including X, i.e. $conv\ X = \bigcap_{C \supset X} C$. Obviously, *conv* X is itself a convex set, and it includes X. In fact, it is the smallest convex set including X. It can also be verified that the set *conv* X is bounded and compact together with X, and that

$$conv\ X = \{ x \in R^n \mid x = \sum_{i=1}^{n+1} t_i x^i,\ x^i \in X,\ t_i \geq 0,\ \sum_{i=1}^{n+1} t_i = 1 \}.$$

Let $H = \{ x \in R^n \mid a'x = \alpha \}$ be a hyperplane, and $H^- = \{ x \in R^n \mid a'x < \alpha \}$ and $H^+ = \{ x \in R^n \mid a'x > \alpha \}$ be the two open half-spaces determined by H.

Definition 2.1.4. A set $X \subset R^n$ is said to be situated on the same side of the hyperplane H if $X \subseteq \bar{H}^-$ or $X \subseteq \bar{H}^+$, where $\bar{H}$ denotes the closure of H.

Definition 2.1.5. A hyperplane H is called *supporting hyperplane* for the set $X \subset R^n$ if X is situated on the same side of H and $\inf_{x \in X} (a'x) = \alpha$ or $\sup_{x \in X} (a'x) = \alpha$ depending on whether X is situated in $\bar{H}^+$ or in $\bar{H}^-$.

Definition 2.1.6. A hyperplan H is said to separate the sets X, $Y \subset R^n$ if $X \subseteq \bar{H}^-$ implies $Y \subseteq \bar{H}^+$ i.e.

$$\sup_{x \in X} (a'x) \leq \alpha \leq \inf_{x \in Y} (a'x). \tag{2.1.2}$$

If the inequalities (2.1.2) are strict, then we say that X and Y are strict separated by the hyperplane H.

Let's state, without proof, a few useful results:

Theorem 2.1.2. *(Strict separation of convex sets). If X, $Y \subset R^n$ are convex sets, and if $X \cap Y = \varnothing$, X is closed, and Y is compact, then $a \in R^n$ exists with $\|a\| = 1$ such that*

$$\sup_{y \in Y} (a'y) < \inf_{x \in X} (a'x)$$

Theorem 2.1.3. *(Supporting hyperplane) If x_0 belongs to the boundary of the convex set $X \subset R^n$, then $a \in R^n$ exists with $\|a\| = 1$ such that*

$$\inf_{x \in X} (a'x) = a'x_0$$

Theorem 2.1.4. *(Separation of convex sets). If X, $Y \subset R^n$ are convex sets, and if $X \cap Y = \varnothing$, then $a \in R^n$ exists with $\|a\| = 1$ such that*

$$\inf_{x \in X} (a'x) \geq \sup_{y \in Y} (a'y)$$

Theorem 2.1.5.* *Let H_{n-1} be a supporting hyperplane of a non-empty, compact and convex set S_n. Then the set $S_{n-1} = H_{n-1} \cap S_n$ is non-empty, compact and convex in the $(n - 1)$ - dimensional space.*

Definition 2.1.7. A point x of a convex set X ($x \in X$) is called *extreme point* of X if there is no $x^1, x^2, x^1 \neq x^2$ and no $t \in (0, 1)$ such that $x = tx^1 + (1 - t)x^2$.

Theorem 2.1.6. *Every non-empty, compact and convex set admits at least one extreme point.*

2.2. CONVEX, QUASICONVEX, PSEUDOCONVEX, LOGARITHMIC CONVEX, αm-CONVEX, AND INVEX FUNCTIONS. EQUIVALENT DEFINITIONS

Let $X \subseteq R^n$ be a non-empty convex set, and let $f : X \rightarrow R$ be a numerical function defined on X. Then:

Definition 2.2.1. The function f is *convex* (*cx*) on X if for $\forall\, x^1, x^2 \in X$ and $\forall\, t \in [0, 1]$

$$f[tx^1 + (1 - t)x^2] \leq tf(x^1) + (1 - t)f(x^2). \tag{2.2.1}$$

The function f is *strict convex* (*sx*) on X if $\forall\, x^1, x^2 \in X$, $(x^1 \neq x^2)$ and $\forall\, t \in (0, 1)$ the inequality (2.2.1) is strict.

Observation: The convexity of f may not be required on the entire domain X, but only in a point $x_0 \in X$. The function f is convex in $x_0 \in X$ if $\forall\, x \in X$ and $\forall\, t \in [0, 1]$ we have

$$f[tx + (1 - t)x_0] \leq tf(x) + (1 - t)f(x_0).$$

If f is convex in any point $x_0 \in X$, then it is convex on X. This observation is true for all the definitions to follow.

Definition 2.2.2. The function f is *quasiconvex* (*qx*) on X if $\forall\, x^1, x^2 \in X$ and $\forall\, t \in [0, 1]$

$$f[tx^1 + (1 - t)x^2] \leq \max\, [f(x^1), f(x^2)]. \tag{2.2.2}$$

The function f is *explicit quasiconvex* (*eqx*) on X if for $\forall\ x^1, x^2 \in X(f(x^1) \neq f(x^2))$ and $\forall\ t \in (0, 1)$ the inequality (2.2.2) is strict. The function f is *strict explicit*

---

*C. Berge, *Topological Spaces*, The Macmillan Company, New York, 1963 *quasiconvex* (*seqx*) on X if $\forall\, x^1, x^2 \in X(x^1 \neq x^2)$ and $\forall\, t \in (0, 1)$ the inequality (2.2.2) is strict.

Definition 2.2.3. The function f is *pseudoconvex* (*pcx*) on X if X is open, f is differentiable, and if $\forall\, x^1, x^2 \in X$, $(x^1 - x^2)' \nabla f(x^2) \geq 0 \Rightarrow f(x^1) \geq f(x^2)$
or equivalently

$$f(x^1) < f(x^2) \Rightarrow (x^1 - x^2)' \nabla f(x^2) < 0.$$

The function f is *strict pseudoconvex* (*spcx*) on X if X is open, f is differentiable, and if $\forall\, x^1, x^2 \in X(x^1 \neq x^2)$

$$(x^1 - x^2)' \nabla f(x^2) \geq 0 \Rightarrow f(x^1) > f(x^2).$$

Definition 2.2.4. The function f is *logarithmic-convex* (*L-convex*) on X if X is open, $f(x) > 0$ $(\forall\, x \in X)$, and if $\ln f$ is convex, i.e. if $\forall\, x^1, x^2 \in X$ and $t \in [0, 1]$:

$$f[tx^1 + (1-t)x^2] \leq [f(x^1)]^t\, [f(x^2)]^{1-t}\,. \tag{2.2.3}$$

The function f is *strict logarithmic-convex* (*strict L-convex*) on X if X is open, $f(x) > 0$ $(\forall\, x \in X)$, and if $\ln f$ is strict convex i.e. $\forall\, x^1, x^2 \in X(x^1 \neq x^2)$ and $\forall\, t \in (0, 1)$ the inequality (2.2.3) is strict.

Definition 2.2.5. Let $\omega \in (0, \infty)$. The function f is *ω-convex* (*ω-concave*) on X if for $\forall\, x^1, x^2 \in X$ and $\forall\, t \in [0, 1]$:

$$f[tx^1 + (1-t)x^2] \underset{(\geq)}{\leq} \frac{1}{\omega} ln[te^{\omega f(x^1)} + (1-t)e^{\omega f(x^2)}]. \tag{2.2.4}$$

Definition 2.2.6. Let $f : X \to R_+$ be a function, and let $\alpha \in R$, $\alpha \neq 0$. The function f is *αm-convex* (*αm-concave*) on X if $\forall\, x^1, x^2 \in X$ and $\forall\, t \in [0, 1]$:

$$f[tx^1 + (1-t)x^2] \underset{(\geq)}{\leq} [tf(x^1)^\alpha + (1-t)f(x^2)^\alpha]^{1/\alpha}. \tag{2.2.5}$$

The function f is *strict αm-convex* (*strict αm-concave*) on X if for $\forall\, x^1, x^2 \in X$ $(x^1 \neq x^2)$ and $\forall\, t \in (0, 1)$ the inequality (2.2.5) is strict.

The function f is *Om-convex* (*Om-concave*) on X if $\forall\, x^1, x^2 \in X$ and $\forall\, t \in (0, 1)$:

$$f[tx^1 + (1-t)x^2] \underset{(\geq)}{\leq} \lim_{\alpha\to 0}[tf(x^1)^\alpha + (1-t)f(x^2)^\alpha]^{1/\alpha}. \tag{2.2.6}$$

The function f is *∞m-convex* (*$-\infty m$-concave*) on X if $\forall\, x^1, x^2 \in X$ and $\forall\, t \in [0, 1]$:

$$f[tx^1 + (1-t)x^2] \underset{(\geq)}{\leq} \lim_{\substack{\alpha\to\infty \\ (\alpha\to-\infty)}} [tf(x^1)^\alpha + (1-t)f(x^2)^\alpha]^{1/\alpha}. \tag{2.2.7}$$

The function f is called *concave (cv), strict concave (sv), explicit quasiconcave (eqv), strict explicit quasiconcave (seqv), pseudoconcave (pcv), strict pseudoconcave (spcv)* on X if $(-f)$ is convex, strict convex, explicit quasiconvex, strict explicit quasiconvex, pseudoconvex, strict pseudoconvex on X, respectively.

Definition 2.2.7. The function f is *L-concave*, and *strict L-concave* on X if lnf is, respectively, concave, and strict concave on X.

Definition 2.2.8. The function f is *linear* if it is both convex and concave. The function f is *pseudolinear* on X if it is both pseudoconvex and pseudoconcave.

It can be easily verified that the set of convex functions is closed with respect to addition and multiplication with positive scalars.

Definition 2.2.9. The function f is *monotonic* if it is both quasiconcave and quasiconvex i.e. for $\forall\ x^1, x^2 \in X$ and $\forall\ t \in [0, 1]$ we have

$$\min\,[f(x^1), f(x^2)] \leq f[tx^1 + (1 - t)x^2] \leq \max\,[f(x^1), f(x^2)].$$

Definition 2.2.10. A differentiable function f is called *invex* on X if a vector function $\eta : X \times X \to R^n$ exists such that

$$\forall\ x, u \in X : f(x) - f(u) \geq [\eta(x, u)]'\ \nabla f(u)$$

One can consider generalizations of the concept of invexity (pseudoinvexity, quasiinvexity). In [656] Mititelu and Stancu-Minasian present several types of invexity, pseudoinvexity and quasiinvexity at a point of nonlinear functions using Clarke's generalized directional derivative. Direct implications of the relations existing between various types of invexity, and the concept of generalized invexity, are presented.

Let $\Lambda \subseteq R^m \times R^k$ be an arbitrary set and let $\varphi : \Lambda \to R$ be a numerical function.

Definition 2.2.11. The function φ is on Λ:

(i) *increasing-decreasing,* if $\forall\ (y^1, z^1), (y^2, z^2) \in \Lambda,\ y^1 \geq y^2,\ z^1 \leq z^2$ we have $\varphi(y^1, z^1) \geq \varphi(y^2, z^2)$;

(ii) *increasing* - ———, if $\forall\ (y^1, z)$ and $(y^2, z) \in \Lambda,\ y^1 \geq y^2$ we have $\varphi(y^1, z) \geq \varphi(y^2, z)$;

(iii) *decreasing* - ———, if $\forall (y^1, z)$ and $(y^2, z) \in \Lambda,\ y^1 \leq y^2$ we have $\varphi(y^1, z) \geq \varphi(y^2, z)$;

(iv) *strict increasing-decreasing,* if $\forall\ (y^1, z^1)$ and $(y^2, z^2) \in \Lambda, y^1 > y^2,\ z^1 \leq z^2$ we have $\varphi(y^1, z^1) > \varphi(y^2, z^2)$.

Similarly, one defines on Λ functions that are strict increasing - strict decreasing, increasing - strict decreasing, increasing - strict decreasing, strict increasing - ———, strict decreasing - ———.

It can be easily verified that for φ differentiable, and for Λ an open and convex set, the conditions (i) - (iv) are equivalent with the following conditions:

(i') $\nabla_1\varphi(y, z) \geq 0$ and $\nabla_2\varphi(y, z) \leq 0$

(ii') $\nabla_1\varphi(y, z) \geq 0$

(iii') $\nabla_1\varphi(y, z) \leq 0$

(iv') $\nabla_1\varphi(y, z) > 0$ and $\nabla_2\varphi(y, z) \leq 0$.

When the nature of a concrete function must be established, one cannot always easily verify the corresponding condition from definitions 2.2.1. - 2.2.10. Hence it is found useful to have alternative equivalent definitions for certain classes of functions:

Theorem 2.2.1. *(Equivalent definitions for convex functions). Let $X \subseteq R^n$ be a convex set and let $f : X \to R$. The following definitions are equivalent :*

a) *f is convex on X;*

b) $\forall\, x^i \in X$ *and* $\forall\, t_i \in [0, 1]$, $i = 1, ..., m$, $\sum_{i=1}^{m} t_i = 1$ *Jensen's inequality holds*

$$f\left(\sum_{i=1}^{m} t_i x^i\right) \leq \sum_{i=1}^{m} t_i f(x^i);$$

c) $\forall\, x \in X$ *and* $y \in R^n$, *the function* $g_{x,y}(t) = f(x + ty) : T(x, y) \to R$ *is convex ($T(x, y)$ is defined in Theorem 2.1.1) ;*

d) $\forall\, x^1, x^2 \in X$, *the function* $h_{x^1,x^2}(\lambda) = f(\lambda x^1 + (1-\lambda)x^2) : [0, 1] \to R$ *is convex on* $[0, 1]$ *;*

e) *the set* $(X, f) = \{ (x, z) \in X \times R \mid f(x) \leq z \}$, *called the epigraph of f, is convex;*

f) $\forall\, x^0 \in X$ *(X open)* $u^0 \in R^n$ *exists such that*

$$f(x) - f(x^0) \geq u^{0\prime}(x - x^0), \quad \forall\, x \in X.$$

Proof. a) $\Rightarrow$ b) is shown by induction through m using Theorem 2.1.1 point b). Implication b) $\Rightarrow$ a) is trivial.

a) $\Rightarrow$ c). Let $t_1, t_2 \in T(x, y)$ and $\lambda \in [0, 1]$. According to Theorem 2.1.1, point c), $T(x, y)$ is convex, hence $\lambda t_1 + (1 - \lambda)t_2 \in T(x, y)$. Using the convexity of f we get:
$g_{x,y}[\lambda t_1 + (1 - \lambda)t_2] = f[x + (\lambda t_1 + (1 - \lambda)t_2)y] = f[\lambda(x + t_1 y) + (1 - \lambda)(x + t_2 y)] \leq$
$\leq \lambda f(x + t_1 y) + (1 - \lambda)f(x + t_2 y) = \lambda g_{x,y}(t_1) + (1 - \lambda)g_{x,y}(t_2)$.

c) $\Rightarrow$ d). Let $x^1, x^2 \in X$ and $\lambda \in [0, 1]$. We have

$$h_{x^1,x^2}(\lambda) = f[\lambda x^1 + (1 - \lambda)x^2] = f[x^2 + \lambda(x^1 - x^2)] = g_{x^2,x^1-x^2}(\lambda).$$

Since the function g_{x^2,x^1-x^2} is convex, it follows that the function h_{x^1,x^2} is also convex.

d) $\Rightarrow$ e). Let $(x^1, z^1), (x^2, z^2) \in (X, f)$ and $\lambda \in (0, 1)$. We have $f(x^1) \leq z^1$, $f(x^2) \leq z^2$ and $\lambda = \lambda \cdot 1 + (1 - \lambda) \cdot 0$. Using the convexity of h_{x^1,x^2}, we get

$$f[\lambda x^1 + (1 - \lambda)x^2] = h_{x^1,x^2}(\lambda) \leq \lambda\, h_{x^1,x^2}(1) + (1 - \lambda)\, h_{x^1,x^2}(0) =$$
$$\lambda f(x^1) + (1 - \lambda)f(x^2) \leq \lambda z^1 + (1 - \lambda)z^2.$$

Hence $\lambda(x^1, z^1) + (1 - \lambda)(x^2, z^2) \in (X, f)$.

e) $\Rightarrow$ f). Let X be an open set, and $x^0 \in X$. The point $(x^0, f(x^0))$ is a boundary point of the convex set (X, f); according to Theorem 2.1.3, $(v, v^0) \neq \mathbf{0}$ exists such that

$$v'x + v^0 z \geq v'x^0 + v^0 f(x^0) \quad \forall \quad (x, z) \in (X, f) \tag{2.2.8}$$

If $v^0 = 0$, then $v'(x - x^0) \geq 0$ for all $x \in X$, and, in particular, in a neighborhood of x^0, implying $v = 0$, and hence $(v, v^0) = \mathbf{0}$ which is absurd. If $v^0 < 0$, then one can take z sufficiently large to have $v'x + v^0 z < v'x^0 + v^0 f(x^0)$ thus contradicting (2.2.8). It follows that $v^0 > 0$. We choose $z = f(x)$ and $u^0 = -v/v^0$; hence (2.2.8) yield the inequality $f(x) - f(x^0) \geq u^{0\prime}(x - x_0)$.

f) ⇒ a). Let $x^1, x^2 \in X$ and $t \in [0, 1]$. According to f), $\forall \; x^0 \in X$ there is $u^0 \in R^n$ such that $f(x^1) - f(x^0) \geq u^{0\prime}(x^1 - x^0)$ and $f(x^2) - f(x^0) \geq u^{0\prime}(x^2 - x^0)$. Multiplying these inequalities by t and $1 - t$ respectively, and then adding up, we get $tf(x^1) + (1 - t)f(x^2) - f(x^0) \geq u^{0\prime}[tx^1 + (1 - t)x^2 - x^0]$. Taking $x^0 = tx^1 + (1 - t)x^2$, the previous inequality yields (2.2.1), i.e. f is convex.

Theorem 2.2.2. *(Equivalent definitions for convex and differentiable functions). Let $X \subseteq R^n$ be an open and convex set and let $f \in C^1(X)$. The following statements are equivalent:*

a) *f is convex on X;*

b) *$\forall\, x \in X,\; y \in R^n$, the function $g'_{x,y}(t) = y' \nabla f(x + ty) : T(x, y) \to R$ is increasing;*

c) *$\forall\, x^1, x^2 \in X$, the function $h'_{x^1, x^2}(\lambda) = (x^1 - x^2)' \nabla f(\lambda x^1 + (1 - \lambda)x^2)$ is increasing on $[0,1]$;*

d) *$\forall\, x^1, x^2 \in X$, we have $f(x^1) - f(x^2) \geq (x^1 - x^2)' \nabla f(x^2)$;*

e) *$\forall\, x^1, x^2 \in X$ we have $f(x^1) - f(x^2) \leq (x^1 - x^2)' \nabla f(x^1)$;*

f) *$\forall\, x^1, x^2 \in X$, we have $(x^1 - x^2)'(\nabla f(x^1) - \nabla f(x^2)) \geq 0$;*

g) *$(f \in C^2(X))$ The Hessian matrix $Hf(x)$ is positive semi-definite for all $x \in X$.*

Proof. a) ⇔ b). According to Theorem 2.2.1, f is convex if and only if the function $g_{x,y} : T(x, y) \to R$ is convex, i.e. its derivative $g'_{x,y} = y' \nabla f(x + ty)$ is increasing.

a) ⇔ c). According to Theorem 2.2.1, f is convex if and only if $\forall\, x^1, x^2 \in X$, the function $h_{x^1,x^2} : [0,1] \to R$ is convex on $[0, 1]$, i.e. its derivative $h'_{x^1,x^2} = (x^1 - x^2)' \nabla f(\lambda x^1 + (1 - \lambda)x^2)$ is increasing on $[0, 1]$.

c) ⇒ d). Since the function h'_{x^1,x^2} is increasing on $[0, 1]$, it follows that $h'_{x^1,x^2}(\lambda) \geq h'_{x^2,x^2}(0)$ for all $\lambda > 0$, i.e.

$$(x^1 - x^2)' \nabla f(\lambda x^1 + (1 - \lambda)x^2) \geq (x^1 - x^2)' \nabla f(x^2). \tag{2.2.9}$$

Then d) follows from (2.2.9), and first order Taylor's formula:

$$f(x^1) - f(x^2) = (x^1 - x^2)' \nabla f(\theta x^1 + (1 - \theta)x^2), \tag{2.2.10}$$

$0 \leq \theta \leq 1$, where we took $\lambda = \theta$.

d) ⇒ e). Indeed, if d) is true, then

$$f(x^1) - f(x^2) = -[f(x^2) - f(x^1)] \leq -(x^2 - x^1)' \nabla f(x^1) = (x^1 - x^2)' \nabla f(x^1).$$

e) ⇒ f). We write e) for x^1 and x^2, and then for x^2 and x^1. By addition we get f).

f) ⇒ c). Let $x^1, x^2 \in X$, $\lambda_1, \lambda_2 \in [0, 1]$, $(\lambda_1 < \lambda_2)$. Since X is convex it follows that $y_1 = \lambda_1 x^1 + (1-\lambda_1)x^2 \in X$, $y_2 = \lambda_2 x^1 + (1-\lambda_2)x^2 \in X$ and hence $y_1 - y_2 = (\lambda_1 - \lambda_2)(x^1 - x^2)$. We have

$$h'_{x^1,x^2}(\lambda_2) - h'_{x^1,x^2}(\lambda_1) = (x^1-x^2)' \nabla f(\lambda_2 x^1 + (1-\lambda_2)x^2) - (x^1 - x^2)' \nabla f(\lambda_1 x^1 + (1-\lambda_1)x^2)$$
$$= [1/(\lambda_1-\lambda_2)](y_1-y_2)' (\nabla f(y_2)-\nabla f(y_1)) = [1/(\lambda_2-\lambda_1)](y_2-y_1)' (\nabla f(y_2)-\nabla f(y_1)) \geq 0.$$

Hence $h'_{x^1,x^2}(\lambda_2) \geq h'_{x^1,x^2}(\lambda_1)$, i.e. $h'_{x^1,x^2}(\lambda)$ is increasing on $[0,1]$.

g) ⇒ d). Let $x^1, x^2 \in X$ and $f \in C^2(X)$. The implication follows immediately from second order Taylor's formula:

$$f(x^1) - f(x^2) = (x^1 - x^2)' \nabla f(x^2) + (1/2)(x^1 - x^2)Hf(\theta x^2 +$$
$$+ (1 - \theta)x^1)(x^1 - x^2), \quad \theta \in (0, 1).$$

b) ⇒ g). From $f \in C^2(X)$ it follows that $g_{x,y}$ is twice differentiable on $T(x, y)$, and from b) it follows that $g''_{x,y}(t) \geq 0$, $t \in T(x, y)$, i.e., $y'Hf(x + ty)y \geq 0$ $(t \in T(x, y))$. For $t = 0 \in T(x, y)$, we have $y'Hf(x)y \geq 0$, $y \in R^n$, i.e., $Hf(x)$ is positive semi-definite for all $x \in X$.

Equivalent definitions for strict convex and differentiable functions can be similarly formulated.

Theorem 2.2.3. *(Equivalent definitions for quasiconvex functions). Let $X \subseteq R^n$ be a convex set, and let $f : X \rightarrow R$. The following definitions are equivalent:*

a) *f is quasiconvex on X;*

b) *$\forall\, x \in X$ and $y \in R^n$, the function $g_{x,y}(t) = f(x + ty) : T(x, y) \rightarrow R$ is quasiconvex on $T(x, y)$;*

c) *$\forall\, x^1, x^2 \in X$, the function $h_{x^1,x^2}(\lambda) = f(\lambda x^1 + (1 - \lambda)x^2) : [0, 1] \rightarrow R$ is quasiconvex on $[0, 1]$;*

d) *$\forall\, \alpha \in R$, the set $X_\alpha = \{ x \in X \mid f(x) \leq \alpha \}$ is convex or empty;*

e) *(X open, $f \in C^1(X)$), $\forall\, x^1, x^2 \in X$,*

$$f(x^1) \leq f(x^2) \Rightarrow (x^1 - x^2)' \nabla f(x^2) \leq 0.$$

Proof. The proof of implications a) ⇒ b) ⇒ c) is similar to the proof of implications a) ⇒ c) ⇒ d) of Theorem 2.2.1.

c) ⇒ d). Let $x^1, x^2 \in X_\alpha$ and $\lambda \in [0, 1]$. Since the function h_{x^1,x^2} is quasiconvex it follows that $h_{x^1,x^2}(\lambda)) \leq \max\, (h_{x^1,x^2}(1),\ h_{x^1,x^2}(0)) = \max\, (f(x^1),\ f(x^2))$, hence $f(\lambda x^1 + (1 - \lambda)x^2) \leq \max\, (f(x^1), f(x^2)) \leq \alpha$ i.e., $\lambda x^1 + (1-\lambda)x^2 \in X_\alpha$, and hence X_α

is convex.

d) ⇒ a). Let x^1, $x^2 \in X$, $\lambda \in [0,1]$ and $\alpha = \max\,(f(x^1), f(x^2))$. One has x^1, $x^2 \in X_\alpha$. Since X_α is convex, it follows $\lambda x^1 + (1-\lambda)x^2 \in X_\alpha$, i.e., $f(\lambda x^1 + (1-\lambda)x^2) \le \alpha$. Hence f is quasiconvex.

c) ⇒ e). Let x^1, $x^2 \in X$ such that $f(x^1) \le f(x^2)$ and let $\lambda \in [0,1]$. The function h_{x^1,x^2} is quasiconvex on $[0, 1]$. For $\lambda \in [0, 1]$, one has

$$h_{x^1,x^2}(\lambda) \le \max(h_{x^1,x^2}(1),\, h_{x^1,x^2}(0)) = \max(f(x^1), f(x^2)) = f(x^2) = h_{x^1,x^2}(0).$$

It follows that h_{x^1,x^2} has a maximum on $[0, 1]$ in the point $\lambda = 0$. Since $f \in C^1(X)$, it follows $h'_{x^1,x^2}(0) \le 0$ i.e., $(x^1 - x^2)' \nabla f(x^2) \le 0$.

e) ⇒ c). Let x^1, $x^2 \in X$ be such that $f(x^1) \le f(x^2)$. It is sufficient to show that $h_{x^1,x^2}(\lambda) \le h_{x^1,x^2}(0)$ for all $\lambda \in [0, 1]$, i.e., h_{x^1,x^2} reaches its global maximum in point $\lambda = 0$. For $\lambda = 0$ and $\lambda = 1$, the previous relation is obvious.

Let

$$\Lambda = \{\, \lambda \in (0, 1) \mid h_{x^1,x^2}(\lambda) > h_{x^1,x^2}(0) \,\}.$$

Since $f \in C^1(\Lambda)$, it follows that h_{x^1,x^2} is differentiable and hence continuous on $[0, 1]$ whereas Λ is open. Assume $\Lambda \ne \emptyset$ and let $\lambda_0 = \mathit{sup}\, \{\, \lambda \mid \lambda \in \Lambda \,\}$. Since Λ is open and $\lambda_0 \notin \Lambda$ we have

$$h_{x^1,x^2}(\lambda_0) \le h_{x^1,x^2}(0). \tag{2.2.11}$$

We shall show that h_{x^1,x^2} is constant on Λ. Let $\lambda \in \Lambda$ and $x_\lambda = \lambda x^1 + (1-\lambda)x^2$. Since $h_{x^1,x^2}(\lambda) > h_{x^1,x^2}(0)$ we have $f(x^2) \le f(x_\lambda)$ hence, according to e),

$$(x^2 - x_\lambda)' \nabla f(x_\lambda) = -\lambda(x^1 - x^2)' \nabla f(x_\lambda) \le 0. \tag{2.2.12}$$

On the other hand, since $f(x^1) \le f(x^2) \le f(x_\lambda)$ it follows

$$(x^1 - x_\lambda)' \nabla f(x_\lambda) = (1-\lambda)(x^1 - x^2)' \nabla f(x_\lambda) \le 0. \tag{2.2.13}$$

From (2.2.12) and (2.2.13) it follows that

$$h'_{x^1,x^2}(\lambda) = (x^1 - x^2)' \nabla f(x_\lambda) = 0.$$

i.e., h_{x^1,x^2} is constant on Λ. Hence,

$$\lim_{\substack{\lambda \to \lambda_0 \\ \lambda \in \Lambda}} h_{x^1,x^2}(\lambda) > h_{x^1,x^2}(0).$$

This, taken together with relation (2.2.11), contradicts the continuity of $h_{x^1,x^2}(\lambda)$ in λ_0.

Notes:

1) Equivalent definitions for strict quasiconvex functions can be formulated in a similar manner.
2) Statement e) can also be read as

$$\forall\, x^1, x^2 \in X,\ (x^1 - x^2)' \nabla f(x^2) > 0 \Rightarrow f(x^1) > f(x^2).$$

3) Implication d) ⇒ a) does not happen if f is only convex.

2.3. PROPERTIES OF CONVEX, QUASICONVEX, PSEUDOCONVEX, LOGARITHMIC CONVEX, αm-CONVEX FUNCTIONS. RELATIONS BETWEEN THEM

Theorem 2.3.1. *Let $Y \subseteq R^n$ be an open set, $f: Y \to R$, $X \subseteq Y$ a convex set and $X_f^* = \{x \in X \mid f(x) = \inf_{y \in X} f(y)\}$ the set of the points of global minimum of functionf on the set X.*

a) *If f is differentiable in a point $x^* \in X_f^*$, then*

$$(x - x^*)' \nabla f(x^*) \geq 0, \qquad \forall\, x \in X\,; \tag{2.3.1}$$

b) *If f is pseudoconvex, then $x^* \in X_f^*$ if and only if (2.3.1) is fulfilled.*

c) *If f is quasiconvex on X, then X_f^* is a convex set;*

d) *If f is strict convex on X, then X_f^* contains at most one point.*

Proof. a) Let $x \in X$, $x^* \in X_f^*$ and $x^t = tx + (1 - t)x^* \in X$, $t \in [0, 1]$. From $x^* \in X_f^*$ it follows $f(x^*) \leq f(x^t)$, whereas from the differentiability of f in x^* it follows that a function $\omega(x^*, t(x - x^*))$ exists such that

$$0 \leq f(x^t) - f(x^*) = t(x - x^*)' \nabla f(x^*) + t\|x - x^*\|\,\omega(x^*, t(x - x^*))$$

with $\lim_{t \to 0} \omega(x^*, t(x - x^*)) = 0$.

Taking into the limit $t \to 0$ yields inequality (2.3.1).

b) If f is pseudoconvex, then it also is differentiable in $x^* \in X_f^*$. According to a), the inequality (2.3.1) is true. Conversely, if (2.3.1) is true, then the pseudoconvexity of f implies $f(x) \geq f(x^*)$, i.e., $x^* \in X_f^*$.

c) The proof is obvious if X_f^* is either empty or a line segment. Or, let $x^1, x^2 \in X_f^*$, $x^1 \neq x^2$ and $t \in [0, 1]$. Let's prove that $x^t = tx^1 + (1 - t)x^2 \in X_f^*$. Indeed, from $x^1, x^2 \in X_f^*$ it follows $f(x^1) = f(x^2) \leq f(x)$, $\forall\, x \in X$. But the function f is quasiconvex on X i.e., $f(x^t) \leq f(x^1) = f(x^2)$. We cannot have $f(x^t) < f(x^1)$ since $x^1 \in X_f^*$ i.e., $f(x^t) = f(x^1) = f(x^2)$ which means $x^t \in X_f^*$.

d) If $x^1, x^2 \in X_f^*$, $x^1 \neq x^2$ exist, then from f being strict convex it follows $f(x^t) < tf(x^1) + (1 - t)f(x^2) = f(x^1) = f(x^2)$ $\forall\, t \in [0, 1]$. But this contradicts $x^1 \in X_f^*$.

Let's now establish relations between certain types of convexity defined in preceding section.

Theorem 2.3.2. *Let $X \subseteq R^n$ be convex and let $f : X \to R$.*

a) *If f is strict convex on X, then f is convex on X;*

b) *If f is L-convex on X, then f is convex on X;*

c) *If f is convex and differentiable on X (open), then f is pseudoconvex on X;*

d) *If f is pseudoconvex on X, then f is explicit quasiconvex on X;*

e) *If f is explicit quasiconvex and lower semicontinuous on X, then f is quasiconvex on X.*

Proof. a) The proof is trivial and is done from the definition.
b) If f is L-convex on X, then from relation (2.2.3), and from the relation between the arithmetic mean and the geometric mean it follows

$$f(tx^1 + (1-t)x^2) \leq [f(x^1)]^t\,[f(x^2)]^{1-t} \leq tf(x^1) + (1-t)f(x^2)$$

i.e., f is convex on X.
c) Indeed, let $x^1, x^2 \in X$, f be convex and differentiable such that $f(x^1) < f(x^2)$. Using Theorem 2.2.2., point d), it follows

$$0 > f(x^1) - f(x^2) \geq (x^1 - x^2)'\,\nabla f(x^2)$$

i.e., f is pseudoconvex.
d) Let $x^1, x^2 \in X$ with $f(x^1) < f(x^2)$. Let's prove that $f(x^t) < f(x^2), \forall\, t \in (0, 1)$ $(x^t = tx^1 + (1-t)x^2)$. Assume that $t \in (0, 1)$ exists such that $f(x^t) \geq f(x^2)$, i.e., $t_0 \in (0, 1)$ exists such that

$$\max_{t \in [0,1]} f(x^t) = f(x_0)\ ;\quad x_0 = t_0x^1 + (1-t_0)x^2.$$

Using Theorem 2.3.1, point a), for x^1 and x^2, it follows $(x^1 - x_0)'\,\nabla f(x_0) \leq 0$ and $(x^2 - x_0)'\,\nabla f(x_0) \leq 0$. Taking into account the value of x_0, yields

$$(x^1 - x^2)'\,\nabla f(x_0) = 0$$

i.e., $(x^1 - x_0)'\,\nabla f(x_0) = 0$. Using the fact that f is pseudoconvex, it follows that $f(x^1) \geq f(x_0) \geq f(x^t)$. From this inequality, and from $f(x^t) \geq f(x^2)$, it follows $f(x^1) \geq f(x^2)$. But this contradicts $f(x^2) < f(x^1)$. i.e., for all $t \in (0, 1)$, we have $f(x^t) < f(x^2)$.
e) Let $x^1, x^2 \in X$ be such that $f(x^1) < f(x^2)$. The proof is trivial considering that f is explicit quasiconvex. Let $f(x^1) = f(x^2)$. Assume that $t \in [0, 1]$ exists such that $f(x^t) > f(x^2)$ i.e., the set $A = \{y \in (x^1,x^2) \mid f(y) > f(x^2)\} \neq \varnothing$. But f is explicit quasiconvex hence $\forall\, x \in [x^1, x^2]$ and $\forall\, y \in A - \{x\}$

$$f(y) > f(x) \tag{2.3.2}$$

From f being lower semicontinuous, it follows that A is open relative to the line segment (x^1, x^2), and A has at least two different points $(y_1 \neq y_2)$ for which, according to (2.3.2), $f(y_1) > f(y_2)$ and $f(y_2) > f(y_1)$. But this is absurd.

Notes:
1) The converse of statements a) - e) are not generally true.
2) A similar theorem can be formulated and proved for concave functions. We shall only state them and leave the proof to the reader:

Theorem 2.3.3. *Let $X \subseteq R^n$ convex and let $f : X \to R$. The following statements are true:*

a) If f is strict concave on X, then f is concave on X;
b) If f is concave and positive on X, then f is L-concave on X;
c) If f is L-concave and differentiable on X, then f is pseudoconcave on X;
d) If f is pseudoconcave on X, then f is explicit quasiconcave on X;
e) If f is explicit quasiconcave and upper semicontinuous on X, then f is quasiconcave on X.

As in Theorem 2.3.2, we note that the converse statements a) - e) are not generally true.

Theorem 2.3.4. *Let $X \subseteq R^n$ be a convex set and let $f : X \to R$, $f(x) > 0$, $\forall x \in X$, $f \in C^2(X)$. The following statements are equivalent : a) f is L-convex; b) $1/f$ is L-concave; c) $f = e^h$, (h convex on X) ; d) f^α is convex on X for all $\alpha \in (0, 1)$; e) $\forall x^i \in X$ and $\forall t_i \geq 0$, $\sum_{i=1}^{m} t_i = 1$, the following inequality is true:*

$$f\left[\sum_{i=1}^{m} t_i x^i\right] \leq \prod_{i=1}^{m} (f(x^i))^{t_i}$$

Proof. a) ⇔ b) The function f is L-convex ⇔ $\ln f$ is convex on X ⇔ $-\ln f$ is concave on X ⇔ $1/f$ is L-concave on X.

a) ⇔ c) If f is L-convex on X, then $h = \ln f$ is convex on X and thus $f = e^h$. Conversely, if $f = e^h$, with h convex on X, it follows that $\ln f = h$ is convex i.e., f is L-convex.

c) ⇒ d) If $f = e^h$, with h convex on X, then, for $\alpha > 0$, the function αh is convex and hence $f^\alpha = e^{\alpha h}$. From implication c) ⇒ a) it follows that f^α is L-convex. According to Theorem 2.3.2. b) f^α is convex on X.

a) ⇔ e). The function f is L-convex ⇔ $\ln f$ is convex. According to Jensen's inequality (Theorem 2.2.1, b)) it follows that $\forall x^i \in X$ and $\forall t_i \geq 0$, with $\sum_i t_i = 1$, we have

$$\ln\left[\sum_{i=1}^{m} t_i x^i\right] \leq \sum_{i=1}^{m} t_i \ln f(x^i),$$

which is equivalent to e).

d) ⇒ a). If $f^\alpha(\alpha \in (0, 1))$ is convex on X, then, $\forall x, y \in X$ and $\forall t \in [0, 1]$,

$$(f(tx + (1 - t)y))^\alpha \leq tf^\alpha(x) + (1 - t)f^\alpha(y).$$

Since $\alpha > 0$,

$$f(tx + (1 - t)y)) \le (tf^{\alpha}(x) + (1 - t)f^{\alpha}(y))^{1/\alpha}$$

Even more, for $\alpha \searrow 0$,

$$f(tx + (1 - t)y) \le \lim_{\alpha \searrow 0}(tf^{\alpha}(x) + (1 - t)f^{\alpha}(y))^{1/\alpha} = f(x)^t f(y)^{1-t}$$

i.e., f is L-convex.

The convex functions have a number of properties, e.g., those of continuity on all open sets, of boundedness on all compact sets, of lower boundedness on all bounded sets, of subdifferentiability, of differentiability and of lateral differentiability on a direction. In the following we shall consider the extremum properties of convex functions and of generalized-convex functions.

Theorem 2.3.5. *If $f : X \subseteq R^n \to R$ is explicit quasiconvex on the convex set X, then any local minimum of function f is a global minimum of f on X.*

Proof. Let $x^0 \in X$ be a point of local minimum of function f, i.e., $\varepsilon > 0$ exists such that $\forall\, x \in X \cap S(x^*, \varepsilon)$, ($S(x^*, \varepsilon)$ open sphere of center x^* and radius ε) we have $f(x^0) \le f(x)$. Assume that there is $x^* \in X$, point of global minimum of f i.e., $f(x^*) < f(x^0)$. Since X is convex, we have $x^t = tx^0 + (1 - t)x^* \in X$, $\forall\, t \in (0, 1)$. But if we choose $t < \delta / \|x^* - x^0\|$, then $x^t \in X \cap S(x^*, \varepsilon)$, i.e., $f(x^t) \ge f(x^0)(*)$. On the other hand, from f being explicit quasiconvex on X it follows

$$f(x^t) < \max\,(f(x^0), f(x^*)) = f(x^0)$$

which contradicts $(*)$.

Notes:

a) The property is not true for quasiconvex functions.

b) Considering Theorem 2.3.2, it follows that the property shown in Theorem 2.3.5 is also true for strict convex, L-convex, convex, pseudoconvex functions, and also for those strict explicit quasiconvex.

Theorem 2.3.6. *(Theorem of the uniqueness of the minimum). Let $Y \subseteq R^n$ be an open set and $f : Y \to R$, $X \subset Y$, a convex set.*

a) If f is strict convex on X, then the global minimum of the function f on X is unique; b) If f is strict pseudoconvex on X, then the global minimum of the function f on X is unique;

c) If f is strict explicit quasiconvex on X, then the global minimum of the functionf on X is unique.

Proof. a) Assume that $x^1, x^2 \in X_f^*$ exist such that $x^1 \neq x^2$. Since f is strict convex on X, it follows that for all $t \in (0, 1)$ we have

$$f(tx^1 + (1 - t)x^2) < tf(x^1) + (1 - t)f(x^2) = f(x^2) = f(x^1)$$

But this contradicts x^1 and $x^2 \in X_f^*$. i.e., X_f^* has at most one point.

b) But if the function f is strict pseudoconvex, then $f(x^1) = f(x^2)$ implies $(x^1 - x^2)' \nabla f(x^2) < 0$, which shows that a direction $x^1 - x^2$ from x^2 exists along which the function f drops into a neighborhood $V(x^2)$ of x^2. Since $x^1, x^2 \in X$ convex, it follows that we can select a $t \in (0, 1)$ such that $x^t = tx^2 + (1 - t)x^1 \in V(x^2)$ i.e., $f(x^t) < f(x^1) = f(x^2)$. But this contradicts x^1 and $x^2 \in X_f^*$.
c) If f is strict explicit quasiconvex, then $f(x^1) = f(x^2)$ implies that for all $t \in (0, 1)$ we have $f(x^t) < f(x^1) = f(x^2)$ which contradicts x^1 and $x^2 \in X_f^*$.

Theorem 2.3.7. *If an explicit quasiconvex function, on a convex set $X \subset R^n$, has a point of global maximum in the interior of X, then the function is a constant.*

Proof. Accept that f is not a constant on X. Let x^* be the point of global maximum of the function f placed in the interior of X i.e., $f(x^*) = \max_{x \in X} f(x)$. Since f is not a constant on X, it follows that there is at least one point $x^1 \in X$ $(x^1 \neq x^*)$ such that

$$f(x^1) < f(x^*) \tag{2.3.3}$$

Let $x^2 \in X$, $x^2 \neq x^1$, $x^2 \neq x^*$, $f(x^2) \neq f(x^1)$, and consider $x^t = tx^1 + (1 - t)x^2$, $(t \in (0, 1))$ such that for a certain $t^* \in (0, 1)$ we have $x^{t^*} = x^*$. Since $x^2 \in X$, either a) $f(x^2) = f(x^*)$ or b) $f(x^2) < f(x^*)$ happens. If a), then (2.3.3) implies $f(x^1) < f(x^2)$. If b), then we may have either $f(x^1) < f(x^2)$, or $f(x^1) > f(x^2)$. In conclusion, we get the following mutually exclusive situations:

$$f(x^1) < f(x^2) \text{ if } f(x^2) \leq f(x^*) \tag{2.3.4}$$

and

$$f(x^1) > f(x^2) \text{ if } f(x^2) < f(x^*) \tag{2.3.5}$$

In the case (2.3.4), f being explicit quasiconvex on X implies $f(x^t) < f(x^2)$ for all $t \in (0, 1)$. In particular, for $t = t^*$ we get $f(x^{t^*}) < f(x^2)$ which contradicts (2.3.4). In the case (2.3.5), we get $f(x^*) < f(x^1)$ which contradicts (2.3.3). The contradictions stem from the assumption that f is not constant on X.

Theorem 2.3.8. *Let $X \subset R^n$ be a non-empty, compact and convex set. If $f : X \to R$ is an explicit quasiconvex function, then f reaches its global maximum in one or more extreme points of X.*

Proof. Since X is non-empty, compact (closed and bounded) and convex, Theorem 2.1.6 implies that it has at least one extreme point. If the point of global maximum of f is in the interior of X, then, according to Theorem 2.3.7, f is constant on X. In the extreme point will have the same maximum value. Assume that f reaches its global maximum global on X in a point x^* on the boundary of X. The theorem is proved if x^* is an extreme point. Otherwise, let H_{n-1}^* be a supporting hyperplane of X in x^*, and let

$X_{n-1} = H^*_{n-1} \cap X$. Theorem 2.1.5 implies that X_{n-1} is a non-empty, compact and convex set in an $(n - 1)$-dimensional space. Then, according to Theorem 2.1.6, X_{n-1} admits at least one extreme point. The argument is continued for X_{n-1}. Thus, if x^* is an interior point of X_{n-1}, then f is constant on X_{n-1} and hence it reaches its global maximum in all points of X_{n-1}, in particular in an extreme point of X_{n-1} (and hence of X). If x^* is on boundary of X_{n-1} and is not extreme point, let H^*_{n-2} be a supporting hyperplane of X_{n-1} in x^* and let $X_{n-2} = H^*_{n-2} \cap X_{n-1}$. According to Theorem 2.1.6, the set X_{n-2} is non-empty, compact and convex and is in the $(n - 2)$-dimensional space. This reasoning is continued for X_{n-2}. The process finishes in a finite number of steps since X_0 is a non-empty, compact and convex set in a 0-dimensional space and hence contains only the point x^*, which, according to Theorem 2.1.6, must be an extreme point of X_0,and hence of X. The proof is complete.

2.4. CONVEXITY, QUASICONVEXITY, PSEUDOCONVEXITY OF COMPOSITE FUNCTIONS

We shall show that, under certain conditions, convexity, quasiconvexity and pseudoconvexity are maintained during functions composition. These results will help us to establish the convexity type of various functions appearing in fractional programming.

Theorem 2.4.1. *Let $X \subseteq R^n$ be a convex set, $f : X \to R^m$ an m-dimensional vector function, $g : X \to R^k$ a k-dimensional vector function, $\varphi : R^m \times R^k \to R$ a real function, and let*

$$\theta(x) = \varphi(f(x),\ g(x)).$$

a) If one of the statements given in Table 2.4.1 with respect to f, g and φ, on their definition domains is true, then the following implications are true :

1. *If φ is strict convex on $R^m \times R^k$, then θ is strict convex on X ;*
2. *If X is open, f and g are differentiable on X, and φ is strict pseudoconvex on $R^m \times R^k$, then θ is strict pseudoconvex on X ;*
3. *If φ is strict explicit quasiconvex on $R^m \times R^k$, then θ is strict explicit quasiconvex on X ;*
4. *If φ is explicit quasiconvex on $R^m \times R^k$, then θ is explicit quasiconvex on X.*

Table 2.4.1

	f (on X)	g (on X)	φ (on $R^m \times R^k$)
(i)	convex	concave	increasing-decreasing
(ii)	linear	linear	——— - ———
(iii)	convex	linear	increasing - ———
(iv)	concave	linear	decreasing - ———

b) If one of the statements given in Table 2.4.2 is true then the following implications are true :

1. *If φ is convex (or strict convex) on $R^m \times R^k$, then θ is strict convex on X ;*

2. *If X is open, f and g are differentiable on X, then*

i) *if φ is quasiconvex on $R^m \times R^k$, then θ is strict pseudoconvex on X ;*

ii) *if φ is pseudoconvex on $R^m \times R^k$, then θ is pseudoconvex on X ;*

3. *If φ is explicit quasiconvex on $R^m \times R^k$, then θ is explicit quasiconvex on X;*

4. *If φ is quasiconvex on $R^m \times R^k$, then θ is strict explicit quasiconvex on X.*

Table 2.4.2

	f (on X)	g (on X)	φ (on $R^m \times R^k$)
(i)	strict convex	concave	strict increasing - decreasing
(ii)	convex	strict concave	increasing - strict decreasing
(iii)	strict convex	strict concave	strict increasing - strict decreasing
(iv)	strict convex	linear	strict increasing - ———
(v)	linear	strict concave	——— - strict decreasing

c) If f and g are linear functions, then the following implications are true:

1. *If φ is strict convex on $R^m \times R^k$, then θ is strict convex on X ;*

2. *If φ is strict pseudoconvex on $R^m \times R^k$, then θ is strict pseudoconvex on X ;*

3. *If φ is strict explicit quasiconvex on $R^m \times R^k$, then θ is strict explicit quasiconvex on X ;*

4. *If φ is explicit quasiconvex on $R^m \times R^k$, then θ is explicit quasiconvex on X.*

Proof. a) - 1) Assume **(i)** is true.
Let $x^1, x^2 \in X$, $t \in (0, 1)$, $x^1 \neq x^2$ and $(f(x^1), g(x^1)) \neq (f(x^2), g(x^2))$. From **(i)** it follows

$$\theta(x^t) = \varphi(f(x^t), g(x^t)) \le \varphi(tf(x^1) + (1 - t)f(x^2), \; tg(x^1) + (1 - t)g(x^2)). \quad (2.4.1)$$

Since φ is strict convex on $R^m \times R^k$, it follows that

$$\varphi(tf(x^1) + (1 - t)f(x^2),\ \ tg(x^1) + (1 - t)g(x^2)) \leq$$
$$\leq t\varphi(f(x^1),\ g(x^1)) + (1 - t)\varphi(f(x^2),\ g(x^2)) = t\theta(x^1) + (1 - t)\theta(x^2). \qquad (2.4.2)$$

From (2.4.1) and (2.4.2) it follows that θ is strict convex on X. In the cases **(ii)**, **(iii)** or **(iv)** the proof is similar, with the exception that in the case **(ii)** the inequality (2.4.1) is replaced by an equality.

a) - 2) We are also in the case **(i)**. Let $x^1,\ x^2 \in X$ be such that

$$(x^1 - x^2)'\ \nabla\theta(x^2) \geq 0 \qquad (2.4.3)$$

Differentiation of composite functions yields

$$(x^1 - x^2)'\ \nabla\theta(x^2) = (\nabla_1\varphi(f(x^2),\ g(x^2)))(x^1 - x^2)'\ \nabla f(x^2) +$$
$$+ (\nabla_2\varphi(f(x^2),\ g(x^2)))(x^1 - x^2)'\ \nabla g(x^2) \geq 0 \qquad (2.4.4)$$

Using **(i)**, inequality (2.4.4) becomes

$$0 \leq (\nabla_1\varphi(f(x^2),\ g(x^2)))(x^1 - x^2)'\ \nabla f(x^2) + (\nabla_2\varphi(f(x^2),$$
$$g(x^2)))(x^1 - x^2)'\ \nabla g(x^2) \leq \nabla_1\varphi(f(x^2),\ g(x^2))(f(x^1) - f(x^2)) +$$
$$+ \nabla_2\varphi(f(x^2),\ g(x^2))(g(x^1) - g(x^2)). \qquad (2.4.5)$$

Since the function φ is strict pseudoconvex on $R^m \times R^k$, expression (2.4.5) yields

$$\varphi(f(x^1),\ g(x^1)) > \varphi(f(x^2),\ g(x^2)) \qquad (2.4.6)$$

i.e.,

$$\theta(x^1) > \theta(x^2) \qquad (2.4.7)$$

From (2.4.3) and (2.4.7) it follows that θ is strict pseudoconvex on X.

In the cases **(ii)**-**(iv)**, the proof is similar.

a) - 3) Let **(i)** be true, and let $x^1,\ x^2 \in X$ $(x^1 \neq x^2)$ and $t \in (0,\ 1)$. Assume $\theta(x^1) \leq \theta(x^2)$, and let's prove $\theta(x^t) < \theta(x^2)$. From $\theta(x^1) \leq \theta(x^2)$ it follows that

$$\varphi(f(x^1),\ g(x^1)) \leq \varphi(f(x^2),\ g(x^2)). \qquad (2.4.8)$$

But the function φ is strict explicit quasiconvex, hence (2.4.8) implies

$$\varphi[tf(x^1) + (1-t)f(x^2),\ tg(x^1) + (1-t)g(x^2)] < \varphi(f(x^2),\ g(x^2)) \qquad (2.4.9)$$

Using **(i)**, the inequality (2.4.9) becomes

$$\varphi[f(tx^1 + (1-t)x^2),\ g(tx^1 + (1-t)x^2)] < \varphi(f(x^2),\ g(x^2)) \qquad (2.4.10)$$

i.e., $\theta(x^t) < \theta(x^2)$. The proof is complete. The proof is similar for **(ii)**-**(iv)**.

a) - 4) The proof is identical with a) - 3), with the exception that we take $\theta(x^1) < \theta(x^2)$, and that inequality (2.4.8) is strict. The proof of cases b) and c) does not pose special problems and is done similarly.

A result similar to Theorem 2.4.1 can be stated and proved by replacing "convex" with "concave" and *vice-versa*. We shall give it as a corollary, leaving the proof to the reader.

Corollary 2.4.1. *Let functions f, g, φ and θ be as defined in Theorem 2.4.1.*
a) If one of the statements given in Table 2.4.3 is true, then the following implications are true:

1. *If φ is strict concave on $R^m \times R^k$, then θ is strict concave on X ;*
2. *If X is open, f and g are differentiable on X, and φ is strict pseudoconcave on $R^m \times R^k$, then θ is strict pseudoconcave on X ;*
3. *If φ is strict explicit quasiconcave on $R^m \times R^k$, then θ is strict explicit quasiconcave on X ;*
4. *If φ is strict quasiconcave on $R^m \times R^k$, then θ is explicit quasiconcave on X.*

Table 2.4.3

	f (on X)	g (on X)	φ (on $R^m \times R^k$)
(i)	concave	convex	increasing -decreasing
(ii)	linear	linear	——— - ———
(iii)	concave	linear	increasing - ———
(iv)	convex	linear	decreasing - ———

b) If one of the statements given in Table 2.4.4 is true, then the following implications are true:

1. *If φ is concave (or strict concave) on $R^m \times R^k$, then θ is strict concave on X;*
2. *If X is open, f and g are differentiable on X, then*
i) if φ is quasiconcave on $R^m \times R^k$, then θ is strict pseudoconcave on X;
ii) if φ is pseudoconcave on $R^m \times R^k$, then θ is pseudoconcave on X;
3. *If φ is explicit quasiconcave on $R^m \times R^k$, then θ is explicit quasiconcave on X;*
4. *If φ is quasiconcave on $R^m \times R^k$, then θ is strict explicit quasiconcave on X.*

Table 2.4.4

	f (on X)	g (on X)	φ (on $R^m \times R^k$)
(i)	strict concave	convex	strict increasing - decreasing
(ii)	concave	strict convex	increasing - strict decreasing
(iii)	strict concave	strict convex	strict increasing - strict decreasing
(iv)	strict concave	linear	strict increasing - ———
(v)	linear	strict convex	——— - strict decreasing

c) If f and g are linear functions, then the following implications are true:

1. *If φ is strict concave on $R^m \times R^k$, then θ is strict concave on X ;*
2. *If φ is strict pseudoconcave on $R^m \times R^k$, then θ is strict pseudoconcave on X;*
3. *If φ is strict explicit quasiconcave on $R^m \times R^k$, then θ is strict explicit quasiconcave on X ;*
4. *If φ is explicit quasiconcave on $R^m \times R^k$, then θ is explicit quasiconcave on X.*

Theorem 2.4.2. *Let $X \subseteq R^n$ be a convex and open set, $f : X \to R^m$ an m-dimensional vector function, $g : X \to R^k$ a k-dimensional vector function (f and g differentiable), $\varphi : R^m \times R^k \to R$ a real function, $l : X \to R$ a linear strict positive function, and let $p(x) = f(x)/l(x)$, $q(x) = g(x)/l(x)$ and $\theta(x) = \varphi(p(x), q(x))$. If one of the statements given in Table 2.4.5 relative to f, g and φ is true, then the following implications are true :*

1. *If φ is strict pseudoconvex on $R^m \times R^k$, then θ is strict pseudoconvex on X;*
2. *If φ is pseudoconvex on $R^m \times R^k$, then θ is pseudoconvex on X.*

Table 2.4.5

	f (on X)	g (on X)	φ (on $R^m \times R^k$)
(i)	convex	concave	increasing - decreasing
(ii)	linear	linear	——— - ———
(iii)	convex	linear	increasing - ———
(iv)	concave	linear	decreasing - ———

Proof. 1. Let (i) be true. Let $x^1, x^2 \in X(x^1 \neq x^2)$, and let

$$(x^1 - x^2)\ \nabla\theta(x^2) \geq 0 \tag{2.4.11}$$

Using (2.4.11), and the chain rule for differentiation of composite functions yields:

$$(x^1 - x^2)\ \nabla\theta(x^2) = \nabla_1\varphi(p(x^2),q(x^2))(x^1 - x^2)\ \nabla p(x) + \\ + \nabla_2\varphi(p(x^2),\ q(x^2),\ q(x^2))(x^1 - x^2)\ \nabla q(x) \geq 0. \tag{2.4.12}$$

But f is convex, g is concave, $l > 0$ and φ is strict increasing-decreasing. Hence (2.4.12) yields

$$0 \leq \nabla_1\varphi(p(x^2),\ q(x^2))(x^1 - x^2)\nabla p(x) + \nabla_2\varphi(p(x^2), \\ q(x^2))(x^1 - x^2)\nabla q(x) \leq l(x^1)/l(x^2)(\nabla_1\varphi(p(x^2), \\ q(x^2))(p(x^1) - p(x^2)) + \nabla_2\varphi(p(x^2),q(x^2))(q(x^1) - q(x^2))). \tag{2.4.13}$$

Since φ is strict pseudoconvex, (2.4.13) yields

$$\varphi(p(x^1),q(x^1)) > \varphi(p(x^2),q(x^2)) \tag{2.4.14}$$

i.e.,

$$\theta(x^1) > \theta(x^2) \tag{2.4.15}$$

From (2.4.11) and (2.4.15) it follows that θ is strict pseudoconvex on X. In the cases (ii)-(iv), the proof is similar. For Case **2** the proof is similar to Case **1** with the exception of inequalities (2.4.14) and (2.4.15) which are of the form "$\geq$" since φ is only pseudoconvex on $R^m \times R^k$.

Similar properties are found by replacing "convex" with "concave" in Theorem 2.4.2, and *vice-versa.* Thus we can formulate

Corollary 2.4.2. *Let functions f, g, φ, θ, p and q be as defined in Theorem 2.4.2. If one of the statements of Table 2.4.6 is true, then the following implications are true :*
1. If φ is strict pseudoconcave on $R^m \times R^k$, then θ is strict pseudoconcave on X ;
2. If φ is pseudoconcave on $R^m \times R^k$, then θ is pseudoconcave on X.

Table 2.4.6

	f (on X)	g (on X)	φ (on $R^m \times R^k$)
(i)	concave	convex	increasing - decreasing
(ii)	linear	linear	——— - ———
(iii)	concave	linear	increasing - ———
(iv)	convex	linear	decreasing - ———

2.5. CONVEXITY OF SOME PARTICULAR FUNCTIONS

As an application of the results presented in Section 2.4, let's consider the convexity of several functions met in fractional programming.

A. *Nonlinear Fractional Functions*

Let $X \subseteq R^n$ be a convex set, $\rho : X \to R$, $\sigma : X \to R$ and $\theta(x) = \rho(x)/\sigma(x)$. Then:

a) If ρ and σ fulfill one of the conditions stated in Table 2.5.1, then θ is explicit quasiconvex on X ;

Table 2.5.1

	1	2	3	4	5
ρ σ	cx ≥ 0 cv > 0	cv ≤ 0 cx < 0	cx ≤ 0 cx > 0	cv ≥ 0 cv < 0	linear linear ≥ 0
	6	7	8	9	
ρ σ	linear ≤ 0 cx > 0	linear ≥ 0 cv > 0	cx linear ≥ 0	cv linear ≤ 0	

b) If ρ and σ fulfill one of the conditions stated in Table 2.5.2, then θ is explicit quasiconcave on X ;

Table 2.5.2

	1	2	3	4	5
ρ σ	cv > 0 cx > 0	cx ≤ 0 cv < 0	cv ≤ 0 cv > 0	cx ≥ 0 cx < 0	linear linear>0
	6	7	8	9	
ρ σ	linear ≤ 0 cv > 0	linear ≥ 0 cv > 0	cv linear > 0	cv linear < 0	

c) If ρ and σ fulfill one of conditions of Table 2.5.3, then :
1) θ is strict pseudoconvex if X is open, and if ρ and σ are differentiable;
2) θ is otherwise strict explicit quasiconvex.

Table 2.5.3

	1	2	3	4	5	6
ρ σ	sx$\geq$0 cv or sv$>$0	cx$\geq$0 sv$>$0	cv$\leq$0 sx$<$0	sv$\leq$0 cx or sx$<$0	sx$\leq$0 cx or sx$>$0	cx$\leq$0 sx$>$0
	7	8	9	10	11	12
ρ σ	sv$\geq$0 cv or sv$<$0	cv$\geq$0 sv$<$0	linear$\leq$0 sx$>$0	linear$\geq$0 sv$>$0	sx linear$>$0	sv linear$<$0

d) If ρ and σ fulfill one of the conditions stated in Table 2.5.4, then:
1) θ is strict pseudoconcave if X is open, and if ρ and σ are differentiable;
2) θ is otherwise strict explicit quasiconcave.

Table 2.5.4

	1	2	3	4	5	6
ρ σ	sv ≥ 0 cx or sx >0	cv ≥ 0 sx >0	cx ≤ 0 cv or sv <0	sx ≤ 0 cv or sv <0	sv ≤ 0 cv or sv >0	cv ≤ 0 sv >0
	7	8	9	10	11	12
ρ σ	sv ≥ 0 cx or sx <0	cx ≥ 0 sx <0	linear ≤ 0 sv >0	linear ≥ 0 sv >0	sv linear >0	sx linear <0

The proof results from Theorem 2.4.1 and Corollary 2.4.1. Consider the function $\varphi(y, z) : R \times R \rightarrow R$, $\varphi(y, z) = y/z$. Note that, on any of the convex sets $\{ (y, z) \in R \times R \mid z > 0 \}$ or $\{ (y, z) \in R \times R \mid z < 0 \}$ we get: in cases a) and b), φ is pseudoconvex (hence also explicit quasiconvex); in cases c) and d) it is strict pseudoconvex (hence also strict explicit quasiconvex). The corresponding identifications for functions f, g, ρ, σ and φ is made.

B. *Reciprocal Functions*

Let $X \subseteq R^n$ be a convex set, $\sigma : X \rightarrow R$, and $\theta(x) = 1/\sigma(x)$.
a) If σ is concave and strict positive on X, then θ is convex and strict positive on X;
b) If σ is convex and strict negative on X, then θ is concave and strict negative on X.

The proof results from Theorem 2.4.1 and Corollary 2.4.1, considering the function $\varphi(y, z) : R \times R \rightarrow R$, $\varphi(y, z) = 1/y$ and using the fact that $\nabla_1\varphi(y, z) = -1/y^2$, $\nabla_2\varphi(y, z) = 0$, and that $\varphi(y, z)$ is convex for $y > 0$ and concave for $y < 0$.

C. *Product Functions*

Let $X \subseteq R^n$ be a convex set, $\rho : X \rightarrow R$, $\sigma : X \rightarrow R$ and $\theta(x) = \rho(x) \cdot \sigma(x)$. Then:
a) If ρ and σ fulfill one of the conditions stated in Table 2.5.5, then θ is explicit quasiconcave on X ;
b) If ρ and σ fulfill one of the conditions stated in Table 2.5.6, then :
1) θ is strict pseudoconcave if X is open, and if ρ and σ are differentiable;
2) θ is otherwise strict explicit quasiconcave on X.
For proof is sufficient to write

$$\theta(x) = \rho(x)/(1/\sigma(x))$$

or

$$\theta(x) = \sigma(x)/(1/\rho(x))$$

and to use the results of A and B.

Table 2.5.5

	1	2	3	4
ρ	cv ≥ 0	cv ≥ 0	cx ≤ 0	cx ≤ 0
σ	cv ≥ 0	cx ≤ 0	cv ≥ 0	cx ≤ 0

Table 2.5.6

	1	2	3	4	5	6
ρ	sv > 0	cv ≥ 0	sv > 0	sx < 0	cx ≤ 0	sx < 0
σ	cv ≥ 0	sv > 0	sv > 0	cx ≤ 0	sx < 0	sx < 0
	7	8	9	10	11	12
ρ	cv ≥ 0	sv > 0	sv > 0	cx ≤ 0	sx < 0	sx < 0
σ	cx < 0	cx ≤ 0	sx < 0	sv > 0	cv ≥ 0	sv > 0

D. *Special Fractional Functions*

Let $X \subseteq R^n$ be a convex set, $\rho : X \to R$, $\sigma : X \to R$ and $\theta(x) = \rho^2(x)/\sigma(x)$. Then:
a) If ρ and σ fulfill one of the conditions stated in Table 2.5.7, then θ is strict convex and is positive, or non-negative, on X, as ρ is positive (negative), or non-negative (non-positive), on X.

Table 2.5.7

	1	2	3	4	5	6	7
ρ	sx > 0	cx ≥ 0	sx > 0	cv ≤ 0	sv < 0	sv < 0	linear
σ	cv > 0	sv > 0	sv > 0	sv > 0	cv > 0	sv > 0	sv > 0

b) If ρ and σ fulfill one of the conditions stated in Table 2.5.8, then θ is strict concave and is negative, or non-positive, as ρ is negative (positive), or non-positive (non-negative), on X.

Table 2.5.8

	1	2	3	4	5	6	7
ρ	sv < 0	cv ≤ 0	sv < 0	cx ≥ 0	sx > 0	sx > 0	linear
σ	cx < 0	sx < 0	sx < 0	sx < 0	cx < 0	sx < 0	sx < 0

For proof is sufficient to consider Theorem 2.4.1 and Corollary 2.4.1 with $\varphi(y, z) : R \times R \to R$; $\varphi(y, z) = y^2/z$, $z \neq 0$. Next note that the function φ is convex on the convex set $\{ (y, z) \mid (y, z) \in R \times R,\ z > 0 \}$, and concave on the convex set $\{ (y, z) \mid (y, z) \in R \times R,\ z < 0 \}$. Corresponding identifications for the functions p, q, ρ, σ and φ must be made.

E. *Bi-linear Special Fractional Functions*

Let $X \subseteq R^n$ be a convex and open set, $\rho : X \to R$, $\sigma : X \to R$, ρ and σ differentiable, $l : X \to R$ a linear and strict positive function, and let

$$\theta(x) = \frac{\rho(x)\sigma(x)}{(l(x))^2}.$$

If ρ and σ fulfill one of the conditions stated in Table 2.5.9, then θ is pseudoconcave on X.

Table 2.5.9

	1	2	3	4
ρ	cv ≥ 0	cv > 0	cx ≤ 0	cx < 0
σ	cv > 0	cv ≥ 0	cx < 0	cx ≤ 0

The proof results from Theorem 2.4.2. and Corollary 2.4.2 through a corresponding identification of the functions p, q, ρ, σ and φ.

F. *Fractional Functions of the form* $F_1(x) = \prod_{i=1}^{m} (f_i(x)^{a_i} \,/\, (f_0(x))^{a_0}$

where $f_i : X \to R$, $i = 0, \ldots, m$, $X \subseteq R^n$, X is convex, f_0 is convex and positive on X, f_i , $i = 1, \ldots, m$ are concave and non-negative on X, and a_i, $i = 0, \ldots, m$ are positive numbers such that

$$a_0 \geq \sum_{i=1}^{m} a_i \tag{2.5.1}$$

Theorem 2.5.1. [788], [791] *If condition (2.5.1) is fulfilled, then the function F_1 is explicit quasiconcave on X.*

Proof. Let $f : X \to R^m$, $f = (f_1, f_2, ..., f_m)$, $g = f_0 : X \to R$, and

$$\varphi : R^m \times R^p, \quad \varphi(y,z) = \prod_{i=1}^{m} y_i^{a_i} \,/\, \prod_{h=1}^{p} z_k^{e_k}$$

(a_i, e_k positive numbers)

The theorem results from Corollary 2.4.1 taking $F_1(x) = \varphi(f(x), g(x))$, and recalling that the function φ is explicit quasiconvex on domain
$D^* = \{ (y, z) \in R^m \times R^p \mid y_i \geq 0, \ z_j > 0, \ i = 1, ..., m, \ j = 1, ..., p \}$
for $p = 1$ and $e_1 = \sum_{i=1}^{m} a_i$. In particular, take $f_i = c'x + c_0 > 0$, $i = 1, 2$ and $f_0 = d'x + d_0 > 0$ for any x from the domain. With $a_1 = a_2 = 1$ and $a_0 = 2$ we get the function $F(x) = ((c'x + c_0)/(dx + d_0))^2$ to be both explicit quasiconvex and explicit quasiconcave.

G. *Fractional Functions of the form* $\theta(x) = (a'x + \alpha)(b'x + \beta)/(c'x + \gamma)$.

Let $X \subseteq R^n$, $X = \{ x \in R^n \mid a'x + \alpha \geq 0, \ b'x + \beta \geq 0, \ c'x + \gamma > 0 \}$ and $\theta : X \to R$. Duca [365] showed that, under certain conditions, the function θ is quasiconcave.

H. *Fractional Functions of the form* $\theta(x) = a'x + b'x/c'x$ [483], [488], [492], [615], [643], [801].

Take $a \neq 0$, $b \neq 0$, b and c linear independent, and $c'x > 0$ for any $x \in X$ (X convex). Consider the following cases :

I. a, b, c linear independent ;

II. a, b linear dependent, i.e., $a = \mu b$ $(\mu \neq 0)$;

III. a, c linear dependent, i.e., $a = \lambda c$ $(\lambda \neq 0)$.

The study of function θ was done by numerous authors starting from the erroneous assumption of Teterev [972], [973], regarding the (explicit) quasiconcavity or (explicit) quasiconvexity of θ. We shall state without proof a result obtained by Schaible [801] regarding the generalized concavity or convexity of θ. (One can, also, consult [488].)

Theorem 2.5.2. *Corresponding to cases I - III, one gets the following situations:*

I. *θ is neither quasiconvex nor quasiconcave on X ;*

II. *If $\mu > 0$, then θ is (explicit) quasiconcave on $\{ x \in X \mid a'x \leq 0 \}$ and (explicit) quasiconvex on $\{ x \in X \mid a'x \geq 0\}$.*
If $\mu < 0$, then θ is (explicit) quasiconcave on $\{x \in X \mid a'x \leq 0,\ c'x \leq (-1)/\mu\}$ and $\{x \in X \mid a'x \geq 0,\ c'x \geq (-1)/\mu\}$, but (explicit) quasiconvex on $\{x \in X \mid a'x \geq 0, c'x \leq (-1)/\mu\}$ and $\{x \in X \mid a'x \leq 0, c'x \geq (-1)/\mu\}$;

III. *If $\lambda < 0$, then θ is explicit quasiconcave on X, and if $\lambda > 0$, then θ is explicit quasiconvex on X.*

CHAPTER 3

METHODS FOR SOLVING LINEAR FRACTIONAL PROGRAMMING PROBLEMS

3.1. PRIMAL SIMPLEX ALGORITHM

Consider the linear fractional programming problem given in the standard form:

Maximize the function

$$F(\boldsymbol{x}) = f(\boldsymbol{x})/g(\boldsymbol{x}) = (\boldsymbol{c}'\boldsymbol{x} + c_0)/(\boldsymbol{d}'\boldsymbol{x} + d_0) \qquad (3.1.1)$$

subject to

$$A\boldsymbol{x} = \boldsymbol{b} \; ; \quad \boldsymbol{x} \geq \boldsymbol{0} \qquad (3.1.2)$$

where:

(i) $A = \|a_{ij}\|$ is a matrix m $\times$ n (m $<$ n),

(ii) $\boldsymbol{x}$, $\boldsymbol{c}$ and $\boldsymbol{d}$ are n dimensional column vectors,

(iii) $\boldsymbol{b}$ is an m dimensional column vector,

(iv) c_0 and d_0 are scalars,

(v) $\boldsymbol{x}'$ means the transpose of vector $\boldsymbol{x}$.

We assume that the domain of feasible solutions $D = \{ \boldsymbol{x} \mid A\boldsymbol{x} = \boldsymbol{b}, \boldsymbol{x} \geq \boldsymbol{0} \}$ represents a regular set i.e., a non-empty and bounded set. Regarding the function F, make the assumption that its denominator maintains a constant sign on D (say positive) i.e.,

$$\boldsymbol{d}'\boldsymbol{x} + d_0 > 0 \quad \forall\, \boldsymbol{x} \in D \qquad (3.1.3)$$

If there are points in D for which $\boldsymbol{d}'\boldsymbol{x} + d_0 = 0$, those will also be points of optimum if $\boldsymbol{c}'\boldsymbol{x} + c_0 > 0$. Assume that the problem (3.1.1) - (3.1.2) does not reduce to a linear programming problem, i.e., assume that $g(\boldsymbol{x}) \neq$ constant on D and that **c** and **d** are linearly independent. Also assume that the system (3.1.2) is non-redundant (i.e., it does not contain redundant equations) and that the rank of the matrix A is m $(r(A) = m)$.

As shown in Chapter 2, Section 2.5 , Case A, the function (3.1.1) is both explicit quasiconvex and explicit quasiconcave. This implies two important properties illustrated in Theorem 2.3.5 and Theorem 2.3.8:

a) Any point of local maximum is point of global maximum;

b) The maximum value is reached in a vertex of D.

Due to these properties, it is possible to solve the problem (3.1.1) - (3.1.2) using a technique similar to the Simplex method of conventional linear programming. The extension of the Simplex algorithm to the fractional case is due to Swarup [942] and Martos [640], [641].

One starts from a vertex x^* of D which is an extreme point. The vertices neighboring x^* are examined. Movement takes place along a side adjacent to x^*, towards a neighboring vertex x^{**} such that the function value increases (for a problem of minimum, it must decrease). The process continues until an extreme point is reached from which onwards one cannot find a point for which the function value increases. The process is finite since the number of vertices of D is finite (equal to C_n^m), and since at each step we have an effective increase of the function value. The Simplex algorithm belongs to the class of adjacent vertex method. Similar to the conventional linear programming, movement rules are given in the neighboring vertices for which the value of the objective function increases. Also, a stopping criterion is established i.e., it can recognize when an optimum vertex is reached. Basic entrance and exit criteria are also established.

Denote by $\boldsymbol{a}_j$ ($j = 1, \dots, n$) the column vectors of matrix $\boldsymbol{A}$. Let $\boldsymbol{B}$ be a base of $\boldsymbol{A}$, *i.e.*, a set of m linear-independent vectors from $\boldsymbol{A}$. Without loss of generality, we can accept that the basis $\boldsymbol{B}$ is formed of the first m columns of $\boldsymbol{A}$, i.e.,

$$\boldsymbol{B} = (\boldsymbol{b}_1 , \boldsymbol{b}_2 , \dots , \boldsymbol{b}_m)$$

The columns of $\boldsymbol{A}$ which are not in the base form a matrix denoted by $\boldsymbol{R}$. Hence $\boldsymbol{A} = (\boldsymbol{B}, \boldsymbol{R})$. The components of the vectors $\boldsymbol{x}$, $\boldsymbol{d}$ and $\boldsymbol{c}$ are similarly partitioned, *i.e.*, $\boldsymbol{x} = (\boldsymbol{x}_B , \boldsymbol{x}_R)$, $\boldsymbol{d} = (\boldsymbol{d}_B , \boldsymbol{d}_R)$ and $\boldsymbol{c} = (\boldsymbol{c}_B , \boldsymbol{c}_R)$, respectively. With these notations, the problem (3.1.1) - (3.1.2) becomes:

$$Max\ F(x) = (\boldsymbol{c}_B'\boldsymbol{x}_B + \boldsymbol{c}_R'\boldsymbol{x}_R + c_0) \,/\, (\boldsymbol{d}_B'\boldsymbol{x}_B + \boldsymbol{d}_R'\boldsymbol{x}_R + d_0) \tag{3.1.1'}$$

subject to

$$\boldsymbol{B}\boldsymbol{x}_B + \boldsymbol{R}\boldsymbol{x}_R = \boldsymbol{b} \tag{3.1.2'}$$

$$\boldsymbol{x}_B \geq \boldsymbol{0} , \quad \boldsymbol{x}_R \geq \boldsymbol{0} \tag{3.1.2''}$$

Denote by $N = \{ 1, 2 , \dots , n \}$ the set of indices of the column vectors of $\boldsymbol{A}$, and with I, the set of indices of the column vectors of the base $\boldsymbol{B}$, and with J those of the matrix $\boldsymbol{R}$. If we take $\boldsymbol{x}_R = \boldsymbol{0}$, then (3.1.2′) yields a basic solution for the problem under consideration, $\boldsymbol{x}_B = \boldsymbol{B}^{-1}\boldsymbol{b}$, which will be feasible if all the components are non-negative $\boldsymbol{x}_B \geq \boldsymbol{0}$. Let $z^1 = \boldsymbol{c}_B'\boldsymbol{x}_B + c_0$ and $z^2 = \boldsymbol{d}_B'\boldsymbol{x}_B + d_0$ be the values of the numerator and denominator, respectively, for this basic solution. For the basic solution $\boldsymbol{x}_B$, the objective function takes the value $F = z^1/z^2$. For the basic feasible solution $\boldsymbol{x}_B$, we also assume that the following elements $u_j = B^{-1}a_j = (u_{1j}, u_{2j}, \dots, u_{mj})'$, $z_j^1 = c_B'u_j$, $z_j^2 = d_B'u_j$ are known $\forall j \in J$. Let's see in what conditions we can use a new base $\hat{B} = (\hat{b}_1, \hat{b}_2, \dots, \hat{b}_m)$, i.e., a new basic feasible solution $\hat{x}_B = \hat{B}^{-1}b$ for which the value of F increases. Assume that the matrix $\hat{B}$ is obtained from the matrix B by replacing the column b_r with the column a_j from outside the base, i.e., the columns of the new base will be: $\hat{b}_i = b_i \quad i \neq r$, $\hat{b}_r = a_j \quad i = r$. The components of the new solution $\hat{x}_B$ are written as functions of those of the old solution just as in linear programming, i.e.,

$$\hat{x}_{Bi} = x_{Bi} - x_{Br}\,(u_{ij} / u_{rj}) \qquad (i \neq r)$$

$$\hat{x}_{Br} = x_{Br} / u_{rj} = \theta \text{ (notation), where } a_j = \sum_{i=1}^{m} u_{ij} b_i$$

Hence, the values of the numerator and of the denominator of the objective function become [942]:

$$\hat{z}^1 = z^1 + \theta\,(c_j - z_j^1) \text{ and } \hat{z}^2 = z^2 + \theta\,(d_j - z_j^2)\,.$$

Here, z_j^1 and z_j^2 refers to the previous basic feasible solution. The new value of the objective function will be $\hat{F} = \dfrac{\hat{z}^1}{\hat{z}^2}$. This value will be improved if $\hat{F} = \hat{z}^1/\hat{z}^2 > z^1/z^2$

$$\text{i.e., } [\, z^1 + \theta(c_j - z_j^1)\,] \,/\, [\, z^2 + \theta(d_j - z_j^2)\,] > z^1/z^2.$$

Eliminating the denominators, which are positive since we assume that the denominator of the objective function has a constant sign (positive) on D, we get

$$z^2\,[\, z^1 + \theta(c_j - z_j^1)\,] - z^1\,[\, z^2 + \theta(d_j - z_j^2)\,] > 0$$

i.e., using $\theta > 0$,

$$\Delta_j = z^2\left(c_j - z_j^1\right) - z^1\left(d_j - z_j^2\right) > 0$$

We get $\Delta_j > 0$ if:

a) $d_j - z_j^2 > 0$ and $(c_j - z_j^1) / (d_j - z_j^2) > z^1/z^2$

b) $d_j - z_j^2 < 0$ and $(c_j - z_j^1) / (d_j - z_j^2) < z^1/z^2$

c) $d_j - z_j^2 = 0$ and $c_j - z_j^1 > 0$

In the case of non-degenerated solutions we have $\theta > 0$ and $\bar{F} > F$, and for degenerated solutions, $\theta = 0$ and $\bar{F} = F$. In conclusion, being given a basic feasible solution $x_B = B^{-1}b$, if there is a column a_j of A outside the base for which $\Delta_j > 0$, and if at least a $u_{ij} > 0$ $(i = 1, \dots, m)$, then it is possible to obtain a new basic feasible solution if we replace with a_j one of the columns of B, and we get $\hat{F} \geq F$. Next, we shall show, under the assumption that D is a regular set, that, for each column a_j of A outside the base, the vector $u_j = (u_{ij})'$ has at least a strict positive component. Indeed, assume that for a_j we have

$$u_{ij} \leq 0 \,, \quad \forall\, i = 1 \,, \dots \,, m \tag{3.1.4}$$

We have $\sum_{i=1}^{m} x_{Bi} b_i = b$ or, by addition and subtraction of $\bar{\theta} a_j$ ($\bar{\theta}$ is any scalar), $\sum_{i=1}^{m} x_{Bi} b_i - \bar{\theta} a_j + \bar{\theta} a_j = b$. Using $\bar{\theta} a_j = \bar{\theta} \sum_{i=1}^{m} u_{ij} b_i$, we also get:

$$\sum_{i=1}^{m} (x_{Bi} - \bar{\theta} u_{ij})\, b_i + \bar{\theta} a_j = b \tag{3.1.5}$$

Taking $\bar{\theta} > 0$ and using (3.1.4) yields $x_{Bi} - \bar{\theta} u_{ij} \geq 0$. The relation (3.1.5) shows that $\tilde{x} = \left(x_{B1} - \bar{\theta} u_{1j} \,, \dots \,, x_{Bm} - \bar{\theta} u_{mj} \,, \bar{\theta}\right)$ is a basic feasible solution. If we take larger and larger values for $\bar{\theta}$, we get that the vector $\tilde{x}$ can have larger and larger components, thus contradicting the assumption that the domain D is bounded. Hence the relation (3.1.4) cannot apply for all the columns a_j ($j \in J$). Thus, is possible to move from vertex to vertex as long as there is a_j ($j \in J$) for which $\Delta_j > 0$, and, at each step, the value of the objective function increases. This process cannot continue indefinitely since there is a finite number of vertices, and since, if degeneration is not present, no basis repeats itself. For the columns of A from the base we get

$$z_j^1 = c_B' u_j = c_B' B^{-1} a_j = c_B' B^{-1} b_i = c_j, \text{ and}$$
$$z_j^2 = d_B' u_j = d_B' B^{-1} a_j = d_B' B^{-1} b_i = d_j,$$

and hence,

$$\Delta_j = z^2 \left(c_j - z_j^1\right) - z^1 \left(d_j - z_j^2\right) = 0.$$

Thus, the search process for an optimum solution will end if $\Delta_j = z^2 \left(c_j - z_j^1\right) - z^1 \left(d_j - z_j^2\right) \leq 0$ for all the columns of A. The method for determining the variables which leave and enter the base is the same as in the conventional linear programming, i.e.,

a) Compute $\Delta_r = \max\limits_{i \in J} \left\{ \Delta_i \mid \Delta_i > 0 \right\}$; the variable x_r enter the base.

b) Compute $x_j \,/\, u_{jr} = \min\limits_{s \in I} \left\{ x_s \,/\, u_{sr} \;;\; u_{sr} > 0 \right\}$; the variable x_j leaves the base.

In the case of fractional programming, the Simplex tables are similar with those of linear programming, with the exception that Δ_j has the previously presented, more complicated formula, and that in these tables we must add a row for $d = (d_j)$ and a column for d_B. Eventually, when calculating Δ_j, we can reveal a row for each terms, i.e., a row for $c_j - z_j^1$ and a row for $d_j - z_j^2$. The initial Simplex table has the form of Table 3.1.1.

Table 3.1.1

	d_j →		$d_1\ d_2\ \dots\ d_r\ \dots\ d_m$	d_{m+1}	$\dots$	d_j	$\dots$	d_n
	c_j →		$c_1\ c_2\ \dots\ c_r\ \dots\ c_m$	c_{m+1}	$\dots$	c_j	$\dots$	c_n
$\boldsymbol{d_B}$ ↓	↓ $\boldsymbol{c_B}$	↓ $\boldsymbol{x_B}$	$a_1\ a_2\ \dots\ a_r\ \dots\ a_m$	a_{m+1}	$\dots$	a_j	$\dots$	a_n
d_1	c_1	$x_1 = b_1$	$1\ 0\ \dots\ 0\ \dots\ 0$	a_{1m+1}	$\dots$	a_{1j}	$\dots$	a_m
d_2	c_2	$x_2 = b_2$	$0\ 1\ \dots\ 0\ \dots\ 0$	a_{2m+1}	$\dots$	a_{2j}	$\dots$	a_{2n}
$\vdots$	$\vdots$	$\vdots$	$\vdots\ \vdots\ \ \vdots\ \ \vdots$	$\vdots$		$\vdots$		$\vdots$
d_r	c_r	$x_r = b_r$	$0\ 0\ \dots\ 1\ \dots\ 0$	a_{rm+1}	$\dots$	a_{rj}	$\dots$	a_{rn}
$\vdots$	$\vdots$	$\vdots$	$\vdots\ \vdots\ \ \vdots\ \ \vdots$	$\vdots$		$\vdots$		$\vdots$
d_m	c_m	$x_m = b_m$	$0\ 0\ \dots\ 0\ \dots\ 1$	a_{mm+1}	$\dots$	a_{mj}	$\dots$	a_{mn}
z^2	z^1	$F(\boldsymbol{x_B})$						
		$c_j - z_j^1$	$0\ 0\ \dots\ 0\ \dots\ 0$	$c_{m+1}-z_{m+1}^1$	$\dots$	$c_j-z_j^1$	$\dots$	$c_n-z_n^1$
		$d_j - z_j^2$	$0\ 0\ \dots\ 0\ \dots\ 0$	$d_{m+1}-z_{m+1}^2$		$d_j-z_j^2$		$d_n-z_n^2$
$\Delta_j = z^2\left(c_j-z_j^1\right) - z^1\left(d_j-z_j^2\right)$			$0\ 0\ \dots\ 0\ \dots\ 0$	Δ_{m+1}	$\dots$	Δ_j	$\dots$	Δ_n

For using Primal Simplex algorithm, we need to know a basic feasible solution of the constraints system, i.e., to know a base $\boldsymbol{B}$ of $\boldsymbol{A}$. This means that in the matrix $\boldsymbol{A}$ we outline the identity matrix. If the constraints are of the form "$\leq$", one can outline the identity matrix by introducing slack variables. If the constraints are of the form "$=$" the identity matrix is outlined by introducing artificial non-negative variables, one for each constraint. We shall apply the Two-phase method. In Phase I we solve a problem of linear programming with the constraints obtained by adding artificial variables, and by taking the objective function as the sum of the artificial variables:

$$\min F_1 = x_1^a + x_2^a + \dots + x_m^a$$

After a number of iterations, one reaches the optimum solution, and one of the following situations may apply:

a) $\min F_1 > 0$, i.e., the initial problem is incompatible; or

b) $\min F_1 = 0$ hence $x_i^a = 0 \quad (\forall\ i = 1\ ,\ \dots\ ,\ m\ )$:

- b_1) If the artificial variables do not enter the base, then the optimum solution is a starting solution of the initial problem, and we enter Phase II; or
- b_2) If, however, the artificial variables enter the base, but with zero values, the optimal solution is again a degenerated solution of the initial problem and we enter Phase II.

In Phase II, we apply the Primal Simplex algorithm to the initial problem with starting base just found. If the constraints are of the form "$\geq$", then we bring them first to the form "=", and then apply the previous case.

Numerical example. Consider the following fractional programming problem:

$$\max F(x) = (3x_1 + 6x_2 + 2x_3 - 11)/(x_1 + 4x_2 + x_3 + 1) ,$$

subject to

$$3x_1 + 4x_2 + x_3 \le 2 ,$$
$$x_1 + 4x_2 + x_3 \le 1 ,$$
$$x_1 , x_2 , x_3 \ge 0 .$$

We bring the problem to standard form by adding the artificial variables x_4 and x_5 having zero coefficients in the objective function:

$$\max F(x) =$$
$$= (3x_1 + 6x_2 + 2x_3 + 0 \cdot x_4 + 0 \cdot x_5 - 11)/(x_1 + 4x_2 + x_3 + 0 \cdot x_4 + 0 \cdot x_5 + 1)$$

subject to

$$3x_1 + 4x_2 + x_3 + x_4 = 2 ,$$
$$x_1 + 4x_2 + x_3 + x_5 = 1 ,$$
$$x_1 , x_2 , x_3 , x_4 , x_5 \ge 0.$$

A starting basic solution is obtained immediately by taking into the base the variables x_4 and x_5. Hence $c_B = \begin{bmatrix} 0 \\ 0 \end{bmatrix}$, $d_B = \begin{bmatrix} 0 \\ 0 \end{bmatrix}$, $x_B = \begin{bmatrix} 2 \\ 1 \end{bmatrix}$

Table 3.1.2.

	$d_j \rightarrow$		1	4	1	0	0
	$c_j \rightarrow$		3	6	2	0	0
$d_B \downarrow$	$c_B \downarrow$	$x_B \downarrow$	a_1	a_2	a_3	a_4	a_5
0	0	$x_4 = 2$	3	4	1	1	0
0	0	$x_5 = 1$	1	4	1	0	1
$z^2 = 1$	$z^1 = -11$	F = -11					
		$c_j - z_j^1$	+3	+6	+2	0	0
		$d_j - z_j^2$	+1	+4	+1	0	0
$\Delta_j = z^2 (c_j - z_j^1) - z^1 - (d_j - z_j^2)$			14	50	13	0	0

The first iteration of the Simplex table (Table 3.1.2) yields:

$$x_4 = 2 , \quad x_5 = 1$$

Choose max $\{ \Delta_1 , \Delta_2 , \Delta_3 \}$ i.e., $\Delta_2 = 50$. Following the Simplex algorithm, the variable x_5 leaves the base and the variable a_2 enters the base. In the Simplex table (Table 3.1.3) not all the differences are non-positive, hence the solution is not optimum. Choose $\max \{ \Delta_1 , \Delta_3 \} = \Delta_1 = 3$, hence a_1 will enter the base and will replace x_4. Table 3.1.4 is obtained. The variable a_3 enters the base, replacing x_2. The new solution is given in Table 3.1.5.

Table 3.1.3

	$d_j \rightarrow$		1	4	1	0	0
	$c_j \rightarrow$		3	6	2	0	0
$d_B \downarrow$	$c_B \downarrow$	$x_B \downarrow$	a_1	a_2	a_3	a_4	a_5
0	0	$x_4 = 1$	2	0	0	1	-1
4	6	$x_2 = 1/4$	1/4	1	1/4	0	1/4
$z^2 = 2$	$z^1 = -19/2$	$F = -19/4$					
		$c_j - z_j^1$	3/2	0	1/2	0	-3/2
		$d_j - z_j^2$	0	0	0	0	-1
	$\Delta_j = z^2 (c_j - z_j^1) - z^1 (d_j - z_j^2)$		3	0	1	0	-25/2

Table 3.1.4

	$d_j \rightarrow$		1	4	1	0	0
	$c_j \rightarrow$		3	6	2	0	0
$d_B \downarrow$	$c_B \downarrow$	$x_B \downarrow$	a_1	a_2	a_3	a_4	a_5
1	3	$x_1 = 1/2$	1	0	0	1/2	-1/2
4	6	$x_2 = 1/8$	0	1	1/4	-1/8	3/8
$z^2 = 2$	$z^1 = -35/4$	$F = -35/8$					
		$x_j - z_j^1$	0	0	1/2	-3/4	-3/4
		$d_j - z_j^2$	0	0	0	0	-1
	$\Delta_j = z^2 (c_j - z_j^1) - z^1 (d_j - z_j^2)$		0	0	1	-3/2	-41/4

Table 3.1.5

		d_j →	1	4	1	0	0
		c_j →	3	6	2	0	0
d_B ↓	c_B ↓	x_B ↓	a_1	a_2	a_3	a_4	a_5
1	3	$x_1 = 1/2$	1	0	0	1/2	-1/2
1	2	$x_3 = 1/2$	0	4	1	-1/2	3/2
$z^2 = 2$	$z^1 = -17/2$	$F = -17/4$					
		$c_j - z_j^1$	0	-2	0	-1/2	-9/2
		$d_j - z_j^2$	0	0	0	0	-1
$\Delta_j = z^2\left(c_j - z_j^1\right) - z^1\left(d_j - z_j^2\right)$			0	-7	0	-7/4	-89/4

In Table 3.1.5, we get $\Delta_j \leq 0$ for all the columns of A. Thus the optimum value $F = -17/4$ for $x_1 = x_3 = 1/2$, $x_2 = 0$ has been obtained.

We presented the Primal Simplex algorithm following Swarup [942]. We find the first ideas regarding this algorithm in the works of Munteanu and Radó [686]. Also, Kucher [588] gave a Simplex type algorithm, by considering the problem of fractional programming as a problem of monotone programming. Yamada [1065] gave, under certain conditions a Simplex algorithm for the fractional problem. Seiji Nabeya [687] generalized and enhanced the Simplex algorithm given by Yamada by discarding some constraints. Rani and Shivpuri [745] brought some improvements to the Primal Simplex algorithm for linear fractional programming. Improvement consisted in that at each iteration, $p(p \neq 1)$ variables from the base (and not one as in the Primal Simplex algorithm) are replaced by just as many variables from outside the base, such that the value of the objective function increases (if the problem is of maximum). The start is made with a basic feasible solution. The determination of a new basic feasible solution requires solving an auxiliary fractional program. The case when $p = 2$ was considered by Kanchan [527], whereas the case $p = 3$ was treated by Sharma, Gupta and Gupta [842].

Note that, if x_j is a non-basic variable of a basic feasible solution and $\Delta_j < 0$, then x_j will not become a basic variable in the next iteration of the Simplex algorithm. Hence, the condition $\Delta_j < 0$ is a necessary condition for a non-basic variable x_j to remain a non-basic variable in the optimal solution.

Chadha [239], assuming that the problem (3.1.1) - (3.1.2) is non-degenerated and that

$c_0 = d_0 = 0$, obtained new sufficient conditions which guarantee that a non-basic variable of a basic feasible solution will remain a non-basic variable of an optimal basic feasible solution.

3.2. NECESSARY AND SUFFICIENT OPTIMALITY CONDITIONS VIA DIFFERENTIAL CALCULUS

The necessary and sufficient optimality conditions, obtained in preceding section will be recovered here in terms of partial derivatives using the classical method of Lagrange multipliers. The derivation follows the work of Swarup [953]. Reconsider the problem (3.1.1) - (3.1.2) for which we know a base $\boldsymbol{B} = (x_1 , x_2 , \ldots , x_m)$ assumed to be, without loss of generality, made up of the first m columns of $\boldsymbol{A}$. We shall stay within the assumption that the denominator keeps the positive sign constant on D. Using the partition of the matrices $\boldsymbol{A} = (\boldsymbol{B} , \boldsymbol{R})$, $\boldsymbol{x} = (\boldsymbol{x}_B , \boldsymbol{x}_R)$, $\boldsymbol{c} = (\boldsymbol{c}_B , \boldsymbol{c}_R)$, $\boldsymbol{d} = (\boldsymbol{d}_B , \boldsymbol{d}_R)$ defined in Section 3.1 yields:

$$\boldsymbol{x}_B = \boldsymbol{B}^{-1}\boldsymbol{b} - \boldsymbol{B}^{-1}\boldsymbol{R}\, \boldsymbol{x}_R \tag{3.2.1}$$

Equation (3.2.1) provides an expression for variables inside the base in terms of variables outside the base. To establish the optimum criterion, express the function F in terms of the variables outside the base. Thus:

$$F(\boldsymbol{x}) = \frac{\boldsymbol{c}'_B\boldsymbol{x}_B + \boldsymbol{c}'_R\boldsymbol{x}_R + c_0}{\boldsymbol{d}'_B\boldsymbol{x}_B + \boldsymbol{d}'_R\boldsymbol{x}_R + d_0} = \frac{\boldsymbol{c}'_B\boldsymbol{B}^{-1}\boldsymbol{b} - \boldsymbol{c}'_B\boldsymbol{B}^{-1}\boldsymbol{R}\boldsymbol{x}_R + \boldsymbol{c}'_R\boldsymbol{x}_R + c_0}{\boldsymbol{d}'_B\boldsymbol{B}^{-1}\boldsymbol{b} - \boldsymbol{d}'_B\boldsymbol{B}^{-1}\boldsymbol{R}\boldsymbol{x}_R + \boldsymbol{d}'_R\boldsymbol{x}_R + d_0} = \tag{3.2.2}$$

$$= \frac{z^1 - \sum_{j=m+1}^{n} z_j^1 x_j + \sum_{j=m+1}^{n} c_j x_j}{z^2 - \sum_{j=m+1}^{n} z_j^2 x_j + \sum_{j=m+1}^{n} d_j x_j} = \frac{z^1 - \sum_{j=m+1}^{n} (z_j^1 - c_j)x_j}{z^2 - \sum_{j=m+1}^{n} (z_j^2 - d_j)x_j}$$

where the following notations and formulae were used:

$$z^1 = \boldsymbol{c}'_B\boldsymbol{x}_B + c_0 \ ; \quad z^2 = \boldsymbol{d}'_B\boldsymbol{x}_B + d_0 \ ,$$
$$z_j^1 = \boldsymbol{c}'_B\boldsymbol{B}^{-1}\boldsymbol{R} = \boldsymbol{c}'_B\boldsymbol{u}_j \ ,$$
$$z_j^2 = \boldsymbol{d}'_B\boldsymbol{B}^{-1}\boldsymbol{R} = \boldsymbol{d}'_B\boldsymbol{u}_j .$$

Theorem 3.2.1. *The necessary and sufficient condition for a non-degenerate basic feasible solution* $\boldsymbol{x}^0 = (x_1^0 , x_2^0 , \ldots , x_n^0)$ *to be a local maximum for the problem* (3.1.1) - (3.1.2) *is that*

$$\frac{\partial F}{\partial x_j} \Big|_{x_j = x_j^0} \le 0 , \quad j = m + 1 , \ldots , n. \tag{3.2.3}$$

Proof. *Necessity:* Expanding the function F in Taylor's series about the point $x^0 = (x_j^0)$ using (3.2.2) yields:

$$F(x_1\,,\,x_2\,,\,\dots\,,\,x_n) - F(x_1^0\,,\,x_2^0\,,\,\dots\,,\,x_n^0) =$$

$$= \sum_{j=m+1}^{n} \frac{\partial F}{\partial x_j}\Big|_{x=x^0} x_j + \theta(x_j^2)$$

Since the solution is non-degenerated, the points $x = (x_j)$ placed in a sufficiently small neighborhood of $x^0 = (x_j^0)$ remain feasible solutions. Assuming $x^0 = (x_j^0)$ to be a local maximum for the problem (3.1.1) - (3.1.2) implies (3.2.3). Indeed, if we had $\frac{\partial F}{\partial x_k}\Big|_{x_k=x_k^0} > 0$ for $j = k$, then taking $x_k \neq 0$ and $x_j = 0$ $(j \neq k)$ yields

$$F(0\,,\,\dots\,,\,0\,,\,x_k\,,\,0\,,\,\dots\,,\,0) - F(x_1^0\,,\,\dots\,,\,x_n^0) > 0.$$

Since we can take x_k sufficiently small for the terms in $\theta(x_j^2)$ to negligible with respect to $\frac{\partial F}{\partial x_k}\Big|_{x=x^0} x_k$. But this relation contradicts x^0 being a local maximum.

Sufficiency. Assume that the relations (3.2.3) take place. Let's prove that $x^0 = (x_j^0)$ is a local maximum for the problem (3.1.1) - (3.1.2). Through differentiation, expression (3.2.2) yields:

$$\frac{\partial F}{\partial x_j} = \frac{-\,[z^2-\sum_{j=m+1}^{n}(z_j^2-d_j)x_j](z_j^1-c_j)+(z_j^2-d_j)[z^1-\sum_{j=m+1}^{n}(z_j^1-c_j)x_j]}{[z^2-\sum_{j=m+1}^{n}(z_j^2-d_j)x_j]^2}$$

The partial derivatives of F evaluated in $x^0 = (x_j) = (x_j^0)$ are

$$\frac{\partial F}{\partial x_j}\Big|_{x_j=x_j^0} = \frac{-\,z^2(z_j^1 - c_j) + z^1(z_j^2 - d_j)}{(z^2)^2}$$

Using relations (3.2.3) yields

$$-\,z^2(z_j^1 - c_j) + z^1(z_j^2 - d_j) \le 0, \text{ or}$$

$$z^2(c_j - z_j^1) - z^1(d_j - z_j^2) \le 0\,, \quad j = m+1\,,\,\dots\,,\,n \tag{3.2.4}$$

Section 3.1 tells us that the conditions (3.2.4) are local optimality conditions for the solution

$$x^0 = (x_j^0), \quad j = 1\,,\,\dots\,,\,n.$$

But these are also conditions for global optimum since a local optimum is a global optimum. The proof is complete.

Finally, we must show that from the optimality conditions (3.2.4) one obtains the optimality conditions for linear programming. This is obvious, since linear programming is a particular case of linear fractional programming, *viz* that when the denominator is constant. In this case $z_j^2 - d_j = 0$, $j = m+1, \dots, n$, and relations (3.2.4) become $z^2(c_j - z_j^1) \leq 0$, $j = m+1, \dots, n$, or, using $z^2 > 0$, $c_j - z_j^1 \leq 0$, $j = 1, 2, \dots, n$, which are the usual maximum conditions of linear programming.

3.3. LINEAR FRACTIONAL PROGRAMMING WITH NONLINEAR CONSTRAINTS. OPTIMALITY CONDITIONS

3.3.1. Consider the following linear fractional programming problem with differentiable nonlinear constraints

$$\max f(x) = (c'x + c_0)/(d'x + d_0), \tag{3.3.1}$$

subject to

$$g_i(x) \leq 0, \quad i = 1, \dots, m, \tag{3.3.2}$$

where, besides the usual notations, we have:

(i) g_i are convex or quasiconvex functions of x, on the domain of feasible solutions D satisfying the Kuhn-Tucker constraints qualification.

(ii) the function f is well defined on D, which means that $c'x + c_0$ and $d'x + d_0$ are not simultaneously zero.

Assume that problem (3.3.1) - (3.3.2) has a finite maximum at a finite point, i.e.,

(a) the denominator does not change sign on D, hence, either $d'x + d_0 \geq 0$ or $d'x + d_0 \leq 0$ and $d'x + d_0 \neq 0$ for at least one $x \in D$.

(b) if $d'x_0 + d_0 = 0$ for x_0, and if $d'x + d_0 \geq 0$, $\forall x \in D$, then $c'x_0 + c_0 < 0$.

We have already shown (Cap.II, Section 2.3) that the function (3.3.1) reaches its optimum value in a vertex of D, and that any local maximum is a global maximum. We shall show that for the problem (3.3.1) - (3.3.2) the Kuhn-Tucker conditions are necessary and sufficient for the existence of an optimum solution. Two cases are depicted:

Case I. The functions $g_i (i = 1, \dots, m)$ are convex.

Theorem 3.3.1. *The necessary and sufficient conditions for* x_0 *to be an optimum solution of the problem* (3.3.1) - (3.3.2) *is that the* m *scalars* $u_i (i = 1, \dots, m)$ *exist such that:*

$$g_i(x_0) \leq 0 \,, \quad i = 1 \,, \dots \,, m \tag{3.3.3}$$

$$\nabla f(x_0) = \sum_{i=1}^{m} u_i \nabla g_i(x_0) \tag{3.3.4}$$

$$\sum_{i=1}^{m} u_i g_i(x_0) = 0 \tag{3.3.5}$$

$$u_i \geq 0 \quad i = 1 \,, \dots \,, m \tag{3.3.6}$$

where ∇ is the gradient operator.

Proof. Since the necessity of conditions (3.3.3) - (3.3.6) results from the general theory of nonlinear programming problems, we shall only prove the sufficiency. In general, the sufficiency only takes place if f is of a certain type. For the case when f and g_i are convex functions, the sufficiency conditions (3.3.3) - (3.3.6) were proved by Kuhn and Tucker ("Non-linear programming", *Proceedings of the Second Berkeley Symposium on Mathematical Statistics and Probability*, University of California, 1951, p. 481 - 492). Let then $x_0 \in D$ and m be scalars u_1 , u_2 , ... , u_m verifying the conditions (3.3.3) - (3.3.6). We have

$$\nabla f(x_0) = c/(d'x_0 + d_0) - f(x_0)\cdot d/(d'x_0 + d_0) \,,$$

or, upon multiplication by $(x - x_0)'$,

$$(x - x_0)' \, \nabla f(x_0) = \frac{c'x - c'x_0}{d'x_0 + d_0} - \frac{f(x_0)}{d'x_0 + d_0}(d'x - d'x_0) =$$

$$= \frac{(c'x + c_0) - (c'x_0 + c_0)}{d'x_0 + d_0} - \frac{f(x_0)}{d'x_0 + d_0}[(d'x + d_0) - (d'x_0 + d_0)] =$$

$$= \frac{d'x + d_0}{d'x_0 + d_0}[(f(x) - f(x_0)] .$$

Thus

$$f(x) - f(x_0) = \frac{d'x_0 + d_0}{d'x + d_0}(x - x_0)' \, \nabla f(x_0) . \tag{3.3.7}$$

Multiply relation (3.3.4) by $(x - x_0)'$ to get

$$(x - x_0)' \, \nabla f(x_0) = \sum_{i=1}^{m} u_i[(x - x_0)' \, \nabla g_i(x_0)] . \tag{3.3.8}$$

Since the functions g_i are assumed to be convex, we have

$$g_i(x) \geq g_i(x_0) + (x - x_0)' \, \nabla g_i(x_0)$$

or

$$(x - x_0)' \, \nabla g_i(x_0) \leq g_i(x) - g_i(x_0) .$$

Multiplying by $u_i \geq 0$ and summating, yields

$$\sum_{i=1}^{m} u_i[(x - x_0)' \, \nabla g_i(x_0)] \leq \sum_{i=1}^{m} u_i[g_i(x) - g_i(x_0)] . \tag{3.3.9}$$

The point x verifies (3.3.2), thus

$$\sum_{i=1}^{m} u_i g_i(x) \leq 0. \tag{3.3.10}$$

Use relations (3.3.9), (3.3.10) and (3.3.5), to transform relation (3.3.8) into

$$(x - x_0)' \nabla f(x_0) \leq \sum_{i=1}^{m} u_i[g_i(x) - g_i(x_0)] =$$

(3.3.11)

$$= \sum_{i=1}^{m} u_i g_i(x) - \sum_{i=1}^{m} u_i g_i(x_0) = \sum_{i=1}^{m} u_i g_i(x) \leq 0.$$

If the point x is chosen sufficiently close to x_0, then, due to the continuity of $d'x + d_0$, we have

$$(d'x_0 + d_0)/(d'x + d_0) > 0. \tag{3.3.12}$$

Relations (3.3.7), (3.3.10) and (3.3.12) yield $f(x) - f(x_0) \leq 0$ for x in a neighborhood of x_0. This means that x_0 is a local maximum. Using a result similar to that of Theorem 2.3.5, Chapter 2, yields that x_0 is also a point of global maximum. The proof is complete.

Case II. The functions $g_i (i = 1, \dots, m)$ are quasiconvex. We shall prove that, again, the Kuhn-Tucker conditions (3.3.3) - (3.3.6) are sufficient for the existence of an optimum solution of the problem (3.3.1) - (3.3.2). Relations (3.3.7) and (3.3.12) apply here too. Let $I = \{ i | g_i(x_0) < 0 \}$, i.e., $g_i(x_0) = 0$ for $i \notin I$, and hence

$$g_i(x) \leq g_i(x_0) \quad i \notin I \text{ and } x \in D.$$

Relations (3.3.3), (3.3.5), (3.3.6) yield

$$u_i = 0, \quad i \in I. \tag{3.3.13}$$

Since the functions g_i are quasiconvex, the relation $g_i(x) \leq g_i(x_0)$, yields

$$(x - x_0)' \nabla g_i(x_0) \leq 0, \quad i \notin I, \quad x \in D. \tag{3.3.14}$$

Upon multiplication by $u_i > 0 \quad (i \notin I)$ and addition, we get

$$(x - x_0)' \nabla \sum_{i \notin I} u_i g_i(x_0) \leq 0. \tag{3.3.15}$$

In the case of $i \in I$, we have $u_i = 0$, thus

$$(x - x_0)' \nabla \sum_{i \in I} u_i g_i(x_0) = 0. \tag{3.3.16}$$

Relations (3.3.15) and (3.3.16) yield

$$(x - x_0)' \sum_{i=1}^{m} u_i \nabla g_i(x_0) \leq 0. \tag{3.3.17}$$

We have:

$$(x - x_0)' \nabla f(x_0) = (x - x_0)' \sum_{i=1}^{m} u_i \nabla g_i(x_0) \leq 0$$

which is the relation (3.3.11). The reasoning follows now the line of that used in *Case* I, and hence, x_0 is a point de global maximum of f.

3.3.2. We present next a method for solving the linear fractional programming problems with a quadratic constraint and a series de linear constraints:

$$max\ f(x) = \frac{c'x + c_0}{d'x + d_0}, \tag{3.3.18}$$

subject to

$$p'x + (1/2)x'Hx \leq \gamma, \tag{3.3.19a}$$

$$Ax \leq b\ , \quad x \geq 0, \tag{3.3.19b}$$

where, besides the usual notations, we have:

a) p a column vector $n \times 1$,

b) H a matrix $n \times n$, positive semi-definite,

c) γ a scalar.

The method is a combination between the Simplex method for linear fractional programming which was presented in Section 3.1, and a parametric version of the Simplex method for quadratic programming. The method has two phases. In the first phase, a solution of the problem is determined without activating the quadratic constraint.

Phase I. Solve the linear fractional programming problem

$$max\ \{\ f(x) = \frac{c'x + c_0}{d'x + d_0}\ \mid Ax \leq b\ ,\ x \geq 0\ \}$$

obtained from the initial problem by neglecting the quadratic constraint. The usual assumptions apply, i.e.: the set of feasible solutions (D) is regular, the denominator of the objective function is positive and finite on D, and the optimum solution is unique. Let x_0 be the optimum solution. If x_0 also verifies the constraint $p'x + (1/2)x'Hx \leq \gamma$, then x_0 is the optimum solution of the problem (3.3.18) - (3.3.19a,b). If it does not verify, then $p'x_0 + (1/2)x^{0\prime}Hx^0 > \gamma$. Proceed with Phase II.

Phase II. To the original system of constraints add the constraint

$$(c'x + c_0)/(d'x + d_0) \geq \mu,$$

i.e.,

$$(c' - \mu d')x \geq \mu d_0 - c_0,$$

and then solve the problem

$$min\ (p'x + (1/2)x'Hx), \tag{3.3.20}$$

subject to

$$Ax \leq b, \tag{3.3.21}$$

$$x \geq 0, \tag{3.3.22}$$

$$(c' - \mu d')x \geq \mu d_0 - c_0. \tag{3.3.23}$$

Note that if we take

$$\mu = \mu_0 = \frac{c'x_0 + c_0}{d'x_0 + d_0},$$

then x_0 is a feasible solution of the problem (3.3.20) - (3.3.23). Thus, the problem (3.3.20) - (3.3.23) is solved beginning with $\mu = \mu_0$, and then the value of μ is parametrically decreased until a value $\mu = \mu^*$ is obtained for which the value of the objective function (3.3.20) is equal with γ. In this case the optimum solution of problem (3.3.20) - (3.3.23) for $\mu = \mu^*$ is also the optimum solution for the initial problem (3.3.18) - (3.3.19 a and b), as shown in the theorem below.

If we cannot obtain a value equal with γ for the function (3.3.20), but is all the time strictly greater than γ, whatever small we make the value of μ, then there are no feasible solutions for the original problem (3.3.18) - (3.3.19). Note that decreasing the value ofμ means relaxing the constraint (3.3.23), which means that the value of the objective function (3.3.20) decreases, or stays the same.

Theorem 3.3.2. *The optimum solution of the problem:*

$$\min\ (p'x + (1/2)x'Hx), \tag{3.3.24}$$

subject to

$$A\,x \le b, \tag{3.3.25}$$

$$(c' - \mu d')x \ge \mu d_0 - c_0, \tag{3.3.26}$$

$$x \ge 0, \tag{3.3.27}$$

for which the value of the objective function is equal to γ is an optimum solution of the problem:

$$\max\ (c'x + c_0)/(d'x + d_0), \tag{3.3.28}$$

subject to

$$p'x + (1/2)x'Hx \le \gamma, \tag{3.3.29}$$

$$A\,x \le b, \tag{3.3.30}$$

$$x \ge 0. \tag{3.3.31}$$

Proof. We can use Theorem 3.3.1 to write the Kuhn-Tucker conditions for the optimum solutions of problems (3.3.24) - (3.3.27) or (3.3.28) - (3.3.31). Thus, the Kuhn-Tucker conditions for the optimum solution of the problem (3.3.28) - (3.3.31) are:

$$\frac{c}{d'x + d_0} - \frac{c'x + c_0}{(d'x + d_0)^2} - (p + Hx)v_1 - A'v + u = 0, \tag{3.3.32}$$

$$p'x + (1/2)x'Hx + y_1 = \gamma, \tag{3.3.33}$$

$$Ax + y = b, \tag{3.3.34}$$

$$u'x + v'y + v_1y_1 = 0, \tag{3.3.35}$$

$$x\,,\,y\,,\,y_1\,,\,u\,,\,v\,,\,v_1 \ge 0, \tag{3.3.36}$$

where y_1 is a slack variable, and v_1 is the Lagrange variable attached to y_1, while u and

$\boldsymbol{v}$ are the vectors of the Lagrange multipliers associated to the vectors $\boldsymbol{x}$ and $\boldsymbol{y}$ of dimensions n and m, respectively. The Kuhn-Tucker conditions for the optimum solution of the problem (3.3.24) - (3.3.27) obtained for $\mu = \mu^*$ are:

$$- \boldsymbol{p} - \boldsymbol{Hx} - \boldsymbol{A'v} + (\boldsymbol{c} - \mu \boldsymbol{d}) v_2 + \boldsymbol{u} = 0 \,, \tag{3.3.37}$$

$$\boldsymbol{A}\,\boldsymbol{x} + \boldsymbol{y} = \boldsymbol{b} \,, \tag{3.3.38}$$

$$(\boldsymbol{c'x} + c_0) - \mu^*(\boldsymbol{d'x} + d_0) - y_2 = 0 \,, \tag{3.3.39}$$

$$\boldsymbol{u'x} + \boldsymbol{v'y} + v_2 y_2 = 0 \,, \tag{3.3.40}$$

$$\boldsymbol{x} \,, \boldsymbol{y} \,, y_2 \,, \boldsymbol{u} \,, \boldsymbol{v} \,, v_2 \geq 0 \,. \tag{3.3.41}$$

The fact that, for $\mu = \mu^*$, the optimum of function (3.3.24) is equal with γ means that $y_1 = 0$. Consider now that $\hat{\boldsymbol{x}}$, $\hat{\boldsymbol{y}}$, $\hat{y}_2 = 0$, $\hat{\boldsymbol{u}}$, $\hat{\boldsymbol{v}}$, $\hat{v}_2$ is a optimum solution of the problem (3.3.24) - (3.3.27), i.e. it verifies the Kuhn-Tucker conditions (3.3.27) - (3.3.41).

$$- \boldsymbol{p} - \boldsymbol{H\hat{x}} - \boldsymbol{A'\hat{v}} + (\boldsymbol{c} - \mu^* \boldsymbol{d}) \hat{v}_2 + \hat{\boldsymbol{u}} = 0 \,, \tag{3.3.42}$$

$$\boldsymbol{A\hat{x}} + \hat{\boldsymbol{y}} = \boldsymbol{b} \,, \tag{3.3.43}$$

$$(\boldsymbol{c'\hat{x}} + c_0) - \mu^*(\boldsymbol{d'\hat{x}} + d_0) = 0 \,, \tag{3.3.44}$$

$$\hat{\boldsymbol{u}}'\hat{\boldsymbol{x}} + \hat{\boldsymbol{v}}'\hat{\boldsymbol{y}} = 0 \,, \tag{3.3.45}$$

$$\boldsymbol{p'\hat{x}} + (1/2)\hat{\boldsymbol{x}}'\boldsymbol{H\hat{x}} = \gamma \,, \tag{3.3.46}$$

$$\hat{\boldsymbol{x}} \,, \hat{\boldsymbol{y}} \,, \hat{\boldsymbol{u}} \,, \hat{\boldsymbol{v}} \,, \hat{v}_2 \geq 0 \,. \tag{3.3.47}$$

Let

$$k = \frac{1}{\boldsymbol{d'\hat{x}} + d_0} > 0 \,, \quad \boldsymbol{u}^* = \left[\frac{k}{\hat{v}_2}\right] \hat{\boldsymbol{u}} \,, \quad \boldsymbol{v}^* = \left[\frac{k}{\hat{v}_2}\right] \hat{\boldsymbol{v}} \,, \quad \hat{v}_1 = \frac{k}{\hat{v}_2} \,.$$

Then $\hat{\boldsymbol{x}}$,$\hat{\boldsymbol{y}}$, $\hat{y}_1 = 0$, $\boldsymbol{u}^*$, $\boldsymbol{v}^*$, $\hat{v}_1$ constitutes an optimum solution of the problem (3.3.28) - (3.3.31) since it verifies the Kuhn-Tucker conditions (3.3.32) - (3.3.36), i.e.,

$$\frac{\boldsymbol{c}}{\boldsymbol{d'\hat{x}} + d_0} - \frac{\boldsymbol{c'\hat{x}} + c_0}{(\boldsymbol{d'\hat{x}} + d_0)}\boldsymbol{d} - (\boldsymbol{p} + \boldsymbol{H\hat{x}})\hat{v}_1 - \boldsymbol{A'v}^* + \boldsymbol{u}^* = 0 \,, \tag{3.3.48}$$

$$\boldsymbol{p'\hat{x}} + (1/2)\hat{\boldsymbol{x}}'\boldsymbol{H\hat{x}} = \gamma \,, \tag{3.3.49}$$

$$\boldsymbol{A\hat{x}} + \hat{\boldsymbol{y}} = \boldsymbol{b} \,, \tag{3.3.50}$$

$$\boldsymbol{u}^{*\prime}\hat{\boldsymbol{x}} + \boldsymbol{v}^{*\prime}\hat{\boldsymbol{y}} = 0 \,, \tag{3.3.51}$$

$$\hat{\boldsymbol{x}} \,, \hat{\boldsymbol{y}} \,, \boldsymbol{u}^*, \boldsymbol{v}^*, \hat{v}_1 \geq 0 \,. \tag{3.3.52}$$

Indeed, the conditions (3.3.49) - (3.3.52) are evident satisfied due to the relations (3.3.43) - (3.3.46). It is left to show that the relation (3.3.48) is true. We take the value of μ^* from relation (3.3.39) and then introduce it in relation (3.3.32) to obtain

$$- \boldsymbol{p} - \boldsymbol{H\hat{x}} - \boldsymbol{A'\hat{v}} + \hat{\boldsymbol{u}} + \left[\boldsymbol{c} - \frac{\boldsymbol{c'\hat{x}} + c_0}{\boldsymbol{d'x} + d_0} \cdot \boldsymbol{d}\right] \hat{v}_2 = 0 \,.$$

By dividing by $\hat{v}_2(\boldsymbol{d'\hat{x}} + d_0)$ and using the definition of $\boldsymbol{u}^*$, $\boldsymbol{v}^*$, $\hat{v}_1$ we get relation (3.3.48). The proof is complete.

3.4. CHANGE OF VARIABLE METHOD

Charnes and Cooper [281] uses the method of variable transformation to solve the problem of linear fractional programming. Thus, the solution of the fractional programming problem is reduced to the solution of at most two linear programming problems which differ among themselves through a sign in a constraint and in the objective function. Consider the linear fractional programming problem

$$\max \{ F(x) = (c'x + c_0)/(d'x + d_0) |\ Ax \leq b\ ;\ x \geq 0 \} \tag{3.4.1}$$

in which we assume that the set D of feasible solutions is regular. To solve the problem (3.4.1), utilize the variable transformation:

$$y = tx\ , \tag{3.4.2}$$

where $t \geq 0$ is a scalar such that we have

$$d'y + d_0 t = \gamma \tag{3.4.3}$$

with $\gamma \neq 0$ a given number. Usually take $\gamma = 1$. Multiplying the numerator and the denominator of F, as well as the constraints of the problem (3.4.1), by t, and using relation (3.4.3), the problem (3.4.1) becomes

$$\max G(y, t) = c'y + c_0 t\ , \tag{3.4.4}$$

subject to

$$\begin{aligned} Ay - bt &\leq 0\ , \\ d'y + d_0 t &= \gamma\ , \\ y \geq 0\ , \quad t &\geq 0\ . \end{aligned} \tag{3.4.5}$$

Let

$$D^* = \{(y, t) \in R^n \times R^1\ |Ay - bt \leq 0\ ;\ d'y + d_0 t = \gamma\ ,\ y \geq 0\ ,\ t \geq 0\ \}$$

be the domain of feasible solutions for the problem (3.4.4) - (3.4.5).

Lemma. *For any feasible solution* $(y, t) \in D^*$ *of the problem* (3.4.4) - (3.4.5) *we have* $t > 0$.

Proof. Assume that $(\hat{y}, 0) \in D^*$ and let $\hat{x} \in D$. Since $A\hat{y} \leq 0$, $\hat{y} \geq 0$, it follows that $x_r = \hat{x} + r\hat{y} \in D$. But r can be as large as required, and hence it follows that D is unbounded, which contradicts its regularity assumption.

The transformation $y = tx$ establishes a one-to-one correspondence between the sets D and D^*. The following theorem shows the connection between the optimum solutions of problems (3.4.1) and (3.4.4) - (3.4.5).

Theorem 3.4.1. *If*

(i) $0 < sgn\ (\gamma) = sgn\ (d'x^* + d_0)$ *for* x^*, *the optimum solution of the problem* (3.4.1), *and if*

(ii) (y^*, t^*) *is an optimum solution of the problem* (3.4.4) - (3.4.5),

then $x = y^*/t^*$ *is an optimum solution of the problem* (3.4.1).

Proof. From the fact that (y^*, t^*) is an optimum solution of the problem (3.4.4) - (3.4.5) it follows that:

$$c'y^* + d_0t^* \geq c'y + d_0t \,, \quad \forall\, (y, t) \in D^* \tag{3.4.6}$$

Assume that $x = y^*/t^*$ were not an optimum solution of the problem (3.4.1). Hence, there is another point $x^0 \in D$ which is optimum, i.e.,

$$\frac{c'x^0 + c_0}{d'x^0 + d_0} > \frac{c'(y^*/t^*) + c_0}{d'(y^*/t^*) + d_0} \,.$$

But

$$\frac{c'(y^*/t^*) + c_0}{d'(y^*/t^*) + d_0} = \frac{c'y^* + c_0t^*}{d'y^* + d_0t^*} = \frac{c'y^* + c_0t^*}{\gamma}$$

since

$$d'y^* + d_0t^* = \gamma \,, \quad (y^*, t^*) \in D^* \,.$$

Hence

$$\frac{c'x^0 + c_0}{d'x^0 + d_0} > \frac{c'y^* + c_0t^*}{\gamma} \tag{3.4.7}$$

Condition (i) implies

$$d'x^0 + d_0 = \rho\gamma \quad (\rho > 0) \,.$$

Let $\hat{y} = x^0/\rho$, $\hat{t} = 1/\rho$. The point $(\hat{y}, \hat{t})$ is a feasible solution of the problem (3.4.4)-(3.4.5) since:

$$\hat{y} \geq 0 \,, \quad t \geq 0 \,, \quad A\hat{y} = A\frac{x^0}{\rho} = \frac{Ax^0}{\rho} \leq \frac{b}{\rho} = b\hat{t} \,,$$

$$d'\hat{y} + d_0\hat{t} = d'\left(\frac{x^0}{\rho}\right) + d_0\left(\frac{1}{\rho}\right) = \frac{d'x^0 + d_0}{\rho} = \gamma \,.$$

We have

$$\frac{c'x + c_0}{d'x^0 + d_0} = \frac{c'x^0/\rho + c_0/\rho}{d'x^0/\rho + d_0/\rho} =$$

$$= \frac{c'\hat{y} + c_0\hat{t}}{d'\hat{y} + d_0\hat{t}} = \frac{c'\hat{y} + c_0\hat{t}}{\gamma} \,. \tag{3.4.8}$$

Relations (3.4.7) and (3.4.8) imply

$$c'\hat{y} + c_0\hat{t} > c'y^* + c_0t^* \,,$$

which contradicts the relation (3.4.6). The proof is complete.

Note that Theorem 3.4.1 was proved under the assumption (i). But if

$$sgn\,(d'x^* + d_0) < 0$$

where x^* is the optimum solution of problem (3.4.1), then, substituting (c', c_0) and (d', d_0) by $(-c', -c_0)$ and $(-d', -d_0)$, respectively, the value of the objective function will not change and we get

$$sgn\,(-d'x^* - d_0) > 0$$

i.e., the assumption i). Thus the results of Theorem 3.4.1 remain valid in this case too. Hence we can state the following result:

Theorem 3.4.2. *If D is a regular set, then for solving the problem* (3.4.1) *is sufficient to solve the following two linear programming problems*:

$$max \ \{(c'y + c_0t) \ |Ay - bt \leq 0 \ , d'y + d_0t = 1 \ , \ y \geq 0 \ , \ \ t \geq 0\} \tag{3.4.9}$$

$$max \ \{(-c'y - c_0t) \ |Ay - bt \leq 0 \ , - d'y - d_0t = 1 \ , \ \ y \geq 0 \ , \ \ t \geq 0\} \tag{3.4.10}$$

If the sign of the denominator at the optimum solution is not known, then both problems, (3.4.9) and (3.4.10), must be solved; and the solution which gives the largest value for F is selected. But if the sign of the denominator at the optimum solution is known, then, according to Theorem 3.4.2, it is sufficient to solve only one of the problems (3.4.9) or (3.4.10), depending on the sign of the denominator. In fact, it is sufficient to know *a priori* that the denominator has the same sign on D, and then to solve only one of the problems (3.4.9) or (3.4.10). Since

$$max \ \frac{c'x + c_0}{d'x + d_0} \approx max \ (-1) \ \frac{d'x + d_0}{c'x + c_0}$$

it results that the same reduction as above can be made also depending on the sign of the numerator at the optimum solution. Thus, if we know the sign of the numerator or of the denominator at the optimum solution, then it is sufficient to solve only one of the problems (3.4.9) or (3.4.10). Let see what happens if the denominator becomes zero on D.

Theorem 3.4.3. *If*

$$d'x + d_0 = 0 \ , \quad \forall \ x \in D \ ,$$

then the problems (3.4.9) and (3.4.10) *are both inconsistent*.

Proof. Indeed, if

$$d'x + d_0 = 0 \ , \ \ \forall \ x \in D \ ,$$

then we cannot have

$$\pm \ t(d'x + d_0) = \pm \ (d'y + d_0t) = 1$$

which means that the problems (3.4.9) and (3.4.10) are both inconsistent. But if $d'x + d_0 = 0$ for $x \in D_1 \subset D$ (which means that $d'x + d_0 \neq 0$ for $x \in D_2 = D - D_1$), then any $x \in D_1$ is the limit of a sequence $\{x_n\}$ of points from D_2 for which $d'x_n + d_0 = \varepsilon_n \neq 0$ and $\varepsilon_n \to 0$. For the corresponding sequence $\{ \ y_n \ , \ t_n \ \}$, we have

$$\pm \ t_n(d'x_n + d_0) = \pm \ d'y_n \pm \ d_0t_n = t_n\varepsilon_n = 1 \ .$$

Thus $t_n = 1/\varepsilon_n \to \infty$.

Zionts [1084] showed that if the problem (3.4.1) has a finite optimum solution, then, for all practical cases, the denominator $d'x + d_0$ cannot have two different signs on D. Consequently, only one of the problems (3.4.9) and (3.4.10) must be solved.

If the sign of the denominator $g(x) = d'x + d_0$ is unknown and if points in D exist were the denominator is zero, then one add to the constraints system a pair of disjunctive constraints $g(x) \geq \varepsilon$ or $g(x) \leq -\varepsilon$ (ε is a sufficiently small constant) that are equivalent to

$$(2z - 1)g(x) \geq \varepsilon$$

where z is a 0-1 variable.

Numerical example. Reconsider the numerical example of Section 3.1 :

$$max \left\{ F(x) = \frac{3x_1 + 6x_2 + 2x_3 - 11}{x_1 + 4x_2 + x_3 + 1} \mid 3x_1 + 4x_2 + x_3 \leq 2 \right.,$$
$$\left. x_1 + 4x_2 + x_3 \leq 1 ,\ x_1, x_2, x_3 \geq 0 \right\}$$

Let's make the change of variable $y_i = tx_i$ $(i = 1, 2, 3)$. Since the denominator has positive coefficients, it will keep a positive sign on D. Solve the problem (3.4.9) in the form

$$max \left\{ G(y, t) = 3y_1 + 6y_2 + 2y_3 - 11t \mid 3y_1 + 4y_2 + y_3 - 2t \leq 0 \right.,$$
$$\left. y_1 + 4y_2 + y_3 - t \leq 0 ,\ y_1 + 4y_2 + y_3 + t = 1 ,\ y_1, y_2, y_3, t \geq 0 \right\}.$$

The optimum solution is $y_1 = 1/4$; $y_2 = 0$; $y_3 = 1/4$; $t = 1/2$, and $max\ G(y, t) = -17/4$. According to Theorem 3.4.1, the optimum solution of the linear fractional programming problem is $x_1 = y_1/t = (1/4)/(1/2) = 1/2$; $x_2 = 0$, $x_3 = (1/4)/(1/2) = 1/2$, and the optimum value of F is $-17/4$.

Remark: In some practical problems, it is possible that, for some feasible solutions, the denominator, $g(x) = d'x + d_0$, takes very large values and hence t takes very small values, practically zero. In such situations, the value of t can be calculated using the formula $t = \dfrac{\gamma}{g(x) - \varepsilon}$, where γ and ε are positive numbers such that $t \geq 0$ for all $x \in D$. The change of variable (3.4.1) takes now the form $y = (t + \varepsilon)x$, while the linear problem (3.4.4) - (3.4.5) will have the form

$$\max\ c'y + c_0t + c_0\varepsilon$$

subject to

$$Ay - bt \leq \varepsilon b$$

$$d'y + d_0t = \gamma - \varepsilon d_0$$

$$y \geq 0, \quad t \geq 0 .$$

3.5. APPROXIMATE METHODS AND ITERATIVE METHODS

Consider the linear fractional programming problem

$$\max \ \{ \ F(x) = \frac{c'x}{d'x} \ | \ Ax = b, \ \ x \geq 0 \ \} \tag{3.5.1}$$

We describe an approximate method for determining the optimal solution of Problem (3.5.1).

To describe this method, we assume that the domain of feasible solutions is a bounded set, and that on this domain $d'x > 0$. These assumptions ensure that the optimal value of Problem (3.5.1) is finite. The method results from a geometrical interpretation of the problem. Consider, in the plane (z_1, z_2), S, the image of the feasible solution domain through the transformation given by the equations

$$z_1 = d'x, \quad z_2 = c'x$$

(For other discussions about S, and for the interpretation of various situations see reference [892]).

The set S (Figure 3.5.1) constitutes a convex polyhedral which, we assume, does not reduce to a straight line segment. Define in S a sequence of points using the following procedure: Choose a point P_1 in S (not necessarily on the boundary of S). Having the point P_i defined, denote by P_i^* the intersection point (different from P_i if this is possible, otherwise consider P_i) of the half-line OP_i, with the boundary of S. Denote by P_i^{**} the midpoint of segment $P_iP_i^*$. From this point, draw a perpendicular on $P_iP_i^*$. Let P_{i+1} be the point on the boundary of S and on the perpendicular, with the ordinate z_2 greater than the ordinate of P_i^*, if such a thing exists, otherwise take P_i^* etc.

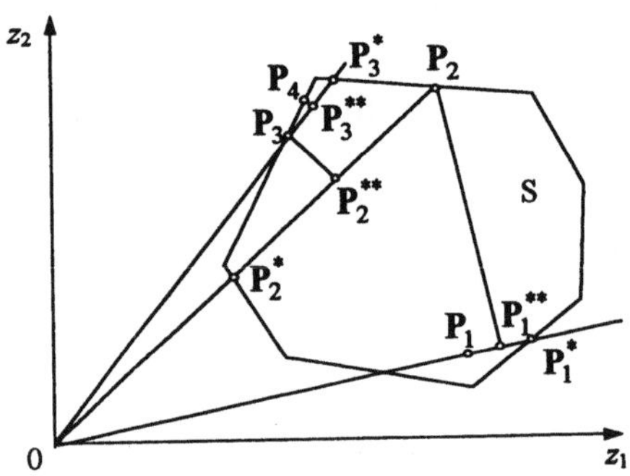

Figure 3.5.1

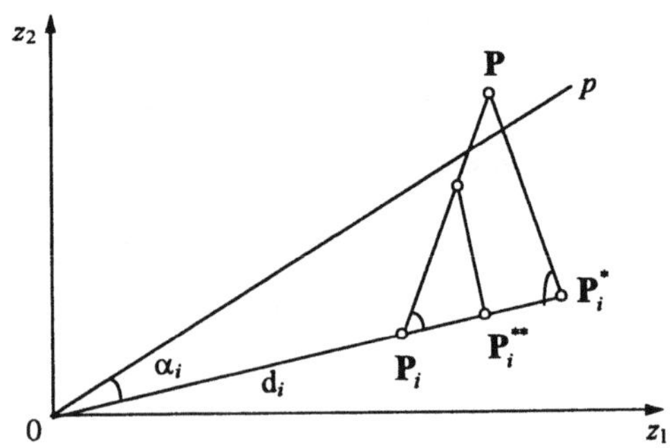

Figure 3.5.2

Theorem 3.5.1. *The sequence of points previously defined converges towards a point $P \in S$, which is the image of the optimal solution of the problem under consideration.*

Proof. The gradients of the half-lines OP_i form a sequence which is bounded and increasing monotone, i.e., convergent. The fact that the sequence is increasing is obvious and stems from its definition. To show that it is bounded, it is sufficient to show that above the line p, whose gradient corresponds to this limit, there is no point belonging to S. Assume the contrary, and let P be the point of S situated above the line p (Figure 3.5.2). Denote by

$$\alpha_i = \sphericalangle(OP_i, p), \quad \beta_i = \min\,[\sphericalangle P_i^* P_i P, \ \sphericalangle PP_i^* P_i], \ OP_i \equiv d_i, \ P_i P_i^* = 2a_i.$$

If P_i^* is situated between O and P_i, then P_i, and P_i^* changes roles. Since all the elements of sequence $\{P_i\}$ are below the line p, it follows that

$$a_i tg\ \beta_i \leq (a_i + d_i) tg\ \alpha_i \text{ or } tg\ \alpha_i \geq (a_i/(a_i + d_i)) tg\ \beta_i$$

From the initial assumptions, and from the fact that S does not reduce to a straight line segment, it follows that $a_i \geq a > 0$, $tg\ \beta_i \geq tg\ \beta > 0$ i.e., $tg\ \alpha_i > 0$, which contradicts the fact that $\alpha_i \to 0$.

Using these facts and Theorem 3.5.1, one can define the following solution algorithm due to Stahl [892]. The algorithm is relatively simple and requires the successive solution of some linear programming problems.

Step 1. Determine x_1 an arbitrary feasible solution of Problem (3.5.1), and let $\lambda_1 = c'x_1/d'x_1$.

Step 2. For a feasible solution x_i, and the corresponding $\lambda_i = c'x_i/d'x_i$, consider the following linear programming problems :

$$\max\ \{d'x \mid Ax = b;\ x \geq 0;\ (c' - \lambda_i d')x = 0\} \tag{3.5.2}$$

and, respectively,

$$\min\ \{d'x \mid Ax = b;\ x \geq 0;\ (c' - \lambda_i d')x = 0\} \tag{3.5.3}$$

Step 3. Denote by x_i^* the optimal solution of one of the above problems such that $d'x_i^* \neq d'x_i$, and let $x_i^{**} = (x_i + x_i^*)/2$.

Step 4. Solve the linear programming problem

$$\max\ \{c'x \mid Ax = b;\ x \geq 0;\ (\lambda_i c' + d')x = (\lambda_i c' + d')x_i^{**}\} \tag{3.5.4}$$

For its optimal solution x_{i+1}, determine $\lambda_{i+1} = c'x_{i+1}/d'x_{i+1}$.

Step 5. Compare λ_i with λ_{i+1}. If $\lambda_i = \lambda_{i+1}$, then x_i is the optimal solution of Problem (3.5.1). If not, go to Step 2 with x_{i+1} instead of x_i.

According to Theorem 3.5.1. the sequence $\{\lambda_i\}$ is increasing monotone and converges towards the optimum value of Problem (3.5.1).

Note that in Step 2 we need not solve Problem (3.5.2) and Problem (3.5.3), but only one of them. If the solution $\boldsymbol{x}^*$ of Problem (3.5.2) verifies the relation

$$\boldsymbol{d}'\boldsymbol{x}_i^* \neq \boldsymbol{d}'\boldsymbol{x}_i$$

then it is no longer necessary to also solve Problem (3.5.3). Also, we can have $\lambda_i = \lambda_{i+1}$, and if the image of S is a straight line segment, which is a possibility that we eliminated from the beginning, since in this case the objective function is a constant. The solution of the linear programming problems (3.5.2), (3.5.3), (3.5.4) is not difficult since these are almost identical, differentiating only through one restriction.

The algorithm described above brings us close to the optimal solution of the linear fractional programming problem. However, remain still open the question of the iteration stopping criterion, and the question of the distance between the optimal solution and the solution at which we stopped.

Numerical example. Solve the following linear fractional programming problem:

$$\max\ (x_1 - x_2)/(x_1 + x_2),$$

subject to

$$\begin{aligned} 2x_1 + x_2 &\geq 2, \\ -x_1 + x_2 &\leq 2, \\ 2x_1 + x_2 &\leq 7, \\ x_1 - 2x_2 &\leq 1, \\ x_2 &\leq 3, \\ x_1, x_2 &\geq 0. \end{aligned}$$

The domain of feasible solutions can be seen in Figure 3.5.3. The optimal solution is $x_1 = 1$, $x_2 = 0$ whereas the optimal value of the objective function is equal to 1. Start from the feasible solution $\boldsymbol{X}_1(2, 2)$ for which $\lambda_1 = 0$, and solve the linear programming problem :

$$\max\ (x_1 + x_2),$$

subject to

$$\begin{aligned} 2x_1 + x_2 &\geq 2, \\ -x_1 + x_2 &\leq 2, \\ 2x_1 + x_2 &\leq 7, \\ x_1 - 2x_2 &\leq 1, \\ x_2 &\leq 3, \\ x_1 - x_2 &= 0, \\ x_1, x_2 &\geq 0. \end{aligned}$$

having the solution $X_1^*(7/3,\ 7/3)$. Determine

$$X_1^{**} = (X_1 + X_1^*)/2 = (13/6,\ 13/6).$$

Solve the linear programming problem:

$$\max\ (x_1 - x_2),$$

subject to

$$2x_1 + x_2 \geq 2,$$
$$-x_1 + x_2 \leq 2,$$
$$2x_1 + x_2 \leq 7,$$
$$x_1 - 2x_2 \leq 1,$$
$$x_2 \leq 3,$$
$$x_1 + x_2 = 13/3,$$
$$x_1,\ x_2 \geq 0.$$

having the solution $X_2 = (8/3,\ 5/3)$ for which $\lambda_2 = 3/13 = 0.23076$.

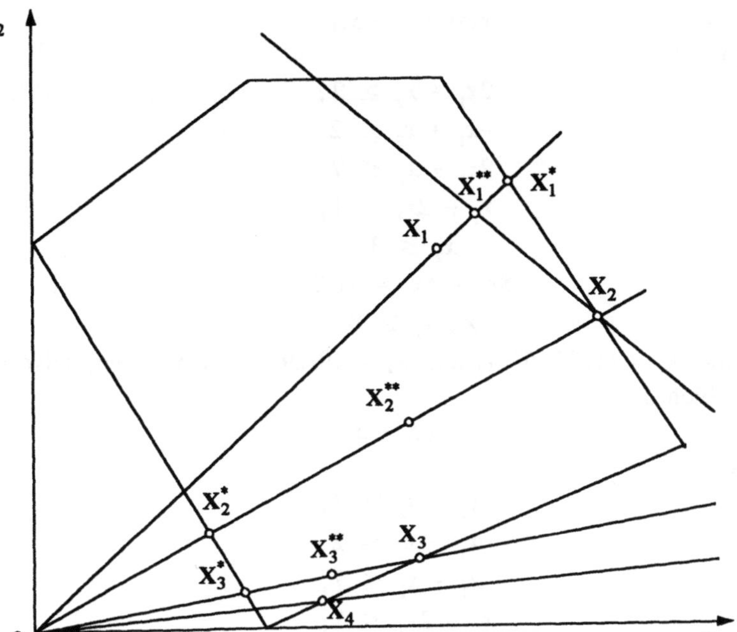

Figure 3.5.3

Now we solve a problem of type (3.5.3) since solving the problem of type (3.5.2) would lead to the same solution $X_2^*(8/3,\ 5/3)$ and we would get $d'X_2^* = d'X_2$. Hence, solve the linear programming problem:

$$\min\ (x_1 + x_2),$$

subject to

$$\begin{aligned} 2x_1 + x_2 &\geq 2, \\ -x_1 + x_2 &\leq 2, \\ 2x_1 + x_2 &\leq 7, \\ x_1 - 2x_2 &\leq 1, \\ x_2 &\leq 3, \\ 5x_1 - 8x_2 &= 0, \\ x_1,\ x_2 &\geq 0. \end{aligned}$$

and get the solution $X_2^* = (16/21,\ 10/21)$. Determine

$$X_2^{**} = (X_2 + X_2^*)/2 = (12/7,\ 15/14).$$

The following problem is to be solved:

$$\max\ (x_1 - x_2),$$

subject to

$$\begin{aligned} 2x_1 + x_2 &\geq 2, \\ -x_1 + x_2 &\leq 2, \\ 2x_1 + x_2 &\leq 7, \\ x_1 - 2x_2 &\leq 1, \\ x_2 &\leq 3, \\ 8x_1 + 5x_2 &= 31/2, \\ x_1,\ x_2 &\geq 0. \end{aligned}$$

having the solution $X_3 = (12/7,\ 5/14)$ and $\lambda_3 = 19/29 = 0.655$. Then, solve the linear programming problem :

$$\min\ (x_1 + x_2),$$

subject to

$$\begin{aligned} 2x_1 + x_2 &\geq 2, \\ -x_1 + x_2 &\leq 2, \\ 2x_1 + x_2 &\leq 7, \\ x_1 - 2x_2 &\leq 1, \\ x_2 &\leq 3, \\ 5x_1 - 24x_2 &= 0, \\ x_1,\ x_2 &\geq 0. \end{aligned}$$

having the solution $X_3^* = (48/53,\ 10/53)$. Determine

$$X_3^{**} = (X_3 + X_3^*)/2 = (486/371,\ 405/742).$$

Solve the linear programming problem:

$$\max\ (x_1 - x_2),$$

subject to

$$\begin{aligned} 2x_1 + x_2 &\geq 2, \\ -x_1 + x_2 &\leq 2. \\ 2x_1 + x_2 &\leq 7, \\ x_1 - 2x_2 &\leq 1, \\ x_2 &\leq 3, \\ 24x_1 + 5x_2 &= 25353/742, \\ x_1,\ x_2 &\geq 0. \end{aligned}$$

having the solution $X_4 = (54416/39326 = 1.38;\ 7545/39326 = 0.19)$ and $\lambda_4 = 46871/61961 = 0.756$, and so on. Continuing in the same way one gets the optimal solution. The values of X approximating the optimal solution are shown in Figure 3.5.3.

Dinkelbach [359] reduces the solution of a linear fractional programming problem to the solution of a sequence of linear programming problems. The method is general in as much as it can be applied even when we have a ratio of functions that are not necessarily linear. We will reopen extend this method in the following Chapter when we tackle the nonlinear case. Let

$$f(x) = \sum_{j=1}^{n} c_j x_j + c_0; \quad g(x) = \sum_{j=1}^{n} d_j x_j + d_0$$

and the following linear fractional programming problem

$$\max\ \{z = \frac{f(x)}{g(x)} \mid Ax = b,\ x \geq 0\} \tag{3.5.5}$$

Make the assumption that the domain of feasible solutions D is non-empty, that the denominator does not reduce to a constant and that it is strictly positive on D. Consider the auxiliary function

$$F(z) = \max_{x \in D}\ [f(x) - zg(x)],\quad z \in R.$$

The following theorem establishes a connection between the fractional programming problem (3.5.5) and a certain parametric linear programming problem.

Theorem 3.5.2. *The vector x_0 is optimal solution for the fractional programming problem* (3.5.5) *if and only if*

$$\max_{x \in D}\ [f(x) - z_0 g(x)] = F(z_0) = 0,$$

where

$$z_0 = \frac{f(x_0)}{g(x_0)}.$$

Proof. If $\boldsymbol{x}_0$ is optimal solution for the fractional problem, then

$$z_0 = \frac{f(\boldsymbol{x}_0)}{g(\boldsymbol{x}_0)} \geq \frac{f(\boldsymbol{x})}{g(\boldsymbol{x})}, \quad \forall \ \boldsymbol{x} \in D$$

which means (since $g(\boldsymbol{x}) > 0, \ \boldsymbol{x} \in D$) that

$$f(\boldsymbol{x}) - z_0 g(\boldsymbol{x}) \leq 0, \quad \forall \ \boldsymbol{x} \in D$$

But, for $\boldsymbol{x}_0$ we have equality, thus

$$\max_{\boldsymbol{x} \in D} \ [f(\boldsymbol{x}) - z_0 g(\boldsymbol{x})] = 0.$$

Conversely, if $\boldsymbol{x}_0$ is the optimal solution of problem

$$\max_{\boldsymbol{x} \in D} \ [f(\boldsymbol{x}) - z_0 g(\boldsymbol{x})] = F(z_0) = 0.$$

This means that

$$f(\boldsymbol{x}) - z_0 g(\boldsymbol{x}) \leq f(\boldsymbol{x}_0) - z_0 g(\boldsymbol{x}_0) = 0, \quad \forall \ \boldsymbol{x} \in D.$$

This implies that $\boldsymbol{x}_0$ is solution of the fractional programming problem (3.5.5). This theorem also gives a procedure for calculating the optimal solution of the fractional programming problem (3.5.5).

If we choose two numbers, z_1 and z_2, such that $z_2 > z_1$, and if, for $z_i (i = 1, 2)$ fixed, $\boldsymbol{x}_i$ maximizes $F(z_i)$, then

$$F(z_2) = \max_{\boldsymbol{x}} [f(\boldsymbol{x}) - z_2 g(\boldsymbol{x})] = f(\boldsymbol{x}_2) - z_2 g(\boldsymbol{x}_2) <$$

$$< f(\boldsymbol{x}_2) - z_1 g(\boldsymbol{x}_2) \leq \max_{\boldsymbol{x}} \ [(f(\boldsymbol{x}) - z_1 g(\boldsymbol{x})] = F(z_1).$$

This means that $F(z)$ is strictly decreasing in z, and consequently $F(z) = 0$ has o unique solution.

The algorithm for the solution of the fractional programming problem (3.5.5) consists in the following steps:

Step 1. Consider $z = z_1 = 0$ and determine $\boldsymbol{X}_1$ the optimal solution of problem

$$F(z_1) = F(0) = \max_{\boldsymbol{x}} f(\boldsymbol{x})$$

Let $z_2 = f(\boldsymbol{X}_1)/g(\boldsymbol{X}_1)$.

Step 2. Determine $\boldsymbol{X}_2$ the optimal solution of the mathematical programming problem

$$F(z_2) = \max_{\boldsymbol{x} \in D} \ [f(\boldsymbol{x}) - z_2 g(\boldsymbol{x})].$$

Let $z_3 = f(\boldsymbol{X}_2)/g(\boldsymbol{X}_2)$. Continue the iterative process.

Step 3. We stop at the iteration of order k if the value $F(z_k)$ is either zero (according to Theorem 3.5.2) or is less than a quantity $\delta > 0$ sufficiently close to zero.

Numerical example. Maximize the function

$$z = \frac{x_1 - 3x_2 + 3x_3}{x_1 + 2x_2 + x_3 + 1} = \frac{f(x)}{g(x)} \tag{3.5.6}$$

subject to

$$\begin{aligned} x_1 + x_2 - x_3 &\leq 6, \\ -2x_1 - x_2 + 2x_3 &\leq 2, \\ 2x_1 - 3x_2 + x_3 &\leq 3, \\ x_1, x_2, x_3 &\geq 0. \end{aligned} \tag{3.5.7}$$

Add the slack variables x_4, x_5, x_6 in the constraints (3.5.7) and get

$$\begin{aligned} x_1 + x_2 - x_3 + x_4 &= 6, \\ -2x_1 - x_2 + 2x_3 + x_5 &= 2, \\ 2x_1 - 3x_2 + x_3 + x_6 &= 3, \\ x_1, x_2, x_3, x_4, x_5, x_6 &\geq 0 \end{aligned} \tag{3.5.8}$$

Consider $z_1 = 0$ and solve the linear programming problem

$$F(z_1) = F(0) = \max_{x \in D} f(x) = \max_{x \in D} (x_1 - 3x_2 + 3x_3).$$

The optimal solution is X_1 (37/3; 14; 61/3; 0; 0; 0), $F(z_1) = F_1 = = 94/3 \approx 33.13$. Let $z_2 = f(X_1)/g(X_1) = 94/185 \approx 0.50$. Solve the linear programming problem

$$F_2 = F(z_2) = F(0.5) = \max_{x \in D} (f(x) - 0.5 \cdot g(x)) =$$

$$= \max_{x \in D}[(1 - 94/185)x_1 + (-3 - 183/185)x_2 + (3 - 94/185)x_3 - 94/185] =$$

$$= \max_{x \in D}(91/185x_1 - 743/185x_2 + 461/185x_3 - 94/185)$$

The optimal solution is X_2(2/3; 0; 5/3; 7; 0; 0) whereas $z_3 = f(X_2)/g(X_2) = 17/10 = 1.7$. Also, $F_2 = 2205/555 \approx 3.98$.

Solve the problem

$$F_3 = F(z_3) = F(1.7) = \max_{x \in D}(f(x) - 1.7g(x)) =$$

$$= \max_{x \in D}[(1 - 17/10)x_1 - (3 + 34/10)x_2 + (3 - 17/10)x_3 - 17/10] =$$

$$= \max_{x \in D}(-7/10x_1 - 64/10x_2 + 13/10x_3 - 17/10).$$

The optimal solution is X_3(2/3; 0; 5/3; 7; 0; 0; 0) and $F_3 = 0$.

According to the presented algorithm, the optimal solution of the fractional programming problem (3.5.6) - (3.5.7) has been obtained.

Elmaghraby and Arisawa [379] applied this algorithm in a particular case of Problem (3.5.5), i.e., when the matrix $\boldsymbol{A}$ is reduced to just a row vector. Very close to the method of Dinkelbach is also the method propose by Bitran and Novaes [192]. In their method, one first solves the problem

$$\max z = \bar{c}'x \text{ with } \bar{c}_i = c_i - [(\sum_{j=1}^{n} c_j d_j)/(\sum_{j=1}^{n} d_j^2)]\, d_i \ .$$

The remaining steps are identical with the method of Dinkelbach.

Bhatt [188] showed that the linearization methods (see also subsection 4.4) proposed by Isbell and Marlow [504], Mangasarian [623], Bitran and Novaes [192] and Bhatt [186] are algorithmically equivalent in the sense that they all generate the same sequence of the points leading to an optimal solution of a linear fractional problem and that the manner in which this sequence is generated remains the same.

3.6. PARAMETRIC METHODS

In this section, we present several methods for the solution of the linear fractional programming problems that we named parametric methods. Such methods are named parametric since the optimal solution is obtained in a number of iterations considering either the objective function, or one of the functions in the numerator or denominator as a real parameter. One parametric method known in the literature is the method of Dinkelbach as presented in Section 3.5. Another parametric method belongs to Frankel, Novaes and Pollack [399] and also consists in considering the value of the objective function as a parameter. Their method was applied by Kydland [590] to the solution of a maritime transportation problem. Consider the problem:

$$\max z = N(\boldsymbol{x})/D(\boldsymbol{x}) = (\boldsymbol{c}'\boldsymbol{x} + c_0)/(\boldsymbol{d}'\boldsymbol{x} + d_0) \tag{3.6.1}$$

subject to

$$\boldsymbol{A}\boldsymbol{x} \le \boldsymbol{b};\ \boldsymbol{x} \ge \boldsymbol{0} \tag{3.6.2}$$

The constraints (3.6.2) are transformed in equalities by adding the slack variables $x_{n+1}, ..., x_{n+m} \ge 0$

$$\begin{aligned} a_{11}x_1 + ... + a_{1n}x_n + x_{n+1} &= b_1 \\ ... \quad ... \quad ... \quad ... \quad ... &\quad ... \\ a_{m1}x_1 + ... + a_{mn}x_n + x_{n+m} &= b_m \\ x_i \ge 0,\ i = 1, ..., n+m & \end{aligned} \tag{3.6.3}$$

Consider z as a parameter, and construct a new function

$$F = (-zd_0 + c_0) + (-z\boldsymbol{d}'\boldsymbol{x} + \boldsymbol{c}'\boldsymbol{x}) \tag{3.6.4}$$

which leads to (3.6.1) if $F = 0$. The problem is to maximize z such that to have the relation

$$(zd_0 - c_0) + (zd'x - c'x) = 0. \tag{3.6.5}$$

Since, in (3.6.4), the term $zd_0 - c_0$ is a constant with respect to x, we are going to neglect it. Hence, we consider and maximize the function

$$f = \sum_{j=1}^{n} (c_j - zd_j)x_j, \quad z \text{ parameter} \tag{3.6.6}$$

instead of the function F. The theory of linear parametric programming indicates that there are values of the parameter z:

$$-\infty < z_1 < z_2 < \dots < z_p < \infty$$

such that, for every interval $[z_i, z_{i+1}]$, exists an optimal solution x^* of the problem max f, and the value the objective function is f^*.

For the initial solution $x^0(x_1 = x_2 = \dots = x_n = 0,\ x_{n+1} = b_1, \dots x_{n+m} = b_m)$, given in Table 3.6.0, we calculated the differences $z_j - c_j$.

Table 3.6.0

			c_1-zd_1	c_2-zd_2	...	c_n-zd_n	0	0	...	0
c_B	Base	P_0	P_1	P_2	...	P_n	P_{n+1}	P_{n+2}	...	P_{n+m}
0	x_{n+1}	b_1	a_{11}	a_{12}	...	a_{1n}	1	0	...	0
0	x_{n+2}	b_2	a_{21}	a_{22}	...	a_{2n}	0	1	...	0
⋮	⋮	⋮	⋮	⋮	...	⋮	⋮	⋮	⋮	⋮
0	x_{n+m}	b_m	a_{m1}	a_{m2}	...	a_{mn}	0	0	...	1
z_j-c_j		f	zd_1-c_1	zd_2-c_2	...	zd_n-c_n	0	0	...	0

For this solution to be optimal solution of the problem max f, we need $z_j - c_j \geq 0$ for all j i.e.,

$$zd_1 - c_1 \geq 0, \dots, zd_n - c_n \geq 0$$

Let $[\underline{z}_1, \overline{z}_1]$ be the domain of z resulting from the above inequalities. Calculate the value of z, let this be z^0, for the solution x^0,

$$z^0 = \frac{c'x^0 + c_0}{d'x^0 + d_0}$$

If $z^0 \in [\underline{z}_1, \overline{z}_1]$, then z^0 is the optimal value for z and is obtained for x^0. If $z^0 \notin [\underline{z}_1, \overline{z}_1]$, then x^0 is not optimal solution of problem (3.6.1) - (3.6.2). Then, the variable x_k from outside the base is introduced in base instead of x_r and we get a new optimal solution x^1 for $z \in [\underline{z}_2, \overline{z}_2]$. The interval $[\underline{z}_2, \overline{z}_2]$ is adjacent to the right or to the left of the interval $[\underline{z}_1, \overline{z}_1]$. Let

$$z^1 = \frac{c'x^1 + c_0}{d'x^1 + d_0}$$

Similarly, if $z^1 \in [\underline{z}_2, \overline{z}_2]$, then x^1 is the optimal solution of Problem (3.6.1) - (3.6.2). Otherwise, determine a new solution x^2 and a new interval $[\underline{z}_3, \overline{z}_3]$ for z etc. The procedure is continued until we find a solution x^s that maximizes the function f, to which corresponds an interval $[\underline{z}_{s+1}, \overline{z}_{s+1}]$ such that

$$z^s = \frac{c'x^s + c_0}{d'x^s + d_0} \in [\underline{z}_{s+1}, \overline{z}_{s+1}]$$

Numerical example. Maximize the function

$$F = \frac{x_1 + 2x_2 + 3x_3 - 24}{2x_1 + 3x_2 + x_3 + 4}$$

subject to

$$\begin{gathered} x_1 \le 5,\ x_2 \le 4,\ x_3 \le 3, \\ x_1 + 2x_2 + 3x_3 \le 15, \\ 2x_1 + 3x_2 + x_3 \le 20, \\ x_i \ge 0, \quad i = 1,2,3. \end{gathered}$$

Introduce the slack variables x_4, x_5, x_6, x_7, x_8 and the constraints system becomes

$$\begin{gathered} x_1 + x_4 = 5, \\ x_2 + x_5 = 4, \\ x_3 + x_6 = 3, \\ x_1 + 2x_2 + 3x_3 + x_7 = 15, \\ 2x_1 + 3x_2 + x_3 + x_8 = 20, \\ x_i \ge 0, \quad i = 1,2, ..., 8. \end{gathered}$$

The objective function becomes "max F" such that the following relation is satisfied:

$$(-4-4F) + (1-2F)x_1 + (2-3F)x_2 + (3-F)x_3 = 0$$

Modify the objective function such that we maximize the function

$$f = (1-2\overline{F})x_1 + (2-3\overline{F})x_2 + (3-\overline{F})x_3$$

with the constraints as above. The initial solution (Table 3.6.1) is

$$x_1 = x_2 = x_3 = 0$$

Table 3.6.1

			$1-2\bar{F}$	$2-3\bar{F}$	$3-\bar{F}$	0	0	0	0	0
c_B	Base	P_0	P_1	P_2	P_3	P_4	P_5	P_6	P_7	P_8
0	x_4	5	1	0	0	1	0	0	0	0
0	x_5	4	0	1	0	0	1	0	0	0
0	x_6	3	0	0	[1]	0	0	1	0	0
0	x_7	15	1	2	3	0	0	0	1	0
0	x_8	20	2	3	1	0	0	0	0	1
$z_j - c_j$		0	$2\bar{F}-1$	$3\bar{F}-2$	$\bar{F}-3$	0	0	0	0	0

Let see if this solution is optimal or not. Put the condition that $z_j - c_j \geq 0$ for the variables outside the base. We have,

$$\left.\begin{aligned} 2\bar{F} - 1 \geq 0 &\Rightarrow \bar{F} \geq 1/2 \\ 3\bar{F} - 2 \geq 0 &\Rightarrow \bar{F} \geq 2/3 \\ \bar{F} - 3 \geq 0 &\Rightarrow \bar{F} \geq 3 \end{aligned}\right\} \Rightarrow 3 \leq \bar{F}$$

For the solution found we have $F = -24/4 = -6$, which is outside the interval $[3,\infty)$, i.e., the solution found is not optimal. For $\bar{F} = 3$, we have $z_3 - c_3 = 0$ hence x_3 will enter the base instead of x_6. The new Simplex table is Table 3.6.2.

Table 3.6.2

			$1-2\bar{F}$	$2-3\bar{F}$	$3-\bar{F}$	0	0	0	0	0
c_B	Base	P_0	P_1	P_2	P_3	P_4	P_5	P_6	P_7	P_8
0	x_4	5	1	0	0	1	0	0	0	0
0	x_5	4	0	1	0	0	1	0	0	0
$3-\bar{F}$	x_3	3	0	0	1	0	0	1	0	0
0	x_7	6	1	[2]	0	0	0	-3	1	0
0	x_8	17	2	3	0	0	0	-1	0	1
$z_j - c_j$		$9-3\bar{F}$	$2\bar{F}-1$	$3\bar{F}-2$	0	0	0	$3-\bar{F}$	0	0

The new solution is $x_1 = 0$, $x_2 = 0$, $x_3 = 3$. The conditions $z_j - c_j \geq 0$ give

$$\left.\begin{array}{l} 2\bar{F} - 1 \geq 0 \Rightarrow \bar{F} \geq 1/2, \quad 3\bar{F} - 2 \geq 0 \Rightarrow \bar{F} \geq 2/3 \\ 3 - \bar{F} \geq 0 \Rightarrow \bar{F} \leq 3 \end{array}\right\} \Rightarrow 2/3 \leq \bar{F} \leq 3$$

For this solution, we have $F = (-15)/7 = -2.14$, thus this solution is not optimal either. Since, for $\bar{F} = 2/3$, we have $z_2 - c_2 = 0$, the variable x_2 will enter instead of the variable x_7 (Table 3.6.3).

Table 3.6.3

			$1-2\bar{F}$	$2-3\bar{F}$	$3-\bar{F}$	0	0	0	0	0
c_B	Base	P_0	P_1	P_2	P_3	P_4	P_5	P_6	P_7	P_8
0	x_4	5	[1]	0	0	1	0	0	0	0
0	x_5	1	-1/2	0	0	0	1	3/2	-1/2	0
$3-\bar{F}$	x_3	3	0	0	1	0	0	1	0	0
$2-3\bar{F}$	x_2	3	1/2	1	0	0	0	-3/2	1/2	0
0	x_8	8	1/2	0	0	0	0	7/2	-3/2	1
$z_j - c_j$		$15-12\bar{F}$	$\bar{F}/2$	0	0	0	0	$\frac{7}{2}\bar{F}$	$1-\frac{3}{2}\bar{F}$	0

Table 3.6.4

			$1-2\bar{F}$	$2-3\bar{F}$	$3-\bar{F}$	0	0	0	0	0
c_B	Base	P_0	P_1	P_2	P_3	P_4	P_5	P_6	P_7	P_8
$1-2\bar{F}$	x_1	5	1	0	0	1	0	0	0	0
0	x_5	7/2	0	0	0	1/2	1	3/2	-1/2	0
$3-\bar{F}$	x_3	3	0	0	1	0	0	1	0	0
$2-3\bar{F}$	x_2	1/2	0	1	0	-1/2	0	-3/2	1/2	0
0	x_8	11/2	0	0	0	-1/2	0	[7/2]	-3/2	1
$z_j - c_j$		$15-\frac{29}{2}\bar{F}$	0	0	0	$-\frac{1}{2}\bar{F}$	0	$\frac{7}{2}\bar{F}$	$1-\frac{3}{2}\bar{F}$	0

The new solution is $x_1 = 0$, $x_2 = 3$, $x_3 = 3$. The conditions $z_j - c_j \geq 0$ give

$$\left.\begin{array}{r} \bar{F}/2 \geq 0 \Rightarrow \bar{F} \geq 0 \\ (7/2)\bar{F} \geq 0 \Rightarrow \bar{F} \geq 0 \\ 1 - (3/2)\bar{F} \geq 0 \Rightarrow \bar{F} \leq 2/3 \end{array}\right\} \Rightarrow 0 \leq \bar{F} \leq 2/3$$

We have $F = (-9)/15 = (-3)/5 < 0$ and hence this solution is not optimal either. Either x_1 or x_6 can enter the base. We choose x_1 to enter the base instead of x_4 (Table 3.6.4). The new solution is $x_1 = 5$; $x_2 = 1/2$; $x_3 = 3$. The conditions $z_j - c_j \geq 0$ give

$$\left.\begin{array}{r} -\bar{F}/2 \geq 0 \Rightarrow \bar{F} \leq 0 \\ (7/2)\bar{F} \geq 0 \Rightarrow \bar{F} \geq 0 \\ 1 - (3/2)\bar{F} \geq 0 \Rightarrow \bar{F} \leq 2/3 \end{array}\right\} \Rightarrow \bar{F} = 0 \text{ and } F = (-18)/37 < 0$$

Make a pivoting of pivot 7/2 in Table 3.6.4 and get Table 3.6.5.

Table 3.6.5

			$1-2\bar{F}$	$2-3\bar{F}$	$3-\bar{F}$	0	0	0	0	0
c_B	Base	P_0	P_1	P_2	P_3	P_4	P_5	P_6	P_7	P_8
$1-2\bar{F}$	x_1	5	1	0	0	1	0	0	0	0
0	x_5	8/7	0	0	0	5/7	1	0	1/7	-3/7
$3-\bar{F}$	x_3	10/7	0	0	1	1/7	0	0	3/7	-2/7
$2-3\bar{F}$	x_2	20/7	0	1	0	-5/7	0	0	-1/7	3/7
0	x_6	11/7	0	0	0	-1/7	0	1	-3/7	2/7
$z_j - c_j$		$15-\frac{160}{7}\bar{F}$	0	0	0	0	0	0	1	$-\bar{F}$

The new solution is $x_1 = 5$; $x_2 = 20/7$; $x_3 = 10/7$. The conditions $z_j - c_j \geq 0$ give $-\bar{F} \geq 0 \Rightarrow \bar{F} \leq 0$, thus $-\infty < \bar{F} \leq 0$. We have $F = -3/8 < 0$, i.e., F has reached its maximum value. Since $z_4 - c_4 = 0$, it follows that problem admits another optimal solution.

Joksch [519] reduces the solution of the linear fractional programming problem (3.6.1) - (3.6.2) to the parametric linear programming problem by considering the value of one of the linear functions from the numerator or from the denominator of the objective function as a parameter λ. For example, if we assume that the value of the denominator is equal to a parameter $\lambda (D(x) = \lambda)$, then a parametric linear programming problem, of parameter λ follows:

$$N^*(\lambda) = \max N(x) \tag{3.6.7}$$

subject to

$$Ax \leq b;\ D(x) = \lambda;\ x \geq 0. \tag{3.6.8}$$

Its optimal solution x^* will depend on λ. Cambini and Martein [219] call x^* an optimal level solution. Also, z will depend on λ. The problem is now to determine the maximum value of z as a function of the real variable λ.

Wolf [1051], [1053], [1054] also make the parametric analysis of Problem (3.6.1) - (3.6.2), in the case when the domain of feasible solutions D is a compact set. In the first step, solve the problem (3.6.7) - (3.6.8) with

$$\lambda \in \Lambda = \{ \lambda \in R \mid D \cap \{x \in R^n \mid D(x) = \lambda\} \neq \emptyset \}$$

and determine the function $N^*(\lambda)$ which is continuous, piecewise linear and concave. In the second step, solve the problem

$$\max_{\lambda \in \Lambda} N^*(\lambda)/\lambda \tag{3.6.9}$$

Wolf shows that, if λ^0 is the optimal solution of Problem (3.6.9), then the optimal value of the objective function of Problem (3.6.1) - (3.6.2) is given by $N^*(\lambda^0)/\lambda^0$.

3.7. CONVEX SIMPLEX METHOD

The convex simplex method extends the simplex method for linear programming to nonlinear functions. The method was developed by Zangwill[1], initially to minimize a convex function on a domain defined by linear constraints. Later, it was applied to more general functions with continuous partial derivatives. In the application of this method we follow, as much as possible, the simplex method from linear programming which we modify only if the nonlinearity of the function requires it. A summary description of the convex simplex method follows.

Consider the nonlinear programming problem with linear constraints

$$\min \{f(x) \mid Ax = b;\ \ x \geq 0\} \tag{3.7.1}$$

[1] Zangwill, W.I., *The convex simplex method*. Management Sci., 14(3),221-238, 1967.

where f is a continuous function with the first partial derivatives continuous. Consider the iteration of order k of the method. In the application of the method, we start with step I of iteration $k = 1$.

<u>The iteration k</u>. Let B be a base of matrix $A, x^k = (x_i^k)'$, a feasible solution having the corresponding table $T^k = (t_{ij}^k)'$, and let $x_B^k = (x_{Bj}^k)'$ be a basic feasible solution and $b^k = (b_i^k)'$.

Step I. Calculate the relative cost vector

$$c^k = (c_i^k)' = (\nabla f(x^k) - (T^k)' \nabla f(x^k)_B),$$

where

$$\nabla f(x^k) = \left[\frac{\partial f(x)}{\partial x_1^k}, \ldots, \frac{\partial f(x)}{\partial x_n^k}\right]'$$

and

$$\nabla f(x^k)_B = \left[\frac{\partial f(x)}{\partial x_{B1}^k}, \ldots, \frac{\partial f(x)}{\partial x_{Bm}^k}\right]'$$

Let $c_{s1}^k = \min \{c_i^k \mid i \in N\}$ and $c_{s2}^k x_{s2}^k = \max \{c_i^k x_i^k \mid i \in N\}$, where $N = \{1,2, \ldots, n\}$. If $c_{s1}^k = c_{s2}^k x_{s2}^k = 0$ the algorithm is finished, x^k is optimal solution. Otherwise, go to Step II.

Step II. Determine the non-basic variables to change their value. If $|c_{s1}^k| > c_{s2}^k x_{s2}^k$, then $x_s^k = x_{s1}^k$ increases, and only the basic variables are modified. If $|c_{s1}^k| < c_{s2}^k x_{s2}^k$, then $x_s^k = x_{s2}^k$ decreases, modifying only the basic variables.

Step III. Calculate x^{k+1} and T^{k+1}.

Case 1. x_s^k must be increased, and, for a certain $i, t_{is}^k > 0$. The increasing of x_s^k means that a basic variable becomes zero. Let $y^k = (y_i^k)'$ be the new value of x. We have

$$y_i^k = x_i^k,\ i \in D-s;\ y_s^k = x_s^k + \Delta^k;\ y_{Bi}^k = x_{Bi}^k - t_{is}^k\Delta^k,\ i \in M \tag{3.7.2}$$

where $D-s$ is the set of indices non-basic variables less the index s, M is the set of indices of the basic variables, and

$$\Delta^k = x_{Br}^k / t_{rs}^k = \min \{x_{Bi}^k / t_{is}^k \mid t_{is}^k > 0\}.$$

Determine x^{k+1} such that

$$f(x^{k+1}) = \min \{f(x) \mid x = \lambda x^k + (1 - \lambda) y^k,\ 0 \leq \lambda \leq 1\} \tag{3.7.3}$$

If $x^{k+1} \neq y^k$, consider $T^{k+1} = T^k$ and go to the iteration k with $k+1$ instead of k.

The base remains unchanged. If $x^{k+1} = y^k$, make a pivoting of pivot t^k_{rs}, form T^{k+1}, go to iteration k with $k+1$ instead of k and substitute x^k_{Br} with x^k_s in the base.

Case 2. x^k_s must be increased and $t^k_{is} \leq 0, \forall\, i$. Define the vector y^k through the equations (3.7.2) with the only exception that $\Delta^k = 1$. Determine x^{k+1} from problem

$$f(x^{k+1}) = \min \{f(x) \mid x = x^k + \lambda(y^k - x^k),\ \lambda \geq 0\}$$

If x^{k+1} does not exist, the optimal value is $-\infty$. If x^{k+1} exists, consider $T^{k+1} = T^k$ and go to iteration k with $k+1$ instead of k, the base remaining the same.

Case 3. x^k_s must be reduced. Determine y^k from equations (3.7.2) with the difference that Δ^k is defined as:

$$\Delta^k = \max\,[\Delta^k_1, \Delta^k_2],$$

where

$$\Delta^k_1 = x^k_{Br}/t^k_{rs} = \max\,\{x^k_{Bi}/t^k_{is} \mid t^k_{is} < 0\} \text{ and } \Delta^k_2 = -x^k_s.$$

If $t^k_{is} \geq 0,\ i \in M$, then $\Delta^k_1 = -\infty$. Determine x^{k+1} from (3.7.3), determine also the table T^{k+1}, and go to the iteration k with $k+1$ instead of k. Continue the calculation process with the new feasible basic solution x^{k+1}_B up to iteration n when $c^n_{s1} = c^n_{s2}x^n_{s2} = 0$. In this case, x^n will be the optimal solution of Problem (3.7.1).

So far, we have presented the convex simplex method for a minimum problem. For a maximum problem, the algorithm is maintained, except some differences that can be easily observed. For example, in *Step* I

$$c^k_{s1} = \max\,\{c^k_i \mid i \in N\} \text{ and } c^k_{s2}x^k_{s2} = \min\,\{c^k_i x^k_i \mid i \in N\}$$

In *Step* II, the sense of the inequalities is changed, whereas in *Step* III, in formula (3.7.3) one must consider max.

After this short presentation of the convex simplex method, let's examine how we can utilize this method to the solution of the linear fractional programming problem.

Consider the following problem:

$$\max f(x) = \frac{\sum_{j=1}^{n} c_j x_j + c_0}{\sum_{j=1}^{n} d_j x_j + d_0} \tag{3.7.4}$$

subject to

$$\sum_{j=1}^{n} a_{ij}x_j \leq b_i, \quad i = 1, ..., m; \quad x_j \geq 0, \quad j = 1, ..., n \tag{3.7.5}$$

Assume that the domain of feasible solutions is a regular set, and that the denominator is positive on this domain. If we start with a basic feasible solution, then the convex simplex method coincides with the simplex method from linear fractional programming. The two methods differ if we start with a non-basic feasible solution, at least until we obtain a basic solution, but in the following iterations they coincide. In the constraints (3.7.5), add the slack variables $x_{n+1}, ..., x_{n+m}$, and then

$$x^1 = (0, 0, ..., 0, x_{n+1}, ..., x_{n+m})' = (0, 0, ..., 0, b_1, ..., b_m)'$$

constitutes a starting basic feasible solution which has the table T^1.

	0	0	...	0	b_1	b_2	...	b_m
	--	--	--	--	--	--	--	--
	a_{11}	a_{12}	...	a_{1n}	1	0	...	0
T^1	a_{21}	a_{22}	...	a_{2n}	0	1	...	0
	...	...	...	...	...	...	...	...
	a_{m1}	a_{m2}	...	a_{mn}	0	0	...	1
	--	--	--	--	--	--	--	--

Note that

$$x_B^1 = (x_{n+1}^1, ..., x_{n+m}^1)' = (b_1, b_2, ..., b_m)'$$

Step I. We have

$$\nabla f(x) = (Z_1, Z_2, ..., Z_n, 0, ..., 0)$$

in which

$$Z_j = (\Delta_j + \sum_{k=1}^{n} \delta_{kj}x_k)/D^2, \; D = \sum_{j=1}^{n} d_jx_j + d_0, \; \Delta_j = d_0c_j - c_0d_j,$$

$$\delta_{kj} = (d_kc_j - c_kd_j)$$

Hence,

$$\nabla f(x^1) = (\Delta_1, \Delta_2, ..., \Delta_n, 0, ..., 0)'$$

and

$$\nabla f(x^1)_B = (0, 0, ..., 0)'$$

Calculate:

$$c^1 = (\nabla f(x^1) - (T^1)' \nabla f(x^1)_B) =$$

$$= \begin{bmatrix} \Delta_1 \\ \Delta_2 \\ \vdots \\ \Delta_n \\ 0 \\ 0 \\ \vdots \\ 0 \end{bmatrix} - \begin{bmatrix} a_{11} & a_{21} & \dots & a_{m1} \\ a_{12} & a_{22} & \dots & a_{m2} \\ \vdots & \vdots & \vdots & \vdots \\ a_{1n} & a_{2n} & \dots & a_{mn} \\ 1 & 0 & \dots & 0 \\ 0 & 1 & \dots & 0 \\ \vdots & \vdots & \vdots & \vdots \\ 0 & 0 & \dots & 1 \end{bmatrix} \begin{bmatrix} 0 \\ 0 \\ \vdots \\ 0 \\ 0 \\ 0 \\ \vdots \\ 0 \end{bmatrix} = \begin{bmatrix} \Delta_1 \\ \Delta_2 \\ \vdots \\ \Delta_n \\ 0 \\ 0 \\ \vdots \\ 0 \end{bmatrix}$$

Let $c_{s1}^1 = \max\{\Delta_1, \Delta_2, ..., \Delta_n, 0, ..., 0\}$ and $c_{s2}^1 x_{s2}^1 = \max\{0, 0, ... 0\}$.

Step II. Let $c_{s1}^1 > |c_{s2}^1 x_{s2}^1|$ corresponding to $s_1 = j$. Thus, $x_s^1 = x_j^1$; $j = 1, ..., n$. Hence, x_j^1 increases.

Step III. Let

$$\Delta^1 = \min\{b_1/a_{1j}, b_2/a_{2j}, ..., b_m/a_{mj}\} = b_r/a_{rj},$$

accepting that $a_{ij} > 0$.

Define the vector $\boldsymbol{y}^1$ such that:

$$y_1^1 = ... = y_{j-1}^1 = y_{j+1}^1 = ... = y_n^1 = 0;$$

$$y_j^1 = x_j^1 + \Delta^1 = \Delta^1 = b_r/a_{rj};\ y_{n+1}^1 = b_1 - (b_r/a_{rj})a_{1j};$$

$$y_{n+2}^1 = b_2 - (b_r/a_{rj})a_{2j};\ ...\ ;\ y_{n+r}^1 = b_r - (b_r/a_{rj})a_{rj} = 0;\ ...\ ;$$

$$y_{n+m}^1 = b_m - (b_r/a_{rj})a_{mj}.$$

Thus

$$\boldsymbol{y}^1 = [0, ..., 0, b_r/a_{rj}, 0, ..., 0, b_1 - b_r \cdot a_{1j}/a_{rj}, ..., b_{r-1} - \\ -b_r \cdot a_{r-1,j}/a_{rj}, 0, b_{r+1} - b_r \cdot a_{r+1,j}/a_{rj}, ..., b_m - b_r \cdot a_{mj}/a_{rj}]'$$

which shows that the r-th basic variable has become zero and the j-th variable outside the base has increased from 0 to b_r/a_{rj}. To find $\boldsymbol{x}^2$ solve the problem

$$f(\boldsymbol{x}^2) = \max\{f(\boldsymbol{x}) \mid \boldsymbol{x} = \lambda\boldsymbol{x}^1 + (1 - \lambda)\boldsymbol{y}^1,\ 0 \le \lambda \le 1\}.$$

But for f a linear fractional function, a finite value for λ cannot be found. Indeed, f is a function of λ in the point

$$\boldsymbol{x} = \lambda\boldsymbol{x}^1 + (1 - \lambda)\boldsymbol{y}^1$$

and if we consider $\partial f(\boldsymbol{x})/\partial\lambda$, then its numerator will not depend on λ, i.e., one cannot find a finite value of λ such that

$$\partial f(\boldsymbol{x})/\partial\lambda = 0$$

Hence, $x^2 = y^1$. Furthermore, $x^{k+1} = y^k$ for any $k = 1,2, \ldots$. Continue the calculation process with the new feasible solution x^2 and with the new basic feasible solution x_B^2 in which x_{Br}^2 is now substituted by x_j^1, up to iteration k in which we have $c_{s1}^k = c_{s2}^k x_{s2}^k = 0$. The optimal solution will be x^k, whereas $f(x^k)$ is the corresponding optimal value.

Numerical example:

$$\max \{f(x) = (2x_1 + x_2)/(5x_1 + x_2 + 1) \mid x_1 + 2x_2 \le 8;\ 5x_1 + 3x_2 \le 15;\ x_1, x_2 \ge 0\}$$

Add the slack variables in the constraints and get:

$$x_1 + 2x_2 + x_3 = 8,$$
$$5x_1 + 3x_2 + x_4 = 15.$$

Iteration 1. We have $x^1 = (0, 0, 8, 15)'$, and its table T^1

	x^1	0	0	8	15
T^1		1	2	1	0
		5	3	0	1

Hence

$$x_B^1 = (x_{B1}^1, x_{B2}^1)' = (x_3^1, x_4^1) = (8, 15)'.$$

Step 1. A simple calculation yields

$$\nabla f(x^1) = (2, 1, 0, 0);\quad \nabla f(x^1)_B = (0, 0)'$$

Thus,

$$c^1 = \begin{bmatrix} 2 \\ 1 \\ 0 \\ 0 \end{bmatrix},\quad c_{s1}^1 = \max\{2, 1, 0, 0\} = 2,\quad c_{s2}^1 x_{s2}^1 = \min\{0, 0, 0, 0\} = 0$$

Step 2. We have

$$c_{s1}^1 > |c_{s2}^1 x_{s2}^1|,\quad x_s^1 = x_1^1$$

Thus, the value of x_1^1 will rise.

Step 3. $\Delta^1 = \min\{8/1, 15/5\} = 3$, which corresponds to $r = 2$. Consequently,

$$y^1 = [3, 0, 5, 0]' \text{ and } x^2 = y^1$$

<u>Iteration 2</u>.

	x^2	3	0	5	0
T^2		0	**7/5**	1	-1/5
		1	3/5	0	1/5

$$x_B^2 = (x_3^2,\ x_1^2)' = (5,\ 3)'$$

Step 1.

$$\nabla f(x^2) = (1/128,\ 5/128,\ 0,\ 0)',\quad \nabla f(x^2)_B = (0,\ 1/128).$$

Thus

$$c^2 = \begin{bmatrix} 1/128 \\ 5/128 \\ 0 \\ 0 \end{bmatrix} - \begin{bmatrix} 0 & 7/5 & 1 & -1/2 \\ 1 & 3/5 & 0 & 1/5 \end{bmatrix}' \begin{bmatrix} 0 \\ 1/128 \end{bmatrix} = \begin{bmatrix} 0 \\ 22/(5\times128) \\ 0 \\ -1/(5\times128) \end{bmatrix}$$

$$c_{s1}^2 = 22/(5\times12),\quad c_{s2}^2 x_{s2}^2 = 0.$$

Step 2.

$$c_{s1}^2 > |c_{s2}^2 x_{s2}^2|,\quad x_s^2 = x_2^2$$

Hence, x_2^2 will rise
$\Delta^2 = \min\{5\times5/7,\ 3\times5/3\} = 25/7$,
$y^2 = [6/7,\ 25/7,\ 0,\ 0]'$ and $x^3 = y^2$.

<u>Iteration 3</u>.

	x^3	6/7	25/7	0	0
T^3		0	1	5/7	-1/7
		1	0	-3/7	**2/7**

and

$$x_B^3 = (x_2^3,\ x_1^3)' = (25/7,\ 6/7)'$$

Step 1.

$$\nabla f(x^3) = \left[\frac{-61/7}{(62/7)^2},\ \frac{25/7}{(62/7)^2},\ 0,\ 0,\right]'$$

$$\nabla f(x^3)_B = \left[\frac{25/7}{(62/7)^2}, \frac{-61/7}{(62/7)^2}\right]'$$

Thus

$$c^3 = \begin{bmatrix} -\frac{61}{7} \times \frac{7^2}{62^2} \\ \frac{25}{7} \times \frac{7^2}{62^2} \\ 0 \\ 0 \end{bmatrix} - \begin{bmatrix} 0 & 1 & 5/7 & -1/7 \\ 1 & 0 & -3/7 & 2/7 \end{bmatrix}' \begin{bmatrix} \frac{25}{7} \times \frac{7^2}{62^2} \\ -\frac{61}{7} \times \frac{7^2}{62^2} \end{bmatrix} = \begin{bmatrix} 0 \\ 0 \\ -\frac{308}{62^2} \\ \frac{147}{62^2} \end{bmatrix}$$

$$c^3_{s1} = \frac{147}{62^2}, \quad c^3_{s2}x^3_{s2} = 0.$$

Step 2.

$$c^3_{s1} > |c^3_{s2}x^3_{s2}|, \quad x^3_s = x^3_4.$$

Thus the component x^3_4 will rise.

Step 3. $\Delta^3 = 3$ corresponding to $r = 2$, $y^3 = [0, 4, 0, 3]'$ and $x^4 = x^3$

<u>Iteration 4</u>.

		x^4	0	4	0	3
T^4	$x_2 = 4$		1/2	1	1/2	0
	$x_4 = 3$		7/2	0	-3/2	1

and

$$x^4_B = (x^4_2, x^4_4)' = (4, 3)'$$

Step 1.

$$\nabla f(x^4) = (-2/5, 1/25, 0, 0)', \quad \nabla f(x^4)_B = (1/25, 0)'.$$

Thus

$$c^4 = \begin{bmatrix} -2/5 \\ 1/25 \\ 0 \\ 0 \end{bmatrix} - \begin{bmatrix} 1/2 & 1 & 1/2 & 0 \\ 7/2 & 0 & -3/2 & 1 \end{bmatrix}' \begin{bmatrix} 1/25 \\ 0 \end{bmatrix} =$$

$$= \begin{bmatrix} -2/5 \\ 1/25 \\ 0 \\ 0 \end{bmatrix} - \begin{bmatrix} 1/(2\times 25) \\ 1/25 \\ 1/(2\times 25) \\ 0 \end{bmatrix} = \begin{bmatrix} -21/50 \\ 0 \\ -1/50 \\ 0 \end{bmatrix}$$

and

$$c^4_{s1} = 0 \quad c^4_{s2}x^4_{s2} = 0.$$

We note that the optimality condition is satisfied and the optimal solution is $x = (0, 4, 0, 3)'$ for which the optimal value of the objective function is equal to 0.8.

3.8. MULTIPLEX METHOD (PROJECTED GRADIENT METHOD)

The multiplex method was formulated by Frisch[2] to solve linear and quadratic programming problems. Here, we present the adaptation of this method to the solution of the linear fractional programming problems following the treatment of Gogia [412]. The multiplex method is an iterative method. Different from the usual simplex method, where we start from a vertex of the feasible solutions domain and then we move from vertex to vertex along its boundary until we find the optimal solution, in this method we start from a general feasible point and then we move towards the optimal solution (which is on the boundary, regularly in a vertex) and maintains always in the feasible solutions domain. Thus, in this method, we move towards or on the boundary of the feasible domain until we reach the optimal solution.

Consider the following linear fractional programming problem:

$$\max F(x_1, ..., x_n) = \frac{\sum_{j=1}^{n} c_j x_j + c_0}{\sum_{j=1}^{n} d_j x_j + d_0} \tag{3.8.1}$$

subject to

$$\sum_{j=1}^{n} a_{ij}x_j = a_{i0}, \quad i = 1, ..., m \tag{3.8.2}$$

$$\underline{x}_j \le x_j \le \bar{x}_j, \quad j = 1, ..., n, \tag{3.8.3}$$

---

[2]R. Frisch, *Multiplex method of linear programming*. Sankhya. Indian Statistical Institute, Calcutta, India, 1955.

R. Frisch, *Multiplex method of linear programming*. Memorandum dated 21st January, 1957, from University of Oslo, Norway.

where $\underline{x}_j$ and $\overline{x}_j$ are limits for the variable x_j. Obviously, $\underline{x}_j \leq \overline{x}_j$. If we take $\underline{x}_j = 0$ and $\overline{x}_j = \infty$, the inequalities (3.8.3) reduce to the usual conditions of variables non-negativity. As usually, assume that the denominator of the objective function is strictly positive on the domain of feasible solutions.

Accept that the matrix of the constraints system has rank m, and let x_B be a basic solution of the system (3.8.2) - (3.8.3). Then, we can express the variables from the base as functions of those outside the base. Denote by I, the set of indices of the vectors from the base and with J the set of those outside the base. Then

$$x_j = b_{j0} + \sum_{k \in J} b_{jk} x_k, \; j \in I, \tag{3.8.4}$$

where b_{j0} and b_{jk} are expressed as functions of a_{ij} and a_{i0}. Relations (3.8.4) can be also written for the variables outside the base by taking:

$$b_{j0} = 0, \; j \in J, \; b_{jk} = \begin{cases} 1 & j = k \\ 0 & j \neq k \quad j \in J \end{cases}$$

The objective function can be written as function of the variables outside the base

$$F(x_j \mid j \in J) = \tag{3.8.5}$$

$$\left(\sum_{j \in I} c_j x_j + \sum_{j \in J} c_j x_j + c_0\right) \Big/ \left(\sum_{j \in I} d_j x_j + \sum_{j \in J} d_j x_j + d_0\right) =$$

$$\left[\sum_{j \in I} c_j \left(b_{j0} + \sum_{k \in J} b_{jk} x_j\right) + \sum_{k \in J} c_k x_k + c_0\right] \Big/ \left[\sum_{j \in I} d_j \left(b_{j0} + \sum_{k \in J} b_{jk} x_j\right) + \right.$$

$$\left. + \sum_{k \in J} d_k x_k + d_0\right)\Bigg] = \left[\left(\sum_{j \in I} c_j b_{j0} + c_0\right) + \sum_{k \in J}\left(c_k + \sum_{j \in I} b_{jk} c_j\right) x_k\right] \Big/$$

$$\left[\left(\sum_{j \in I} d_j b_{j0} + d_0\right) + \sum_{k \in J}\left(d_k + \sum_{j \in J} b_{jk} d_j\right) x_k\right] =$$

$$\left(\sum_{k \in j} C_k x_k + C_0\right) \Big/ \left(\sum_{k \in J} D_k x_k + D_0\right)$$

with the notations:

$$C_0 = \sum_{j \in I} c_j b_{j0} + c_0, \; D_0 = \sum_{j \in I} d_j b_{j0} + d_0, \; C_k = c_k + \sum_{j \in I} b_{jk} c_j, \; D_k = d_k + \sum_{j \in I} b_{jk} d_j.$$

Let

$$f_k = \frac{\partial F}{\partial x_k} = \frac{\sum_{l \in J} (C_k D_l - C_l D_k) x_l + C_k D_0 - C_0 D_k}{\left(\sum_{l \in J} D_l x_l + D_0\right)^2}, \quad k \in J \tag{3.8.6}$$

and let x_k^0 be the starting point. From this point we move to point x_k such that the value of the objective function increases. We have

$$x_k = x_k^0 + \lambda d_k^0, \quad k \in I \tag{3.8.7}$$

For the other variables, we have

$$x_j = x_j^0 + \lambda \sum_{k \in J} b_{jk} d_k^0, \quad j \in J$$

Denoting by

$$d_j^0 = d_k^0 \ (k \in J), \quad d_j^0 = \sum_{k \in J} b_{jk} d_k^0 \quad (j \in I) \tag{3.8.8}$$

then we obtain the coordinates of the new point:

$$x_j = x_j^0 + \lambda d_j^0, \quad j = 1,2, \ldots, n \tag{3.8.9}$$

The vector (d_j^0) represents the exploration direction, whereas λ, assumed to be always positive, represents the step length. Initially, the exploration direction is taken equal to f_k^0, the value of f_k in the point x_k^0, i.e., $d_k^0 = f_k^0$. The parameter λ is determined such that the new point is also in the domain of feasible solutions, i.e.,

$$\lambda_j^1 = (\bar{x}_j - x_j^0)/d_j \ \text{ if } \ d_j > 0;$$

$$\lambda_j^2 = (\underline{x}_j - x_j^0)/d_j \text{ if } d_j < 0 \quad (j = 1,2, \ldots, n) \tag{3.8.10}$$

Let

$$\lambda_{\min} = \lambda_s = \min \{\lambda_j^1; \lambda_j^2\}, \quad j = 1, \ldots, n.$$

In these conditions, the new point will be

$$x_j = x_j^0 + \lambda_{\min} d_j^0.$$

The index s corresponding to $\lambda_{\min}$ will correspond to variable x_s which is at maximum limit. Introduce a set M which will contain all the variables which are at the upper bound. The variable x_s being at the upper bound, choose $d_s = 0$.

Let's see what is the criterion for stopping or for continuation. If F^0 is the value of the objective function at x_k^0 and F at x_k, then the process will continue as long as $F - F^0 \geq 0$, or, considering (3.8.5):

$$\frac{\sum_{k \in J} C_k x_k + C_0}{\sum_{k \in J} D_k x_k + D_0} - \frac{\sum_{k \in J} C_k x_k^0 + C_0}{\sum_{k \in J} D_k x_k^0 + D_0} \geq 0$$

and even more,

$$\frac{\left(\sum_{k \in J} C_k x_k + C_0\right)\left(\sum_{k \in J} D_k x_k^0 + D_0\right) - \left(\sum_{k \in J} C_k x_k^0 + C_0\right)\left(\sum_{k \in J} D_k x_k + D_0\right)}{\left(\sum_{k \in J} D_k x_k + D_0\right)\left(\sum_{k \in J} D_k x_k^0 + D_0\right)} \geq 0$$

Considering only the numerator, in which we have substituted x_k from relation (3.8.7), we have :

$$\Delta_j = \sum_{k \in J} \sum_{l \in J} (C_k D_l - C_l D_k) d_k^0 x_l^0 + \sum_{k \in J} (C_k Q_0 - C_0 Q_k) d_k \geq 0 \tag{3.8.11}$$

Thus, the process of moving to another point will continue if condition (3.8.11) is satisfied.

If the set M contains a number of variables whose indices constitutes the set T, then from a point on the boundary we move to another point in the so called *regressive direction* by taking

$$d_k = f_k^0 + \sum_{i \in T} B_i^0 b_{ik}, \quad k \in J \tag{3.8.12}$$

where B_i^0, called *regression coefficients*, are solutions of the system of equations:

$$M_{i0}^0 + \sum_{h \in T} B_h^0 M_{hi} = 0, \quad i \in T \tag{3.8.13}$$

with

$$M_{i0}^0 = \sum_{k \in J} b_{ik} f_k^0 \text{ and } M_{ih} = \sum_{k \in J} b_{ik} b_{hk}$$

Using the following formulae, we also determine d_k for $k \in I$.

$$\begin{aligned} d_j &= \sum_{k \in J} b_{jk} d_k^0 = \sum_{k \in J} b_{jk} \left[f_k^0 + \sum_{i \in T} B_i^0 b_{ik} \right] = \\ &= \sum_{k \in J} b_{jk} f_k^0 + \sum_{i \in T} \sum_{k \in J} b_{jk} B_i^0 b_{ik} = M_{j0}^0 + \sum_{i \in T} B_i^0 M_{ij}, \; j = 1, \ldots, n \end{aligned} \tag{3.8.14}$$

The direction d_j is *feasible* if it is non-negative when x_j is at the lower bound, and non-positive when x_j is at the upper bound. Otherwise d_j is *infeasible*. Bringing a variable to the upper bound, hence it inclusion in the set M, means touching a constraints hyperplane. Using the direction d_j, given by the formulae (3.8.14) we move along the boundary until we meet another hyperplane. In this way, other variables will be at the upper bound, hence will enter in M. The variables from M must be linearly independent. To verify this thing, calculate B_i^0 with (3.8.12) and put the condition that the matrix M_{ih}^0 is non-singular. The matrix M_{ij} is called a moment matrix. Note that this matrix is symmetric. The regression coefficient B_i^0 has the *correct sign* if is non-negative when x_i is at lower bound, or non-positive when x_i is at the upper bound. Otherwise, B_i^0 has the *incorrect sign*. In the set M, keep the variables with the correct sign and exclude those with incorrect sign. In conclusion, the process of annexing or deleting variables from the set M is continued until the optimal solution is obtained.

Numerical example. For understanding the gradient projection method we will solve in detail the following example:

$$\max \{F(x_1, x_2) = (5x_1 + 2x_2)/(5x_1 + x_2 + 1) \mid 2x_1 + 3x_2 \leq 12,$$
$$5x_1 + 3x_2 \leq 15, \ x_1, x_2 \geq 0\}. \tag{3.8.15}$$

Introduce the slack variables x_3 and x_4 and write the constraints as

$$x_3 = 12 - 2x_1 - 3x_2,$$
$$x_4 = 15 - 5x_1 - 3x_2, \tag{3.8.16}$$
$$x_i \geq 0, \ i = 1, ..., 4.$$

A starting base is $\{x_3, x_4\}$, hence the initial solution is $X^0(x_1 = x_2 = 0, x_3 = 12, x_4 = 15)$, whereas the value of F is $F(X^0) = 0$. Note that $\underline{x}_i = 0$ and $\overline{x}_i = \infty$ $i = 1, ..., 4$. Also, (3.8.16) yields $b_{11} = b_{22} = 1$, $b_{12} = b_{21} = 0$, $b_{31} = -2$, $b_{32} = -3$, $b_{41} = -5$; $b_{42} = -3$, $I = \{3,4\}$, $J = \{1,2\}$. From point X^0 we move towards another point X^1 in the direction (d_1, d_2, d_3, d_4), where $d_i (i = 1, ..., 4)$ are obtained using the formulae (3.8.8) and (3.8.9):

$$d_1 = \frac{\partial F}{\partial x_1}\Big|_{X = X^0} = \frac{5 - 5x_2}{(5x_1 + x_2 + 1)^2}\Big|_{X = X^0} = 5,$$

$$d_2 = \frac{\partial F}{\partial x_2}\Big|_{X = X^0} = \frac{2 + 5x_1}{(5x_1 + x_2 + 1)}\Big|_{X = X^0} = 2,$$

$$d_3 = b_{31}d_1 + b_{32}d_2 = (-2)5 + (-3)2 = -16,$$
$$d_4 = b_{41}d_1 + b_{42}d_2 = (-5)5 + (-3)2 = -31.$$

According to relations (3.8.10), we have $\lambda_1^1 = \infty$, $\lambda_2^1 = \infty$, $\lambda_3^2 = (0-12)/(-16) = 0.7500$, $\lambda_4^2 = (0-15)/(-31) = 0.4838$ hence $\lambda_{\min} = \lambda_4^2 = 0.4838$ and we get $M = \{x_4\}$. The new point, X^1, is obtained with formulae (3.8.9).

$$x_1 = 0 + 0.4838 \times 5 = 2.419;$$
$$x_2 = 0 + 0.4838 \times 2 = 0.9676;$$
$$x_3 = 12 + 0.4838 \times (-16) = 4.2592, \ x_4 = 0.$$

Hence $X^1(x_1 = 2.419; x_2 = 0.9676; x_3 = 4.2592; x_4 = 0)$ and consequently $F(X^1) = 0.997696$. For X^1, we determine

$$f_1 = \frac{\partial F}{\partial x_1}\Big|_{X = X^1} = 0.000895; \ f_2 = \frac{\partial F}{\partial x_2}\Big|_{X = X^1} = 1.0023.$$

We have :

$$M_{44} = b_{41}^2 + b_{42}^2 = 34; \ M_{40}^0 = b_{41}f_1^0 + b_{42}f_1^0 =$$
$$= (-5) \times 0.000895 + (-3) \times 1.0023 = -0.004475 - 3.006900 = -3.011375$$

Equations (3.8.13) yield

$$B_4^0 = \frac{3.011375}{34} = 0.0885698$$

The regressive moving direction from point X^1 is determined with formulae (3.8.12) and (3.8.14):

$$d_1 = f_1 + B_4^0 b_{41} = 0.000895 + (-5)\times 0.0885698 =$$
$$= 0.000895 - 0.442849 = -0.441954;$$
$$d_2 = f_2 + B_4^0 b_{42} = 1.0023 + (-3) \times 0.0885698 = 1.0023 - 0.2657 = 0.7366;$$
$$d_3 = b_{31}d_1 + b_{32}d_2 = (-2)(-0.441954) + (-3)0.7366 =$$
$$= 0.883908 - 2.2098 = -1.325892;$$
$$d_4 = 0 \text{ (since } x_4 \in M)$$

According to relations (3.8.10), we have

$$\lambda_1^2 = (0 - 2.419)/(-0.441954) = 5.4734,$$
$$\lambda_2^2 = \infty,$$
$$\lambda_3^2 = (0 - 4.2592)/(-1.325892) = 3.212328$$

and hence

$$\lambda_{min} = \lambda_3^2 = 3.212328$$

whereas $M = \{x_3, x_4\}$. The new point X^2, obtained with formulae (3.8.9), is

$$x_1 = 2.419 + 3.212328 \times (-0.441954) = 2.4197 - 1.41970 = 1;$$
$$x_2 = 0.9676 + 3.212328 \times 0.7366 = 0.9676 + 2.3662 = 3.333;$$
$$x_3 = 4.2592 + 3.212328 \times (-1.325892) = 4.2592 - 4.2592 = 0;$$
$$x_4 = 0.$$

Hence $X^2(x_1 = 1;\ x_2 = 3.333;\ x_3 = x_4 = 0)$ and $F(X^2) = 1.346$. For X^2, we determine again $f_1 = 0$, $f_2 = 0.2485207$. Also,

$$M_{34} = b_{31}b_{41} + b_{32}b_{42} = 19,\ M_{33} = b_{31}^2 + b_{32}^2 = 13,$$
$$M_{43} = b_{41}b_{31} + b_{42}b_{32} = 10 + 9 = 19,\ M_{44} = b_{41}^2 + b_{42}^2 = 34,$$
$$M_{30}^0 = b_{31}f_1 + b_{32}f_2 = (-2) \times 0 + (-3) \times 0.2485207 = -0.7455621$$
$$M_{40}^0 = b_{41}f_1 + b_{42}f_2 = (-5) \times 0 + (-3) \times 0.2485207 = -0.7455621$$

Equations (3.8.13) gives

$$M_{40}^0 + B_4^0 M_{44} + B_3^0 M_{34} = 0,\ M_{30}^0 + B_4^0 M_{43} + B_3^0 M_{33} = 0$$

and hence

$$\begin{bmatrix} B_4^0 \\ B_3^0 \end{bmatrix} = \begin{bmatrix} M_{44} & M_{34} \\ M_{43} & M_{33} \end{bmatrix}^{-1} \begin{bmatrix} -M_{40}^0 \\ -M_{30}^0 \end{bmatrix} = \begin{bmatrix} 34 & 19 \\ 19 & 13 \end{bmatrix}^{-1} \begin{bmatrix} 0.7455621 \\ 0.7455621 \end{bmatrix} =$$

$$= \begin{bmatrix} 0.1604938 & -0.2345679 \\ -0.2345679 & 0.419753 \end{bmatrix} \begin{bmatrix} 0.7455621 \\ 0.7455621 \end{bmatrix} = \begin{bmatrix} 0.5591715 \\ 0.1371834 \end{bmatrix}$$

Since B_4^0 is negative, and since the variable x_4 is at the lower bound, B_4^0 is of incorrect sign and x_4 will exit M. Hence $M = \{x_3\}$. Rewrite the equations (3.8.13) for the new set M, and get

$$M_{30}^0 + B_3^0 M_{33} = 0 \text{ and hence } B_3^0 = -\frac{M_{30}^0}{M_{33}} = \frac{0.7455621}{13} = 0.0573508.$$

Using formulae (3.8.12) and (3.8.14) determine the regressive directions

$$d_1 = f_1 + B_3^0 b_{31} = 0 + 0.0573508(-2) = -0.1147016$$

$$d_2 = f_1 + B_3^0 b_{32} = 0.2485207 + (-3)0.0573508$$

$$= 0.2485207 - 0.1720524 = 0.0764683, \; d_3 = 0,$$

$$d_4 = b_{41}d_1 + b_{42}d_2 = (-5)(-0.1147016) + (-3)(0.0764683) = 0.3441$$

According to relations (3.8.10) we have $\lambda_1^2 = (0-1)/(-0.1147016) = 8.7184$; $\lambda_2^1 = \infty$; $\lambda_4^1 = \infty$, hence $\lambda_{min} = \lambda_1^2 = 8.7184$ and the new point X^3 is $x_1 = 1 + (-0.1147)(8.7184) = 1 - 1 = 0$, $x_2 = 3.33 + 0.0764683 \times 8.7184 = 4$, $x_3 = 0$, $x_4 = 3.001$. Hence

$$X^3(x_1 = 0; x_2 = 4; x_3 = 0; x_4 = 3) \text{ and } F(X^3) = 1.6.$$

Again, calculate

$$f_1 = -3, \; f_2 = 0.4$$

The variable x_1 will be included in M, hence $M = \{x_1, x_3\}$. Calculate the moments

$$M_{10}^0 = -3; \; M_{30}^0 = b_{31}f_1 + b_{32}f_2 = (-2)(-3) + (-3)(0.4) = 4.8$$

$$M_{11} = 1; \; M_{13} = M_{31} = -2; \; M_{33} = 13.$$

For the new set M, the system (3.8.13) is written

$$M_{30}^0 + B_3^0 M_{33} + B_1^0 M_{13} = 0$$

$$M_{10}^0 + B_3^0 M_{31} + B_1^0 M_{11} = 0$$

and hence

$$\begin{bmatrix} B_3^0 \\ B_1^0 \end{bmatrix} = \begin{bmatrix} M_{33} & M_{13} \\ M_{31} & M_{11} \end{bmatrix}^{-1} \begin{bmatrix} -M_{30}^0 \\ -M_{10}^0 \end{bmatrix} = \begin{bmatrix} 13 & -2 \\ -2 & 1 \end{bmatrix}^{-1} \begin{bmatrix} -4.8 \\ 3 \end{bmatrix} =$$

$$= \begin{bmatrix} \frac{1}{9} & \frac{2}{9} \\ \frac{2}{9} & \frac{13}{9} \end{bmatrix} \begin{bmatrix} -4.8 \\ 3 \end{bmatrix} = \begin{bmatrix} 0.13334 \\ 3.26660 \end{bmatrix}$$

Both values for B_3^0 and B_1^0 are positive, whereas the variables x_1 and x_3 are at the lower bounds. Hence, both B_1^0 and B_3^0 are of correct sign, which means that we maintain both variables, x_1 and x_3, in M.
Calculate the regressive directions:

$$d_1 = f_1 + B_3^0 b_{31} + B_1^0 b_{11} = -3 + (-2)0.13334 + 3.2666 = 0,$$

$$d_2 = f_2 + B_3^0 b_{32} + B_1^0 b_{12} = 0.4 + (-3) \times 0.13334 = 0.$$

The value of F will no longer increase, which means that we have obtained the optimal solution, which is X^3. The successive values of X are shown in Figure 3.8.1.

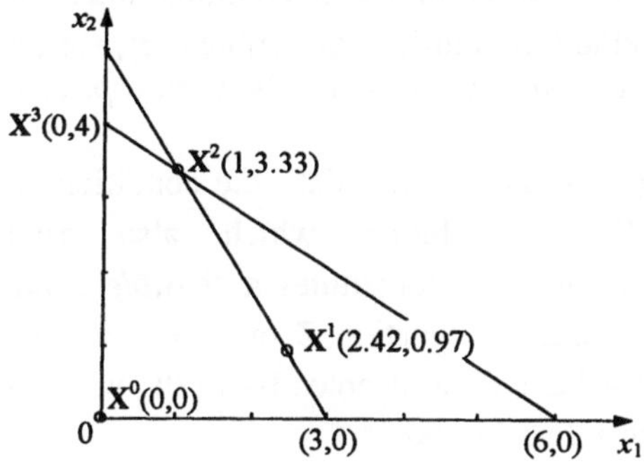

Figure 3.8.1

3.9. COMPLEMENTARY PROGRAMMING

The general problem of complementary programming with linear fractional objective function is as follows:

$$\max\{Z = \frac{dx + eu + fv + g}{\alpha x + \beta u + \gamma v + \delta} \mid Ax + Bu + Cv \leq h,\ u \cdot v = 0,\ x, u, v \geq 0\} \quad (3.9.1)$$

where $\boldsymbol{x}$, $\boldsymbol{u}$ and $\boldsymbol{v}$ are unknown vectors of size n, m, and m whereas $\boldsymbol{d}$, $\boldsymbol{e}$ and $\boldsymbol{f}$ from the numerator of the objective function are constant vectors of size n, m, and m, respectively, $\boldsymbol{\alpha}$, $\boldsymbol{\beta}$ and γ from the denominator of the objective function are constant vectors of size n, m, and m, respectively, g and δ are constants, $\boldsymbol{A}$,$\boldsymbol{B}$ and $\boldsymbol{C}$ are constant matrices of size $p\times n$, $p\times m$ and $p\times m$, respectively, whereas $\boldsymbol{h}$ is a constant vector of size p. We call problem (3.9.1) Problem $\boldsymbol{P}$. Without condition $\boldsymbol{u} \cdot \boldsymbol{v} = 0$, Problem $\boldsymbol{P}$ constitutes a common linear fractional programming problem and, for this, the previous sections have given various solution methods. Due to condition $\boldsymbol{u} \cdot \boldsymbol{v} = 0$, (called complementary condition) Problem $\boldsymbol{P}$ accepts only solutions for which we have $u_j = 0$ or $v_j = 0$ or both zero for any $j = 1, ..., m$. Its solution requires a special method. Define C^* as the set of the variable pairs (u_j,v_j), $j = 1, ..., m$. If the objective function Z of (3.9.1) is linear, one gets the complementary problem studied by Ibaraki[3]. For its solution, Ibaraki gave a branch-and-bound method. In the following, we present the branch-and-bound method for the solution of Problem $\boldsymbol{P}$ [403], the method which is based on the technique of Ibaraki and on the method of solution for a common linear fractional programming problem. To present the method of solution for Problem $\boldsymbol{P}$, we make a few clarifications in the form of some definitions following Ibaraki's work (*op. cit*).

Any set de variables $\boldsymbol{t} = (\boldsymbol{x}, \boldsymbol{u}, \boldsymbol{v})$ which satisfy the condition $\boldsymbol{u} \cdot \boldsymbol{v} = 0$ constitutes a *solution* of Problem $\boldsymbol{P}$. A solution which also satisfies the constraints $\boldsymbol{Ax} + \boldsymbol{Bu} + \boldsymbol{Cv} \leq \boldsymbol{h}$, $\boldsymbol{x}, \boldsymbol{u}, \boldsymbol{v} \geq \boldsymbol{0}$ constitutes a *feasible solution* of Problem $\boldsymbol{P}$. A feasible solution which maximizes the function Z of (3.9.1) represents an *optimal solution* of Problem $\boldsymbol{P}$. The optimal value will be denoted by $Z(P)$, whereas $z(\boldsymbol{t})$ denotes the value of the efficiency function in a feasible point $\boldsymbol{t}$.

By a *partial solution* $\boldsymbol{S}$, we understand one which is affecting zero to certain variables u_j and v_j. We adopt the convention that j represents $u_j = 0$ whereas $-j$ represents $v_j = 0$. For example, the partial solution $\boldsymbol{S} = \{-3, 2, -1\}$ means $v_3 = 0$, $u_2 = 0$, $v_1 = 0$. The variables u_j and v_j are called *free variables* if $j \in \boldsymbol{S}$ or $-j \in \boldsymbol{S}$. A partial solution $\boldsymbol{T}$ constitutes a continuation of the partial solution $\boldsymbol{S}$ if $\boldsymbol{S} \subset \boldsymbol{T}$, where $\boldsymbol{S}$ and $\boldsymbol{T}$ are seen as index sets with signs. $\boldsymbol{T}$ constitutes a completion of $\boldsymbol{S}$ if for any j we have j or $-j$ in $\boldsymbol{T}$. To problem $\boldsymbol{P}$ and to the partial solution $\boldsymbol{S}$ we associate the *partial problem* $\boldsymbol{P(S)}$.

Problem $\boldsymbol{P(S)}$.

$$\max Z = (\boldsymbol{dx} + \boldsymbol{eu} + \boldsymbol{fv} + g)/(\boldsymbol{\alpha x} + \boldsymbol{\beta u} + \boldsymbol{\gamma u} + \delta) \tag{3.9.2}$$

subject to

$$\boldsymbol{Ax} + \boldsymbol{Bu} + \boldsymbol{Cv} \leq \boldsymbol{h}, \quad \boldsymbol{u} \cdot \boldsymbol{v} = 0, \quad \boldsymbol{x}, \boldsymbol{u}, \boldsymbol{v} \geq \boldsymbol{0} \tag{3.9.3}$$
$$u_j = 0 \text{ if } j \in \boldsymbol{S}, \quad v_j = 0 \text{ if } -j \in \boldsymbol{S}$$

[3]T. Ibaraki, *Complementary programming*. Operations Res. **19**(6), 1523-1529, 1971.

Hence, Problem ***P(S)*** is obtained from problem ***P*** by giving zero values to all the variables contained in ***S***. Since Problem ***P(S)*** has a smaller dimension than Problem ***P***, it is expected that it will be easier to solve. The method of solution for Problem ***P*** consists in generating, following a certain rule, the partial solutions ***S*** and the solution to problems ***P(S)***. The obtained results can be represented as a tree (Figure 3.9.1). Initially, the tree consists of node ***P*** which corresponds to problem ***P***. If this can be solved, the algorithm is finished and the tree of Figure 3.9.1 is no longer necessary. If ***P*** cannot be solved, then choose a pair de free variables (u_j, v_j) (following a rule to be explained later) and consider two partial problems, ***P(j)*** and ***P(-j)***, of which only one has to be solved, etc.

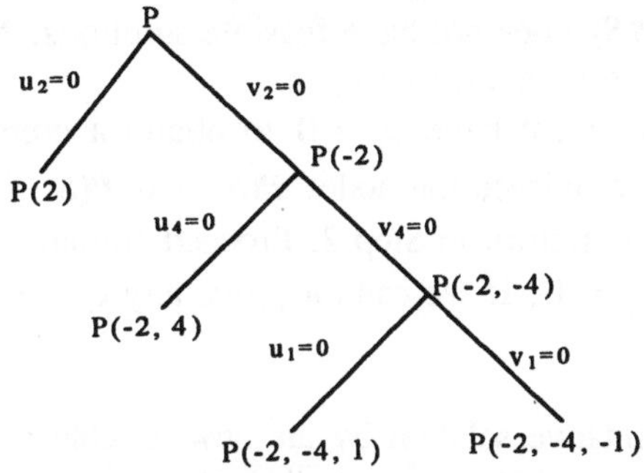

Figure 3.9.1

Although optimal solutions can be obtained for the partial problems ***P(S)***, these solutions are not necessarily optimal solutions for problem ***P***, but only feasible solutions. From among the feasible solutions resulting from various nodes of the tree, choose the best and call it *representative solution*. Among the tree nodes, we distinguish active nodes and finished nodes. A node is called *active* if, starting from it, one can get further branches of the tree, i.e., better and better solutions. The other nodes are called *finished*. If a node is declared finished, it will be neglected for subsequent branching. A node is considered finished if:

a) The corresponding problem ***P(S)*** is solved.

b) It is proved that further branching will generate worse solution than the representative solution.

c) ***P(S)*** is already partitioned into two smaller partial problems, i.e., from the respective node already stem two branches.

The calculation algorithm stops when all nodes are finished. The optimal solution is given by the representative solution. If denote by Z^* the value of the representative solution of which value will obviously modify in the calculation process, then we will exclude those feasible solutions for which we have

$$\frac{dx + eu + fv + g}{\alpha x + \beta u + \gamma v + \delta} \leq Z^* \tag{3.9.4}$$

The branch-and-bound algorithm for the solution of problem $\boldsymbol{P}$ consists of the following steps:

Step 1. Consider the tree which consists of only one active node $\boldsymbol{P}$.

Step 2. Select, by a certain rule, an active node of the tree. Let this be $P(S)$. If there is no such a node, the algorithm stops. For node $P(S)$ verify which of the following situations apply:

i) An optimal solution of Problem $P(S)$ can be obtained. Node $P(S)$ is considered finished and go to la Step 3.

ii) Note that $P(S)$ does not have feasible solutions. Node $P(S)$ is considered finished and go, again, to Step 2.

iii) Note that we must have $u_j = 0$ to obtain a completion of $P(S)$. In this case, consider finished the nodes $P(S)$ and $P(S,-j)$, and generate an active node $P(S, j)$. Return to Step 2. Proceed similarly if we can prove that we must have $v_j = 0$. If we cannot prove any of the situations i) - iii), go to Step 4.

Step 3. Replace the representative solution by the new feasible solution. Consider finished all the active nodes in the tree that are worse than the representative solution. Return to Step 2.

Step 4. Select a pair de free variables (u_j, v_j) of S called *branching variables*, and generate two active nodes $P(S, j)$ and $P(S,-j)$ from $P(S)$. Consider finished node$P(S)$ and return to Step 2.

In the application of the branch-and-bound algorithm, in each node, it is necessary to either solve a problem $P(S)$, or to prove that this does not have solution. We shall show how to solve problem $P(S)$. Start from it, we will construct Problem $P'(S)$ which is obtained from Problem $P(S)$ neglecting condition $\boldsymbol{u} \cdot \boldsymbol{v} = 0$.

Problem $P'(S)$

$$\max Z = \frac{dx + eu + fv + g}{\alpha x + \beta u + \gamma v + \delta} \tag{3.9.5}$$

subject to

$$\begin{aligned} &Ax + Bu + Cv \leq h,\ x,\ u,\ v \geq 0 \\ &u_j = 0 \text{ if } j \in S,\ v_j = 0 \text{ if } -j \in S. \end{aligned} \tag{3.9.6}$$

Problem $P'(S)$ constitutes a common problem of linear fractional programming and for its solution we can use any of the methods presented in previous sections, for example, the simplex method of Section 3.1. Let, to this purpose, $s = (s_1, s_2, ..., s_p)' \geq \mathbf{0}$ be the vector of the slack variables added to give

$$Ax + Bu + Cv + s = h.$$

Solve problem $P'(S)$. Its optimal solution and the simplex tables constructed for obtaining the optimal solution give useful information about the optimal solution of Problem $P(S)$.

We have the following possibilities:

a) The problem $P'(S)$ does not have feasible solutions. Then, problem $P(S)$ does not have feasible solutions either.

b) The optimal solution of Problem $P'(S)$ satisfies condition $u \cdot v = 0$, in which case this is also the optimal solution for Problem $P(S)$.

c) The optimal solution of Problem $P'(S)$ does not verify the condition $u \cdot v = 0$. In this case, the following obvious relation exist among the optimal values of problems $P(S)$ and $P'(S)$:

$$Z_P(S) \leq Z_{P'}(S)$$

In this case, we proceed as follows: Consider the rth row in the simplex table of Problem $P'(S)$, which corresponds to the basic variable u_j. Define a penalty function $p(u_j)$ such that:

$$p(u_j) = \frac{x_{Hr}}{Z^{(2)}} \frac{1}{Z^{(2)}} \frac{\bar{\Delta}_s}{\bar{u}_{rs}},$$

where

$$\bar{\Delta}_s = -\Delta_s, \quad \frac{\bar{\Delta}_s}{\bar{u}_{rs}} = \max \left\{ \frac{\bar{\Delta}_j}{\bar{u}_{rj}} \mid \bar{u}_{rj} > 0 \right\},$$

$$\bar{u}_{rs} = \frac{x_{Hr}(d_s - z_s^{(2)})}{Z^{(2)}} + u_{rs},$$

$$\bar{u}_s = (u_{1s}, u_{2s}, ..., u_{ps});$$

$$\bar{u}_s = H^{-1}a_s, \ s = 1, ..., n+2m+p,$$

H is the base corresponding to the optimal solution of Problem $P'(S)$.

Similarly, define the function $p(v_j)$ for v_j. Choose $Z_{P'}(S) - p(u_j)$ as a lower bound of the objective function value for $P(S, j)$, and $Z_{P'}(S) - p(v_j)$ for $P(S, -j)$.

d) If $Z_{p'}(S) - p(u_j) < Z^*$, consider the finished node $P(S, j)$, whereas if $Z_{p'}(S) - p(v_j) < Z^*$, consider finished node $P(S,-j)$. Using the values $p(u_j)$ and $p(v_j)$, determine the branching variables (u_m, v_m), where m is that index for which we get max $|p(u_j) - p(v_j)|$. If $p(u_m) < p(v_m)$, then examine the partial problem $P(S, m)$, whereas if $p(u_m) \geq p(v_m)$, then examine the partial problem $P(S, -m)$.

e) If, in the Simplex table for problem $P'(S)$, a row exists such that $\bar{u}_{rk} \leq 0$, $(k = 1,2, ..., n+2m+p)$, and $x_{Hr} > 0$, then if the free variable u_j is associated with the rth row, the partial problem $P(S, j)$ can be considered finished. It follows that $u_j \neq 0$, and we go to the consideration of variable $v_j = 0$. We reason similarly for variable v_j,

f) If, in the Simplex table for $P'(S)$, there is a row such that

$$x_{Hr} < 0, \; \bar{u}_{rk} \geq 0 \quad (k = 1, ..., n + 2m + p), \; k \neq k_0, \; \bar{u}_{rk_0} < 0,$$

assuming that $\bar{u}_{rk_0}$ corresponds to the non-basic free variable u_j, then $P(S, j)$ is considered finished. We reason similarly for variable v_j.

Summarizing the things presented until now, it follows that for a certain partial solution S, solve the problem $P'(S)$, and the penalty functions $p(u_j)$ and $p(v_j)$ are calculated. The information that is obtained assists us in either finishing certain vertices, or to construct additional branches of the tree.

Numerical example.

$$\max Z = \frac{x_1 + x_2}{2x_1 + x_2 + 1}$$

subject to

$$\begin{aligned} -x_1 + x_2 &\leq 1, \\ x_1 + 3x_2 &\leq 11, \\ x_1 + x_2 &\leq 7, \\ x_1 - 2x_2 &\leq 4, \\ x_1 x_2 &= 0, \\ x_1 \geq 0, \; x_2 &\geq 0. \end{aligned}$$

By adding the slack variables $x_3, ..., x_6$, the constraints become

$$\begin{aligned} -x_1 + x_2 + x_3 &= 1, \\ x_1 + 3x_2 + x_4 &= 11, \\ x_1 + x_2 + x_5 &= 7, \\ x_1 - 2x_2 + x_6 &= 4. \end{aligned}$$

Neglect the condition $x_1x_2 = 0$, and, for the resulting problem, construct the initial Simplex table (Table 3.9.1). Apply the Primal Simplex algorithm of Section 3.1. Hencex_1 will enter the base instead of x_6, etc. The optimal solution of the given Problem, without condition $x_1x_2 = 0$, is obtained after four iterations, as shown in Table 3.9.2. This solution is $x_1 = 2$, $x_2 = 3$, $Z = 5/8 = 0.625$. Observe that it does not verify condition $x_1x_2 = 0$.

In order to determine the partial problem to be solve next, calculate the penalty functions$p(x_1)$ and $p(x_2)$. To the basic variable x_1, one associates the 4th row (thus $r = 4$) in Table 3.9.2. $\bar{u}_{43} = 2(1/8)(5/4)-3/4 = -7/16$; $\bar{u}_{44} = 2(1/8)(-3/4) + 1/4 = 1/16$. Since only $\bar{u}_{44} > 0$, we have $\bar{\Delta}_s/\bar{u}_{rs} = \bar{\Delta}_4/\bar{u}_{44} = 4$, hence $p(x_1) = \dfrac{2}{8 \times 8}\, 4 = 1/8$.

Table 3.9.1

		$c_j \rightarrow$		1	1	0	0	0	0
		$d_j \rightarrow$		2	1	0	0	0	0
d_B	c_B	x_B	P_0	x_1	x_2	x_3	x_4	x_5	x_6
0	0	x_3	1	-1	1	1	0	0	0
0	0	x_4	11	1	3	0	1	0	0
0	0	x_5	7	1	1	0	0	1	0
0	0	x_6	4	[1]	-2	0	0	0	1
$Z^{(2)}=1$	$Z^{(1)}=0$	$Z = 0$							
		$c_j - z_j^1$		1	1	0	0	0	0
		$d_j - z_j^2$		2	1	0	0	0	0
		Δ_j		2	1	0	0	0	0

Also, the 3rd row corresponds to the basic variable x_2, (hence, $r = 3$) and we have

$$\bar{u}_{33} = 3(5/4)(1/8) + 1/4 = 23/32$$
$$\bar{u}_{34} = 3(-3/4)(1/8) + 1/4 = (-1)/32$$

We have $\bar{\Delta}_s/\bar{u}_{rs} = \bar{\Delta}_4/\bar{u}_{34} = 72/23$, and hence $p(x_2) = (3/8)(1/8)(72/23) = 27/184$. Since $p(x_1) < p(x_2)$, the next partial problem to be solved is $P(S)$ with $S = \{1\}$.

This means that we must have $x_1 = 0$. To achieve this, we make an iteration in Table 3.9.2 (the pivot is marked) through which the variable x_4 enter the base instead of x_1. We get Table 3.9.3 in which we remove the column corresponding to x_1.

Table 3.9.2

d_B	c_B	x_B	P_0	x_1	x_2	x_3	x_4	x_5	x_6
0	0	x_5	2	0	0	1/2	-1/2	1	0
0	0	x_6	8	0	0	5/4	1/4	0	1
1	1	x_2	3	0	1	1/4	1/4	0	0
2	1	x_1	2	1	0	-3/4	[1/4]	0	0
$Z^{(2)}=8$	$Z^{(1)}=5$	$Z = 0.625$							
		$c_j - z_j^1$				1/2	-1/2		
		$d_j - z_j^2$				5/4	-3/4		
		Δ_j				-9/4	-1/4		

Table 3.9.3

d_B	c_B	x_B	P_0	x_1	x_2	x_3	x_4	x_5	x_6
0	0	x_5	6	2	0	1	0	1	0
0	0	x_6	6	-1	0	2	0	0	1
1	1	x_2	1	-1	1	1	0	0	0
0	0	x_4	8	4	0	-3	1	0	0
$Z^{(2)}=2$	$Z^{(1)}=1$	$Z = 0.5$							
		$c_j - z_j^1$		2		-1			
		$d_j - z_j^2$		3		-1			
		Δ_j		1		-1			

Note that, in the resulting table, we have $\Delta_j \leq 0$, hence the optimal solution $x_1 = 0$, $x_2 = 1$, with $Z^* = 1/2$. Since this optimal solution verifies condition $x_1x_2 = 0$, it is kept as a representative solution. From Table 3.9.2, and from the values $p(x_1)$ and $p(x_2)$, it follows that $Z_{p'}(S) - p(x_1) = 5/8 - 1/8 = 1/2$ constitutes a lower bound of the value of the objective function for $P(1)$, whereas $Z_{p'}(S) - p(x_2) = 5/8 - 27/184 = 11/23$ for $P(-1)$. But $11/23 < 1/2$, which means that $S = \{-1\}$ does not lead to a solution better than the representative solution. In conclusion, the optimal solution of problem is $x_1 = 0$, $x_2 = 1$ and $Z = 1/2$.

3.10. LINEAR FRACTIONAL PROGRAMMING WITH SUPPLEMENTARY CONSTRAINTS

In this section we study the linear fractional programming problem in which, to the usual constraints given as equalities or /and inequalities, we add supplementary constraints. We consider again the problem presented in Section 3.1.

$$\max \{F(x) = (c'x + c_0)/(d'x + d_0) \mid Ax = b; \ x \geq 0\} \tag{3.10.1}$$

where A is a $m \times n$ matrix, x is a column vector with n components, b is an m-dimensional column vector, c and d are n-dimensional vectors, whereas c_0 and d_0 are given constants.

Assume that the domain of feasible solutions is a regular set and that, on this domain, the denominator is positive. Call problem (3.10.1) the *favored problem*. Add to problem (3.10.1) the constraints of the form

$$Sx = 0 \tag{3.10.2}$$

in which S is a matrix of size $p \times n$. Call the problem (3.10.1) - (3.10.2) the *extended problem* of the favored problem. Note that in the case where the function F is linear, the extended problem was studied by Bakes[4]. When F is linear fractional, the problem was treated by Horváth [496]. The method we are going to present for the solution of Problem (3.10.1) - (3.10.2) follows the ideas exposed in reference [496] and is based on the Simplex method as presented by Charnes and Cooper[5].

---

[4]Bakes M.D., *Solution of special linear programming problems with additional constraints.* Operational Research Quarterly, 17(4), 425-445, 1966.

[5]Charnes, A., Cooper, W.W., *Management Models and Industrial Applications of Linear Programming*. John Wiley and Sons, New York and London, 1961.

Let A_F be a base of favored problem, A_E a matrix constituted from p columns of A, different from those of A_F, and chose such that, if we denote by S_F and S_E the columns of S corresponding to A_F and A_E, then the matrix $B = \begin{bmatrix} A_F & A_E \\ S_F & S_E \end{bmatrix}$ constitutes a base of extended problem. Hence, a basic solution of extended problem is obtained from the matrix relation

$$\begin{bmatrix} A_F & A_E \\ S_F & S_E \end{bmatrix} \times \begin{bmatrix} x_F \\ x_E \end{bmatrix} = \begin{bmatrix} b \\ 0 \end{bmatrix}, \quad x_F \geq 0,\ x_E \geq 0$$

Performing the calculations yields

$$A_F x_F + A_E x_E = b \tag{3.10.3}$$

$$S_F x_F + S_E x_E = 0 \tag{3.10.4}$$

Since A_F is a base of the favored problem, an unique matrix will exist T_E such that we have the relation

$$A_E = A_F T_E \tag{3.10.5}$$

Substituting this relation in (3.10.3) we obtain

$$A_F(x_F + T_E x_E) = b$$

or

$$A_F \hat{x} = b \tag{3.10.6}$$

where we denoted

$$\hat{x} = x_F + T_E x_E$$

Since A_F is a base, it inverse A_F^{-1} exists and is unique, and hence relation (3.10.6) becomes

$$\hat{x} = A_F^{-1} b \tag{3.10.7}$$

Substitute in relation (3.10.4) the value

$$x_F = \hat{x} - T_E x_E \tag{3.10.8}$$

and obtain

$$S_F \hat{x} = (S_F T_E - S_E) x_E$$

or, using relation (3.10.7),

$$S_F A_F^{-1} b = (S_F T_E - S_E) x_E. \tag{3.10.8'}$$

Relation (3.10.8') yields the value of x_E if we prove that matrix $(S_F T_E - S_E)$ is non-singular. Indeed, if we assume that matrix $(S_F T_E - S_E)$ is singular, then the matrix equation

$$(S_F T_E - S_E) x = 0$$

would have a non-zero solution. Denoting $y = T_E x$, the matrix equation becomes

$$S_F y - S_E x = 0.$$

Add to it equation

$$A_F y - S_E x = 0$$

to which we can also add equation

$$A_F y - A_F T_E x = 0.$$

These two last equations can be written in matrix form as:

$$\begin{bmatrix} A_F & A_E \\ S_F & S_E \end{bmatrix} \times \begin{bmatrix} y \\ x \end{bmatrix} = 0$$

which shows that the columns of the extended problem are not linearly independent, and this thing contradicts the fact that B is a base of the extended problem. Hence, the matrix $(S_F T_E - S_E)$ is non-singular and relation (3.10.8') yields

$$x_E = (S_F T_E - S_E)^{-1} S_F A_F^{-1} b \tag{3.10.9}$$

The value of x_F is obtained from relations (3.10.5), (3.10.7), (3.10.8) :

$$x_F = A_F^{-1}b - A_F^{-1}A_E(S_F T_E - S_E)^{-1} S_F A_F^{-1} b = A_F^{-1}b - A_F^{-1}A_E x_E \tag{3.10.10}$$

Again, denoting by I_m the unit matrix $m \times m$,

$$x_F = [I_m - A_F^{-1}A_E(S_F T_E - S_E)^{-1}]A_F^{-1}b.$$

Let's now see what are the conditions that the basic solution of the extended problem must satisfy for optimality. Let $I = \{1,2, ..., m + p\}$ be the set of indices of the basic vectors, and let $J = \{m + p + 1, ..., n\}$ be the set of indices of the non-basic vectors, and let's used the notations

$$A^i = \begin{bmatrix} a^i \\ s^i \end{bmatrix}, \quad A^0 = \begin{bmatrix} b \\ 0 \end{bmatrix}, \text{ where } a^i = \begin{bmatrix} a_{1i} \\ \vdots \\ a_{mi} \end{bmatrix} \text{ and } s^i = \begin{bmatrix} a_{m+1,\, i} \\ \vdots \\ a_{m+p,\, i} \end{bmatrix}$$

For the basic solution $\begin{bmatrix} x_F \\ x_E \end{bmatrix}$ of the extended problem, we write the optimality conditions as

$$\begin{aligned} \Delta_j = (d_F x_F + d_E x_E + d_0)(c_j - c_F a^j - c_E s^j) - \\ - (c_F x_F + c_E x_E + c_0)(d_j - d_F a^j - d_E s^j) \le 0, \quad j \in J \end{aligned} \tag{3.10.11}$$

where the vectors c_F, d_F, c_E and d_E have as components the elements of c and d corresponding to the vectors that make up the base A_F, and those corresponding to the vectors contained in A_E, respectively. The basic solution can be improved if the relation (3.10.11) is not satisfied for all $j \in J$. Let $\Delta_j > 0$ for $j \in J' \subseteq J$. Determine

$$\Delta_r = \max_{j \in J} \Delta_j$$

and the vector x_r will enter the base. The vector which exits the base is determined in the usual way. The vector x_k exists the base since

$$\frac{x_k}{w_{kr}} = \min_{\substack{i \in I \\ w_{jr} > 0}} \frac{x_j}{w_{jr}}$$

where w_{ir} are the components of vector

$$w_i = B^{-1}A^i, \quad i \in J.$$

Note that, due to relation (3.10.5), the matrix $S_F T_E - S_E$ which intervenes in relation (3.10.9), can be written as $S_F A_F^{-1} A_E - S_E$, which means that, if $k \in \{m + 1, ..., m + p\}$, i.e., a^k does not belong to base A_F, to obtain the new solution x_F, x_E the calculation is reduced substantially since the products $S_F A_F^{-1}$ and $S_F A_F^{-1} b$ are known from the previous iteration.

Numerical example. Maximize the function

$$F(x) = (3x_1 + 5x_2 - 5x_3 + x_4 + 1)/(2x_1 + 4x_2 + 7x_5 + 10)$$

on the set of solutions of the system :

$$\begin{gathered} 3x_1 + 5x_2 - 6x_3 - x_4 + 5x_5 \leq 12, \\ -x_1 + 2x_2 + 3x_3 + 3x_5 \leq 9, \\ 2x_1 + 8x_2 + x_3 + 4x_4 - x_5 \leq 5, \\ x_i \geq 0, \quad i = 1, ..., 5. \end{gathered}$$

with the supplementary constraint

$$x_1 + 2x_2 - x_3 - 4x_4 + x_5 = 0.$$

In the above constraints system, add the slack variables x_6, x_7, x_8 and get the system

$$\begin{gathered} 3x_1 + 5x_2 - 6x_3 - x_4 + 5x_5 + x_6 = 12, \\ -x_1 + 2x_2 + 3x_3 + 3x_5 + x_7 = 9, \\ 2x_1 + 8x_2 + x_3 + 4x_4 - x_5 + x_8 = 5, \\ x_i \geq 0, \quad i = 1, ..., 8. \end{gathered}$$

A base of the favored problem is

$$A_F = (a^6, a^7, a^8).$$

Consider $A_E = a^5$, and hence the base of the extended problem is

$$B_1 = \begin{bmatrix} A_F & A_E \\ S_F & S_E \end{bmatrix} = \left[\begin{array}{ccc|c} a^6 & a^7 & a^8 & a^5 \\ \hline s^6 & s^7 & s^8 & s^5 \end{array}\right] = \left[\begin{array}{ccc|c} 1 & 0 & 0 & 5 \\ 0 & 1 & 0 & 3 \\ 0 & 0 & 1 & -1 \\ \hline 0 & 0 & 0 & 1 \end{array}\right]$$

For details see Table 3.10.1. Since $S_F = (0,0,0)$, formula (3.10.9) yields $x_E = x_5 = 0$, whereas relation (3.10.10) gives:

$$x_F = (x_6, x_7, x_8)' = A_F^{-1}b = [10; 9; 5]'$$

The values of x_F and x_E are written in Table 3.10.2. Note that the values of x_F and x_E can be easily obtained be performing a Simplex iteration in Table 3.10.1 with pivot 1 (marked in the table). One gets Table 3.10.2.

Table 3.10.1

c_j				3	5	-5	1	0	0	0	0
	d_j			2	4	0	0	7	0	0	0
		Base	P_0	P_1	P_2	P_3	P_4	P_5	P_6	P_7	P_8
0	0	x_6	12	3	5	-6	-1	5	1	0	0
0	0	x_7	9	-1	2	3	0	3	0	1	0
0	0	x_8	5	2	8	1	4	-1	0	0	1
0	7	x_5	0	1	2	-1	-4	[1]	0	0	0

Table 3.10.2

c_j				3	5	-5	1	0	0	0	0
	d_j			2	4	0	0	7	0	0	0
		Base	P_0	P_1	P_2	P_3	P_4	P_5	P_6	P_7	P_8
0	0	x_6	12	-2	-5	-1	19	0	1	0	0
0	0	x_7	9	-4	-4	6	12	0	0	1	0
0	0	x_8	5	3	10	0	0	0	0	0	1
0	7	x_5	0	0	[2]	-1	-4	1	0	0	0
		$z_j^{(1)}$	1	0	0	0	0	0	0	0	0
		$z_j^{(2)}$	10	7	14	-7	-28	7	0	0	0
		$c_j - z_j^{(1)}$		3	5	-5	1	0	0	0	0
		$d_j - z_j^{(2)}$		-5	-10	7	28	0	0	0	0
$z^2(c_j-z_j^{(1)})-z^1(d_j-z_j^{(2)})$			1/10	35	0	-57	-18	0	0	0	0

Table 3.10.2 indicates that

$$x^1 = (x_6, x_7, x_8, x_5)' = (12, 9, 5, 0)' \text{ and } F(x^1) = 1/10$$

Table 3.10.2 also shows the values Δ_j calculated with formulae (3.10.11). The solution is not optimal since there are positive values Δ_j. The vector P_2 will enter the base instead of vector P_5, and we get the base

$$B_2 = \begin{bmatrix} A_F & A_E^* \\ \\ S_F & S_E^* \end{bmatrix} = \left[\begin{array}{ccc|c} a^6 & a^7 & a^8 & a^2 \\ -- & -- & -- & -- \\ s^6 & s^7 & s^8 & s^2 \end{array}\right]$$

The formulae (3.10.9) and (3.10.10) gives $x_E = x_2 = 0, x_F = (x_6, x_7, x_8)'$ $= A_F^{-1}b = [12, 9, 5]'$ (Table 3.10.3). Hence, $x^2 = (x_6, x_7, x_8, x_2)' = (12, 9, 5, 0)$ and $F(x^2) = 1/10$. Table 3.10.3 is obtained from Table 3.10.2 through a pivoting of pivot 2.

Table 3.10.3

c_j				3	5	-5	1	0	0	0	0
	d_j			2	4	0	0	7	0	0	0
		Base	P_0	P_1	P_2	P_3	P_4	P_5	P_6	P_7	P_8
0	0	x_6	12	1/2	0	-7/2	9	5/2	1	0	0
0	0	x_7	9	-2	0	4	4	2	0	1	0
0	0	x_8	5	-2	0	5	[20]	-5	0	0	1
5	4	x_2	0	1/2	1	-1/2	-2	1/2	0	0	0
		$z_j^{(1)}$	1	5/2	5	-5/2	-10	5/2	0	0	0
		$z_j^{(2)}$	10	2	4	-2	-8	2	0	0	0
		$c_j - z_j^{(1)}$		1/2	0	-5/2	11	-5/2	0	0	0
		$d_j - z_j^{(2)}$		0	0	2	8	5	0	0	0
		Δ_j	1/10	5	0	-27	102	30	0	0	0

There are still positive values for Δ_j. The vector P_4 will enter the base instead of P_8. The new base is

$$B_3 = \begin{bmatrix} A_F^* & A_E^* \\ \\ S_F^* & S_E^* \end{bmatrix} = \left[\begin{array}{ccc|c} a^6 & a^7 & a^4 & a^2 \\ -- & -- & -- & -- \\ s^6 & s^7 & s^4 & s^2 \end{array}\right]$$

whereas the solution is

$$x_E = x_2 = (S_F^* A_F^{*-1} A_E^* - S_E^*)^{-1} S_F^* A_F^{*-1} b =$$

$$= \left[(0\ \ 0\ \ -4) \begin{pmatrix} 1 & 0 & 1/4 \\ 0 & 1 & 0 \\ 0 & 0 & 1/4 \end{pmatrix} \begin{pmatrix} 5 \\ 2 \\ 8 \end{pmatrix} - 2 \right]^{-1} (0\ \ 0\ \ -4) \begin{pmatrix} 1 & 0 & 1/4 \\ 0 & 1 & 0 \\ 0 & 0 & 1/4 \end{pmatrix} \begin{pmatrix} 12 \\ 9 \\ 5 \end{pmatrix} =$$

$$= \left[(0\ \ 0\ \ -1) \begin{pmatrix} 5 \\ 2 \\ 8 \end{pmatrix} - 2 \right]^{-1} (0\ \ 0\ \ -1) \begin{pmatrix} 12 \\ 9 \\ 5 \end{pmatrix} = (-10)^{-1}(-5) = \frac{1}{2}$$

$$x_F = (x_6, x_7, x_4)' = A_F^{*-1} b - A_F^{*-1} A_E^* x_E =$$

$$= \begin{pmatrix} 1 & 0 & 1/4 \\ 0 & 1 & 0 \\ 0 & 0 & 1/4 \end{pmatrix} \left[\begin{pmatrix} 12 \\ 9 \\ 5 \end{pmatrix} - \begin{pmatrix} 5 \\ 2 \\ 8 \end{pmatrix} \left(\frac{1}{2} \right) \right] = \begin{pmatrix} 1 & 0 & 1/4 \\ 0 & 1 & 0 \\ 0 & 0 & 1/4 \end{pmatrix} \begin{pmatrix} 19/2 \\ 8 \\ 1 \end{pmatrix} = \begin{pmatrix} 39/4 \\ 8 \\ 1/4 \end{pmatrix}$$

In the above calculations we had

$$A_F = \begin{pmatrix} 1 & 0 & -1 \\ 0 & 1 & 0 \\ 0 & 0 & 4 \end{pmatrix}, \quad A_E = \begin{pmatrix} 5 \\ 2 \\ 8 \end{pmatrix}, \quad S_F = (0\ \ 0\ \ -4), \quad S_E = 2$$

Hence,

$$x^3 = (x_6, x_7, x_4, x_2)' = (39/4,\ 8,\ 1/4,\ 1/2)' \text{ and } F(x^3) = 5/16$$

The new values for the solution x^3 are written in Table 3.10.4. As we have already mentioned, the values of x^3 can be obtained through a simple pivoting of pivot 20 in Table 3.10.3.

Table 3.10.4

c_j	d_j	Base	P_0	P_1	P_2	P_3	P_4	P_5	P_6	P_7	P_8
c_j				3	5	-5	1	0	0	0	0
	d_j			2	4	0	0	7	0	0	0
		Base	P_0	P_1	P_2	P_3	P_4	P_5	P_6	P_7	P_8
0	0	x_6	39/4	7/5	0	-23/4	0	19/4	1	0	-9/20
0	0	x_7	8	-8/5	0	3	0	3	0	1	-1/5
1	0	x_4	1/4	-1/10	0	1/4	1	-1/4	0	0	1/20
5	4	x_2	1/2	$\boxed{3/10}$	1	0	0	0	0	0	1/10
		$z_j^{(1)}$	15/4	7/5	5	1/4	1	-1/4	0	0	11/20
		$z_j^{(2)}$	12	6/5	4	0	0	0	0	0	2/5
		$c_j - z_j^{(1)}$		8/5	0	-21/4	0	1/4	0	0	-11/20
		$d_j - z_j^{(2)}$		4/5	0	0	0	7	0	0	-2/5
		Δ_j	5/16	81/5	0	-63	0	-93/4	0	0	-51/10

The solution x^3 is not optimal. The vector P_1 will enter the base instead of P_2. The new base is

$$B_4 = \begin{bmatrix} A_F^* & A_E^{**} \\ S_F^* & S_E^{**} \end{bmatrix} = \begin{bmatrix} A_F^* & a^1 \\ S_F^* & s^1 \end{bmatrix}$$

whereas the new solution

$$x_E = x_1 =$$

$$= \left[(0\ \ 0\ \ -4) \begin{pmatrix} 1 & 0 & 1/4 \\ 0 & 1 & 0 \\ 0 & 0 & 1/4 \end{pmatrix} \begin{pmatrix} 3 \\ -1 \\ 2 \end{pmatrix} - 1 \right]^{-1} (0\ \ 0\ \ -4) \begin{pmatrix} 1 & 0 & 1/4 \\ 0 & 1 & 0 \\ 0 & 0 & 1/4 \end{pmatrix} \begin{pmatrix} 12 \\ 9 \\ 5 \end{pmatrix} =$$

$$= \left[(0\ \ 0\ \ -1) \begin{pmatrix} 3 \\ -1 \\ 2 \end{pmatrix} - 1 \right]^{-1} (0\ \ 0\ \ -1) \begin{pmatrix} 12 \\ 9 \\ 5 \end{pmatrix} = [(-2)-1]^{-1}(-5) = \frac{5}{3}$$

$$x_F = (x_6, x_7, x_4)' = \begin{bmatrix} 1 & 0 & 1/4 \\ 0 & 1 & 0 \\ 0 & 0 & 1/4 \end{bmatrix} \left[\begin{bmatrix} 12 \\ 9 \\ 5 \end{bmatrix} - \begin{bmatrix} 3 \\ -1 \\ 2 \end{bmatrix} \left(\frac{5}{3} \right) \right] = \begin{bmatrix} 89/12 \\ 32/3 \\ 5/12 \end{bmatrix}$$

Hence, $x_4 = (x_6, x_7, x_4, x_1)' = (89/12,\ 32/3,\ 5/12,\ 5/3)'$ and $F(x^4) = 77/120$. Make a pivoting of pivot 3/10 and get Table 3.10.5.

Table 3.10.5

c_j				3	5	-5	1	0	0	0	0
	d_j			2	4	0	0	7	0	0	0
		Base	P_0	P_1	P_2	P_3	P_4	P_5	P_6	P_7	P_8
0	0	x_6	89/12	0	-14/3	-25/4	0	19/4	1	0	-11/12
0	0	x_7	32/3	0	16/3	3	0	3	0	1	1/3
1	0	x_4	5/12	0	1/3	1/4	1	-1/4	0	0	1/12
3	2	x_1	5/3	1	10/3	0	0	0	0	0	1/3
		$z_j^{(1)}$	77/12	3	31/3	1/4	1	-1/4	0	0	13/12
		$z_j^{(2)}$	40/3	2	20/3	0	0	0	0	0	2/3
		$c_j - z_j^{(1)}$		0	-16/3	-21/4	0	1/4	0	0	-13/12
		$d_j - z_j^{(2)}$		0	-8/3	0	0	7	0	0	-2/3
		Δ_j	$\frac{77}{160}$	0	-54	-70	0	$-\frac{499}{12}$	0	0	$-\frac{61}{6}$

Note that $\Delta_j \leq 0,\ j \in \{1, ..., 8\}$ hence x^4 is the optimal solution of the problem, i.e., $x_1 = 5/3,\ x_4 = 5/12,\ x_6 = 89/12,\ x_7 = 32/3,\ x_2 = x_3 = x_5 = x_8 = 0$ whereas the optimal value of the objective function is $F = 77/160$.

3.11. HISTORICAL NOTES. COMMENTARIES

1. As in the case of linear programming, the revised simplex method can be considered for linear fractional programming. This was signaled by Gogia [413]. Starting with

problem (3.1.1) - (3.1.2), add to the initial constraints system, two more constraints, the numerator and the denominator of the objective function denoted by $z^{(1)}$ and $z^{(2)}$. The constraints system becomes :

$$Ax = b, \quad -c'x + z^{(1)} = c_0, \quad -d'x + z^{(2)} = d_0$$

or

$$A^*x^{(1)} = b^{(1)}, \text{ where } x^{(1)} = (x, z^{(1)}, z^{(2)})', \; b^{(1)} = (b, c_0, d_0)'$$

and

$$A^* = \begin{bmatrix} A & 0 & 0 \\ -c & 1 & 0 \\ -d & 0 & 1 \end{bmatrix}.$$

If a_j are the columns of matrix A, whereas $a_j^{(1)}$ are the columns of matrix A^*, we have

$$a_j^{(1)} = (a_j, -c_j, -d_j)', \quad j = 1, ..., n,$$
$$a_{n+1}^{(1)} = (0, 0, ..., 1, 0)'$$
$$a_{n+2}^{(1)} = (0, 0, ..., 0, 1)'$$

Assuming that rank $A = m$, then matrix A^* will have rank $m + 2$. Next, proceed as in linear programming. Bitran and Magnanti [191] developed a primal dual algorithm for solving fractional programs.

2. The Primal Simplex algorithm presented in Section 3.1 can be also applied to other fractional objective functions if they have the property that the optimal value is attained in a vertex of the feasible solutions domain. Several such cases are given next.

a) Aggarwal [17] considered the problem

$$\max \{ f(x) = \frac{(c'x + c_0)^2}{(d'x + d_0)^2} \mid Ax = b, \; x \geq 0 \}$$

The function $f(x)$ is explicit quasi-monotone (Chapter II, Section 2.5, case F). Let

$$\Delta_j = [z^2(c_j - z_j^{(1)}) - z^1(d_j - z_j^{(2)})][\xi_j z^2(c_j - z_j^{(1)}) + z^1(d_j - z_j^{(2)}) + 2z^1z^2]$$

A basic feasible solution $x = (x_B, 0)$ is optimal if $\Delta_j \leq 0$, for all the columns of A. Schaible [794] showed that, if $c'x + c_0$ does not change its sign on D, the above problem is equivalent with the linear fractional programming problem

$$\max \{ g(x) = \frac{c'x + c_0}{d'x + d_0} \mid Ax = b, \; x \geq 0 \}$$

b) Teterev [972], [973] considered the problem

$$\max \{ f(x) = c'x + d'x/e'x \mid Ax = b, \; x \geq 0 \}$$

In certain conditions (Chapter 2, Section 2.5., Theorem 2.2.3), the function f is explicit quasiconcave or explicit quasiconvex. A basic feasible solution $x = (x_B, 0)$ is optimal if $\Delta_j = c_j - z_j^{(0)} + (Z^2(d_j - z_j^{(1)}) - Z^1(e_j - z_j^{(2)}))/(Z^2)^2 \leq 0,\ j=1, ..., n$.

c) Bector [122] and Aggarwal [14] considered the problem

$$\min \{f(x) = (c'x + c_0)(d'x + d_0)/(e'x + e_0)^2 \mid Ax = b,\ x \geq 0\}$$

Assuming that the domain of feasible solutions D is a regular set and that $c'x + c_0$, $d'x + d_0$, $e'x + e_0$ are strict positive on D, then f is an explicit quasiconcave and pseudoconcave function, hence the optimal solution is attained in a vertex of D.

Let

$$\Delta_j = T_3[T_1T_3(d_j - z_j^{(2)}) + T_2T_3(c_j - z_j^{(1)}) - 2T_1T_2(e_j - z_j^{(3)})] + \\ + \xi_j[(c_j - z_j^{(1)})(d_j - z_j^{(2)})T_3^2 - (e_j - z_j^{(3)})^2\ T_1T_2]$$

A basic feasible solution $x = (x_B, 0)$ is optimal if $\Delta_j \geq 0$ for all the columns of A, where:

$$T_1 = c_B'x_B + c_0,\ T_2 = d_B'x_B + d_0,\ \ T_3 = e_B'x_B + e_0$$
$$y_j = B^{-1}a_j,\ \ z_j^{(1)} = c_B'y_j,\ \ z_j^{(2)} = d_B'y_j,\ \ z_j^{(3)} = e_B'y_j$$

d) Aggarwal [6] considered the problem

$$\max \{ f(x) = \frac{(c'x)^2}{d'x + d_0} \mid Ax \geq b,\ \ x \geq 0 \}.$$

Assuming $d'x + d_0 > 0$, the function f is convex (Chapter II, Section 2.5, case D). Let

$$\Delta_j = V_1[2V_2(c_j - z_j^{(1)}) - V_1(d_j - z_j^{(2)})] + V_2\xi_j(c_j - z_j^{(1)})^2.$$

A basic feasible solution $x = (x_B, 0)$ is optimal if $\Delta_j \leq 0$ for all the columns of A, where

$$V_1 = c_B'x_B \text{ and } V_2 = d'x_B + d_0$$

e) Stancu-Minasian [900] proposed a simplex-type algorithm for solving the following pseudomonotonic programming problem

$$\min \{ f(x) = \sum_{i=1}^{r} \left[\frac{c'x + c_0}{d'x + d_0} \right]^i \mid Ax = b,\ \ x \geq 0 \}$$

Hirche [486] considered a more general case i.e. the problem of minimizing a composite objective function with linear constraints.

$$\min \{ f(x) = \sum_{i=1}^{k} a_i q^i(x) \mid Ax = b,\ \ x \geq 0 \}$$

where a_i are real numbers and $q(x)$ is a linear fractional function.
He proposed a solving method based on examination of a polynomial and on a linear fractional programming.

f) Aggarwal [16] considered the problem

$$\max \left\{ f(x) = \frac{(c'x + c_0)^2}{(d'x + d_0)^2} \mid Ax = b,\ x \geq 0,\ L_j \leq x_j \leq U_j \right\}$$

and developed a computational procedure.

3. Swarup [743] showed that the Dual Simplex algorithm applied to the linear programming problem obtained from the fractional programming problem through the change of variable method, generates a Dual Simplex algorithm for the linear fractional programming problem. The basic idea is that of starting with a dual feasible base (i.e., a non-feasible base satisfying conditions $\Delta_j \leq 0$) and then obtain in the following iterations other basic solutions by always keeping satisfied conditions $\Delta_j \leq 0$ for all the columns of A. One gets the optimal solution when a basic feasible solution is obtained.

4. The method presented in Section 3.3.1 for the solution of the linear fractional programming problem with one non-linear constraint of form $p'x + (1/2)x'Hx \leq \gamma$ was applied by Aggarwal [18] for solving the problem

$$\max \left\{ f(x) = \frac{(c'x + c_0)(d'x + d_0)}{(e'x + e_0)^2} \mid p'x + (1/2)x'Hx \leq \gamma,\ \ Ax = b,\ \ x \geq 0 \right\}$$

5. Horváth [497] gave a criterion for determining the optimum of the linear fractional programming problem with supplementary constraints (3.10.1) - (3.10.2) using duality considerations.

6. Stahl [892] gave a finite method for the solution of the problem

$$\max \left\{ F(x) = \frac{c'x}{d'x} \mid Ax = b;\ \ x \geq 0 \right\}$$

The method reduces to the determination of the upper bound λ^* of the values of λ for which the problem below has the optimal value equal to zero

$$\min \{ I'y + y_{mn} \mid Ax + I'y = b,\ (c' - \lambda d')x + y_{mn} = 0,\ \ x, y \geq 0,\ y_{mn} \geq 0 \},$$

where I and I' are unit matrices of the corresponding dimensions.

7. For the solution of the complementary programming problem with linear objective function, Garg and Swarup [405] examine the introduction of a new type of constraints using the complementarity condition $\boldsymbol{u} \cdot \boldsymbol{v} = \mathbf{0}$. These constraints are called C-cuts (complementarity cuts) and can be also use in the branch-and-bound algorithm presented in Section 3.9.

8. A special case of the linear fractional programming is that of fractional interval programming of the form

$$\max \{ f(x) = \frac{c'x + c_0}{d'x + d_0} \mid b^- \leq Ax \leq b^+ \} \qquad (3.11.1)$$

It can be easily observed that the set of constraints defines a parallelipiped. If we apply the method of change of variable presented in Section 3.4, the constraints of the resulting linear programming problem will not be of interval form. The fractional interval programming was studied by numerous authors. Charnes and Cooper [285] applied the change of variable method and developed a solution algorithm for Problem (3.11.1) using the duality theory of linear programming. Charnes, Granot and Frieda Granot [295], [296] transformed problem (3.11.1) into an equivalent problem of same format using the generalized inverse of A. Next, the constraints of form $a \leq y \leq b$ of the equivalent problem were reduced to the form $0 \leq y \leq 1$, followed by explicit solutions of problem (3.11.1). Bühler [205] transformed problem (3.11.1) into an equivalent one of the same form using the generalized inverse of A, to which he associated a linear parametric program used to obtain solution for Problem (3.11.1). Agrawal and Verma [47] also transformed the problem into an equivalent one, but their solution to the equivalent problem is different from other authors. Lata and Mittal [598] gave a solution method for Problem (3.11.1) using the principle of decomposition for linear fractional programming [230]. Charnes, Granot and Frieda Granot [297] also gave an algorithm for the solution of Problem (3.11.1) using the special structure of the constraints. The algorithm starts with a feasible solution and finds the optimal solution (if such a solution exists) after a certain number of iterations. Agrawal and Verma [46] presented a method for the solution of the following interval programming problem

$$\max \{ f(x) = c'x + p'x/q'x \mid b^- \leq Ax \leq b^+ \}$$

where

$$x,\ c',\ p',\ q' \in R^n, \quad b^-,\ b^+ \in R^m \text{ and } A \in R^{m \times n}$$

9. Wagner and Yuan [1029] showed that the methods presented in Section 3.1 and Section 3.4 are identical, in as much that they generate the same sequences of bases during the iterations. More precisely, for a given non-optimal basic solution for $Ax = b,\ x \geq 0$, both methods will select the same variable to enter the base and the same variable to exit the base. Mond [672] showed that this equivalence ceases if the set of feasible solutions is unbounded.

10. Bitran [189] presented the results of some numerical experiments using six methods for the solution of linear fractional programming problems. The methods were tested on a set of 20 problems randomly generated.

11. Storøy [935] showed that most of the vertex ranking algorithm for the linear programming problem can be used with a minor modification to rank the vertices of the linear fractional programming problem.

12. Karmarkar N. [A new polynomial-time algorithm for linear programming, *Combinatorica* 4(1984)4, 373 - 395] proposed a gradient projection method for linear programming problems which is worst case polynomial with complexity $0(n^{3.5} L^2)$, where n is the number of variables and L is the number of bits in the input data.
We recall that an algorithm is classified as a polynomial time algorithm if the number of computational steps required to solve a problem using it is bounded above by a polynomial in terms of the size of the problem.
Anstreicher [62] showed that Karmarkar's projective algorithm is fundamentally an algorithm for linear fractional programming on the simplex.
Also, Tang and Li [971] proposed two polynomial-time algorithms for solving linear fractional programming problem, using Karmarkar's algorithm. The computational complexity of the two algorithms is $0(n^{3.5} L^2)$.

13. Swarup [956] considered bounded-variable linear fractional programs i.e. linear fractional programs in which some or all the variables are restricted to lie within specified finite lower and upper bounds

$$L_j \leq x_j \leq U_j$$

He developed a computational procedure in the framework of linear fractional programs without bounds of the variables.

CHAPTER 4

NONLINEAR FRACTIONAL PROGRAMMING

In this chapter we deal with the following nonlinear fractional programming problem:

$$P : \max_{x \in S} q(x) = (f(x) + \alpha)/(g(x) + \beta)$$

where f, g: $R^n \to R$, α, $\beta \in R$, $S \subseteq R^n$. To simplify things, and without restricting the generality of the problem, it is usually assumed that, $g(x) + \beta > 0$ for all $x \in S, S$ is non-empty and that the objective function has a finite optimal value. As we will see, the methods for solving the problem P are generalizations to the nonlinear case of the methods presented in Chapter 3. Such methods can be grouped in the following categories:

a) *Methods based on the change of variables*. The problem P is transformed in a problem P' which can be solved easier than P. Such methods will be described in Section 4.2 and in Section 4.3 for homogeneous functions;

b) *Direct methods*. View P as a nonlinear programming problem which, in principle, can be solved through any of the known methods. If f and g satisfy certain concavity or convexity hypotheses, then the function q has a series of properties, which facilitates the finding of the optimal solution. For example, if f and g are differentiable on S, $f(x) + \alpha \geq 0$ and concave on S, $g(x) + \beta > 0$ and convex on S, then, according to Chapter 2, Section 2.5, Case A, the function q is pseudoconcave, which means that every local maximum is global maximum. The global optimum can be obtained using, for example, any gradient method. Also, we can use the method of linearization. Such a method will be described in Section 4.4. Among direct methods we also mention the SUMT method [185].

c) *Parametric methods*. Starting with problem P, we associate the parametric problem $\max_{x \in S} \{[f(x) + \alpha] - \lambda[g(x) + \beta]\}$. Neglecting the constant term $(\alpha - \lambda\beta)$, we consider the parametric problem:

$$Q(\lambda) : \max_{x \in S} f(x) - \lambda g(x) , \quad \lambda \in R.$$

The optimal solution of problem P is determined by solving the parametric problem for various values of the parameter (Section 4.5).

4.1. NECESSARY AND SUFFICIENT OPTIMALITY CONDITIONS

Consider the nonlinear fractional programming problem with nonlinear constraints

$$\max \left\{ q(x) = \frac{f(x) + \alpha}{g(x) + \beta} \mid g_i(x) \le 0, \quad i = 1 , \dots , m \right\} , \tag{4.1.1}$$

where

(i) $g_i \in C^1 (i = 1 , \dots , m)$ are quasiconvex functions on a (feasibility) domain S which satisfies the Kuhn-Tucker constraint qualifications,
(ii) α and β are arbitrary scalar constants,
(ii) The set of feasible solutions for problem (4.1.1),

$$D = \{ x \mid g_i(x) \le 0, \quad i = 1 , \dots , m, \quad x \in S \}$$

is a convex set,
(iv) $f \in C^1$ is concave on S and $f(x) + \alpha \ge 0, \quad \forall\, x \in D$,
(v) $g \in C^1$ is convex on S and $g(x) + \beta > 0, \quad \forall\, x \in D$.

Using the results of Chapter 2, Section 2.5, Case A, the function q is explicit quasiconvex and consequently has the property that a point of local maximum is also of global maximum (for a direct proof, one can consult Bector [119]).

Theorem 4.1.1. *The necessary and sufficient condition for* x_0 *to be an optimal solution of problem* (4.1.1) *is that m scalars* u_i $(i = 1 , \dots , m)$ *exist such that:*

$$g_i(x_0) \le 0, \quad i = 1 , \dots , m , \tag{4.1.2}$$

$$\nabla q(x_0) = \sum_{i=1}^{m} u_i \nabla g_i(x_0) , \tag{4.1.3}$$

$$\sum_{i=1}^{m} u_i g_i(x_0) = 0 , \tag{4.1.4}$$

$$u_i \ge 0 , \quad i = 1 , \dots , m . \tag{4.1.5}$$

Proof. As in the linear case, we only prove the sufficiency. For $x_0 \in D$, assume that m scalars $u_1 , \dots , u_m$ exist which verify conditions (4.1.2) - (4.1.5). We have

$$\nabla q(x_0) = \frac{\nabla f(x_0)}{g(x_0) + \beta} - \frac{f(x_0) + \alpha}{[g(x_0) + \beta]^2} \nabla g(x_0)$$

or, multiplying by $(x - x_0)$,

$$(x - x_0)' \nabla q(x_0) =$$

$$= \frac{(x - x_0)' \nabla f(x_0)}{g(x_0) + \beta} - \frac{f(x_0) + \alpha}{[g(x_0) + \beta]^2} (x - x_0)' \nabla g(x_0) .$$

Using conditions (iv) and (v), we get

$$(x - x_0)' \nabla q(x_0) \geq \frac{f(x) - f(x_0)}{g(x_0) + \beta} - \frac{f(x_0) + \alpha}{[g(x_0) + \beta]^2}[g(x) - g(x_0)] =$$

$$\frac{f(x) + \alpha}{g(x_0) + \beta} - \frac{f(x_0) + \alpha}{g(x_0) + \beta} - \frac{[f(x_0) + \alpha][g(x) + \beta]}{[g(x_0) + \beta]^2} + q(x_0) = \frac{[g(x) + \beta]}{[g(x_0) + \beta]}[q(x) - q(x_0)].$$

Hence,

$$q(x) - q(x_0) \leq \frac{g(x_0) + \beta}{g(x) + \beta}[(x - x_0)' \nabla q(x_0)] \ . \tag{4.1.6}$$

Multiply relation (4.1.3) by $(x - x_0)$ and get

$$(x - x_0)' \nabla q(x_0) = \sum_{i=1}^{m} u_i (x - x_0)' \nabla g_i(x_0) \tag{4.1.7}$$

Let $K = \{ i \mid g_i(x_0) < 0 \}$. Relations (4.1.2), (4.1.4) and (4.1.5) yield

$$u_i = 0 \text{ for } i \in K. \tag{4.1.8}$$

Also, $g_i(x) \leq g_i(x_0) = 0$ for all $x \in D$ and $i \notin K$, and hence, using the quasi-convexity of g_i on D, we get

$$(x - x_0)' \nabla g_i(x_0) \leq 0 \ , \quad x \in D \ , \quad i \notin K. \tag{4.1.9}$$

Multiplication by $u_i \geq 0$ $(i \notin K)$ and summation gives

$$\sum_{i \notin K} u_i (x - x_0)' \nabla g_i(x_0) \leq 0 \ , \quad x \in D \tag{4.1.10}$$

Due to relation (4.1.8), the summation can be extended to the whole set of indices $\{ 1 , \ldots , m \}$ since

$$\sum_{i \in K} u_i (x - x_0)' \nabla g_i(x_0) = 0 \ , \quad \forall \ x \in D \tag{4.1.11}$$

and hence

$$\sum_{i=1}^{m} u_i (x - x_0)' \nabla g_i(x_0) \leq 0 \ . \tag{4.1.12}$$

Relations (4.1.6), (4.1.7) and (4.1.12), and conditions (v), give $q(x) \leq q(x_0)$ i.e., x_0 is a local maximum (hence, also global maximum) of function q. The proof is complete. The proof given above can be adapted to the case when either $K = \varnothing$ or $K = \{ 1 , \ldots , m \}$. If $K = \varnothing$, then omit relations (4.1.8) and (4.1.11), whereas if $K = \{ 1 , \ldots , m \}$, we omit relations (4.1.9) and (4.1.10).

4.2 CHANGE OF VARIABLE METHOD

Consider the fractional programming problem

$$\max\ q(x) = f(x)/g(x) \tag{4.2.1}$$

subject to

$$h(x) \leq \mathbf{0}\ , \tag{4.2.2}$$

where $f,\ g : R^n \to R$ are nonlinear scalar functions, and $h : R^n \to R^m$ is a nonlinear vector function. Let $S = \{\ x \in R^n \mid h(x) \leq \mathbf{0}\ \}$ be the domain of feasible solutions for problem (4.2.1) - (4.2.2). It is assumed that on S, f and g are not simultaneously zero. We present the change of variable method using the work of Mond and Craven [675]. Let $\Phi_0 : R \to R$ a monotonic function, strictly increasing, with $\Phi_0(t) > 0$ for $t > 0$. Let $\Phi_i : R \to R$, $(i = 1, \ldots, m)$, be m positive functions i.e., $t \geq 0$ implies $\Phi_i(t) \geq 0$. For every $i = 1, \ldots, m$, $t \in R_+$ and $y \in R^n$ we define the functions

$$\begin{aligned} F(y,\ t) &= f(y/t)\Phi_0(t), \quad G(y,\ t) = g(y/t)\Phi_0(t)\ , \\ H_i(y,\ t) &= h_i(y/t)\Phi_i(t)\ , \quad i = 1, \ldots, m\ . \end{aligned} \tag{4.2.3}$$

For the case when the limits given below exist, consider

$$\begin{aligned} F(y,\ 0) = \lim_{t\to 0} F(y,\ t)\ ,\quad G(y,\ 0) = \lim_{t\to 0} G(y,\ t)\ ,\quad H_i(y,\ 0) = \\ = \lim_{t\to 0} H_i(y,\ t)\ ,\quad i = 1, \ldots, m\ . \end{aligned} \tag{4.2.4}$$

Assume that $G(\mathbf{0},\ 0) = 0$. Let $H(y,\ t) = (H_i(y,\ t))$, $i = 1, \ldots, m$. To solve the problem (4.2.1) - (4.2.2), use the change of variable

$$y = tx\ , \tag{4.2.5}$$

where $t \geq 0$ is a scalar chosen such that, for a given function Φ_0 and $\gamma \in R, \gamma \neq 0$ we have

$$G(y,\ t) = \gamma \tag{4.2.6}$$

Multiply f and g by Φ_0 and h_i by $\Phi_i (i = 1, \ldots, m)$ and use relations (4.2.3), to reduce the problem (4.2.1) - (4.2.2) to the following nonlinear programming problem

$$\max\ F(y,\ t)\ , \tag{4.2.7}$$

subject to

$$G(y,\ t) = \gamma\ , \tag{4.2.8}$$

$$H(y,\ t) \leq \mathbf{0}\ , \tag{4.2.9}$$

$$t \geq 0\ . \tag{4.2.10}$$

Let $T = \{(y,\ t) \mid G(y,\ t) = \gamma,\ \ H(y,\ t) \leq \mathbf{0}\ ,\ t \geq 0\}$ be the domain of feasible solutions for problem (4.2.7) - (4.2.10). The following theorem establishes a connection between the optimal solutions of problems (4.2.1) - (4.2.2) and (4.2.7) - (4.2.10).

Theorem 4.2.1. *If*

(i) *the point* $(y, 0)$ *is not feasible for problem* (4.2.7) - (4.2.10),

(ii) $0 < sgn\ \gamma = sgn\ g(x^*)$ *for* x^* *the optimal solution of the problem* (4.2.1) - (4.2.2) *and*

(iii) (y^*, t^*) *is an optimal solution of the problem* (4.2.7) - (4.2.10),

then y^*/t^* *is an optimal solution for problem* (4.2.1) - (4.2.2).

Proof. The fact that (y^*, t^*) is an optimal solution of the problem (4.2.7) - (4.2.10) implies that

$$F(y^*, t^*) \geq F(y, t)\ , \quad \forall\ (y, t) \in T\ . \tag{4.2.11}$$

Assume that $x = y^*/t^*$ is not optimal solution of problem (4.2.1) - (4.2.2). Hence, another point $x^* \neq y^*/t^*$ exists with this property, i.e.

$$\frac{f(x^*)}{g(x^*)} > f\left[\frac{y^*}{t^*}\right] / g\left[\frac{y^*}{t^*}\right]\ . \tag{4.2.12}$$

Condition (ii) implies $g(x^*) = \theta\gamma(\theta > 0)$. If we consider

$$y = \Phi_0^{-1}(1/\theta)x^*\ , \quad t = \Phi_0^{-1}(1/\theta)\ , \tag{4.2.13}$$

then (y, t) is a feasible solution for problem (4.2.7) - (4.2.10). Indeed,

$$G(y, t) = \Phi_0(t)g(x^*) = \gamma$$

and

$$H_i(y, t) = h_i(y/t)\Phi_i(t) = h_i(x^*)\Phi_i(t) \leq 0\ , \quad i = 1\ , \ldots\ , m\ .$$

We have

$$\frac{f(x^*)}{g(x^*)} = \frac{\Phi_0(t)f(x^*)}{[\Phi_0(t)g(x^*)]} = \frac{F(y, t)}{G(y, t)} = \frac{F(y, t)}{\gamma}. \tag{4.2.14}$$

Also,

$$\frac{f(y^*/t^*)}{g(y^*/t^*)} = \frac{\Phi_0(t^*)f(y^*/t^*)}{\Phi_0(t^*)g(y^*/t^*)} = \frac{F(y^*, t^*)}{G(y^*, t^*)} = \frac{F(y^*, t^*)}{\gamma} \tag{4.2.15}$$

Relations (4.2.12), (4.2.14) and (4.2.15) imply

$$F(y^*, t^*) < F(y, t) \tag{4.2.16}$$

which contradicts relation (4.2.11). Hence, y^*/t^* is a solution of problem (4.2.7) - (4.2.10) and the proof is complete.

If sgn $g(x^*) < 0$, with x^* optimal solution of problem (4.2.1) - (4.2.2), then replacing f with $-f$, g with $-g$, the objective function does not change and, for the new denominator, $-g(x^*) > 0$ and Theorem 4.2.1 can be applied. Hence, using the change of variable (4.2.5), the solution of the problem (4.2.1) - (4.2.2) reduces to the solution of at most two nonlinear programming problems: problem (4.2.7) - (4.2.10), and problem:

$$max\ \{\ -\ F(y, t)\ |\ -\ G(y, t) = \gamma, \quad H(y, t) \leq \mathbf{0}, \quad t \geq 0\ \}\ .$$

Observation. In problem (4.2.1) - (4.2.2) variable non-negativity constraints were considered included in constraints (4.2.2). But if we consider explicitly the constraint $x \geq 0$, then we must take care to consider the constraint $y \geq 0$ in the equivalent nonlinear problems.

Taking particular forms for the functions f, g, Φ_i $(i = 0, 1, \ldots, m)$ and the set S, we get various cases studied in literature :

a) If f and g are linear, $\Phi_0(t) = t$ and $S = \{x \in R^n \mid Ax \leq b, \quad x \geq 0\}$, we get the problem studied by Charnes and Cooper [281] and presented by us in Chapter 3, Section 3.4.

b) If f and g are quadratic functions, $\Phi_0(t) = t^2$ and $S = \{x \in R^n \mid Ax \leq b, \; x \geq 0\}$ we get the problem studied by Swarup [941].

c) If f and g are polynomials of degree k, h linear, $\Phi_0(t) = t^k$ and $\Phi_i(t) = t$, $(i = 1, \ldots, m)$, then we get the results of Sharma [836].

d) If f and g are arbitrary, h linear and $\Phi_i(t) = t$, $(i = 1, \ldots, m)$ we get the problem studied by Mond and Craven [674].

In the method presented here, although we remove the fractional characteristic of the objective function, other difficulties are introduced in problem (4.2.7) - (4.2.10). First, even if h_i are linear, the problem (4.2.7) - (4.2.10) will have nonlinear constraints. Secondly, even if G is a convex function of y and t, in the case when G is nonlinear, due to constraint (4.2.8) the domain T is not a convex set. Because of this, we replace problem (4.2.7) - (4.2.10) by another problem in which we consider, instead of constraint (4.2.8), a constraint of the form:

$$G(y, t) \leq \gamma . \qquad (4.2.8')$$

In this case, if G is a convex function, then the set

$$T' = \{ (y, t) \mid G(y, t) \leq \gamma, \quad H(y, t) \leq 0, \quad t \geq 0 \}$$

is convex. If (y^*, t^*) is an optimal solution for problem (4.2.7), (4.2.8') - (4.2.10), $t^* > 0$ and $G(y^*, t^*) = 0$, then the maximum of problem (4.2.1) - (4.2.2) can be $+\infty$, since $x^* = y^*/t^*$ is a feasible point but $g(x^*) = 0$. But if

$$G(y^*, t^*) = \gamma_1 (0 < \gamma_1 < \infty) ,$$

then (y^*, t^*) is also an optimal solution for problem (4.2.7) - (4.2.10), taking $\gamma = \gamma_1$ and hence Theorem 4.2.1 can be applied. But if the maximum of problem (4.2.7), (4.2.8') - (4.2.10) appears at $(0, 0)$, then nothing can be said about the optimum of problem (4.2.1) - (4.2.2).

Assumption i) Theorem 4.2.1 is satisfied, for example, if S is a regular set, h is linear and

$$\Phi_i(t) = t , \quad (i = 1, \ldots, m) .$$

This thing is also valid if S is regular, $\Phi_i(t) = t^2$ and h is nonlinear of the form

$$h_i(x) = x'Cx + \alpha \leq 0 \quad (i = 1, \ldots, m) , \qquad (4.2.17)$$

where C is a symmetric positive semi-definite matrix and $\alpha \in R$. In this case, we have

$$H_i(y, t) = y'Cy + \alpha t^2 \leq 0 \quad (i = 1, \ldots, m) \tag{4.2.18}$$

If $(y, 0)$ satisfies (4.2.8) from $y'Cy \leq 0$ and C positive semi-definite then $y'Cy = 0$ i.e., $Cy = 0$. Since $G(y, t) = \gamma \neq 0$ and $G(\mathbf{0}, 0) = 0$ it follows that in $(y, 0)$ we have $y \neq \mathbf{0}$. Let x, which satisfies (4.2.17), and let k be a scalar. We have

$$h_i(x + ky) = (x + ky)'C(x + ky) + \alpha =$$

$$x'Cx + k^2y'Cy + 2kx'Cy + \alpha = x'Cx + 0 + 0 + \alpha \leq 0$$

which means that $x + ky \in S$, which contradicts the assumption that S is regular.

Let's analyze now problem (4.2.1) - (4.2.2) in the case when $f, g : S \to R$ are positive functions, convex or concave on the convex set S. Through the change of variable

$$y = x/g(x), \quad t = 1/g(x), \quad x \in S \tag{4.2.19}$$

the problem $max \{q(x) = f(x)/g(x) \mid x \in S\}$ becomes

$$max \{q'(y, t) = tf(y/t) \mid y \in R^n, t > 0, y/t \in S, tg(y/t) = 1\} \tag{4.2.20}$$

Let

$$T = \left\{ \begin{bmatrix} y \\ t \end{bmatrix} \in R^n \times R \mid t > 0, \ \ y/t \in S, \ \ tg(y, t) = 1 \right\}.$$

First, we prove the following result:

Lemma 4.2.1. *If f is concave on S, then the function*

$$q'(y, t) = tf(y/t)$$

is concave on

$$S^* = \left\{ \begin{bmatrix} y \\ t \end{bmatrix} \in R^n \times R \,\middle|\, t > 0, \ \ y/t \in S \right\}.$$

Proof. The case when $f \in C^2(S)$ was proved by Manas [622] constructing the Hessian matrix $Hq'(y, t)$. Next, we give a direct proof of the lemma. Let $\begin{bmatrix} y' \\ t' \end{bmatrix}, \begin{bmatrix} y'' \\ t'' \end{bmatrix} \in S^*$, $0 < \lambda < 1$,

$$y^\lambda = \lambda y' + (1 - \lambda)y'' \ \text{ and } \ t^\lambda = \lambda t' + (1 - \lambda)t''.$$

Then

$$q'(y^\lambda, t^\lambda) = t^\lambda f(y^\lambda/t^\lambda) = t^\lambda f\left[\frac{\lambda t'}{t^\lambda}\cdot\frac{y'}{t'} + \frac{(1-\lambda)t''}{t^\lambda}\cdot\frac{y''}{t''}\right] =$$

$$t^\lambda f\left[\lambda'\frac{y'}{t'} + (1-\lambda)\frac{y''}{t''}\right] \geq \lambda t' f(y'/t') + (1-\lambda)t'' f(y''/t'') = \tag{4.2.21}$$

$$\lambda q'(y', t') + (1-\lambda)q'(y', t') .$$

We have used the fact that f is concave function on S.
The following cases will be analyzed next:

a) f convex and g concave,
b) f concave and g convex,
c) f convex and g convex,
d) f concave and g concave.

In Case a), the objective function q is explicit quasiconvex whereas in Case b) the function q is quasiconcave. Cases c) and d) are equivalent due to relation $\dfrac{f(x)}{g(x)} = \dfrac{(1/g)}{(1/f)}$ and due to the fact that the inverse of a concave and strict positive function is a convex and strict positive function (Chapter 2, Section 2.5, Case B). Because of this, we analyze only the Case c). Through the change of variable (4.2.19) the initial problem reduces to problem (4.2.20). Since f is convex, the function $q'(y, t) = tf(y/t)$ is convex (due to Lemma 4.2.1) hence, quasiconvex. Consider now

$$T' = \left\{ \begin{bmatrix} y \\ t \end{bmatrix} \in R^n \times R \mid t > 0 , \ y/t \in S , \ tg(y/t) \leq 1 \right\}$$

and the problem

$$\max \left\{ q'(y, t) = tf(y/t) \ \middle|\ \begin{bmatrix} y \\ t \end{bmatrix} \in T' \right\} . \tag{4.2.22}$$

Since $tg(y/t)$ is a concave function, then T' is a convex set. We show now that the optimal solutions of problems (4.2.20) and (4.2.22) coincide. To show this, it is sufficient to show that a point $\begin{bmatrix} y' \\ t' \end{bmatrix} \in T'$ such that $t'g(y'/t) < 1$ cannot be an optimal solution for (4.2.22). Let $\hat{y}, \hat{t}$ be defined through $\hat{y}/\hat{t} = y'/t'$ and $\hat{t}g(y'/t') = 1$. Obviously $\begin{bmatrix} \hat{y} \\ \hat{t} \end{bmatrix} \in T$.

Since $\hat{t} > t'$, we have

$$\hat{t}f(\hat{y}/\hat{t}) > t'f(\hat{y}/\hat{t}) = t'f(y'/t')$$

hence, $\begin{bmatrix} y' \\ t' \end{bmatrix}$ is not an optimal solution for (4.2.22).

For example, consider problem (1.1) presented in Chapter 1. The function $T(x)$ is quasiconcave [26]. Using the change of variable $v = \lambda x (\lambda \geq 0)$ such that $\sum_j t_j v_j = 1$, problem (1.1) becomes

$$Max \left\{ T^*(\lambda, v) = \sum_i \sum_j v_j p_{ij} \log \left[\lambda p_{ij} \bigg/ \sum_k v_k \, p_{ik} \right] \right. \tag{4.2.23}$$

subject to

$$\begin{aligned} \sum_j v_j - \lambda &= 0 \\ \sum_j t_j v_j &= 1 \\ v_j &\geq 0 \end{aligned} \tag{4.2.24}$$

The function $T^*(\lambda, v)$ is concave [26] in $(\lambda, v) > \mathbf{0}$. According to Theorem 4.2.1, $\lambda > 0$ and if (λ^*, v^*) is optimal solution for problem (4.2.23) - (4.2.24), then $x^* = v^*/\lambda^*$ is optimal solution for problem (1.1).

4.3 NONLINEAR FRACTIONAL PROGRAMMING WITH HOMOGENEOUS FUNCTIONS

In this section, we analyze a particular case of the problem presented in Section 4.2, i.e. that in which the functions are homogeneous of degree r . Recall that a function f is homogenous of degree r if

$$f(\lambda x) = \lambda^r f(x) \tag{4.3.1}$$

where r is a non-zero real constant. Hence, consider the problem:

Problem P: $\max\ q(x) = \dfrac{c(x) + c_0}{d(x) + d_0}$

subject to

$$x \in S = \left\{ x \in R^n \,|\, x \in K, \;\; g(x) \leq \mathbf{0}, \; h(x) = 1 \right\}$$

where $K \subseteq R^n_+$, c, d, $h : K \to R$, $g : K \to R^m$ are real functions, $d(x) + d_0 \neq 0, c_0$ and d_0 are arbitrary real constants. Following the work of Suppe [940], we adopt the following assumptions:

A) $K \subseteq R_+^n$ is a convex cone, $c, d, h : K \to R$, $g : K \to R^m$ are continuous and positive homogeneous functions of degree $r (r \neq 0, r \in R), c_0, d_0 \in R$ are arbitrary real constants, $d(x) + d_0 > 0$ for all $x \in S$.

The case in which c, d, h, g are continuous and positive homogeneous functions of degree 1 was considered by Bradley and Frey, Jr. [201].

Also consider:

Problem P_1^+ : $\max q_1(x) = \dfrac{c(x) + c_0}{d(x) + d_0}$

subject to

$$x \in S_1^+ = \{ x \in R^n \mid x \in K,\ g(x) \leq 0,\ h(x) = 1,\ d(x) + d_0 > 0 \}$$

Similarly can be defined problem P_1^- in the case when $d(x) + d_0 < 0$. We associate to problem P_1^+ the following problem:

Problem P_2^+ : $\max q_2(y) = c(y) + c_0 h(y)$

subject to

$$y \in S_2^+ = \{ y \mid y \in K,\ g(y) \leq 0,\ d(y) + d_0 h(y) = 1 \}$$

The problem P_2^- associated to problem P_1^- is similarly defined.

We show the equivalence between problems P_1^+ and P_2^+. To do this, we consider the change of variables:

$$y = t_1(x) = \frac{x}{(d(x) + d_0)^{1/r}}, \quad d(x) + d_0 > 0 \tag{4.3.2}$$

$$x = t_2(y) = \frac{y}{(h(y))^{1/r}}, \quad h(y) > 0 \tag{4.3.3}$$

Theorem 4.3.1. *Under the assumptions* A), *and based on the change of variables* (4.3.2) *and* (4.3.3) *we have*:

a) *If x^* is feasible solution for problem P_1^+, then $y^* = t_1(x^*)$ is feasible solution for P_2^+.*

b) *If y^* is feasible solution for P_2^+, then $x^* = t_2(y^*)$ is feasible solution for P_1^+.*

Proof. a) Let x^* be feasible solution for P_1^+. We have

$$g(y^*) = g(t_1(x^*)) = g\left[\frac{x^*}{(d(x^*) + d_0)^{1/r}}\right] = \frac{1}{d(x^*) + d_0} g(x^*) \leq 0$$

considering that g is a positive homogenous function of degree r, $d(x^*) + d_0 > 0$ and $g(x^*) \leq \mathbf{0}$. On the other hand,

$$d(y^*) + d_0 h(y^*) = (1/(d(x^*) + d_0))d(x^*) + d_0(1/(d(x^*) + d_0))h(x^*) = 1 .$$

Also, $y^* \in K$ since

$$x^* \in K \text{ and } d(x^*) + d_0 > 0.$$

In conclusion, $y^* = t_1(x^*)$ is feasible solution for P_2^+. Point b) of the theorem is proved similarly.

Theorem 4.3.2. *Under the assumptions* A), *and based on the change of variables* (4.3.2) *and* (4.3.3) *we have* :

a) *If* x^* *is feasible solution for* P_1^+, *then*

$$q_1(x^*) = q_2(y^*) = q_2(t_1(x^*))$$

b) *If* y^* *is feasible solution for* P_2^+, *then*

$$q_2(y^*) = q_1(x^*) = q_1(t_2(y^*)).$$

Proof: a) Let $x^* \in S_1^+$. From Theorem 4.3.1 it follows

$$y^* = t_1(y^*) \in S_2^+$$

Based on the assumptions A) we have

$$q_1(x^*) = (c(x^*) + c_0)/(d(x^*) + d_0) = c(x^*/(d(x^*) + d_0)^{1/r}) +$$
$$+ c_0 \cdot h(x^*/(d(x^*) + d_0)^{1/r}) = c(y^*) + c_0 h(y^*) = q_2(y^*).$$

Similarly, we prove point b) of the theorem.

Theorem 4.3.3. *If the assumptions* A) *are true, then*

a) *If* x^0 *is optimal solution of problem* P_1^+, *then* $y^0 = t_1(x^0)$ *is optimal solution of problem* P_2^+ *and*

$$q_2(y^0) = q_1(x^0).$$

b) *If* y^0 *is optimal solution of problem* P_2^+ *then* $x^0 = t_2(y^0)$ *is optimal solution of problem* P_1^+ *and*

$$q_1(x^0) = q_2(y^0)$$

Proof. a) Let x^0 be the optimal solution of problem P_1^+ i.e.,

$$q_1(x^0) \geq q_1(x) \quad \forall\, x \in S_1^+ \tag{4.3.4}$$

If we accept that $y^0 = t_1(x^0)$ is not optimal solution of problem P_2^+, then

$$q_2(y^0) < q_2(y^*) \text{ for a certain } y^* \in S_2^+. \tag{4.3.5}$$

Theorems 4.3.1 and 4.3.2 imply that

$$q_1(x^0) = q_2(y^0) < q_2(y^*) = q_1(x^*).$$

Hence, $q_1(x^0) < q_1(x^*)$ for at least one $x^* \in S_1^+$, which contradicts relation (4.3.4). Hence, $y^0 = t_1(x^0)$ is optimal solution of P_2^+. Similarly, we prove point b).

As an example, reconsider the production efficiency problem presented in Chapter 1, Example 6. The fractional programming model presented there can be put in the form of problem P if we add the variable x_{n+1}, such that:

$$\max \{ q(x) = \prod_{j=1}^{n} a_0 x_j^{a_j} / (\sum_{j=1}^{n} c_j x_j + c_0) \mid \sum_{j=1}^{n} a_{ij} x_j - b_i x_{n+1} \geq 0, \quad i = 1, ..., m$$
$$x_{n+1} = 1, x_j \geq 0, \ j = 1, ..., n+1 \}$$

Under the assumption that the denominator is strict positive, the problem P_2^+ has the form

$$\max \{ q_2(y) = \prod_{j=1}^{n} a_0 y_j^{a_j} \mid \sum_{j=1}^{n} a_{ij} y_j - b_i y_{n+1} \geq 0 \quad (i = 1, ..., m);$$
$$\sum_{j=1}^{n} c_j y_j + c_0 y_{n+1} = 1, \quad y_j \geq 0, \quad (j = 1, ..., n+1)\}. \tag{4.3.6}$$

If we consider the logarithm of the objective function, this problem becomes a concave programming problem with linear constraints. Even more, if $\sum_{i=1}^{n} a_i = 1$, then we get a linear and homogenous programming problem (the objective function and the constraints are homogeneous of degree 1). It is obviously that, if in (4.3.6) the variables are strict positive, then we have a geometric programming problem for the solution of which we can use any algorithm existing in the literature.

4.4. LINEARIZATION METHODS

For solving the nonlinear fractional programming problem one can apply the linearization of the objective function (and/or of the constraints). Since the work of Kortanek and Evans, the problem has been studied by numerous researchers. Bector and Bhatt [133] utilize the linearization method for solving an interval programming problem with linear fractional objective function. Bhatt[1] considers the case when the objective function is pseudoconcave and Bector and Bhatt [136] considers the case of interval programming with the objective function pseudomonotonic.

---

[1]Bhatt, S.K. "A linearization technique for pseudo-concave programming". Congressus Numerantium XX Proc. Seventh Manitoba Conference on Numerical Math. and Computing 1977, 235-249

Bhatt [186] applies the linearization technique to linear fractional or pseudomonotonic programming problems. Recently, Ţigan[2] considers a more general case of quasimonotonic programming. Let's recall the result of Kortanek and Evans[3].

Theorem 4.4.1. *Let f be a continuous and differentiable function defined on a convex and open set $D \subseteq R^n$. Consider the following two problems:*

$$\text{I. } \max \{ f(x) : x \in X \}, \text{ II. } \max \{ x' \nabla f(x^*) : x \in X \},$$

where X is a closed and convex set contained in D and $x^* \in X$. The point x^* is an optimal solution for problem **I** *if and only if x^* is an optimal solution for problem* **II** *with the condition that f is pseudoconcave on D or quasiconcave on D and $\nabla f(x^*) \neq \mathbf{0}$.*

Theorem 4.4.1. suggests that, for solving problem **I,** we can utilize the linear programming problem **II**, equivalent with it, for which the calculation effort is much less than for problem **I**. However, a difficulty arises from the fact that x^* is not known, and hence, we cannot say exactly who $\nabla f(x^*)$ is. The algorithm we are going to present will remove this dilemma and seeks iteratively the point $x^* \in X$ for which Theorem 4.4.1 applies. We consider that f is pseudoconcave on X, a bounded set of the form

$$X = \{x \in R^n \mid Ax = b, x \geq \mathbf{0}\}.$$

Certain fractional problems (linear or nonlinear) are pseudoconcave (see Chapter II, Section 2.5).Let x^1, ..., x^p be some p vertices of the convex set X and consider the problem

$$(\Lambda_p) \max \left\{ \Phi(\lambda) = f\left(\sum_{i=1}^{p} \lambda_i x^i \right) \mid \sum_{i=1}^{p} \lambda_i = 1, \quad \lambda_i \geq 0, \quad i = 1, ..., p \right\}.$$

Let $\lambda^1 = (\lambda_1^1, ..., \lambda_p^1)$ be the solution of problem (Λ_p). Bhatt (*op.cit.*) proposes finding this solution using a simplified version of Zoutendijk's method. Consider

$$x^{\lambda^1} = \sum_{i=1}^{p} \lambda_i^1 x^i,$$

to which we attach the problem

$$(P_{\lambda^1}) \max_{x \in X} x' \nabla f(x^{\lambda^1}).$$

Theorem 4.4.2. *If x^{p+1} is an optimal solution of the problem (P_{λ^1}) and $x^{p+1} = x^i$ for a certain $i \in \{1, ..., p\}$, then x^{λ^1} is optimal solution of the problem* **I**.

---

[2] Ţigan, Şt., "On the linearization technique for quasi-monotonic optimization problems". Anal. Numér. Théor. Approx. **12**(1), (1983), 89-96

[3] Kortanek, K.O., Evans, J.P., "Pseudo-concave programming and Lagrange Regularity". Oper. Res., **15**(1967), 882-391

Proof. Let X_1 = conv [$\{x^1, ..., x^p\}$] and let Ω be the domain of feasible solutions of problem Λ_p. Since λ^1 is optimal solution of problem (Λ_p), we have $\Phi(\lambda^1) \geq \Phi(\lambda)$ for all $\lambda \in \Omega$, and hence $f(x) \leq f(x^{\lambda^1})$ for all $x \in X_1$. From here it follows that x^{λ^1} is optimal solution of problem $\max_{x \in X_1} f(x)$. Applying Theorem 4.4.1, it follows that x^{λ^1} is the optimal solution for problem

$$\max_{x \in X_1} x' \nabla f(x^{\lambda^1})$$

i.e.,

$$(x^{\lambda^1})' \nabla f(x^{\lambda^1}) \geq x' \nabla f(x^{\lambda^1}) \quad \forall\, x \in X.$$

In particular, for $x^{p+1} \in X_1$, we have

$$(x^{\lambda^1})' \nabla f(x^{\lambda^1}) \geq (x^{p+1})' \nabla f(x^{\lambda^1}) \tag{4.4.1}$$

But x^{p+1} is an optimal solution of problem (P_{λ^1}) i.e.

$$(x^{p+1})' \nabla f(x^{\lambda^1}) \geq x' \nabla f(x^{\lambda^1}), \quad \forall\, x \in X \tag{4.4.2}$$

From (4.4.1) and (4.4.2) it follows

$$(x^{\lambda^1})' \nabla f(x^{\lambda^1}) \geq x' \nabla f(x^{\lambda^1}) \quad \forall\, x \in X. \tag{4.4.3}$$

From (4.4.3) it follows that x^{λ^1} is optimal solution of problem $\max_{x \in X} x' \nabla f(x^{\lambda^1})$ and hence, according to Theorem 4.4.1, also of problem **I**. The proof is complete.

The algorithm for solving problem **I** consists of the following steps:

Step 1. Choose $x^0 \in R^n$ such that $\nabla f(x^0) \neq \mathbf{0}$ and solve the problem

(L_0) $$\max_{x \in X} x' \nabla f(x^0)$$

Let x^1 be the optimal solution of problem (L_0). Make $i = 1$.

Step 2. For $x^i \in R^n$ given, solve the problem

(L_i) $$\max_{x \in X} x' \nabla f(x^i),$$

the optimal solution of which we denote by x^{i+1}.

Step 3. If $x^{i+1} \notin \{x^1, ..., x^i\}$, then replace i with $i+1$ and repeat the algorithm starting with Step 2. Continue the application of steps 2 and 3 and assume that we have generated the vertices $x^1, ..., x^p$ of X such that, when we solve the problem (L_i) with $i=p$, we obtain the solution $y \in \{x^1, ..., x^p\}$.

Step 4. a) If $y = x^p$, then, based on Theorem 4.4.1, y is optimal solution of problem **I**.

b) If $y = x^j (j < p)$, go to Step 5.

Step 5. Solve the nonlinear programming problem

$$(\Lambda_p) \max \left\{ \Phi(\lambda) = f\left(\sum_{i=j}^{p} \lambda_i x^i\right) \;\middle|\; \sum_{i=j}^{p} \lambda_i = 1,\ \lambda_i \geq 0,\ i = j, ..., p \right\}$$

and let $\lambda^1 = (\lambda_j^1, ..., \lambda_p^1)$ be its optimal solution. Consider $x^{\lambda^1} = \sum_{i=j}^{p} \lambda_i^1 x^i$.

Step 6. Solve the linear programming problem:

$$(P_{\lambda_1}) \qquad \max_{x \in X} x' \nabla f(x^{\lambda^1})$$

and let y^1 be its optimal solution.

a) If $y^1 \in \{x^j, ..., x^p\}$, then, due to Theorem 4.4.2, x^{λ^1} is the optimal solution of problem **I**.

b) If $y^1 \notin \{y^1, ..., y^p\}$, then take $x^{p+1} = y^1$ and go to Step 5 with $x^j, ..., x^p, x^{p+1}$.

The convergence of the algorithm in a finite number de steps is ensured, under the condition that the solution of problems Λ_p can be obtained in a finite number of iterations. Indeed, since X is a bounded set, a finite number of vertices exists, and each time we get a vertex of X. But, if the solution of problem Λ_p cannot be obtained in a finite number of iterations, then, in Step 5, consider λ^1 an approximate solution of problem (Λ_p). Construct x^{λ^1} and go to Step 6. If we are in Case 6a, then consider x^{λ^1} as approximate optimal solution of problem **I**. If we are in Case 6b, we obtain $y^1 = x^{p+1}$. Solve(Λ_{p+1}) with $x^j, ..., x^p, x^{p+1}$ to obtain a new solution λ^2 (exact or approximate).

Consider $x^{\lambda^2} = \sum_{i=j}^{p+1} \lambda_i^2 x^i$. If $f(x^{\lambda^2}) - f(x^{\lambda^1}) < \delta$ (δ depends on the required degree of approximation), then consider x^{λ^2} as an approximate optimal solution of problem **I**. Otherwise, consider $x^{p+2} = y^2$ and go to Step 5 with $x^j, ..., x^{p+1}, x^{p+2}$.

Consider now the particular case when f is pseudomonotonic (pseudolinear). Such a function is, for example, the linear fractional function. In this case, during the solution of problems $L_1, L_2, ..., L_p$ in Step 2, we get increasing values for the function f, as it results from the following theorem:

Theorem 4.4.3. *If f is pseudomonotonic, then,*

$$f(x^1) \le f(x^2) \le ... \le f(x^p).$$

Proof. Consider the problem $L_k (0 \le k \le p-1)$ of which solution is x^{k+1}. Hence,

$$(x^{k+1} - x^k)' \nabla f(x^k) \ge 0$$

which implies $f(x^{k+1}) \ge f(x^k)$, since f is pseudomonotonic hence, pseudoconvex. The proof is complete.

Since the number of vertices of X is finite, and since, through the solution of problems $L_k (k = 1,2, ...)$, the obtained solutions are also vertices of X, it is possible that certain vertices are met again. In this case, we have the following theorem:

Theorem 4.4.4. *If f is pseudomonotonic and if, at a certain step, say $x^p, x^j, 1 \le j \le p$, solves problem (L_p), then x^j, ..., x^p solve problem* **I**.

Proof. Since x^j solves problem (L_p) we have

$$(x^j - x^p)' \nabla f(x^p) \ge 0, \tag{4.4.4}$$

from where, since f is pseudoconvex,

$$f(x^j) \ge f(x^p).$$

Combining this result with that of Theorem 4.4.3, which states that $f(x^j) \le f(x^{j+1}) \le ... \le f(x^p)$, we get

$$f(x^j) = f(x^{j+1}) = ... = f(x^p). \tag{4.4.5}$$

The function f is, by assumption, pseudoconvex and hence, quasiconvex (Theorem 2.3.2, Chapter II, Section 3.2) and, according to the definition, we have

$$f(x^j) = f(x^p) \Rightarrow (x^j - x^p)' \nabla f(x^p) \le 0 \tag{4.4.6}$$

From (4.4.4) and (4.4.6) it follows that

$$(x^j)' \nabla f(x^p) = (x^p)' \nabla f(x^p)$$

i.e., x^p solves problem (L_p). Using Theorem 4.4.1 it follows that x^p is optimal solution for problem **I**. Due to relations (4.4.5), x^j, x^{j+1}, ..., x^{p-1} are also optimal solutions for problem **I**.

Note that, in this algorithm, the first step can be omitted if the first basic feasible solution can be constituted from only slack or artificial variables. Also, the calculation time can be reduced considerably if the optimal solution of problem (L_i) is considered as initial basic feasible solution of problem (L_{i+1}).

4.5 PARAMETRIC METHODS

Consider the nonlinear fractional programming problem

$$(P) \qquad \max \{ q(x) = \frac{f(x)}{g(x)} \mid x \in S \},$$

where $S \subseteq R^n$ is a compact and non-empty set, $f, g : S \to R$ are continuous functions, $g(x) > 0$ for all $x \in S$ and $f(x) \ge 0$ for at least an $x \in S$. In virtue of this assumption, problem (P) has at least an optimal solution. About the functions f and g, we make no assumptions yet regarding their convexity. We are going to describe a class of methods for solving problem (P) which are based on the solution of the parametric problem:

$$Q(\lambda) \qquad \max \{f(x) - \lambda g(x) \mid x \in S\} \quad \lambda \in R \quad (\lambda \text{ parameter})$$

Such a method was presented in Chapter 3, Section 3.6, for the case when f and g are linear functions. Denote by

$$F(\lambda) = \max_{x \in S} (f(x) - \lambda g(x)),$$

the optimal value of the objective function of problem $Q(\lambda)$.

Theorem 4.5.1. *Let $\bar{x}$ be an optimal solution of problem (P) and let*

$$\bar{\lambda} = \frac{f(\bar{x})}{g(\bar{x})}$$

Then:
a) $F(\lambda) > 0$ *if and only if* $\lambda < \bar{\lambda}$,
b) $F(\lambda) = 0$ *if and only if* $\lambda = \bar{\lambda}$,
c) $F(\lambda) < 0$ *if and only if* $\lambda > \bar{\lambda}$.

Proof. We prove only case a); the other cases are treated similarly. Let $F(\lambda) > 0$. Then, there is at least one solution x' of problem $Q(\lambda)$ for which we have $f(x') - \lambda g(x') > 0$. Considering that $g(x') > 0$, then

$$\frac{f(x')}{g(x')} > \lambda \tag{4.5.1}$$

Also, x' is a feasible solution of problem (P) hence, $\bar{\lambda} \geq f(x')/g(x')$. This and (4.5.1) imply that $\lambda < \bar{\lambda}$. Conversely, if $\lambda < \bar{\lambda}$, at least one solution x'' of (P) exists (for example $x'' = \bar{x}$) such that

$$\frac{f(x'')}{g(x'')} > \lambda$$

But $g(x'') > 0$, hence,

$$F(\lambda) = \max_{x \in S} (f(x) - \lambda g(x)) \geq f(x'') - \lambda g(x'') > 0$$

Corollary 4.5.1. *If $F(\lambda) = 0$, then an optimal solution of problem $Q(\lambda)$ is also an optimal solution of problem (P).*

Proof. If $F(\lambda) = 0$, then through Theorem 4.5.1 $\lambda = \bar{\lambda}$. If x' is an optimal solution of problem $Q(\lambda)$, then $f(x') - \lambda g(x') = 0$ and hence, $\bar{\lambda} = f(x')/g(x')$, i.e., x' is optimal solution of problem (P). The proof is complete.

Note that the result of this corollary is contained in the result of Theorem 3.5.2 of Chapter 3, Section 3.5. The proof given in Theorem 3.5.2 did not depend on the fact that f and g were considered, in that case, linear. From these results, it follows that the solution to problem (P) is equivalent to the determination of $\lambda = \bar{\lambda}$ for which $F(\lambda) = 0$, hence, the determination of the root of the nonlinear equation $F(\lambda) = 0$. It will be shown that this root is unique. The methods used for solving equation $F(\lambda) = 0$ will generate the solution algorithms for the nonlinear fractional programming problems. First, we present several interesting properties of the function F.

Theorem 4.5.2. *The function F is continuous, convex and strict decreasing on R.*

Proof. Since f and g are continuous functions, standard calculus yields that F is a continuous function on R. Let's prove that F is convex. Let $\lambda' < \lambda''$, $t \in [0,1], \lambda' = t\lambda' + (1-t)\lambda''$ and let x' be an optimal solution of problem $Q(\lambda')$. We have,

$$F(\lambda') = f(x') - \lambda' g(x') = f(x') - [t\lambda' + (1-t)\lambda'']g(x') =$$
$$= t[f(x') - \lambda' g(x')] + (1-t)[f(x') - \lambda'' g(x')] \leq tF(\lambda') + (1-t)F(\lambda'')$$

hence, F is convex. To prove that F is strict decreasing on R, let $\lambda' < \lambda''$ and let x' and x'' be the optimal solutions of problems $Q(\lambda')$ and $Q(\lambda'')$. Then

$$F(\lambda'') = \max_{x \in S}[f(x) - \lambda'' g(x)] = f(x'') - \lambda'' g(x'') <$$
$$< f(x'') - \lambda' g(x'') \leq \max_{x \in S}[f(x) - \lambda' g(x)] = F(\lambda')$$

which implies that F is strict decreasing on R. The proof is complete.

Even more, the assumption that $f(x) \geq 0$ for at least one $x \in S$ yields that $F(0) \geq 0$. Also, $F(\lambda) \to -\infty$ when $\lambda \to \infty$, and $F(\lambda) \to +\infty$ when $\lambda \to -\infty$. This and the results of Theorem 4.5.2 imply that the equation $F(\lambda) = 0$ has an unique solution $\lambda = \bar{\lambda}$. The shape of the function F can be seen in Figure 4.5.1. If f and g are linear, then F is a piecewise linear function.

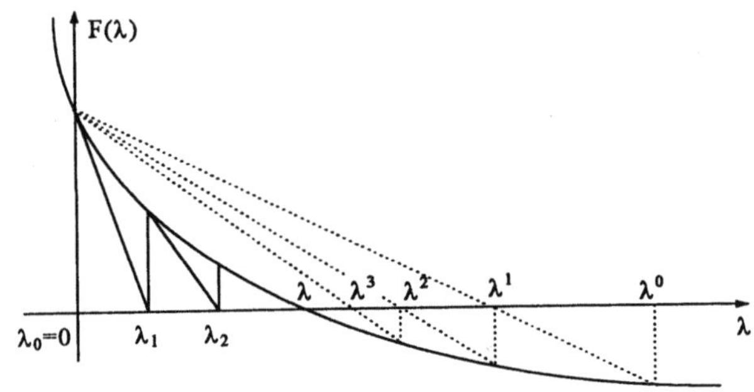

Figure 4.5.1

The most widely known method for solving equation $F(\lambda) = 0$ is the Newton-Raphson method. Isbell and Marlow [504] applied it to the linear case. Method was generalized to the nonlinear case by Dinkelbach [360]. Hence, it is widely known as the "Dinkelbach method". One starts with $\lambda_0 = 0$, and then generates the sequence $\lambda_0, \lambda_1, \lambda_2, \ldots$ as follows: the element λ_{i+1} of the sequence is abscissa of the point where the tangent to $F(\lambda)$ in the point $\lambda = \lambda_i$ meets the axis $O\lambda$ (Figure 4.5.1). The tangent to $F(\lambda)$ in the point $\lambda = \lambda_i$ has the equation

$$y = f(x^i) - \lambda g(x^i)$$

where x^i is the optimal solution of $Q(\lambda_i)$. This tangent crosses the axis $O\lambda$ in the point of abscissa $f(x^i)/g(x^i)$ which can be considered as λ_{i+1} in Newton's method. The tangent mentioned above can be defined even if F is not differentiable in $\lambda = \lambda_i$ but has a subgradient to $\lambda = \lambda_i$ ($-g(x^i)$ is the subgradient of $F(\lambda_i)$ to λ_i, and coincides with the corresponding derivative if $F(\lambda)$ is differentiable in λ_i). The resulting sequence $\{\lambda_i\}$ is convergent towards $\bar{\lambda}$, the solution of equation $F(\lambda)=0$.

In practice, we stop at the term λ_i of the sequence $\{\lambda_i\}$ if $\bar{\lambda}_i$ is sufficiently close to $\bar{\lambda}$ i.e., $|F(\lambda_i)| \leq \delta$ (δ being a non-negative given constant).

Dinkelbach's algorithm consists of the following steps:

Step 1. Take $\lambda = \lambda_1$ such that $F(\lambda_1) \geq 0$,

Step 2. Solve problem $Q(\lambda)$. If $|F(\lambda)| \leq \delta$, stop. Otherwise, go to Step 3.

Step 3. Let $\lambda = f(x^*)/g(x^*)$, where x^* is an optimal solution of problem $Q(\lambda)$ obtained in Step 2. Repeat Step 2.

Note that, in Step 1, $\lambda_1 = 0$ or $\lambda_1 = f(\hat{x})/g(\hat{x})$, where $\hat{x}$ is a feasible solution of the problem. If through some procedure we manage to find a λ_0 sufficiently close to $\bar{\lambda}$, then the convergence rate of the algorithm increases rapidly. Bitran and Magnanti [191] proposed that, in Step 3, one takes

$$\lambda = f(x^*)/g(x^*) + \varepsilon$$

($\varepsilon \geq 0$ is a given constant).

The presented algorithm converges to λ with a superlinear convergence rate, as shown next. First, we prove several intermediate results.

Lemma 4.5.1. *Let λ' and $\lambda'' \in R$, and let x' and x'' be the optimal solutions of problems $Q(\lambda')$ and $Q(\lambda'')$, respectively. Then*

$$q(x'') - q(x') \geq (-F(\lambda''))[1/g(x') - 1/g(x'')] \tag{4.5.2}$$

and

$$q(x'') - q(x') \leq [-F(\lambda'') + (\lambda' - \lambda'')g(x'')][1/g(x') - 1/g(x'')]. \tag{4.5.3}$$

Proof. Since x'' is optimal solution of problem $Q(\lambda'')$, we have

$$f(x'') - \lambda'' g(x'') \geq f(x') - \lambda'' g(x')$$

from where, by division with $g(x')$,

$$f(x'')/g(x') - \lambda'' g(x'')/g(x') \geq f(x')/g(x') - \lambda''.$$

Hence,

$$\frac{f(x'')}{g(x'')} - \frac{f(x')}{g(x')} \geq \frac{f(x'')}{g(x'')} - \frac{f(x'')}{g(x')} + \lambda'' \left[\frac{g(x'')}{g(x')} - 1\right] = -F(\lambda'')\left[\frac{1}{g(x')} - \frac{1}{g(x'')}\right]$$

i.e., (4.5.2). Similarly, one proves (4.5.3). The inequalities (4.5.2) and (4.5.3) imply

Lemma 4.5.2. *If* $\lambda' < \lambda''$, *then* $g(x') \geq g(x'')$.

From lemma 4.5.2, the inequality (4.5.2), and the fact that $F(\lambda'') \geq 0$ for $\lambda'' \leq \bar{\lambda}$, we obtain

Lemma 4.5.3. *If* $\lambda' < \lambda'' \leq \bar{\lambda}$, *then* $q(x') \leq q(x'')$.

Let now $\lambda'' = \bar{\lambda}$. Lemma 4.5.2, the inequality (4.5.3), and the fact that $F(\bar{\lambda}) = 0$ imply

Lemma 4.5.4. *For* $\lambda' < \bar{\lambda}$, *we have* $\bar{\lambda} - q(x') \leq (\bar{\lambda} - \lambda')[1 - g(\bar{x})/g(x')]$, *where* $0 \leq 1 - g(\bar{x})/g(x') \leq 1$.

This lemma implies that, in Dinkelbach's algorithm, we get

$$\bar{\lambda}-\lambda_{i+1} \leq (\bar{\lambda}-\lambda_i)(1-g(\bar{x})/g(x')) \quad \forall\ \lambda_i<\bar{\lambda} \quad (i = 1, 2, ...), \tag{4.5.4}$$

where $0 \leq 1-g(\bar{x})/g(x_i) < 1$. Based on Lemma 4.5.2, the sequence $\{1-g(\bar{x})/g(x_i)\}$ is a non-increasing sequence, and hence $\{\lambda_i\}$ converges at least linearly towards $\bar{\lambda}$. Equation (4.5.4) yields a more general result.

Theorem 4.5.3. *The sequence* $\{\lambda_i\}$, *generated by the Dinkelbach's algorithm, converges for every* $\lambda_1 < \bar{\lambda}$ *superlinear to* $\bar{\lambda}$, *i.e.*

$$\lim_{i \to \infty} \frac{\bar{\lambda}-\lambda_{i+1}}{\bar{\lambda}-\lambda_i} = 0$$

if $\{\lambda_i\}$ *is infinite.*

Another method of solving equation $F(\lambda) = 0$ is the *regula falsi* method. In this method, a sequence $\lambda^0,\lambda^1,\lambda^2, ...\ \lambda^0 > \lambda^1 > \lambda^2 ...$ is constructed such that it converges towards $\bar{\lambda}$ as follows: Start with $\lambda^0 > \bar{\lambda}$. The straight line which joints the points $(\lambda_0, F(\lambda_0))$ and $(\lambda^0, F(\lambda^0))$, intersects the axis $O\lambda$ in the point of abscissa $v(\lambda_0, \lambda^0)$. Take $\lambda^1 = v(\lambda_0, \lambda^0)$. The straight line which joints the points $(\lambda_0, F(\lambda_0))$ and $(\lambda^1, F(\lambda^1))$ intersects the axis $O\lambda$ in the point of abscissa $v(\lambda_0, \lambda^1)$. Take $\lambda^2 = v(\lambda_0, \lambda^1)$ etc. We get a sequence $\{\lambda^i\}_{i=0,1,...}$ (Figure 4.5.1) which converges linearly towards $\bar{\lambda}$ (see Schaible [800]). One obtains the "OUTER" method for the solution of nonlinear fractional programming problems:

Step 1. Choose λ' and λ'' such that $F(\lambda') \geq 0$ and $F(\lambda'') \leq 0$. Let $\lambda = v(\lambda',\lambda'')$.

Step 2. Solve problem $Q(\lambda)$. If $|F(\lambda)| \leq \delta$, Stop. Otherwise, go to Step 3.

Step 3. Let $\lambda = v(\lambda',\lambda)$. Repeat the algorithm starting at Step 2.

Note that, in Step 1, we can consider $\lambda^l = 0$, and can consider that λ^u is a sufficiently large number to ensure that $\bar{\lambda} < \lambda^u$. Schaible [800] suggested that λ^u can be determined using the dual algorithm. Also, to obtain an accelerated algorithm, we can consider

$$\tilde{\lambda}^i = \frac{f(x^i)}{g(x^i)},$$

where x^i is an optimal solution of problem $Q(\lambda^i)$.

Another solution method of equation $F(\lambda) = 0$ is the *binary line search* method or the *bisection* method. This method start with an interval $[\lambda^l, \lambda^u]$ which contains $\bar{\lambda}$. Depending on whether $F(\lambda) \geq 0$ or $F(\lambda) < 0$, where $\lambda = (1/2)(\lambda^l + \lambda^u)$ is the middle of segment $[\lambda^l, \lambda^u]$, one considers the interval $[\lambda, \lambda^u]$ or the interval $[\lambda^l, \lambda]$ as the next interval containing $\bar{\lambda}$, etc. The algorithm corresponding to this method has the following steps:

Step 1. Determine λ' and λ'' such that $F(\lambda') \geq 0$ and $F(\lambda'') < 0$. Let $\lambda^l = \lambda'$ and $\lambda^u = \lambda''$.

Step 2. Let $\lambda = (1/2)(\lambda^l + \lambda^u)$. Solve problem $Q(\lambda)$. If $|F(\lambda)| \leq \delta$ stop; otherwise, go to Step 3.

Step 3. If $F(\lambda) > 0$, then take $\lambda^l = \lambda$ and go to Step 2. Otherwise, take $\lambda^u := \lambda$ and go to Step 2.

It is obvious that rate of convergence of this algorithm is linear, and that the interval $[\lambda^l, \lambda^u]$, after p iterations, satisfies the relation

$$|\lambda^u - \lambda^l| \leq |\lambda'' - \lambda'|/2^p$$

and contains $\bar{\lambda}$. Other methods, which are combinations of the three methods presented above, and which resulted from the continuity and convexity properties of F, can also be mentioned (see Ibaraki [501]). These methods are known under the names MODBIL (modified binary with an improved lower bound) and INTBIN (interpolated binary). The MODBIL method is based on the property that $\bar{\lambda}$ is contained in the interval

$$I = [\max\ (f(x^u)/g(x^u),\ f(x^l)/g(x^l)),\ \ v(\lambda^l, \lambda^u)],$$

where $F(\lambda^l) \geq 0$, $F(\lambda^u) < 0$ and x^l, x^u are the optimal solutions of problems $Q(\lambda^l)$ and $Q(\lambda^u)$ respectively. Consequently, instead of bisecting the interval $[\lambda^l, \lambda^u]$, as in the bisection method, let's consider the bisection of the interval I. The INTBIN method is a variant of the bisection method.

Schaible and Ibaraki [820] presented the results of some numerical experiments performed with the 5 methods mentioned above for the knapsack problem with the fractional objective function. The results of these numerical experiments for this particular problem showed that the INTBIN method is the most rapid, while Dinkelbach's and MODBIL methods are less efficient.

Any other solution method of problem (P) can be expressed as a combination of the 5 methods. In any case, the solution of problem $Q(\lambda)$ to obtain $F(\lambda)$ remains a difficult question which is going to increase the number of operations for the entire algorithm. This difficulty can be removed by using certain techniques such as the sensitivity analysis and the parametric programming. Assume that problem $Q(\lambda)$ is solved with a primal algorithm that generates a sequence of feasible solutions that produce increasing values of the objective function. If, for a feasible solution x' of this sequence, we have $f(x') - \lambda g(x') > 0$, then $F(\lambda) > 0$, and hence $\lambda < \bar{\lambda}$, even if the exact value of $F(\lambda)$ is not known.

Combining this observation with Dinkelbach's algorithm we obtain the PRIMAL algorithm:

Step 1. Take $\lambda := \lambda'$ such that $F(\lambda') \geq 0$,

Step 2. Apply a primal algorithm to determine a sequence of feasible solutions for $Q(\lambda)$ and for which we can have $|F(\lambda)| > \delta$ or $|F(\lambda)| \leq \delta$. In the first case, go to Step 3; the second case, the algorithm stops.

Step 3. Take $\lambda := f(x')/g(x')$, where x' is the best feasible solution of problem $Q(\lambda)$ obtained in Step 2. Return to Step 2.

Another method of solving the problem $Q(\lambda)$ is that of an heuristic algorithm by which one determines a feasible solution for which $f(x') - \lambda g(x') > 0$. After this, one can utilize Dinkelbach's algorithm.

Another way to determine $\bar{\lambda}$, the root of equation $F(\lambda) = 0$, is to draw the graph of $F(\lambda)$. Ibaraki *et al.* [502] applied such an idea to solving of a quadratic fractional programming problem. We also mention Megiddo's method [649] for solving of the equation $F(\lambda) = 0$.

4.6. FRACTIONAL OPTIMAL CONTROL PROBLEM

Consider a dynamic system which is described at every moment by the state vector $x(t) = (x^1(t), \ldots, x^n(t))$, and of which evolution is driven by the control vector $u(t) = (u^1(t), \ldots, u^m(t))$ through the equations

$$\dot{x}^i(t) = dx^i/dt = f^i(t, x(t), u(t)), \quad i = 1, \ldots, n.$$

The initial state of the system, $x(t_0) = (x^1(t_0), \ldots, x^n(t_0))$, the set of feasible controls, $U \subseteq R^m$, and the final set, $G(t) \subseteq R^n$ $(\tau_0 \leq t \leq \tau_1)$, are given. Define

$$\Delta = \Delta\,\{f^1(t, x, u), \ldots, f^n(t, x, u), U, x_0, G(t)\},$$

the set of all controls $u(t)$ on various subintervals $t_0 \leq t \leq t_1$ with $\tau_0 \leq t_0 \leq t_1 \leq \tau_1$ such that $x(t_0) = x_0$ and $x(t_1) \in G(t_1)$. If we associate to the control problem the cost function $C(u)$, then, in Δ, a control vector, $u_*(t)$, is optimal if $C(u_*) \leq C(u)$ for all $u(t) \in \Delta$

(obviously in the case of a minimum problem). Next, let's accept that the cost function de cost $C(u)$ is fractional, and let's formulate the fractional optimal control problem following Bhatt [184].

Problem I. Minimize the cost functional

$$C(u) = \int_{t_0}^{t_1} f^{n+1}(t, x(t), u(t))dt \Big/ \int_{t_0}^{t_1} f^{n+2}(t, x(t), u(t))dt \qquad (4.6.1)$$

with respect to $u(t) \in \Delta$ satisfying the constraints

$$\dot{x}^i(t) = f^i(t, x(t), u(t)) = g^i(t, x) + h^i_j(t, x)u^j, \; i = 1, ..., n, \; j = 1, ..., m \qquad (4.6.2)$$

$$U \subseteq R^m \qquad (4.6.3)$$

$$x(t_0) = x_0 \in R^n \qquad (4.6.4)$$

$$G(t) \subseteq R^n, \quad \tau_0 \leq t \leq \tau_1 \qquad (4.6.5)$$

where : $f^{n+i}(t, x, u) = g^{n+i}(t, x) + h^{n+i}_j(t, x)u^j$, $i = 1, 2$; $g^i(t, x)$, $h^i_j(t, x)$ and $\dfrac{\partial g^i(t, x)}{\partial x^k}$ $\dfrac{\partial h^i_j(t, x)}{\partial x^k}$, $i, k = 1, ..., n, j = 1, ..., m$ are continuous on $R \times R^n$; $g^{n+i}(t, x), h^{n+i}_j(t, x)$, $i = 1, 2$ are continuous on $R \times R^n$; and

$$\int_{t_0}^{t_1} f^{n+2}(t, x(t), u(t))dt > 0 \text{ for all } u(t) \in \Delta, \qquad (4.6.6)$$

U is a convex set, compact and non-empty, $G(t)$ is a compact and non-empty set on the finite interval $\tau_0 \leq t \leq \tau_1$.
We associate to problem I the following non-fractional optimal control problem:

Problem II. $\min\limits_{u(t) \in \Delta} \left[\int_{t_0}^{t_1} \{f^{n+1}(t, x(t), u(t)) - qf^{n+2}(t, x(t), u(t))\}dt\right]$

for $q \in R$, and under the assumptions of Problem I.

E.B. Lee and L. Markus ("Optimal Control for Nonlinear Processes", *Arch. Rational Mech. Anal.* 8(1), 1961, 36-58) showed that, for $q \in R$ and under the assumptions of Problem I, there is an optimal control $u(t) \in \Delta$ for Problem II. We are going to apply Dinkelbach's algorithm, presented in Section 4.5, to solve Problem I.
Consider the function:

$$F(q) = \min_{u(t) \in \Delta} \left[\int_{t_0}^{t_1} \{ f^{n+1}(t, x(t), u(t)) - qf^{n+2}(t, x(t), u(t)) \}dt\right], \; q \in R$$

A number of properties of the function F, similar to those of the corresponding function of Section 4.5, will be used. Let's prove the existence theorem of the optimal control for Problem I and let's find an algorithm for obtaining the corresponding optimal control.

Lemma 4.6.1. *The function F is concave on R.*

Proof. Let $q_1, q_2 \in R$, $q_1 \neq q_2$, $0 \leq \lambda \leq 1$, $x_\lambda(t)$ be a solution of the differential equations system (4.6.2) corresponding to the control $u_\lambda(t)$ and let $(x_\lambda(t), u_\lambda(t))$ solution of Problem II with $q^\lambda = \lambda q_1 + (1-\lambda)q_2$. We have

$$F(q^\lambda) = \int_{t_0}^{t_1} \{ f^{n+1}(t, x_\lambda(t), u_\lambda(t)) - q^\lambda f^{n+2}(t, x_\lambda(t), u_\lambda(t)) \}dt =$$

$$\lambda\int_{t_0}^{t_1} \{ f^{n+1}(t, x_\lambda(t), u_\lambda(t)) - q_1 f^{n+2}(t, x_\lambda(t), u_\lambda(t) \} dt +$$

$$(1-\lambda)\int_{t_0}^{t_1} \{ f^{n+1}(t, x_\lambda(t), u_\lambda(t)) - q_2 f^{n+2}(t, x_\lambda(t), u_\lambda(t)) \} dt \geq$$

$$\lambda \min_{u(t) \in \Delta} [\int_{t_0}^{t_1} \{ f^{n+1}(t, x(t), u(t)) - q_1 f^{n+2}(t, x(t), u(t)) \} dt] +$$

$$(1-\lambda) \min_{u(t) \in \Delta} [\int_{t_0}^{t_1} \{ f^{n+1}(t, x(t), u(t)) - q_2 f^{n+2}(t, x(t), u(t)) \} dt] =$$

$$\lambda F(q_1) + (1 - \lambda)F(q_2)$$

Lemma 4.6.2. *The function F is continuous on R.*
The proof follows that for Theorem 4.5.2, Section 4.5.

Lemma 4.6.3. *F is a monotonic function, strict decreasing i.e., $q_1, q_2 \in R$, $q_1 < q_2$ implies $F(q_2) < F(q_1)$.*

Proof. For $q_1 < q_2$, let $(x_1(t), u_1(t))$ be solution of Problem II obtained for $q = q_1$. Considering (4.6.6), we have $\int_{t_0}^{t_1} f^{n+2}(t, x_1(t), u_1(t))dt > 0$. Hence,

$$F(q_1) = \int_{t_0}^{t_1} \{ f^{n+1}(t, x_1(t), u_1(t)) - q_1 f^{n+2}(t, x_1(t), u_1(t)) \}dt >$$

$$\int_{t_0}^{t_1} \{ f^{n+1}(t, x_1(t), u_1(t)) - q_2 f^{n+2}(t, x_1(t), u_1(t)) \} dt \geq$$

$$\min_{u(t) \in \Delta} [\int_{t_0}^{t_1} \{ f^{n+1}(t, x(t), u(t)) - q_2 f^{n+2}(t, x(t), u(t)) \} dt] = F(q_2)$$

Lemma 4.6.4. *Equation $F(q) = 0$ has an unique real solution.*
The proof is immediate using lemmas 4.6.2 and 4.6.3 and the fact that $\lim_{q \to -\infty} F(q) = +\infty$ and $\lim_{q \to \infty} F(q) = -\infty$.

Lemma 4.6.5. *Let $u(t) \in \Delta$ and*

$$q = \int_{t_0}^{t_1} f^{n+1}(t, x(t), u(t))dt \Big/ \int_{t_0}^{t_1} f^{n+2}(t, x(t), u(t))dt$$

Then $F(q) \leq 0$.

Proof. Indeed,

$$F(q) = \min_{u(t) \in \Delta} \Big[\int_{t_0}^{t_1} \{ f^{n+1}(t, x(t), u(t)) - qf^{n+2}(t, x(t), u(t)) \}dt\Big] \leq$$

$$\int_{t_0}^{t_1} \{ f^{n+1}(t, x(t), u(t)) - qf^{n+2}(t, x(t), u(t)) \} dt = 0$$

We prove now the existence theorem for an optimal control of Problem I. Let

$$F(q_*) = F(q_*, x_*(t), u_*(t)) = \min_{u(t) \in \Delta} \int_{t_0}^{t_1} \{f^{n+1}(t, x(t), u(t)) - q_* f^{n+2}(t, x(t), u(t))\} dt$$

Theorem 4.6.1. *If $F(q_*) = 0$, then*

$$q_* = \min_{u(t) \in \Delta} \Big[\int_{t_0}^{t_1} f^{n+1}(t, x(t), u(t))dt \Big/ \int_{t_0}^{t_1} f^{n+2}(t, x(t), u(t))dt\Big] =$$

$$= \int_{t_0}^{t_1} f^{n+1}(t, x_*(t), u_*(t))dt \Big/ \int_{t_0}^{t_1} f^{n+2}(t, x_*(t), u_*(t))dt$$

and $u_*(t)$ is a solution of Problem I.

Proof. Let q_* be the solution of equation $F(q) = 0$. For $q = q_*$, there is an optimal solution for Problem II (see, Lee and Markus, *op. cit.*) and we have:

$$0 = F(q_*) = \int_{t_0}^{t_1} \{ f^{n+1}(t, x_*(t), u_*(t)) - q_* f^{n+2}(t, x_*(t), u_*(t)) \} dt$$

$$\leq \int_{t_0}^{t_1} \{ f^{n+1}(t, x(t), u(t)) - q_* f^{n+2}(t, x(t), u(t)) \} dt \quad \forall\, u(t) \in \Delta$$

Hence,

$$q_* = \int_{t_0}^{t_1} f^{n+1}(t, x_*(t), u_*(t))dt \Big/ \int_{t_0}^{t_1} f^{n+2}(t, x_*(t), u_*(t))dt \le$$

$$\int_{t_0}^{t_1} f^{n+1}(t, x(t), u(t))dt \Big/ \int_{t_0}^{t_1} f^{n+2}(t, x(t), u(t))dt \quad \forall\, u(t) \in \Delta$$

and $u_*(t)$ is a solution of Problem I with $C(u^*) = q_*$.

If $u_*(t) \in \Delta$ is an optimal solution of Problem I, then, using Theorem 4.6.1, we see that Problem I consists of determining the control $u_m(t)$ such that $|q(u_*) - q(u_m)| < \varepsilon$ ($\varepsilon > 0$, given). Further more, since F is continuous, Problem I consists of determining the control $u_l(t)$ and

$$q_l = \int_{t_0}^{t_1} f^{n+1}(t, x_l(t), u_l(t))dt \Big/ \int_{t_0}^{t_1} f^{n+2}(t, x_l(t), u_l(t))dt = C(u_l)$$

such that $|F(q_*) - F(q_l)| = |F(q_l)| < \delta(\delta > 0$, given).

In the application of the algorithm, one starts with $u_1(t) \in \Delta$. Let $C(u_1) = q_1$ such that $F(q_1) \le 0$. Make $k = 2$.

1) Determine an optimal control $u_k(t)$ which solves Problem II with $q = q_k$. Let

$$F(q_k) = \min_{u(t)\in\Delta} \Big[\int_{t_0}^{t_1} \{ f^{n+1}(t, x(t), u(t)) - q_k \int_{t_0}^{t_1} f^{n+2}(t, x(t), u(t)) \}\, dt\Big]$$

2) If $|F(q_k)| < \delta$, then stop. If $F(q_k) < 0$, then consider $u_k(t) = u_l(t)$.
3) If $F(q_k) = 0$, then $u_k(t) = u_*(t)$.
4) If $|F(q_k)| \ge \delta$, then calculate $q_{k+1} = C(u_k)$ and repeat the algorithm from Step 1) with q_{k+1} instead of q_k.

The algorithm converges towards the optimal solution of Problem I. Indeed, we show that $q_{k+1} < q_k$ for all $k = 1,2, \ldots$ for which $|F(q_k)| \ge \delta$. Lemma 4.6.5 implies that $F(q_k) < 0$, and through the way q_{k+1} was defined, we have

$$\int_{t_0}^{t_1} f^{n+1}(t, x_k(t), u_k(t))dt = q_{k+1} \int_{t_0}^{t_1} f^{n+2}(t, x_k(t), u_k(t))dt$$

Consequently,

$$F(q_k) = \int_{t_0}^{t_1} \{f^{n+1}(t, x_k(t), u_k(t)) - q_k f^{n+2}(t, x_k(t),u_k(t))\}\, dt =$$

$$= (q_{k+1} - q_k) \int_{t_0}^{t_1} f^{n+2}(t, x_k(t), u_k(t))\, dt < 0$$

Using (4.6.6) yields $q_{n+1} < q_k$.

Let's now show that $\lim_{k \to \infty} q_k = q(u_*) = q_*$. Through *reductio ad absurdum*, assume that $\lim_{k \to \infty} q_k = q > q_*$. In this case, we have a sequence $u_k(t)$ with which we determine q_k such that $\lim_{k \to \infty} F(q_k) = F(q) = 0$. But F is monotonic strict decreasing, hence $0 = F(q) < F(q_*) = 0$ which is a contradiction. Hence, $\lim_{k \to \infty} F(q_k) = F(q_*)$, and Lemma 4.6.2 yields $\lim_{k \to \infty} q_k = q_*$.

4.7. HISTORICAL NOTES. COMMENTARIES

We are going to outline some additional aspects of the nonlinear fractional programming. The optimality conditions have been also given for other classes of fractional functions. Thus, Mond [671] gave necessary and sufficient conditions for the existence of an optimal solution of the problem:

$$\min \left\{ \frac{f(x) - (x'Bx)^{1/2}}{g(x) + (x'Dx)^{1/2}} \mid h(x) \le 0 \right\} \tag{4.7.1}$$

where f,g and h are differentiable functions from R^n into R, R and R^m, respectively, B and D are $n \times n$ symmetric positive semi-definite matrices, and $g(x) + (x'Dx)^{1/2} > 0$ for all feasible x.

Mjelde [657] proved the sufficiency of the Kuhn-Tucker conditions for the following problem with each variable occurring in a single constraint

$$\max \left\{ \frac{f(x)}{g(x)} \mid H_j(x_j) \le h_j,\ j = 1, ..., J \right\}, \tag{4.7.2}$$

where $x = (x_{jk}) \ge 0$ is a $J \times K$ matrix and $x_j = (x_{j1}, ..., x_{jk})$ is the j-th row of x. The function f is concave and differentiable with continuous derivative, and the functions g ($g(x) > 0$ for all $x \ge 0$) and $H_j(x_j)$ ($H_j(0) = 0$) for $j = 1, ..., J$ are convex and differentiable and $h_j > 0$ for $j = 1, ..., J$.

Aggarwal and Saxena [25] presented the optimality conditions for the problem

$$\min \{[f(x) + (x'Bx)^{1/2}]/h(x) \mid g(x) \ge 0\}, \tag{4.7.3}$$

where f is convex and differentiable, g and h are concave and differentiable, whereas B is a symmetric positive semi-definite matrix, $x \in S = \{ x: g_i(x) \ge 0,\ i = 1, ..., m \}$ is a non-empty, bounded convex subset of R^n, B is a $n \times n$ symmetric positive semi-definite matrix, f is convex and differentiable, h and g are concave and differentiable, and $h(x) > 0$ for all $x \in S$.

Singh [853] considers the problem

$$\min \left\{ \frac{f(x)}{h(x)} \mid g(x) \le 0 \right\} \tag{4.7.4}$$

and establishes the Fritz John and Kuhn-Tucker optimality conditions with or without the assumption of differentiability for the functions f, g and h. Such conditions are obtained through the change of variable method by which the fractional problem reduces to a non-fractional problem for which such optimality conditions exist. Xu [1058] showed in an example that Kuhn-Tucker Saddle-Point Necessary Optimality Criterion is not true.

Sinha and Aylawadi [864] obtained the necessary and sufficient optimality conditions for the problem

$$\max \left\{\frac{f(x)}{g(x)} \mid h_i(x) \le 0, \quad i = 1, ..., m, \quad p'x + (x'Bx)^{1/2} \le 1\right\} \tag{4.7.5}$$

where f, g and h_i $(i = 1, ..., m)$ are differentiable functions, B is a $n \times n$ symmetric and positive semi-definite matrix, and $g(x) > 0$. Note that problem (4.7.5) has a non-differentiable constraint. Also, the authors extended the analysis to programs with more than one non-differentiable constraint functions.

The change of variable method presented in Section 4.2 was used by various authors to solve some other nonlinear fractional programming problems. First, recall that Craven and Mond [331] considered the problem presented in Section 4.2 under some more general conditions, i.e. they considered that S is a convex cone in R^m, and, for the transformed problem, that the domain T is also a convex cone. Our results are recovered by taking $S = R^m_+$. Swarup [941] applied the change of variable method to the solution of the problem

$$\max \left\{ \frac{x'Cx + d'x + \alpha}{x'Ex + f'x + \beta} \mid Ax \le b, x \ge 0 \right\} \tag{4.7.6}$$

where E is a positive semi-definite matrix. Through such a transformation, the solution of problem (4.7.6) reduces to the solution of, at most, two problems of quadratic programming with a quadratic constraint. Patkar *et al.* [715], through a change of variable, reduced the solution of the quadratic fractional programming problem

$$\max \left\{ \frac{(c_1'x + \alpha)(c_2'x + \beta)}{(c_3'x + \gamma)(c_4'x + \delta)} \mid Ax \le b, \quad x \ge 0 \right\} \tag{4.7.7}$$

to a quadratic programming problem. Aggarwal [8] and Stancu-Minasian [896] utilized the change of variables for solving the problem

$$\max \left\{ \frac{c'x - (x'Bx)^{1/2} + \alpha}{d'x + \beta} \mid Ax \le b, \quad x \ge 0 \right\} \tag{4.7.8}$$

where B is a positive semi-definite matrix. This problem appears in connection with some stochastic programming problems.

Pardalos [702] gives an algorithm for the quasiconcave nonlinear fractional programming problem

$$\min_{x \in P} \frac{c'x + x'Qx + \alpha}{d'x + \beta} \tag{4.7.9}$$

where $P = \{x: Ax \leq b; x \geq 0\}$ is a bounded polyhedron in R^n, A is an $m \times n$ matrix, $b \in R^m$, $\alpha, \beta \in R$, $c, d \in R^n$, Q is a symmetric matrix and $d'x + \beta > 0$ for all $x \in P$.
The algorithm is based on ranking the vertices of a linear fractional programming problem (see, Storøy [935]) and techniques from global optimization.

For other types of nonlinear fractional programming problems reducible to non-fractional nonlinear problems, the reader can consult Stancu-Minasian [910].
Ţigan [995] applied the change of variable for solving the problems

$$\max_x h\left[\frac{ax + e}{dx + r}, \frac{cx + p}{dx + r}\right]$$

and (4.7.10)

$$\max_x \min_i h\left[\frac{a^i x + e^i}{dx + r}, \frac{cx + p}{dx + r}\right]$$

From among the direct methods for solving nonlinear fractional programming problems, we have presented only the linearization method. In the specialized literature, other direct methods also exists.

Gupta and Swarup [453] considered the nonlinear problem

$$\max \left\{\frac{F(x_1, ..., x_n)}{G(x_1, ..., x_n)} \mid g_i(x_1, ..., x_n) = b_i,\ i=1, ..., m,\ L_j \leq x_j \leq U_j,\ j=1, ..., n\right\}.$$

Their proposed method consists in the linearization of the functions F, G, g_i around the fixed point x^i. By Taylor series expansion and solution of the linear programming problems the solution x^{i+1} is found. Then the linearization procedure for the functions F, G, g_i is repeated around x^{i+1}. We can stop at the point x^p which is an approximate solution of the initial problem. The linearization of the both the denominator and the numerator was used by Meister and Oettli [651] for solving a problem of maximization of the transmission speed through an information channel.

Thang [975] considered a minimization problem for a pseudoconvex function on a domain determined by linear constraints using a variant of the Beale algorithm. The algorithm is then illustrated on functions of the form (4.7.6). Other methods for quadratic fractional programming are given in references [537] and [538].

Sharma [836] utilized the method of feasible directions to solve nonlinear fractional programming problems. Aggarwal and Savita Arora [19] also dealt with the problem (4.7.6). Through a change of variable, the problem is transformed in a separable quadratic fractional programming problem for which a linearization method is applied.

The parametric method considered by us in Section 4.5 was used by Ţigan [990] to the solution of some fractional optimization problems in topological spaces and for solving dynamic fractional programming problems [993]. Closely connected with the parametric method is the parametric analysis of nonlinear fractional programming problems. In this case, the problem P is associated with a parametric problem (not necessary of the form of problem $Q(\lambda)$) by considering the numerator or the denominator of the objective function as a parameter. For example, see the solution given by Falk [386] for the maximization problem of the signal-to-noise ratio (Chapter 1, Example 8). Wolf [1051] did a parametric analysis of a fractional program with the solution set depending on a parameters vector. The following problem was considered

$$Q(\alpha) = \max \{q(x) = f(x)/g(x) \mid \alpha \in M, \ x \in X(\alpha)\} \tag{4.7.11}$$

where $X(\alpha) \subseteq R^n$ is a compact and convex set for all $\alpha \in M$, $M = \{\alpha \in R^k \mid X(\alpha) \neq \varnothing\}$ is compact, f continuous, g continuous and convex and $g(x) > 0$ for all $x \in X(\alpha)$ and any $\alpha \in M$.

It is shown that Q is a continuous function on M and that can be determined in two steps. In the first step, one makes the parametric analysis of the auxiliary problem

$$F(\alpha,\gamma) = \max \{ f(x) \mid x \in X(\alpha), (\alpha \in M), g(x) \le \gamma, \gamma \in N_\alpha = \{g(x) \mid x \in X(\alpha)\}\},$$

and, in the second step, one shows that

$$Q(\alpha) = \max_{\gamma \in N_\alpha} F(\beta,\gamma)/\gamma, \quad \alpha \in M.$$

Among other nonlinear fractional programming problems studied in the literature we can mention the problem

$$\max \{ F(x) = \ln[f(x) + \alpha] / \ln[g(x) + \beta] \mid g_i(x) \le 0, \ i = 1, ..., m \}$$

considered by Stancu-Minasian [904],and the problem of maximization of a ratio of optimization problems

$$\max_{x \in R^n} \{\min_y P(x,y) / \max_z Q(x,z) \mid h(x) \le 0, \ y \in C_1, \ z \in C_2\}$$

considered by Craven and Mond [337].

CHAPTER 5

DUALITY IN FRACTIONAL PROGRAMMING

In mathematical programming, the duality theory takes a central place. In this theory, to a given problem, named "primal", one associates another problem, named "dual", and the relationship between the two problems is used to highlight the properties of the optimal solutions of both problems. An important consequence of the duality principle is that if one of the problems has a finite optimum, then the other problem also has a finite optimum, and the values of the two optimums coincide.

For the nonlinear programming problem, the dual problem has been defined in various ways. Thus, we meet the duality according to Wolfe, according to Legendre etc. If we assume that the primal problem is of minimum, the dual problem according to Wolfe consists in maximizing the Lagrangian function.

For the fractional programming problem, the dual constructed according to the definition given by Wolfe is of no practical interest. Although the converse duality theorem takes place, the weak duality theorem and the direct duality theorem do not take place. The following example shows that even for the linear fractional programming problem, the duality theorems are not true. If, for problem max $\{ 1/x \mid x > 0, \quad 1 \le x \le 2 \}$, we construct the dual according to Wolfe then we get the problem

$$\min \left\{ \frac{1}{x} - u_1(-x + 1) - u_2(x - 2) \mid x > 0 \, , \right.$$
$$\left. u_1, \; u_2 \ge 0 \, , \; -\frac{1}{x^2} + u_1 - u_2 = 0 \right\} .$$

Note that the primal problem has the optimal solution $x = 1$, while the dual problem does not have optimal solution being unbounded from below. Thus, neither the weak duality theorem, nor the direct duality theorem take place. Hence, in fractional programming, we need to define another dual mode. Another natural way to construct the dual is that of transforming the fractional problem into a non-fractional problem the dual of which is constructed in the classical way. We are going to exemplify this in Section 5.1, for the linear fractional programming problem, and in Section 5.3 for the nonlinear fractional programming problem.For the problem

$$v = \sup \{ f(x) = \frac{n(x)}{d(x)} \mid x \in X \subseteq R^n,\ g(x) \geq 0 \},$$

Gol'shtein [418] introduced the fractional Lagrangian $L(x, u) = \frac{n(x)}{d(x)} + u\frac{g(x)}{d(x)}$ and used it to define the fractional dual. Bector [125] used the change of variable and the Lagrangian to construct the dual to a linear fractional programming problem.

In Section 5.1 we are going to introduce another dual problem for the linear fractional problem. This dual has the property that is also a linear fractional programming problem. In Section 5.2 we construct the dual and prove the duality theorems for a particular type of linear fractional programming problem. In Section 5.3, dual to a nonlinear fractional programming problem is constructed in two different ways. Section 5.4 shows the dual for a particular nonlinear fractional programming problem. In Section 5.5, the dual problems for linear and nonlinear fractional programming problems are obtained *via* duality in geometric programming.

5.1. DUALITY FOR LINEAR FRACTIONAL PROGRAMMING

In this section, we are going to construct, in several ways, the dual to the linear fractional programming problem:

$$max\ F(x) = \frac{c'x + c_0}{d'x + d_0}, \tag{5.1.1}$$

subject to

$$Ax \leq b, \tag{5.1.2}$$
$$x \geq 0, \tag{5.1.3}$$

where A is a $m \times n$ $(m < n)$ matrix of rank m, x, c, and d are column vectors with n components, b is a column vector with m components, c_0 and d_0 are arbitrary constants. Make the assumption that, on the domain of feasible solutions $S = \{ x \in R^n \mid Ax \leq b,\ x \geq 0 \}$, assumed to be a regular set, the denominator keeps constant sign, for example, positive. Also, we assume that F is not constant on S.

One way to construct the dual of problem (5.1.1) - (5.1.3) is to transform it in a linear programming problem for which the dual can be constructed. Using the change of variable $y = tx$ (Section 3.4, Chapter 3), the problem (5.1.1) - (5.1.3) becomes

$$max\ \{ G(y, t) = c'y + c_0t \mid Ay - bt \leq 0, \tag{5.1.4}$$
$$d'y + d_0t = 1,\quad y \geq 0,\quad t \geq 0 \}$$

According to Lemma 3.4.1, Section 3.4, Chapter 3, we have $t > 0$. The dual of problem (5.1.4) is easily written as

$$min\ H(u, u_{m+1}, u_{m+2}) = u_{m+1} - u_{m+2} ,$$

subject to

$$A'u + d(u_{m+1} - u_{m+2}) \geq c , \quad b'u - d_0(u_{m+1} - u_{m+2}) \leq - c_0 ,$$
$$u = (u_1 , \ldots , u_m) \geq 0 , \quad u_{m+1} \geq 0 , \quad u_{m+2} \geq 0 ,$$

or, denoting $z = u_{m+1} - u_{m+2}$ (z without sign restriction),

$$\min \{ H(u, z) = z \mid A'u + dz \geq c , \ b'u - d_0 z \leq - c_0 , \ u \geq 0 \}. \tag{5.1.5}$$

In this dual, for the complementary slackness theorem to take place at the optimal solutions, the inequality $b'u - d_0 z \leq - c_0$ must be of the form $b'u - d_0 z = - c_0$.

Another way to construct the dual of problem (5.1.1) - (5.1.3) is that of dualizing the linear parametric problem

$$P_\lambda \quad max \{ (c'x + c_0) - \lambda(d'x + d_0) \mid Ax \leq b , \ x \geq 0 \}.$$

According to Theorem 3.5.2, Chapter 3, $\bar{x}$ is a solution of problem (5.1.1) - (5.1.3) and the objective function has the optimal value λ, if and only if $\bar{x}$ is solution of problem P_λ and $H(\lambda) = 0$ ($H(\lambda)$ is the maximum value of the objective function from P_λ). Note that, since $H(\lambda)$ is a monotonically decreasing function, the problem (5.1.1) - (5.1.3) consists of determining λ for which $H(\lambda) = 0$ (or, to minimize λ such that $H(\lambda) \leq 0$).

The dual to problem P_λ is

$$min \{ b'u + c_0 - \lambda d_0 \mid A'u \geq c - \lambda d , \ u \geq 0 \} .$$

Following the above observation, this dual is written as

$$min \{ \lambda \mid A'u \geq c - \lambda d , \ b'u + c_0 - \lambda d_0 \leq 0 , \ u \geq 0 \}$$

which coincide with the dual (5.1.5) constructed through the change of variable.

Now, following reference [831], we are going to construct another dual for problem (5.1.1) - (5.1.3). This dual will have the property to also be a fractional programming problem. The dual has the form

$$min\ I(u, v) = \frac{c'u + c_0}{d'u + d_0} , \tag{5.1.6}$$

subject to

$$c \cdot d'u - d \cdot c'u - A'v \leq c_0 d - d_0 c , \tag{5.1.7}$$

$$c_0 \cdot d'u - d_0 \cdot c'u + b'v \leq 0 , \tag{5.1.8}$$

$$u \geq 0 , \ v \geq 0 , \ u \in R^n , \ v \in R^m . \tag{5.1.9}$$

Let's prove the duality theorems. Let T be the feasible domain for the dual problem.

Theorem 5.1.1. *If* $x \in S$ *and* $(u, v) \in T$, *then*

$$\max_{x \in S} F(x) \le \min_{(u,v) \in T} I(u, v) \quad . \tag{5.1.10}$$

Proof. Assume that if $S = \emptyset$, then $\max_{x \in S} F(x) = -\infty$, and if $T = \emptyset$, then $\min_{(u,v) \in T} I(u, v) = +\infty$. In these conditions the inequality (5.1.10) is true. Let's prove the inequality (5.1.10) in the case in which $S \ne \emptyset$ and $T \ne \emptyset$. Multiply (5.1.7) by $x \ge 0$ and get

$$c'x \cdot d'u - d'x \cdot c'u - x'A'v \le c_0 \cdot d'x - d_0 \cdot c'x \; . \tag{5.1.11}$$

Also, multiply (5.1.2) by v and, using (5.1.8), get

$$c_0 \cdot d'u - d_0 \cdot c'u \le - b'v \le - x'A'v \; . \tag{5.1.12}$$

Equations (5.1.11) and (5.1.12) imply

$$c'x \cdot d'u - d'x \cdot c'u + c_0 \cdot d'u - d_0 \cdot c'u \le c_0 \cdot d'x - d_0 \cdot c'x$$

or,

$$(c'x + c_0)(d'u + d_0) \le (c'u + c_0)(d'x + d_0) \; .$$

Hence,

$$F(x) = \frac{c'x + c_0}{d'x + d_0} \le \frac{c'u + c_0}{d'u + d_0} = I(u, v).$$

Thus,

$$\max_{x \in S} F(x) \le \min_{(u,v) \in T} I(u, v).$$

Corollary 5.1.1. *If* $x \in S$ *and* $(u, v) \in T$ *with the property that* $F(x) = I(u, v)$, *then* x *is optimal solution of problem* (5.1.1) - (5.1.3) *and* (u, v) *is an optimal solution of problem* (5.1.6) - (5.1.9).

Theorem 5.1.2. *If* x^* *is the optimal solution for problem* (5.1.1) - (5.1.3), *then an optimal solution* $(u^*, v^*) \ge 0$ *for problem* (5.1.6) - (5.1.9) *exists, and* $F(x^*) = I(u^*, v^*)$.

Proof. Let $\lambda = \dfrac{c'x^* + c_0}{d'x^* + d_0}$, and construct the auxiliary linear parametric program

$$\max \{ (c'x + c_0) - \lambda(d'x + d_0) \mid Ax \le b \; , \; x \ge 0 \}. \tag{5.1.13}$$

According to Theorem 3.5.2, Chapter 3, x^* is also an optimal solution of problem (5.1.13) and the optimal value is equal to zero. The dual of problem (5.1.13) has the form

$$\min \{ b'v + c_0 - \lambda d_0 \mid A'v \ge c - \lambda d \; , \; v \ge 0 \; , \; v \in R^m \} \tag{5.1.14}$$

Denote its optimal solution by v_1^*. The optimal values of problems (5.1.13) and (5.1.14) are equal, hence

$$b'v_1^* + c_0 - \lambda d_0 = 0 \; . \tag{5.1.15}$$

Consider $u^* = x^*$ and $v^* = v_1^*(d'x^* + d_0)$. Let's prove that $(u^*, v^*) \in T$. Indeed,

$$c \cdot d'u^* - d \cdot c'u^* - A'v^* \leq c \cdot d'x^* - d \cdot c'x^* - (c - \lambda d)(d'x^* + d_0) =$$
$$= c \cdot d'x^* - d \cdot c'x^* - c(d'x^* + d_0) + d(c'x^* + c_0) = c_0 d - d_0 c$$

i.e. inequality (5.1.7). Also, using (5.1.15), through multiplication by $d'x^* + d_0$ one gets

$$b'v^* + c_0(d'x^* + d_0) - d_0(c'x^* + c_0) = b'v^* + c_0 \cdot d'u^* - d_0 \cdot c'u^* = 0$$

i.e. (5.1.8). The definition of u^* and v^* yields $(u^*, v^*) \geq 0$, *i.e.* (5.1.9). Hence, $(u^*, v^*) \in T$. Even more,

$$F(x^*) = \frac{c'x^* + c_0}{d'x^* + d_0} = \frac{c'u^* + c_0}{d'u^* + d_0} = I(u^*, v^*) .$$

Corollary 5.1.1 yields that (u^*, v^*) is optimal solution of the problem (5.1.6) - (5.1.9).

Theorem 5.1.3. *If* (u^*, v^*) *is the optimal solution of problem* (5.1.6) - (5.1.9), *then an optimal solution,* x^*, *of problem (5.1.1) - (5.1.3) exists, and*

$$F(x^*) = I(u^*, v^*) .$$

Proof. Let

$$\lambda = \frac{c'u^* + c_0}{d'u^* + d_0}$$

and construct the auxiliary linear parametric program

$$min\ (c'u + c_0) - \lambda(d'u + d_0) , \tag{5.1.16}$$

subject to

$$c \cdot d'u - d \cdot c'u - A'v \leq c_0 d - d_0 c ,$$
$$c_0 \cdot d'u - d_0 \cdot c'u + b'v \leq 0 \quad u \geq 0 , \quad v \geq 0 . \tag{5.1.17}$$

According to Theorem 3.5.3, Chapter 3, (u^*, v^*) is an optimal solution of problem (5.1.16) - (5.1.17) and the optimal value is zero. The dual of problem (5.1.16) - (5.1.17) has the form

$$max\ ((- c_0 d + d_0 c)'y + c_0 - \lambda d_0) \tag{5.1.18}$$

subject to

$$a) \quad (-cd' + dc')'y + (-c_0 d + d_0 c)t \leq c - \lambda d$$
$$b) \quad Ay - bt \leq 0 , \quad c) \quad y \geq 0, \ t \geq 0, \ y \in R^n, \ t \in R . \tag{5.1.19}$$

Denote by (y^*, t^*) the optimal solution of problem (5.1.18) - (5.1.19). The optimal values of problems (5.1.16) - (5.1.17) and (5.1.18) - (5.1.19) are equal, hence

$$- c_0 \cdot d' y^* + d_0 \cdot c'y^* + c_0 - \lambda d_0 = 0 . \tag{5.1.20}$$

We have $t^* \neq 0$. Indeed, if $t^* = 0$, then (5.1.19) yields $Ay^* \leq 0$, $y^* \geq 0$. This yields that for $x \in S$ and $\mu > 0$ we have $A(x + \mu y^*) \leq b$, $x + \mu y^* \geq 0$ *i.e.* $x + \mu y^* \in S$ which contradicts the fact that S is a bounded set.

We cannot have $t^* = 0$ and $y^* = \mathbf{0}$ since, in this case, (5.1.20) yields $c_0 - \lambda d_0 = 0$, and (5.1.19.a) yields $c \geq \lambda d$. For $x \in S$, we have

$$c'x \geq \lambda d'x \quad \text{and} \quad c_0 = \lambda d_0$$

hence

$$(c'x + c_0) \geq \lambda(d'x + d_0)$$

i.e. $F(x) \geq \lambda$. Theorem 5.1.1 yields

$$F(x) \leq I(u^*, v^*) = \lambda$$

hence $F(x) = \lambda$. Since x was chosen arbitrarily in S, it follows that F is a constant on S, which contradicts the assumption made initially for problem (5.1.1) - (5.1.3). Thus, $t^* > 0$. Consider $x^* = y^*/t^*$. Evidently $x^* > \mathbf{0}$, and (5.1.19.b) by division with t^*, yields $Ax^* \leq b$, which means that $x^* \in S$. The complementary slackness theorem for the dual problems (5.1.16) - (5.1.17) and (5.1.18) - (5.1.19) gives

$$c'y^*d'u^* - d'y^*c'u^* - y^{*\prime}A'v^* - c_0d'y^* + d_0c'y^* = 0\,, \tag{5.1.21}$$

$$c_0t^*d'u^* - d_0t^*c'u^* + t^*b'v^* = 0\,, \tag{5.1.22}$$

$$-d'u^*c'y^* + c'u^*d'y^* - c_0t^*d'u^* + d_0t^*c'u^* - c'u^* + \lambda d'u^* = 0\,, \tag{5.1.23}$$

$$y^{*\prime}A'v^* - t^*b'v^* = 0 \tag{5.1.24}$$

Relations (5.1.21), (5.1.22) and (5.1.24), through addition, give

$$c'y^*d'u^* - d'y^*c'u^* - c_0d'y^* + d_0c'y^* + c_0t^*d'u^* - d_0t^*c'u^* = 0$$

or

$$(c'y^* + c_0t^*)(d'u^* + d_0) = (d'y^* + d_0t^*)(c'u^* + c_0)\,.$$

Hence,

$$\frac{c'y^* + c_0t^*}{d'y^* + d_0t^*} = \frac{c'u^* + c_0}{d'u^* + d_0}\,.$$

It follows that

$$F(x^*) = \frac{c'x^* + c_0}{d'x^* + c_0} = \frac{c'u^* + c_0}{d'u^* + d_0} = I(u^*, v^*)$$

and, according to Corollary 5.1.1, x^* is optimal solution of problem (5.1.1) - (5.1.3).

5.2 DUALITY FOR A PARTICULAR LINEAR FRACTIONAL PROGRAMMING PROBLEM

Consider the problem

$$\min F(x) = \frac{(c'x)^2}{d'x} \tag{5.2.1}$$

subject to

$$Ax \geq b \tag{5.2.2}$$

$$x \geq \mathbf{0} \tag{5.2.3}$$

where A is a $m \times n$ matrix, x, c and d are column vectors with n components, b is a column vector with m components. Let $S = \{x \mid Ax \geq b,\ x \geq 0\}$ be the domain of feasible solutions. On domain S, assume that $c'x \geq 0$ and $d'x > 0$. We show that the dual of problem (5.2.1) - (5.2.3) is the problem

$$\max G(v, t) = b'v \tag{5.2.4}$$

subject to

$$A'v + dt^2 \leq 2ct \tag{5.2.5}$$

$$v \geq 0, \quad t \geq 0 \tag{5.2.6}$$

Let $T = \{(v, t) \mid v \geq 0,\ t \geq 0,\ A'v + dt^2 \leq 2ct\}$ be the domain of feasible solutions. The function F, defined in (5.2.1), is convex (see Section 2.5, Chapter 2, case D).

Theorem 5.2.1. *If $x \in S$ and $(v, t) \in T$, then*

$$\inf_{x \in S} F(x) \geq \sup_{(v,t) \in T} G(v, t) \tag{5.2.7}$$

Proof. As in Section 5.1, assume that, if $S = \emptyset$, then $\inf F(x) = +\infty$, and if $T = \emptyset$, then $\sup G(v, t) = -\infty$. In these conditions, the inequality (5.2.7) is true. Let's prove it in the case when $S \neq \emptyset$ and $T \neq \emptyset$. Right-multiply (5.2.5) by $x \geq 0$ and get

$$x'A'v + x'dt^2 \leq 2x'ct \tag{5.2.8}$$

Left-multiply (5.2.2) by $v \geq 0$ and get

$$x'A'v \geq b'v \tag{5.2.9}$$

Equations (5.2.8) and (5.2.9) yield

$$b'v + d'xt^2 - 2c'xt \leq 0$$

or

$$b'v + [t(d'x)^{1/2} - (c'x)(d'x)^{-1/2}]^2 - \frac{(c'x)^2}{d'x} \leq 0$$

Hence $b'v - \dfrac{(c'x)^2}{d'x} \leq 0$, *i.e.* $b'v \leq \dfrac{(c'v)^2}{d'x}$. Thus, $\inf\limits_{x \in S} F(x) \geq \sup\limits_{(v,t) \in T} G(v, t)$ which is the inequality (5.2.7).

Corollary 5.2.1. *If $x \in S$ and $(v, t) \in T$ such that $F(x) = G(v, t)$, then x is optimal solution of problem (5.2.1) - (5.2.3), and (v, t) is optimal solution of problem (5.2.4) - (5.2.6).*

Theorem 5.2.2. *If x^* is optimal solution of problem (5.2.1) - (5.2.3), then there is $(v^*, t^*) \geq 0$ an optimal solution of problem (5.2.4) - (5.2.6) such that $F(x^*) = G(v^*, t^*)$.*

Proof. Consider x^*, optimal solution for problem (5.2.1) - (5.2.3). For this solution the Kuhn-Tucker conditions are satisfied, hence there is $(w^*, y^*) \geq 0$ such that

$$Ax^* - b \geq 0 \tag{5.2.10}$$

$$(Ax^* - b)'w^* = 0 \tag{5.2.11}$$

$$-A'w^* + \left[\frac{2(c'x^*)(d'x^*)c - (c'x^*)^2 d}{(d'x^*)^2}\right] - y^* = 0 \tag{5.2.12}$$

$$-x^{*\prime}A'w^* + \left[\frac{2(c'x^*)^2(d'x^*) - (c'x^*)^2(d'x^*)}{(d'x^*)^2}\right] - x^{*\prime}y^* = 0 \tag{5.2.13}$$

$$x^{*\prime}y^* = 0 \tag{5.2.14}$$

Since $y^* \geq 0$, Equation (5.2.12) yields

$$A'w^* + \frac{(c'x^*)^2}{(d'x^*)^2}d \leq 2\frac{c'x^*}{d'x^*}c \tag{5.2.15}$$

or, making $t^* = \dfrac{c'x^*}{d'x^*}$,

$$A'w^* + dt^{*2} \leq 2ct^* \tag{5.2.16}$$

Equation (5.2.16) yields that $(v^* = w^*, t^*) \in T$. Equation (5.2.11) yields

$$x^{*\prime}A'w^* = b'w^* = b'v^* \tag{5.2.17}$$

In Equation (5.2.13), use relations (5.2.14) and (5.2.17), to get

$$b'v^* - \frac{(c'x^*)^2(d'x^*)}{(d'x^*)^2} = 0 \text{ or, } b'v^* = \frac{(c'x^*)^2}{d'x^*}$$

The proof is complete.

To prove the converse duality theorem, assume that the constraints (5.2.5) and (5.2.6) verify the Kuhn-Tucker constraint qualification.

Theorem 5.2.3. *If (v^*, t^*) is an optimal solution of problem (5.2.4) - (5.2.6), then there is an optimal solution x^* of problem (5.2.1) - (5.2.3) and $F(x^*) = G(v^*, t^*)$.*

Proof. Since (v^*, t^*) is optimal solution of the dual problem, the Kuhn-Tucker conditions are satisfied, and hence there is $(\bar{x}, \bar{y}) \geq 0$ such that:

$$-b + A\bar{x} = \bar{y} \geq 0 \tag{5.2.18}$$

$$(b - A\bar{x})'v^* = 0, \ \bar{y}'v^* = 0 \tag{5.2.19}$$

$$-2t^*d'\bar{x} + 2c'\bar{x} = 0 \tag{5.2.20}$$

$$(A'v^* + dt^{*2} - 2ct^*)'\bar{x} = 0 \tag{5.2.21}$$

Equation (5.2.18) yields that, by considering $x^* = \bar{x}$, we get a feasible solution for the primal problem. Additionally, Equations (5.2.19) and (5.2.21) yield:

$$b'v^* = \bar{x}'A'v^* = 2c'\bar{x}t^* - t^{*2}d'\bar{x} \tag{5.2.22}$$

Or, using (5.2.20), yields

$$2c'\bar{x}t^* - t^{*2}d'\bar{x} = 2\frac{(c'\bar{x})}{d'\bar{x}}(c'\bar{x}) - \frac{(c'\bar{x})^2}{(d'\bar{x})^2}(d'\bar{x}) = \frac{(c'\bar{x})}{d'\bar{x}} \tag{5.2.23}$$

Equations (5.2.22) and (5.2.23) yield $b'v^* = \frac{(c'x^*)^2}{d'x^*}$. Corollary 5.2.1 yields that x^* is optimal solution of the primal problem, and $F(x^*) = G(v^*, t^*)$. The proof is complete.

5.3 DUALITY FOR NONLINEAR FRACTIONAL PROGRAMMING

We introduce the dual problem for the nonlinear fractional programming problem in two ways. The first way is based on the connection between the nonlinear fractional problem and a certain convex parametric programming problem. The second way is to use the Lagrangian function.

Consider the nonlinear fractional programming problem

$$\min_{x \in S} f(x) = \frac{N(x)}{D(x)} \tag{5.3.1}$$

where : a) $x \in S = \{x \mid g_i(x) \leq 0,\ i=1, ..., m\}$, b) g_i are convex and differentiable functions on R^n, c) N is a convex and differentiable function on S, and D is concave and differentiable on S. Additionally, assume that d) S is a bounded set, e) $D(x) > 0$ for any $x \in S$, f) $\min_{x \in S} N(x) \geq 0$, g) g_i satisfy certain constraint qualifications. Associate to problem (5.3.1) the following parametric programming problem

$$F(\lambda) = \min_{x \in S} N(x) - \lambda D(x) \tag{5.3.2}$$

where λ is a scalar non-negative parameter. According to Theorem 4.5.1, Chapter 4, x^* is an optimal solution for problem (5.3.1) if and only if $\min_{x \in S} N(x) - \lambda^* D(x) = F(\lambda^*) = 0$, where $\lambda^* = \frac{N(x^*)}{D(x^*)}$. Theorem 4.5.2, Chapter 4, yields that $F(\lambda)$ is a continuous function, concave and strict decreasing on R_+. Hence, F^{-1} exists, is continuous and decreasing, and $\lambda^* = F^{-1}(0)$. The dual according to Wolfe of problem (5.3.2) is:

$$\max_{(x,u) \in T_\lambda} N(x) - \lambda D(x) + \sum_{i=1}^{m} u_i g_i(x) \tag{5.3.3}$$

where

$$T_\lambda = \{ (x, u) \mid \nabla N(x) - \lambda \nabla D(x) + \sum_{i=1}^{m} u_i \nabla g_i(x) = 0,\ u_i \geq 0 \}.$$

Define now the following nonlinear program and show that it is the dual of problem (5.3.1):

$$\max \lambda \tag{5.3.4}$$

subject to

$$\nabla N(x) + \sum_{i=1}^{m} u_i \nabla g_i(x) = \lambda \nabla D(x) \tag{5.3.5}$$

$$N(x) - \lambda D(x) + \sum_{i=1}^{m} u_i g_i(x) \geq 0 \tag{5.3.6}$$

$$u_i \geq 0, \ i = 1, ..., m, \ \lambda \geq 0 \tag{5.3.7}$$

Theorem 5.3.1. *If* x^* *is optimal solution of problem (5.3.1), then there is* $(u^*, \lambda^*) \in R^m \times R_+$ *such that* (x^*, u^*, λ^*) *is the optimal solution of problem (5.3.4) - (5.3.7) and* $\lambda^* = \dfrac{N(x^*)}{D(x^*)}$.

Proof. Since (5.3.3) is the dual of problem (5.3.2), it follows that, if x_λ is optimal solution of problem (5.3.2), then there is $u_\lambda \in R^m$ with the property that (x_λ, u_λ) is optimal solution for (5.3.3) and

$$F(\lambda) = \min_{x \in S} (N(x) - \lambda D(x)) = \max_{(x,u) \in T_\lambda} (N(x) - \lambda D(x) + \sum_{i=1}^{m} u_i g_i(x)) \text{ for any } \lambda \geq 0.$$

Let now x^* be the optimal solution of problem (5.3.1). According to Theorem 4.5.1, x^* is optimal solution of problem (5.3.2) for $\lambda = \lambda^* = F^{-1}(0)$ and $\lambda^* = \dfrac{N(x^*)}{D(x^*)}$. The duality between (5.3.2) and (5.3.3) yields that there is $u^* \in R^m$ such that (x^*, u^*) is optimal solution of problem (5.3.3) for $\lambda = \lambda^*$, and we have

$$\max_{(x,u) \in T_{\lambda^*}} (N(x) - \lambda^* D(x) + \sum_{i=1}^{m} u_i g_i(x)) = F(\lambda^*) = 0 \tag{5.3.8}$$

But F is a strict decreasing monotonic function, hence

$$\max_{(x,u) \in T_\lambda} \{ \lambda \mid N(x) - \lambda D(x) + \sum_{i=1}^{m} u_i g_i(x) \geq 0 \} = \lambda^*,$$

i.e., there is (x^*, u^*, λ^*) optimal solution of problem (5.3.4)-(5.3.7) and $\lambda^* = \dfrac{N(x^*)}{D(x^*)}$.

Theorem 5.3.2. *If* x^* *is an optimal solution of problem (5.3.1) and* $(\bar{x}, \bar{u}, \bar{\lambda})$ *is an optimal solution of problem (5.3.4)-(5.3.7), then: a)* $\bar{\lambda} = \dfrac{N(x^*)}{D(x^*)}$, *b) If* $N(x) - \lambda D(x) + \sum_{i=1}^{m} u_i g_i(x)$ *is strict convex in point* $\bar{x}$, *then* $\bar{x} = x^*$.

Proof. Let x^* and $(\bar{x}, \bar{u}, \bar{\lambda})$ be the optimal solutions discussed in the statement of the theorem. Theorem 5.3.1 and the weak duality theorem for problems (5.3.1) and (5.3.4) - (5.3.7) yield:

a$_1$) $\bar{\lambda} = \dfrac{N(x^*)}{D(x^*)}$,

b$_1$) $N(\bar{x}) - \bar{\lambda}D(\bar{x}) + \sum_{i=1}^{m} \bar{u}_i g_i(\bar{x}) = 0$,

c$_1$) $\max_{(x,u) \in T_\lambda} (N(x) - \bar{\lambda}D(x) + \sum_{i=1}^{m} u_i g_i(x)) \leq 0$

Hence, $(\bar{x}, \bar{u})$ is optimal solution of problem (5.3.3) for $\lambda = \bar{\lambda}$. The fact that $N(x) - \lambda D(x) + \sum_{i=1}^{m} u_i g_i(x)$ is strict convex in point $\bar{x}$, and the converse duality theorem yield that $\bar{x} = x^*$.

Restart the consideration of problem (5.3.1). If the function D is only linear, then condition f) can be neglected. Consider problem (5.3.1) in a modified form:

$$\min_{x \in S^*} f(x) = \frac{N(x)}{D(x)}, \tag{5.3.9}$$

where $S^* = \{ x \mid \dfrac{g_i(x)}{D(x)} \leq 0,\ i = 1, ..., m\}$. It is obvious that the sets S and S^* are identical and, if x^* is optimal solution of problem (5.3.1), then x^* is also optimal for problem (5.3.9) and *vice-versa*. Let's show that the dual of problem (5.3.9) is the problem

$$\max_{(x,u) \in T} F(x, u) = \frac{N(x) + u'g(x)}{D(x)} \tag{5.3.10}$$

where $T = \{(x, u) \in R^{n+m} \mid \nabla_x F(x, u) = 0,\ u \geq 0,\ u \in R^m\}$. The set T can also be expressed as $T = \{(x, u) \in R^{n+m} \mid \nabla_x f(x) + u' \nabla_x \dfrac{g(x)}{D(x)} = 0,\ u \geq 0,\ u \in R^m\}$.

Let's prove the duality theorems:

Theorem 5.3.3. *If* $x \in S$ *and* $(x_0, u_0) \in T$, *then* $f(x) \geq F(x_0, u_0)$.

Proof. We have

$$(x - x_0)' \nabla_x f(x_0) = \frac{D(x_0)(x - x_0)' \nabla_x N(x_0) - N(x_0)(x - x_0)' \nabla_x D(x_0)}{(D(x_0))^2}$$

or, using the convexity of N and the linearity (the concavity) of D, and the fact that D is strict positive for each $x \in S$,

$$(x - x_0)' \nabla_x f(x_0) \leq \frac{D(x_0)(N(x) - N(x_0)) - N(x_0)(D(x) - D(x))}{(D(x_0))^2} = \frac{D(x)}{D(x_0)}(f(x) - f(x_0)).$$

Hence,

$$f(x) \geq f(x_0) + \frac{D(x_0)}{D(x)}(x - x_0)' \nabla_x f(x_0) =$$

$$f(x_0) - \frac{D(x_0)}{D(x)} u_0'(x - x_0)' \nabla_x \frac{g(x_0)}{D(x_0)} \geq f(x_0) + u_0' \left[\frac{g(x_0)}{D(x_0)} - \frac{g(x)}{D(x)} \right] =$$

$$F(x_0, u_0) - u_0' \frac{g(x)}{D(x)} \geq F(x_0, u_0)$$

The proof is complete.

Corollary 5.3.1. *If x^* $\in S$ and $(x^*, u^*) \in T$ are such that $f(x^*) = F(x^*, u^*)$, then x^* is optimal solution of problem (5.3.1), and (x^*, u^*) is optimal solution of problem (5.3.10).*

Theorem 5.3.4. *Let $x^* \in S$ be optimal solution of problem (5.3.1) and assume that at x^*, g satisfies certain constraint qualifications. Then, there is $u^* \in R^m$ such that (x^*, u^*) is the optimal solution of problem (5.3.10) and $f(x^*) = F(x^*, u^*)$.*

Proof. Since $x^* \in S$ is optimal solution of problem (5.3.1) and at x^*, g satisfies certain constraint qualifications, it follows that the optimality conditions (4.1.2) - (4.1.5), defined in Chapter 4, Section 4.1, are satisfied. Rewrite these conditions for problem (5.3.9), hence there is $u^* \in R^m$ such that (x^*, u^*) satisfies the constraints

$$\nabla_x f(x^*) + u^{*\prime} \nabla_x \left[\frac{g(x^*)}{D(x^*)} \right] = \mathbf{0} \tag{5.3.11}$$

$$u^{*\prime} \frac{g(x^*)}{D(x^*)} = 0 \tag{5.3.12}$$

$$\frac{g(x^*)}{D(x^*)} \leq \mathbf{0} \tag{5.3.13}$$

$$u^* \geq \mathbf{0} \tag{5.3.14}$$

Equations (5.3.11) and (5.3.14) yield that there is u^* such that $(x^*, u^*) \in T$. Consider $(x, u) \in T$. Using (5.3.12) and Corollary 5.3.1, we have $F(x^*, u^*) = \dfrac{N(x^*) + u^{*\prime} g(x^*)}{D(x^*)} = f(x^*) + u^{*\prime} \dfrac{g(x^*)}{D(x^*)} = f(x^*) \geq F(x, u)$. It follows that (x^*, u^*) is optimal solution of problem (5.3.10). The proof is complete.

Theorem 5.3.5. *Let $x_0 \in S$ be the optimal solution of problem (5.3.1) and assume that, in x_0, g satisfies certain constraint qualifications. If (x^*, u^*) is optimal solution of problem (5.3.10), and if $N(x) + u^{*\prime} g(x)$ is strict convex in x^*, then $x^* = x_0$ i.e., x^* is optimal solution of problem (5.3.1) and $f(x_0) = F(x^*, u^*)$.*

Proof. Note that $N(x) + u^{*\prime}g(x)$ is strict convex in x^* if or N is strict convex in x^* or if, for a certain i, $u_i^* > 0$ and g_i is strict convex in x^*. With this observation, let's prove the converse duality theorem. We assume that $x^* \neq x_0$. Since x_0 is optimal solution of problem (5.3.1), Theorem 5.3.4 yields that there is $u_0 \in R^m$ such that $(x_0, u_0) \in T$ is optimal solution of problem (5.3.10) and

$$F(x_0, u_0) = F(x^*, u^*) = \max_{(x,\, u) \in T} F(x, u) \tag{5.3.15}$$

Since $(x^*, u^*) \in T$, we have

$$\nabla_x F(x^*, u^*) = 0 \tag{5.3.16}$$

The function $N(x) + u^{*\prime}g(x)$ is strict convex in x^*, hence

$$(x - x_0)' \nabla_x F(x^*, u^*) < \frac{D(x_0)}{D(x^*)}(F(x_0, u^*) - F(x^*, u^*)) \tag{5.3.17}$$

Equation (5.3.16) and (5.3.17), and strict positiveness of D yield

$$F(x_0, u^*) > F(x^*, u^*) = F(x_0, u_0)$$

or,

$$u^{*\prime}\frac{g(x_0)}{D(x_0)} > u_0'\frac{g(x_0)}{D(x_0)} \tag{5.3.18}$$

Since x_0 is an optimal solution of problem (5.3.1), it follows that

$$u^{*\prime}\frac{g(x_0)}{D(x_0)} = 0 \tag{5.3.19}$$

Equation (5.3.18) and (5.3.19) yield $u^{*\prime}g(x_0) > 0$, which contradicts the fact that $u^{*\prime} \geq 0$ and $g(x_0) \leq \mathbf{0}$. Hence $x^* = x_0$. We also have

$$f(x_0) = \frac{N(x_0)}{D(x_0)} + u_0'\frac{N(x_0)}{D(x_0)} = F(x_0, u_0) = F(x^*, u^*)$$

The proof is complete.

5.4. DUALITY FOR A PARTICULAR NONLINEAR FRACTIONAL PROGRAMMING PROBLEM

Consider the nonlinear fractional programming problem

$$\max F(x) = \frac{c'x - \sqrt{x'Ux} + c_0}{d'x + d_0} \tag{5.4.1}$$

subject to

$$Ax \leq b, \quad x \geq \mathbf{0} \tag{5.4.2}$$

where A is an $m \times n$ matrix ($m<n$) of rank m, x, c, and d are column vectors with n components, b is a column vector with m components, c_0 and d_0 are positive numbers, and U is a positive semi-definite $n \times n$ matrix. Assume that the denominator keeps constant sign, for example positive, on the whole feasible solutions domain $S = \{x | Ax \leq b, x \geq 0\}$. Also, assume that S is non-empty and bounded, *i.e.*, $Ax \leq 0 \Rightarrow x = 0$. The problem (5.4.1) - (5.4.2) appears in stochastic programming, *e.g.*, in Kataoka's problem for a ratio of linear functions, with the numerator having random coefficients [896], [910].

We show that the dual of problem (5.4.1) - (5.4.2) is of the form

$$\min G(v, z, w) = z \tag{5.4.3}$$

subject to

$$A'v + dz + Uw \geq c \tag{5.4.4}$$

$$-b'v + d_0 z \geq c_0 \tag{5.4.5}$$

$$w'Uw \leq 1 \tag{5.4.6}$$

$$v \geq 0 \tag{5.4.7}$$

where v is a vector with m components, and w is a vector with n components. Let T be the domain defined by the constraints (5.4.4) - (5.4.7).

Theorem 5.4.1. *If* $x \in S$ *and* $(v, z, w) \in T$, *then*

$$\max_{x \in S} F(x) \leq \min_{(v, z, w) \in T} G(v, z, w)$$

Proof. Multiply (5.4.4) by $x \geq 0$ and we get

$$x'A'v + x'dz + x'Uw \geq x'c \tag{5.4.8}$$

Since $x'Av = b'v$, Equation (5.4.8) becomes $b'v + x'dz + x'Uw \geq x'c$. Term-by-term addition of this inequality with Equation (5.4.5) yields

$$(x'd + d_0)z + x'Uw \geq x'c + c_0 \tag{5.4.9}$$

Use now the following result.

Lemma 5.4.1. *If* $x, y \in R^n$ *and* C *is a positive semi-definite matrix, then*

$$(x'Cy) \leq (x'Cx)^{1/2}(y'Cy)^{1/2}$$

The equality takes place if there is $\lambda \geq 0$ *such that* $Cx = \lambda Cy$.

Using this lemma, the inequality (5.4.9) becomes

$$(x'd + d_0)z + (x'Ux)^{1/2}(w'Uw)^{1/2} \geq x'c + c_0$$

or, using (5.4.6),

$$(x'd + d_0)z + (x'Ux)^{1/2} \geq x'c + c_0$$

It follows that

$$z \geq \frac{c'x - \sqrt{x'Ux} + c_0}{d'x + d_0}$$

Thus,

$$\max_{x \in S} F(x) \leq \min_{(v,\, z,\, w) \in T} G(v, z, w)$$

Corollary 5.4.1. *If* $x \in S$ *and* $(v, z, w) \in T$ *are such that* $F(x) = G(v, z, w)$, *then* x *is optimal solution of problem (5.4.1) - (5.4.2), and* (v, z, w) *is optimal solution of problem (5.4.3) - (5.4.7).*

Theorem 5.4.2. *If* x^* *is optimal solution of problem (5.4.1) - (5.4.2), then there is* (v^*, z^*, w^*) *optimal solution of problem (5.4.3) - (5.4.7) and*

$$F(x^*) = G(v^*, z^*, w^*)$$

Proof. Let x^* be an optimal solution of problem (5.4.1) - (5.4.2). Through the change of variable $y = tx$ $(t \geq 0)$, problem (5.4.1) - (5.4.2) becomes the equivalent problem:

$$\max H(y, t) = c'y - \sqrt{y'Uy} + c_0 t \qquad (5.4.10)$$

subject to

$$\begin{aligned} Ay &\leq bt \\ d'y + d_0 t &= 1 \qquad (5.4.11) \\ y \geq \mathbf{0}, &\quad t \geq 0 \end{aligned}$$

According to Theorem 4.2.1, Chapter 4, between the optimal solutions x^* and (y^*, t^*) of problems (5.4.1) - (5.4.2) and (5.4.10) - (5.4.11), one has the relation $x^* = y^*/t^*$, and

$$F(x^*) = H(y^*, t^*). \qquad (5.4.12)$$

The dual of problem (5.4.10) - (5.4.11) is problem (5.4.3) - (5.4.7)[1]. It follows that, if (y^*, t^*) is optimal solution of problem (5.4.10) - (5.4.11), then there is (v^*, z^*, w^*) optimal solution of problem (5.4.10) - (5.4.11) for which we have

$$H(y^*, t^*) = G(v^*, z^*, w^*) \qquad (5.4.13)$$

Equations (5.4.12) and (5.4.13) yield $F(x^*) = G(v^*, z^*, w^*)$. The proof is complete.

Let's now prove the converse duality theorem.

Theorem 5.4.3. *If* (v^*, z^*, w^*) *is an optimal solution for problem (5.4.3) - (5.4.7), then there is* x^* *optimal solution for problem (5.4.1) - (5.4.2), and the corresponding optimal values of the objective functions are equal.*

Proof. Since (v^*, z^*, w^*) is optimal solution of problem (5.4.3) - (5.4.7), the Kuhn-Tucker conditions are satisfied, hence there is a vector $(x_0, y_0) \geq \mathbf{0}$ and $\mu \geq 0$ such that

---

[1] S.M. Sinha; *A duality theorem for non-linear programming*, Management Sci. **12** (1966), 385-390.

$$
\begin{array}{ll}
\text{a)} & x_0'd + y_0 d_0 = 1 \\
\text{b)} & Ax_0 - by_0 \leq 0 \\
\text{c)} & v^{*\prime}(Ax_0 - by_0) = 0 \\
\text{d)} & Ux_0 = 2\mu Uw^* \\
\text{e)} & \mu(1 - w^{*\prime}Uw^*) = 0 \\
\text{f)} & x_0'(A'u^* + dz^* + Uw^* - c) = 0 \\
\text{g)} & -b'v^* + d_0 z^* - c_0 \geq 0 \\
\text{h)} & y_0(-b'v^* + d_0 z^* - c_0) = 0
\end{array}
\tag{5.4.14}
$$

Let $x^* = x_0/y_0$. We have $y_0 > 0$. Hence, using Equation (5.4.14a), we have $x^* \in S$. Relations (5.4.14c) and (5.4.14h) yield $v^{*\prime}Ax_0 = y_0(d_0 z^* - c_0)$. Substituting this latter relation in (5.4.14f), we get

$$c'x_0 - x_0'Uw^* = (d'x_0 + d_0 y_0)z^* - c_0 y_0 \tag{5.4.15}$$

Equation (5.4.14d) and Lemma 5.4.1 imply that $x_0'Ux^* = (x_0'Ux_0)^{1/2}(w^{*\prime}Uw^*)^{1/2}$ or,

$$x_0'Uw^* = (x_0'Ux_0)^{1/2} \tag{5.4.16}$$

since Equation (5.4.14e) yields or, $\mu = 0$ or, $w^{*\prime}Uw = 1$. Equations (5.4.15) and (5.4.16) yield

$$c'x_0 - (x_0Ux_0)^{1/2} = (d'x_0 + d_0 y_0)z^* - c_0 y_0$$

or,

$$\frac{c'x_0 - (x_0'Ux_0)^{1/2} + c_0 y_0}{d'x_0 + d_0 y_0} = z^*$$

or, even more,

$$\frac{c'x^* - (x^{*\prime}Ux^*)^{1/2} + c_0}{d'x^* + d_0} = z^*$$

The proof is complete.

When the matrix U is positive semi-definite, the solution of problem (5.4.1) - (5.4.2) cannot be done using the convex programming methods since the numerator of the function is not differentiable. For this reason, we have to solve the dual problem (5.4.3) - (5.4.7) which is a convex programming problem.

5.5 DUALITY FOR FRACTIONAL PROGRAMMING VIA GEOMETRIC PROGRAMMING DUALITY

I <u>**Linear case**</u>. We will construct the dual of a linear fractional programming problem using the duality theorems of geometric programming. We will show that this dual is a linear programming problem, which offers a series of computational advantages. Consider the linear fractional programming problem

$$Min \quad F(x_1, ..., x_n) = (\sum_{i=1}^{n} c_i x_i + c_0)/(\sum_{i=1}^{n} d_i x_i + d_0)$$

subject to

$$\sum_{j=1}^{n} a_{ij} - b_i \leq 0 \ , \ i = 1, ..., m$$

$$-x_j \leq 0 \ , \ j = 1, ..., n$$

where the denominator is assumed strict positive on the feasible solutions domain. The equivalent linear program, obtained through the transformation $y_i = tx_i$, $i = 1, ..., n$ is

$$Min \ H(y_1, ..., y_n, t) = \sum_{i=1}^{n} c_i y_i + c_0 t$$

subject to

$$\sum_{j=1}^{n} d_j y_j + d_0 t - 1 \leq 0$$

$$-\sum_{j=1}^{n} d_j y_j - d_0 t + 1 \leq 0$$

$$\sum_{j=1}^{n} a_{ij} y_j - b_i t \leq 0, \quad i = 1, ..., m$$

$$-y_j \leq 0, \quad j = 1, ..., n$$

$$-t \leq 0$$

To this program we now apply the Federowicz[2] transformation

$$H(y) = \log h, \ -1 = \log \mu_1, \ +1 = \log \mu_2, \ y_j = \log t_j, \ j = 1, ..., n, \ t = \log t_{n+1}$$

After some calculations, and using the fact that the logarithmic function is monotonic, we get the geometric programming problem

$$Min \ h(t_1, ..., t_{n+1}) = \prod_{j=1}^{n} t_j^{c_j} t_{n+1}^{c_0}$$

subject to

---

[2]Duffin, R.J., Peterson, E.L. and Zener, C.M. : *Geometric Programming*, John Wiley and Sons, New York, 1967.

$$\mu_1 \prod_{j=1}^{n} t_j^{d_j} t_{n+1}^{d_0} \leq 1$$

$$\mu_2 \prod_{j=1}^{n} t_j^{-d_j} t_{n+1}^{-d_0} \leq 1$$

$$\prod_{j=1}^{n} t_j^{a_{ij}} t_{n+1}^{-b_i} \leq 1 \quad i = 1, ..., m$$

$$0 < t_j^{-1} \leq 1 \quad j = 1, ..., n+1$$

Its dual is [Duffin et al., *op. cit.*]

$$Max\ v(\delta_1, ..., \delta_{n+m+4}) = \left(\frac{\mu_1}{\delta_2}\right)^{\delta_2} \left(\frac{\mu_2}{\delta_3}\right)^{\delta_3} \prod_{i=1,\ i\neq 2,3}^{n+m+4} \delta_i^{-\delta_i} \prod_{i=2}^{n+m+4} \delta_i^{\delta_i}$$

subject to

$$\delta_1 = 1$$

$$c_i\delta_1 + d_i\delta_2 - d_i\delta_3 + \sum_{j=4}^{n+3} a_{j-3,i}\delta_j - \delta_{n+i+3} = 0,\ i = 1, ..., n$$

$$c_0\delta_1 + d_0\delta_2 - d_0\delta_3 - \sum_{j=1}^{n} b_j\delta_{j+3} - \delta_{n+m+4} = 0$$

$$\delta_j \geq 0 \quad j = 1, ..., n+m+4$$

Applying the logarithmic operator, with the notation $V(\delta) = \log v(\delta)$, we write the dual as

$$Max\ V(\delta_1, ..., \delta_{n+m+4}) = -\delta_2 + \delta_3$$

subject to

$$c_i + d_1\delta_2 - d_i\delta_3 + \sum_{j=4}^{m+3} a_{j-3,i}\delta_j - \delta_{n+i+3} = 0,\ i = 1, ..., n$$

$$c_0 + d_0\delta_2 - d_0\delta_3 - \sum_{j=1}^{m} b_j\delta_{j+3} - \delta_{n+m+4} = 0$$

$$\delta_j \geq 0 \quad j = 1, ..., n+m+4$$

This problem is the dual of the initial linear fractional programming problem. The duality theorems result from the duality theorems of geometric programming.

Numerical example

$$Min\ F(x_1, x_2) = \frac{5x_1 + 2x_2}{5x_1 + x_2 + 2}$$

subject to

$$2x_1 + 3x_2 \leq 12$$

$$5x_1 + 3x_2 \geq 15$$

$$x_2 \geq 1$$

$$x_1, x_2 \geq 0$$

The optimal solution is $x_1 = \frac{12}{5}$, $x_2 = 1$. The minimum value is $F(x_1, x_2) = 14/15$. Through the transformation $y_i = tx_i (i = 1, 2)$, one gets the following equivalent linear programming problem.

$$\min H(y_1, y_2, t) = 5y_1 + 2y_2$$

subject to

$$2y_1 + 3y_2 - 12t \le 0$$
$$5y_1 + 3y_2 - 15t \ge 0$$
$$y_2 - t \ge 0$$
$$y_1, y_2, t \ge 0$$

Applying the Federowicz transformation, we get the following primal problem of geometric programming:

$$\min h(t_1, t_2, t_3) = t_1^5 t_2^2$$

subject to

$$\mu_1 t_1^5 t_2 t_3^2 \le 1$$
$$\mu_2 t_1^{-5} t_2^{-1} t_3^{-2} \le 1$$
$$t_1^2 t_2^3 t_3^{-12} \le 1$$
$$t_1^{-5} t_2^{-3} t_3^{15} \le 1$$
$$t_2^{-1} t_3^1 \le 1$$
$$0 < t_i^{-1} \le 1, \quad i = 1, 2, 3$$

where $\log \mu_1 = -1$ and $\log \mu_2 = +1$. The dual geometric program is

$$\max v(\delta_1, ..., \delta_9) = \left[\frac{\mu_1}{\delta_2}\right]^{\delta_2} \left[\frac{\mu_2}{\delta_3}\right]^{\delta_3} \prod_{i=1,\, i\ne 2,3}^{9} \delta_i^{-\delta_i} \prod_{i=2}^{9} \delta_i^{\delta_i}$$

subject to

$$\delta_1 = 1$$
$$5\delta_1 + 5\delta_2 - 5\delta_3 + 2\delta_4 - 5\delta_5 - \delta_7 = 0$$
$$2\delta_1 + \delta_2 - \delta_3 + 3\delta_4 - 3\delta_5 - \delta_6 - \delta_8 = 0$$
$$2\delta_2 - 2\delta_3 - 12\delta_4 + 15\delta_5 + \delta_6 - \delta_9 = 0$$
$$\delta_i \ge 0, \ i = 1, ..., 9$$

Or, applying the transformation $V(\delta) = \log v(\delta)$,

$$\mathit{Max}\ V(\delta_1, ..., \delta_9) = -\delta_2 + \delta_3$$

subject to

$$-5\delta_2 + 5\delta_3 - 2\delta_4 + 5\delta_5 + \delta_7 = 5$$
$$-\delta_2 + \delta_3 - 3\delta_4 + 3\delta_5 + \delta_6 + \delta_8 = 2$$
$$-2\delta_2 + 2\delta_3 + 12\delta_4 - 15\delta_5 - \delta_6 + \delta_9 = 0$$
$$\delta_i \ge 0, \quad i = 1, ..., 9$$

The optimal solution is $\delta_2 = 0$, $\delta_3 = 14/15$, $\delta_4 = 0$, $\delta_5 = 1/15$ and $\delta_6 = 13/15$, and the maximum value $V(\delta_1, ..., \delta_9) = 14/15$.

II Nonlinear case. Consider the nonlinear and non-convex fractional programming problem

$$Min\ \{F(x) = \frac{c(x)}{f(x)} \mid Ax \geq b,\ x \in C \subset R^n\} \tag{5.5.1}$$

where c is a convex function defined on a convex closed set C, $f = f_1'x + f_0 > 0$, A is a matrix $m \times n$, $b \in R^m$, $f_1 \in R^n$ and $f_0 \in R$. Before writing the dual of problem (5.5.1), recall that, in geometric programming, the following problems are dual[3]

$$\min\ \{g(x) \mid x \in \chi\}$$
$$\min\ \{g^*(x^*) \mid x^* \in \chi^*\}$$

The function g is defined and convex on the convex closed set C and χ is a convex closed cone from R^n. The function g^*, defined on C^*, is the conjugate of g, χ^* is the dual cone of χ defined as:

$$g^*(x^*) = \sup_{x \in C}(x'x^* - g(x))$$
$$\chi^* = \{x^* \mid x'x^* \geq 0 \text{ for all } x \in \chi\}$$
$$C^* = \{x^* \in R^n \mid g^*(x^*) < \infty\}$$

According to Peterson [*op.cit.*], the following relations are true at optimality:

a) $\displaystyle \min_{x \in C \cap \chi} g(x) + \min_{x^* \in C^* \cap \chi^*} g^*(x^*) = 0$

b) $x \in \partial g^*(x^*)$, $x^* \in \partial g(x)$

c) $x'x^* = 0$

where $\partial g(x)$ is the subgradient of g calculated in point x, $\partial g(x) = \{x^* \mid < x^*, y - x > \leq g(y) - g(x), \text{ for all } y \in C\}$ ($<x, y>$ is the scalar product between x and y). For the convex function c defined on C, its positive and homogeneous extension is

$$c^+(x, \lambda) = \begin{cases} \sup_{x^* \in C^*} x'x^* & \text{if } \lambda = 0 \text{ and } \sup_{x^* \in C^*} x'x^* < \infty \\ \lambda c(x/\lambda) & \text{if } \lambda > 0 \text{ and } x/\lambda \in C \end{cases}$$

which is defined on

$$C^+ = \{(x, \lambda) \mid \lambda = 0 \text{ and } \sup_{x^* \in C^*} x'x^* < \infty\} \cup \{(x, \lambda) \mid \lambda > 0 \text{ and } x/\lambda \in C\}$$

---

[3]E.L. Peterson: *Geometric programming*. SIAM Rev. 18, 1976, 1-52

The conjugate of $c^{+}(x, \lambda)$ is an identically zero function defined on the convex set $c^{*}(x^{*}) + \lambda^{*} \leq 0$.

We return now to problem (5.5.1), and we transform it into a convex problem through the transformation

$$z_0 = 1/f(x), \quad z = z_0 x$$

$$\min \{g(z, z_0) = c(z/z_0)z_0 \mid Az - bz_0 \geq 0, f_1'z + f_0 z_0 = 1, z_0 > 0\}$$

Or, upon closure,

$$\min c^{+}(z, z_0) \text{ over } (z, z_0)' \in C^{+}, \quad \alpha \in \{1\}$$

subject to

$$(z, z_0, \alpha) \in \chi = \{(z, z_0, \alpha) \mid Az - bz_0 \geq 0, f_1'z + f_0 z_0 - \alpha = 0\}$$

In this program, α has been associated to the zero component of the objective function. The dual cone is

$$\chi^{*} = \{(z^{*}, z_0^{*}, \alpha^{*}) \mid \langle z, z^{*} \rangle + z_0 z_0^{*} + \alpha\alpha^{+} \geq 0, (z, z_0, \alpha) \in \chi\}$$

or

$$\begin{Bmatrix} z^{*} \\ \alpha^{*} \\ z_0^{*} \end{Bmatrix} = \begin{Bmatrix} A^{T} \\ 0 \\ -b' \end{Bmatrix} u + \begin{Bmatrix} f_1 \\ -1 \\ f_0 \end{Bmatrix} v$$

$u \in R^{m}, u \geq 0, v \in R$

The dual program is written as:

$$\min \alpha^{*}$$

subject to

$$c^{*}(z^{*}) - b'u - f_0\alpha^{*} \leq 0$$
$$z^{*} = A^{T}u - f_1\alpha^{*}$$
$$u \geq 0$$

where $c^{*}(z^{*})$ is the conjugate of $c(z)$.

5.6. HISTORICAL NOTES. COMMENTARIES

The dual problem (5.1.5) of Section 5.1 was constructed following Bector [127], and the dual problem (5.2.4) - (5.2.6) was constructed following [717]. The development of Section 5.3 follows Jagannthan [513] for the dual (5.3.4) - (5.3.7), and Bector [124] for the dual (5.3.10). The dual problem (5.4.3) - (5.4.7), and the duality theorems of Section 5.4 were introduced and proved by Aggarwal and Saxena [23].

The dual problems constructed through geometric programming were formulated by Aggarwal and Patkar [21] for the linear case, and by Scott and Jefferson [824] for the nonlinear case.

The first treatment of the dual to a linear fractional programming problem was that of Swarup [951] who showed that the dual of the problem

$$\max \left\{ z = \frac{c'x}{d'x} \mid Ax = b,\ x \geq 0 \right\} \tag{5.6.1}$$

is the problem

$$\min \left\{ w = \frac{b'u}{b'v} \mid b'v(A'u - c) - b'u(A'v - d) \geq 0,\ \ b'v \geq 0,\right. \\ \left. u,\ v \text{ without sign constraints},\ |b'u| + |b'v| > 0 \right\} \tag{5.6.2}$$

Swarup proved the weak and direct duality theorem but did not prove the converse duality theorem. In another work, Swarup [951] considered a problem more general than (5.6.1), *i.e.*, when the constraints are defined by convex functions and showed that the Huard's converse duality theorem takes place. Bector [127] showed that the dual problem (5.6.2) can be obtained by dualizing the parametric problem associated to problem (5.6.1). Thus, the converse duality theorem between (5.6.1) and (5.6.2) takes place. Bector [127] also showed that the dual problems obtained by Bector [125] and by Chadha [233] are equivalent to problem (5.1.8) considered in Section 5.1 of this book. Sharma and Swarup [838] constructed the dual of the problem

$$\max \left\{ f(x) = \frac{c'x}{d'x} \mid Ax \leq b,\ x \geq 0 \right\} \tag{5.6.3}$$

having the same form as the primal problem

$$\max \left\{ g(u) = \frac{c'u}{d'u} \mid A'v - c \cdot d'u + d \cdot c'u \geq 0,\ -b'v \geq 0, u, v \geq 0 \right. \\ \left. |c'u| + |d'u| > 0 \right\} \tag{5.6.4}$$

and proved the duality theorems.

Duality for homogeneous fractional programming was considered by Craven and Mond [332], [335], Morris [683], and Aylawadi [88]. A comparative study of various dual problems for linear fractional programming was done by Abrham and Luthra [3]. Kydland [591] constructed the dual of the problem:

$$\max \left\{ f(x) = \frac{c'x + c_0}{d'x + d_0} \mid Ax = b,\ x \geq 0 \right\} \tag{5.6.5}$$

of the form

$$\min \{ f(x) - u(Ax - b) + vIx \mid \nabla f(x) - uA + vI = 0,\ v \geq 0 \} \tag{5.6.6}$$

Craven and Mond [333] constructed, using the change of variable method, the dual to the problem

$$\max \left\{ \frac{c'x + c_0}{d'x + d_0} \mid Ax \le b,\ x \ge 0 \right\} \tag{5.6.7}$$

in the form of the problem

$$\min \left\{ \frac{-\lambda bu - c_0 z - \lambda}{b'u - d_0 z + 1} \mid u \ge 0,\ b'u - d_0 z + 1 > 0, \right. \tag{5.6.8}$$
$$\left. (A' - c^* b')u + (d + c^* d_0)z \ge c^* \right\}$$

where $\lambda = -d_0^{-1}(1 + c_0)$ and $c^* = c + \lambda d$. Aggarwal and Om Parkash [20] constructed the dual to the problem

$$\min \left\{ \frac{c'x + c_0}{d'x + d_0} \mid Ax \le b,\ x \ge 0 \right\} \tag{5.6.9}$$

in the form

$$\max \left\{ \frac{c'u + c_0}{d'u + d_0} \mid A'v + c(d'u + d_0) - d(c'u + c_0) \ge 0, \right. \tag{5.6.10}$$
$$\left. -B'v + c_0(d'u + d_0) - d_0(c'u + c_0) \ge 0,\ u,\ v \ge 0 \right\}$$

These results are similar to those presented by us in Section 5.1, second part.
Important duality results for fractional programming were obtained by Schaible [797], [799]. In particular, Schaible [797] constructed the dual to the problem

$$sup \left\{ \frac{x'Cx + c'x + c_0}{x'Dx + d'x + d_0} \mid Ax \le b,\ x \in S \right\} \tag{5.6.11}$$

where C and D are negative and positive semi-definite matrices, respectively, and $S \subseteq R^n$ is a non-empty open and convex set. Other duality results for fractional programming were obtained by Aggarwal, Saxena [22], Kaska [534], Chandra and Gulati [266], Gulati [434], Parkash, Saxena and Patkar [704], Chadha [235], Jagannthan and Schaible [515], Crouzeix, Ferland and Schaible [340], Rani and Kaul [743], Kaul and Chadha [542], Jeyakumar [518], Bitran and Magnanti [190], [191] and others.

CHAPTER 6

FRACTIONAL PROGRAMMING WITH MULTIPLE OBJECTIVE FUNCTIONS

The theory of mathematical programming (linear or nonlinear) with several objective functions, also called vector programming, is well developed. In the case when several objective functions exist, the optimal solution for a function is not necessarily optimal for the other functions, and hence one introduces the notion of the "best-compromise solution", also known as "non-dominant solution", "efficient solution", "Pareto-optimal solution", etc. Recently, a number of authors have extended various aspects of the mathematical programming with several objective functions to the case when the objective functions are fractional. Some of these recent results will be presented in this chapter. Note that, in general, in fractional programming, the set of efficient points will no longer have a series of properties existing in the linear case.

6.1. DEFINITIONS. PRELIMINARY RESULTS

Let z_i $(i = 1,..., k)$ be k arbitrary objective functions defined on a compact set $D \subseteq R^n$ and let $S \subseteq D$. Consider the mathematical programming problem with several objective functions:

$$(P) \quad \max_{x \in S} \; (z_1(x) \; , \; ... \; , \; z_k(x))$$

Let $I = \{1 \; , \; ... \; , \; k\}$ and for any $i \in I$, denote by $I_i = I - \{i\}$.

Definition 6.1.1. A point $x^0 \in S$ is *efficient point (solution)* for problem (*P*) if and only if there is no $x \in S$ such that $z_r(x) \geq z_r(x^0)$ for any $r \in I$ and for at least a r_0 to have $z_{r_0}(x) > z_{r_0}(x^0)$. In other words, x^0 is efficient if there is no $x \in S$ which dominates weakly on x^0. This definition corresponds to the notion of *strong efficient point.*

Definition 6.1.2. A point $x^0 \in S$ is *weak efficient point (solution)* for problem (*P*) if and only if there is no $x \in S$ such that $z_i(x) > z_i(x^0)$. In other words, there is no $x \in S$

which strongly dominates x^0.

Sometimes, the two definitions, 6.1.1 and 6.1.2, are used in the following form:

Definition 6.1.3. $x^0 \in S$ is called *efficient (weak efficient) solution* for problem (*P*), if the inequality $z_i(x) > z_i(x^0)$ takes place for some $i \in I$ and some $x \in S$, it follows that there is at least an index $j \in I_i$, such that

$$z_j(x) < z_j(x^0) \qquad \text{(respectively, } z_j(x) \leq z_j(x^0))$$

Denote by E, E^w the sets of efficient, weak efficient points, respectively. Obviously $E \subseteq E^w$. To eliminate certain efficient points with unwanted properties (which, in the case of concave functions, are situated on the boundary of *S*), H. W. Kuhn and A.W. Tucker[1] introduced the notion of properly efficient solution.

Definition 6.1.4. $x^0 \in S$ is called *K-T properly efficient solution* for problem (*P*), if and only if there is no $x \in S$ such that $\nabla z_r(x^0)x \geq \nabla z_r(x^0)x^0$ for any $r \in I$ and for at least a r_0 to have $\nabla z_{r_0}(x^0)x > \nabla z_{r_0}(x^0)x^0$ (∇ is the gradient operator).

Geoffrion[2] reformulated the definition of the properly efficient solution given above in order to exclude those efficient solutions for which the ratio of the gain for a function and the loss for another function can be made as large as desired.

Definition 6.1.5. $x^0 \in S$ is called *properly efficient solution* for problem (*P*) if it is efficient and if there is a scalar $M > 0$, such that for any $i \in I$ and any $x \in S$ for which $z_i(x) > z_i(x^0)$ there is at least an index $j \in I_i$ such that $z_j(x) < z_j(x^0)$ and $(z_i(x)-z_i(x^0))/(z_j(x^0)-z_j(x)) \leq M$.

Denote by *PE(KT)* and *PE(G)* the set of efficient points which are properly efficient according to Kuhn and Tucker and Geoffrion, respectively. Subsequently, any properly efficient solution will be considered according to Geoffrion.
H. Isermann[3] showed that, if the functions z_i are linear, then each efficient solution is properly efficient, *i.e.* $PE(G) = E$. Other results existing in literature will also be used

---

[1]H. W. Kuhn and A. W. Tucker, "Nonlinear programming", *Proceedings Second Berkeley Symposium on Mathematical Statistics and Probability*, Univ. of California Press, Berkeley, California, 1950, 481-482

[2]Geoffrion, "Properly efficiency and the theory of vector maximization", *J. Math. Anal. Appl.*, **22** (1968), 618-630

[3]H. Isermann "Properly efficiency and the linear vector maximum problem", *Oper. Res.*, **22** (1974), 189-191

in our exposition. The following theorem and lemma show a connection between the vector programming problems and the mathematical programming problems with only one objective function.

Theorem 6.1.1. (Geoffrion (*op. cit.*)). *Let S be a convex set, and let $z_i (i \in I)$ be concave on X. Then $x^0 \in S$ is properly efficient for problem (P) if and only if x^0 is an optimal solution for problem*

$$\max_{x \in S} \sum_{i=1}^{k} \lambda_i z_i(x) \quad \textit{for a certain } \lambda = (\lambda_1, \ldots, \lambda_k) > \mathbf{0} .$$

Lemma 6.1.1. *$x^0 \in S$ is efficient solution for problem (P) if and only if x^0 maximizes each function z_i on the set*

$$S_i = \{ x \in S, \; z_j(x) \geq z_j(x^0), \; j \neq i \} .$$

Proof. Assume that $x^0 \in S$ is efficient solution for problem (P) but that, for a certain i, x^0 does not maximize the function z_i on set S_i. Hence, there is an $y \in S$ such that $z_j(y) \geq z_j(x^0)$, $j \neq i$ and $z_i(y) > z_i(x^0)$, which contradicts the fact that x^0 is efficient solution for problem (P). Conversely, assume that x^0 maximizes each function z_i on S_i. If x^0 is not efficient solution for problem (P), then there is a $y \in S$ such that $z_j(y) \geq z_j(x^0)$ for any $j \in I$ and $z_i(y) \geq z_i(x^0)$ for a certain i. This contradicts the assumption that x^0 maximizes the function z_i on S_i.

6.2. KUHN-TUCKER CONDITIONS FOR LINEAR FRACTIONAL PROGRAMMING WITH SEVERAL OBJECTIVE FUNCTIONS

Consider the linear fractional programming problem with several objective functions (LFVMP)

$$\max F(x) = \{F_1(x) = (c_1'x + c_{10})/(d_1'x + d_{10}) \; \ldots \; = (c_k'x + c_{k0})/(d_k'x + d_{k0})\} \tag{6.2.1}$$

subject to

$$x \in S = \{x \in R^n \mid Ax \leq b\} \tag{6.2.2}$$

where A is an $m \times n$ matrix, b is a column vector with m components, c_i and d_i, $i \in I = \{1, \ldots, k\}$, are column vectors with n components, and c_{i0} and d_{i0}, $i \in I$, are real constants. Assume that S is bounded and that

$$d_i'x + d_{i0} > 0 \text{ for any } i = 1, \ldots, k \text{ and any } x \in S. \tag{6.2.3}$$

Several properties of functions F_i will be used, and will be stated in the form of a lemma.

Lemma 6.2.1. *For any* $x,\ y \in S$,

$$F_i(x) \leq F_i(y) \text{ if and only if } (y - x)' \nabla F_i(x) \geq 0 \tag{6.2.4}$$

$$F_i(x) < F_i(y) \text{ if and only if } (y - x)' \nabla F_i(x) > 0 \tag{6.2.5}$$

$$F_i \text{ are both pseudoconvex and pseudoconcave} \tag{6.2.6}$$

Lemma 6.2.1 implies that F_i is monotonic on any segment $[x, y]$, $x, y \in S$, which means that

$$F_i(x) < F_i(y) \text{ implies that } F_i(x) < F_i(x^\lambda) < F_i(y),\ \forall\ \lambda \in (0, 1) \tag{6.2.7}$$

where $x^\lambda = \lambda x + (1 - \lambda)y$.

Denote by $p_1, ..., p_m$ the lines of A.

Definition 6.2.1. A subset T of S is called *face* of S if there is $J \subset \{1, ..., m\}$ such that, for the points of T, we have

$$p_i'x = b_i,\quad i \in J \tag{6.2.8}$$

$$p_i'x < b_i,\quad i \notin J \tag{6.2.9}$$

We say that T is the face corresponding to J. The following theorem represents the Kuhn-Tucker conditions for problem LFVMP.

Theorem 6.2.1. *A point* $x^* \in T$ *(the face corresponding to* $J = \{1, ..., r\}$*) is efficient if and only if the real numbers* $a_i \geq 0$, $i = 1, ..., r$, *and* $w_j \geq 1$, $j = 1, ..., k$, *exist such that*

$$a_1p_1 + ... + a_rp_r = w_1\ \nabla F_1(x^*) + ... + w_k\ \nabla F_k(x^*) \tag{6.2.10}$$

Proof. Without loss of generality, we can assume $k = 2$, such that Equation (6.2.10) becomes

$$a_1p_1 + ... + a_rp_r = w_1\ \nabla F_1(x^*) + w_2\ \nabla F_2(x^*) \tag{6.2.11}$$

Necessity. Let x^* efficient and assume that there is x such that

$$p_ix \leq b_i \tag{6.2.12}$$

$$(x - x^*)' \nabla F_1(x^*) \geq 0 \tag{6.2.13}$$

$$(x - x^*)' \nabla F_2(x^*) > 0 \tag{6.2.14}$$

From (6.2.7) and (6.2.9), it follows that there is a point y between x and x^*, sufficiently close to x^*, such that $y \in S$ and $F_1(x^*) \leq F_1(y)$ and $F_2(x^*) < F_2(y)$, which contradicts the fact that x^* is efficient. Hence, the system

$$p_i'(x - x^*) \leq 0,\ i = 1, ..., r;$$

$$(x - x^*)' \nabla F_1(x^*) \geq 0;$$

$$(x - x^*)' \nabla F_2(x^*) > 0 ;$$

does not have a solution $x - x^*$. Applying Farcas' lemma, it follows that the real numbers $a_{1i} \geq 0$, $i = 1, ..., r$, and $u_1 > 0$ exist such that

$$a_{11}p_1 + ... + a_{1r}p_r = u_1 \nabla F_1(x^*) + \nabla F_2(x^*) \tag{6.2.15}$$

Repeating the above argument for F_2 instead of F_1, it follows that the real numbers $a_{2i} \geq 0$, $i = 1, ..., r$, and $u_2 > 0$ exist such that

$$a_{21}p_1 + ... + a_{2r}p_r = \nabla F_1(x^*) + u_2 \nabla F_2(x^*). \tag{6.2.16}$$

Adding Equations (6.2.15) and (6.2.16), and noting that $a_i = a_{1i} + a_{2i} > 0$, $i = 1, ... r$, $w_1 = u_1 + 1 > 1$, $w_2 = u_2 + 1 > 1$ we get Equation (6.2.11).

Sufficiency. Assume that Equation (6.2.11) takes place, but x^* is not efficient, hence, there is x such that $F_1(x^*) \leq F_1(x)$ and $F_2(x^*) < F_2(x)$. From Lemma 6.2.1, it follows that $(x - x^*)' \nabla F_1(x^*) \geq 0$ and $(x - x^*)' \nabla F_2(x^*) > 0$ or

$$w_1(x - x^*)' \nabla F_1(x^*) + w_2(x - x^*)' \nabla F_2(x^*) > 0 \tag{6.2.17}$$

On the other hand, from (6.2.8), it follows that

$$a_1(x - x^*)' p_1 + ... + a_r(x - x^*)' p_r \leq 0 \tag{6.2.18}$$

From (6.2.17) and (6.2.18), it follows that Equation (6.2.11) cannot take place. Contradiction. Hence, x^* is efficient. The proof is complete.

Note that (6.2.10) does not represent a linear system of equations, as shown in the following example. In general, $T \cap E$ is not a set defined by linear equations.

Example 6.2.1. [313]. Let $F_1(x) = x_1/x_2$, $F_2(x) = x_3$, $F_3(x) = (x_1 - x_3)/(x_2 + 1)$, and

$$S = \{x \in R^3 \mid 0.1 \leq x_i \leq 4;\ i = 1, 2, 3\}$$

and let T be the face of S corresponding to $J = \varnothing$. In this case, (6.2.10) is the equation $x_1 = x_2x_3$. Hence,

$$T \cap E = \left\{x \in R^3 \mid 0.1 \leq x_i \leq 4,\ \ i = 1, 2, 3;\ \ x_1 = x_2x_3\right\}.$$

Note that $T \cap E$ is not defined by linear constraints. Even more, $T \cap E$ is not even convex. But, Equation (6.2.11) is equivalent with a system of linear equations and inequalities, as shown by the following lemma.

Lemma 6.2.2. *A point $x^* \in T$ is efficient if and only if the real numbers $a_1, ..., a_r, h_1, h_2, q_1, q_2$ exist such that*

$$a_1p_1 + ... + a_rp_r = q_1c_1 - h_1d_1 + q_2c_2 - h_2d_2,\ \ c_1'x^* + c_{10} = h_1,$$
$$d_1'x^* + d_{10} = q_1,\ \ d_{10}h_1 + d_{20}h_2 - c_{10}q_1 - c_{20}q_2 = b_1a_1 + ... + b_ra_r,$$
$$p_i'x^* = b_i,\ \ i = 1, ..., r,$$
$$p_i'x^* < b_i,\ i = r{+}1, ..., m,\ a_i \geq 0,\ i = 1, ..., r;\ q_i > 0,\ \ i = 1,2.$$

Proof. We have

$$\nabla F_i(x) = [(d_i'x + d_{i0})c_i - (c_i'x + c_{i0})d_i]/(d_i'x + d_{i0})^2$$

From (6.2.3), it follows that, for any $x \in S$, the direction of the gradient of the function is given by the direction of the vector $(d_i'x + d_{i0})c_i - (c_i'x + c_{i0})d_i$, hence, Equation (6.2.11) can be replaced by

$$a_1p_1 + ... + a_rp_r = w_1(d_1'x^* + d_{10})c_1 - w_1(c_1'x^* + c_{10})d_1 + \qquad (6.2.19)$$
$$+ w_2(d_2'x^* + d_{20})c_2 - w_2(c_2'x^* + c_{20})d_2.$$

Equation (6.2.19) is equivalent to the system

$$a_1p_1 + ... + a_rp_r = q_1c_1 - h_1d_1 + q_2c_2 - h_2d_2 \qquad (6.2.20\text{ a})$$
$$F_i(x^*) = h_i/q_i,\ i = 1, 2;\ q_i \geq 0,\ i = 1, 2 \qquad (6.2.20\text{ b})$$

Equation (6.2.20b) will be replaced by

$$c_1'x^* + c_{10} = h_1 \qquad (6.2.21)$$
$$d_1'x^* + d_{10} = q_1 \qquad (6.2.22)$$

Apply the scalar product through x^* to both sides of Equation (6.2.20a) and get

$$b_1a_1 + ... + b_ra_r = q_1c_1'x^* - h_1d_1'x^* + q_2c_2'x^* - q_2d_2'x^*. \qquad (6.2.23)$$

From $F_2(x^*) = h_2/q_2$, it follows that $h_2d_2'x^* + h_2d_{20} = q_2c_2'x^* + q_2c_{20}$ or

$$h_2d_{20} - q_2c_{20} = q_2c_2'x^* - h_2d_2'x^*$$

or, even more

$$h_2d_{20} - q_2c_{20} - q_1(c_1'x^* + c_{10}) = q_2c_2'x^* - h_2d_2'x^* - h_1(d_1'x^* + d_{10})$$

Regrouping the terms, we get

$$h_1d_{10} + h_2d_{20} - q_1c_{10} - q_2c_{20} = q_1c_1'x^* - h_1d_1'x^* + q_2c_2'x^* - h_2c_2'x^*$$

From the latter relation and from (6.2.23), it follows that

$$h_1d_{10} + h_2d_{20} - q_1c_{10} - q_2c_{20} = b_1a_1 + ... + b_ra_r.$$

In conclusion, Equation (6.2.11) is equivalent to the system

$$a_1p_1 + ... + a_rp_r = q_1c_1 - h_1d_1 + q_2c_2 - h_2d_2,\ c_1'x^* + c_{10} = h_1,$$
$$d_1'x^* + d_{10} = q_1,\ h_1d_{10} + d_{20}h_2 - c_{10}q_1 - c_{20}q_0 =$$
$$b_1a_1 + ... + b_ra_r,\ q_i > 0,\ i = 1, 2$$

The proof is complete.

From Lemma 6.2.2, it follows

Corollary 6.2.1. *The set of all the efficient points in any face of S is convex.*

Theorem 6.2.2. *The set $E \cap T$ is defined by linear equations and inequalities and E is a finite reunion of such sets.*

As an illustration, consider

Example 6.2.2. [313] (Figure 6.2.1)

$$\max \{ F_1(x) = x_1/(x_1 + x_2);\ F_2(x) = (-3x_1 + 2x_2)/(x_1 - x_2 + 3) \}$$

subject to

$$x \in S = \{ x \in R^2 \mid 2x_1 - 4x_2 \leq 4,\ -x_1 - 2x_2 \leq -2,\ -x_1 + x_2 \leq 1,\ x_1 \leq 6 \}$$

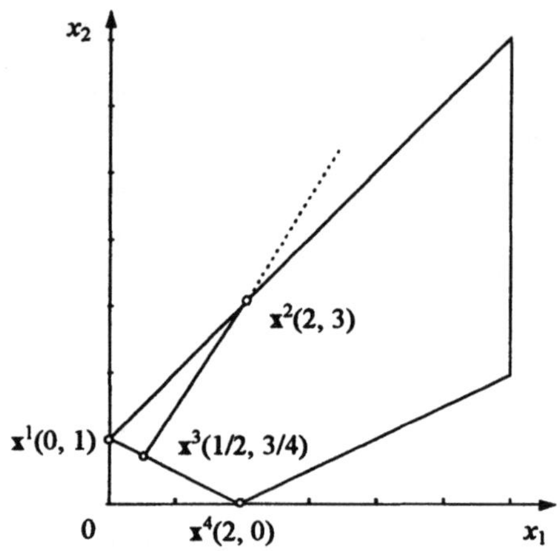

Figure 6.2.1

The sets of efficient points on the faces of S are the points and the straight line segments $\{x^1\}$, $[x^1, x^2]$, (x^2, x^3), $[x^3, x^4)$, $\{x^4\}$. Hence, $E = \{x^1\} \cup (x^1, x^2] \cup (x^2, x^3) \cup [x^3, x^4) \cup \{x^4\}$ is a finite reunion of points or straight line segments.

6.3. PROPERTIES OF THE SET OF EFFICIENT (WEAK EFFICIENT) POINTS FOR THE LINEAR FRACTIONAL PROGRAMMING PROBLEM WITH MULTIPLE OBJECTIVE FUNCTIONS

In this section, we are going to show that, for the problem LFVMP, any efficient solution is properly efficient and that the set E^w is closed. Definition 6.1.5 implies the following lemma, which is true for more than just the linear fractional functions:

Lemma 6.3.1. *If the efficient solution* x^* *of problem LFVMP is not properly efficient, then, for an efficiency function* F_r, *a sequence of points from* S, $\{x^1, x^2, ...\}$ *exists with the properties*

$$F_r(x^n) > F_r(x^*),\ n = 1, 2, \ldots \tag{6.3.1}$$

For every $i = 1, \ldots, k$ *we have*

a) $F_i(x^n) < F_i(x^*),\ n = 1, 2, \ldots$ *or*

b) $F_i(x^n) \geq F_i(x^*),\ n = 1, 2, \ldots$ (6.3.2)

$$J^* = \left\{ i \,\middle|\, F_i(x^n) < F_i(x^*),\ n = 1, 2, \ldots \right\} \neq \emptyset \tag{6.3.3}$$

$$\lim_{n \to \infty} (F_r(x^n) - F_r(x^*)/(F_j(x^*) - F_j(x^n)) = +\infty, \quad \forall\ j \in J^*. \tag{6.3.4}$$

Theorem 6.3.1. *If x^* is efficient solution for problem LFVMP, then x^* is properly efficient solution.*

Proof. By *reductio ad absurdum*, assume that the efficient solution x^* for problem LFVMP is not properly efficient. Assume that $x^* \in T$, the face corresponding to $J = \{1, \ldots, r\}$. According to Lemma 6.3.1, there is a sequence $\{x^1, x^2, \ldots\}$ with the properties (6.3.1) - (6.3.4). Construct the sequence $y^n = (x^n - x^*)/\|x^n - x^*\|$, $n = 1, 2, \ldots$. Since S is compact, we can assume that both sequences, $\{x^1, x^2, \ldots\}$ and $\{y^1, y^2 \ldots\}$, are convergent. Also, we can assume that $F_i(x^*) = 0$, $i = 1, \ldots, k$ (*i.e.* $c_i'x^* = -c_{i0}$) from which it follows that

$$F_i(x^n) - F_i(x^*) = F_i(x^n) = (c_i'x^n - c_i'x^*)/(d_i'y^n + d_{i0}) =$$
$$= (c_i'y^n)\|x^n - x^*\|/(d_i'x^n + d_{i0})$$

If $F_i(x^*) = 0$, then $\nabla F_i(x^*) = (1/(d_i'x^* + d_{i0}))c_i$, and efficiency characterization Equation (6.2.10) becomes

$$\sum_{i=1}^{k} w_i c_i = \sum_{i=1}^{r} a_i p_i \tag{6.3.5}$$

For $j \in J^*$, we have

$$(F_r(x^n) - F_r(x^*))/(F_j(x^*) - F_j(x^n)) = \tag{6.3.6}$$
$$(d_j'x^n + d_{j0})(c_r'y^n)/(d_r'x^n + d_{r0})(-c_j'y^n).$$

From (6.3.6) and (6.3.4), it follows that

$$\lim_{n \to \infty} c_j'y^n/c_r'y^n = 0, \quad j \in J^* \tag{6.3.7}$$

From Lemma 6.2.1 and from the gradient expression $\nabla F_i(x^*)$, it follows that

$$c_r'y^n > 0 \text{ and } c_i'y^n \geq 0 \text{ for any } n = 1, 2, \ldots \text{ and any } i \notin J^* \tag{6.3.8}$$

But x^* is efficient and, according to Theorem 6.2.1 there is $w_i \geq 1$ $(i = 1, \ldots, k)$ and $a_i \geq 0$ $(i = 1, \ldots, r)$ such that Equation (6.3.5) takes place. Multiplying Equation (6.3.5) by y^n and dividing by $c_r'y^n$, we get

$$(1/c_r'y^n)\sum_{i=1}^{k} w_i c_i' y^n = w_r + \sum_{i \in J^*, i \neq r} w_i (c_i' y^n / c_r' y^n) + \sum_{j \in J^*} w_j (c_j y^n / c_r' y^n) \qquad (6.3.9)$$

For n sufficiently large, Equation (6.3.7) implies that, in (6.3.9), the last term is negligible and (6.3.8), (6.3.9) and $w_i \geq 1 \quad (i = 1, ..., k)$, imply that

$$\sum_{i=1}^{k} (w_i c_i)' \, y^n > 0 \qquad (6.3.10)$$

Since $p_i' x^0 = b_i$ and $p_i' x^n \leq b_i$, $i \in J$, we have $p_i' y^n \leq 0$, $\forall \, i \in J$. From here, and from condition $a_i \geq 0 \quad (i \in J)$, it follows that

$$\sum_{i \in J} (a p_i)' \, y^n \leq 0 \qquad (6.3.11)$$

Equations (6.3.10) and (6.3.11) contradict Equation (6.3.5). Hence, x^* is properly efficient solution.

The set E is not, in general, closed even if S is compact. An interior point of S can be efficient without having $E = S$, as it happens in the linear case. Hughes [498] derived general conditions for the bi-criteria linear fractional problem to have interior efficient solutions. He stated that the set of interior efficient solutions (if any) is contained in a hyperplane separating the points where each objective function is optimized.

We show now that E^w, the set of weak efficient points for problem LFVMP, is closed. The following theorem is true if the function $F = (F_1, ..., F_k)$ is continuous and if S is an arbitrary set.

Theorem 6.3.2. *The set E^w of weak efficient solutions for problem LFVMP is closed.*

Proof. Let $\bar{E}^w$ be the closure of set E^w. Let's show that $\bar{E}^w = E^w$. Since $E^w \subseteq \bar{E}^w$, we must show that $\bar{E}^w \subseteq E^w$. Choose $x^* \in \bar{E}^w$. Hence, there is a sequence $\{x^n\}$, $n = 1,2, ...,$ $x^n \in E^w$ such that $\lim_{n \to \infty} x^n = x^*$. By *reductio ad absurdum*, assume that $x^* \notin E^w$ which means, according to Definition 6.1.2, that there is $x \in S$ such that $F(x^*) < F(x)$. From the continuity of F, it follows that $F(x^n) < F(x)$, for sufficiently large n. But this contradicts the fact that $x^n \in E^w$. Hence, $x^* \in E^w$ and the proof is complete.

6.4. DETERMINATION OF EFFICIENT SOLUTIONS FOR LINEAR FRACTIONAL VECTOR PROGRAMMING PROBLEMS

Consider linear fractional vector programming problem

$$\max F(x) = \{ F_1(x) = f_1(x)/g_1(x) = c_1'x/d_1'x, \ldots, F_k(x) = f_k(x)/g_k(x) = c_k'x/d_k'x\} \tag{6.4.1}$$

subject to

$$x \in S = \{ x \in R^n \mid Ax = b, \ x \geq 0 \} \tag{6.4.2}$$

where the matrices and the vectors have the usual dimensions as given in Section 6.2. As usual, make the assumption

$$g_r(x) = d_r'x > 0, \quad \forall \ r = 1, \ldots, k \text{ and } x \in S \tag{6.4.3}$$

With the points x, $x^* \in S$ and $j \in I = \{1, \ldots, k\}$, given, we are going to use the notations

$$S_j(x^*) = \{x \in S \mid F_j(x) > F_j(x^*)\} \ j \in I,$$
$$S(x^*) = \{ j \in I \mid S_j(x^*) \neq \varnothing \},$$
$$S(x, x^*) = \{i \in I \mid F_i(x) < F_i(x^*)\}.$$

Let $x^0 \in S$. Define the matrix $H = (h_1, \ldots, h_k)$ where $h_i = c_i g_i(x^0) - d_i f_i(x^0)$, $(i = 1, \ldots, k)$. From the definition of the coefficients h_i and from (6.4.3), it follows that

Observation 1. $h_i'x^0 = 0$, $i = 1, \ldots, k$.

Observation 2. $(F_1(x), \ldots, F_k(x)) > (F_1(x^0), \ldots, F_k(x^0))$ if and only if $h_i'x \geq 0$, $\forall \ i = 1, \ldots, k$ and for at least an index i_0 we have $h_{i_0}'x > 0$.

Associate to problem (6.4.1) - (6.4.2) a linear programming problem with several objective functions:

$$\max_{x \in S} G(x) = \left\{G_1(x) = h_1'x, \ldots, G_k(x) = h_k'x\right\} \tag{6.4.4}$$

The following theorem reduces the problem of finding the efficient points for problem (6.4.1) - (6.4.2) to that of finding the efficient points for problem (6.4.4) having linear objective functions.

Theorem 6.4.1. *$x^0 \in S$ is an efficient solution for problem (6.4.1) - (6.4.2) if and only if x^0 is efficient solution for problem (6.4.4).*

Proof. Let $x^0 \in S$ be efficient solution for problem (6.4.1) - (6.4.2), and assume that x^0 is not efficient solution for problem (6.4.4). Hence, a $x^1 \in S$ exists such that

$$G_i(x^1) = h_i^1 x^1 \geq h_i'x^0 = G_i(x^0), \quad i = 1, \ldots, k \tag{6.4.5}$$

and, for at least an index i_0, we have

$$G_{i_0}(x^1) > G_{i_0}(x^0) \tag{6.4.6}$$

From the definition of the functions $G_i(x)$, $i = 1, ..., k$, it follows that Equations (6.4.5) - (6.4.6) become

$$c'_i g_i(x^0)x^1 - d'_i f_i(x^0)x^1 \geq c'_i g_i(x^0)x^0 - d'_i f_i(x^0)x^0 = 0, \quad i \in I \tag{6.4.5'}$$

$$c'_{i_0} g_{i_0}(x^0)x^1 - d'_{i_0} f_{i_0}(x^0)x^1 > 0 \tag{6.4.6'}$$

From these relations and from (6.4.3), it follows that $F_i(x^1) \geq F_i(x^0)$, $i \in I$ and $F_{i_0}(x^1) > F_{i_0}(x^0)$, which contradicts the assumption that $x^0 \in S$ is efficient solution for problem (6.4.1) - (6.4.2). The converse of the theorem is similarly proven.

To prove a similar theorem for properly efficient solutions, introduce the additional notations

$$\begin{aligned} \bar{S}_j(x^0) &= \{x \in S \mid h'_j x > h'_j x^0\}, \ j \in I \\ \bar{S}(x^0) &= \{ j \in I \mid \bar{S}_j(x^0) \neq \varnothing\} \\ \bar{S}(x, x^0) &= \{i \in I \mid h'_i x < h'_i x^0\} \end{aligned} \tag{6.4.7}$$

The following relations follow from the definition of vectors h_i, $i = 1, ..., k$,

$$\bar{S}_j(x^0) = S_j(x^0) \ (j \in I), \ \bar{S}(x^0) = S(x^0), \ \bar{S}(x, x^0) = S(x, x^0), \ x \in S \tag{6.4.8}$$

Theorem 6.4.2. *Assume that* $0 < \gamma \leq d'_i x \leq \beta$, $x \in S$, $i \in I$. *Then* $x^0 \in S$ *is properly efficient solution for problem (6.4.1) - (6.4.2) if and only if* x^0 *is properly efficient solution for problem (6.4.4).*

Proof. If x^0 is properly efficient solution for problem (6.4.1) - (6.4.2), it follows that there is a scalar $\alpha > 0$ such that, for any $j \in S(x^0)$ and any $x \in S_j(x^0)$, an index $i \in S(x, x^0)$ exists such that

$$(F_j(x) - F_j(x^0))/(F_i(x^0) - F_i(x)) \leq \alpha \tag{6.4.9}$$

Using Equations (6.4.7) - (6.4.9) and Observation 1, we get

$$\begin{aligned} \alpha &\geq \frac{(h'_j x)/((d'_j x)(d'_j x^0))}{(-h'_i x)/((d'_i x)(d'_i x^0))} = \\ &\frac{(h'_j x - h'_j x^0)}{(h'_i x^0 - h'_i x)} \cdot \frac{(d'_i x)(d'_i x^0)}{(d'_j x)(d'_j x^0)} \geq \\ &((h'_j x) - (h'_j x^0))/((h'_i x^0) - (h'_i x)) \cdot \gamma^2/\beta^2 \end{aligned} \tag{6.4.10}$$

i.e. there is a scalar $r = \alpha\beta^2/\gamma^2 > 0$ such that, for any $j \in \bar{S}(x^0)$ and any $x \in \bar{S}_j(x^0)$, an $i \in \bar{S}(x, x^0)$ exists such that

$$(h'_j x - h'_j x^0)/(h'_i x^0 - h'_i x) \leq r$$

From here, it follows that x^0 is a properly efficient solution for problem (6.4.4).

The converse of the theorem is similarly proven.

The following theorem is identical with Theorem 6.3.1, only that its result was obtained using the equivalence between vector fractional problem and programming problem with several objective functions.

Theorem 6.4.3. *Assume that the following condition is satisfied*

$$0 < \gamma \leq d_i'x \leq \beta, \quad x \in S, \quad i \in I.$$

Then, $x^0 \in S$ is properly efficient solution for problem (6.4.1) - (6.4.2) if and only if x^0 is efficient solution.

Proof. If $x^0 \in S$ is efficient solution for problem (6.4.1) - (6.4.2), then, according to Theorem 6.4.1, x^0 is efficient solution for problem (6.4.4). Since, in the linear case, the sets of the efficient solutions and of the properly efficient solutions coincide, it follows that x^0 is properly efficient solution for problem 6.4.4. Finally, from Theorem 6.4.2, it follows that x^0 is properly efficient solution for problem (6.4.1) - (6.4.2).

Next, based on the things presented so far, we are going to sketch an algorithm for obtaining an efficient solution of problem (6.4.1) - (6.4.2). The algorithm also gives a sets of efficient solutions formed from vertices of S. The algorithm consists of the following steps :

Step 1. Determine a vertex x^0 of the polyhedron S. Let B_0 be the basic matrix corresponding to x^0.

Step 2. Given a basic solution x^{t-1} of problem (6.4.4), *i.e.*, a vertex of S, with the basic matrix B_{t-1}, consider the matrix

$$H^t = (h_1^t, \ldots, h_k^t) \text{ with } h_i^t = c_i(d_i'x^{t-1}) - d_i(c_i'x^{t-1}), \quad i = 1, \ldots k$$

Construct, for each i, the vector $B_{h_i^t}$ having m components, specifically those components of h_i^t corresponding to the basic variables from x^{t-1}.

Step 3. Calculate vectors $\bar{h}_i^t = h_i^t - B_{h_i^t} B_{t-1}^{-1} A$, $i = 1, \ldots, k$ and go to the next step.

Step 4.
a) if condition (C1): there is a $s \in \{1,2, \ldots, n\}$ such that $P_s = (\bar{h}_{s1}^t, \bar{h}_{s2}^t, \ldots, \bar{h}_{sr}^t) > 0$, is satisfied, then x^{t-1} is not efficient solution for problem (6.4.4) corresponding to H^{t-1} and hence, according to Theorem 6.4.1, x^{t-1} is not efficient solution for problem (6.4.1) - (6.4.2). Go to Step 5.
b) if condition (C1) is not satisfied, then x^{t-1} is efficient solution for problem (6.4.4) with $H = H^{t-1}$ and hence, it is efficient solution for problem (6.4.1) - (6.4.2). The algorithm stops and chooses x^{t-1} as efficient solution.

Step 5. Perform a simplex iteration starting with x^{t-1} and obtain the improved solution x^t. For the solution x^t, repeat the algorithm starting with Step 2.

Note that the solution x^t, obtained at Step 5 in the non-degenerate case, has the property that there is an index $i_0 \in I$ such that $h_{i_0}^{\prime\prime\prime} x^{t-1} > h_{i_0}^{\prime\prime\prime} x^{t-1}$, and, for the other indices, one has $h_i^{\prime\prime t} x^t \geq h_i^{\prime\prime t} x^{t-1}$, $i = 1, \ldots, r$, *i.e.*,

$$F(x^t) > F(x^{t-1}) \tag{6.4.11}$$

which means that in Step 5 one gets an improvement of the vector function F. If degeneration of solution x^{t-1} appears, then this can be eliminated as in linear programming.

The algorithm finishes after a finite number of steps. Indeed, from Equation (6.4.11), it follows that in the sequence $x^0, x^1, \ldots, x^t, \ldots$, we do not have two consecutive terms that are equal and also that the sequence is finite since the set of vertices of the polyhedron S is finite.

Next, we are going to associate to problem (6.4.1) - (6.4.2) a linear vector programming problem, different from problem (6.4.4), and will give a necessary condition for x^0 to be efficient for problem (6.4.1) - (6.4.2):

$$\max_{x \in S} (c_1' x, \ldots, c_k' x, -d_1' x, \ldots, -d_k' x) \tag{6.4.12}$$

Theorem 6.4.4. *If x^0 is an efficient solution for problem (6.4.1) - (6.4.2), then x^0 is efficient solution for problem (6.4.12).*

Proof. Let $x^0 \in S$ be efficient solution for problem (6.4.1) - (6.4.2) and assume that x^0 is not efficient solution for problem (6.4.12). Hence, there is a vector $x' \in S$ such that

$$c_i' x' \geq c_i' x^0; \quad d_i' x' \leq d_i' x^0, \quad i = 1, \ldots, k \tag{6.4.13}$$

and, for at least an index i_0, we have

$$c_{i_0}' x' > c_{i_0}' x^0 \text{ or } d_{i_0}' x' < d_{i_0}' x^0. \tag{6.4.14}$$

From Equations (6.4.13) and (6.4.14), it follows that

$$c_i' x' / d_i' x' \geq c_i' x^0 / d_i' x^0, \quad i = 1, \ldots, k \text{ and for } i_0$$
$$c_{i_0}' x' / d_{i_0}' x' > c_{i_0}' x^0 / d_{i_0}' x^0$$

which contradicts the assumption that $x^0 \in S$ is efficient solution for problem (6.4.1) - (6.4.2).

The converse of the theorem is not always true, as shown by the following example.

Example 6.4.1. For simplicity, take $k = 1$. Let max $F(x) = x_1/x_2$ on the domain $S = \{x \in R^2 \mid x_1 + x_2 \geq 4;\ -x_1 + 11x_2 \geq 14,\ 3x_1 - 2x_2 \leq 20,\quad -7x_1 + 59x_2 \leq 225\}$. The domain of the feasible solutions is given in Figure 6.4.1.

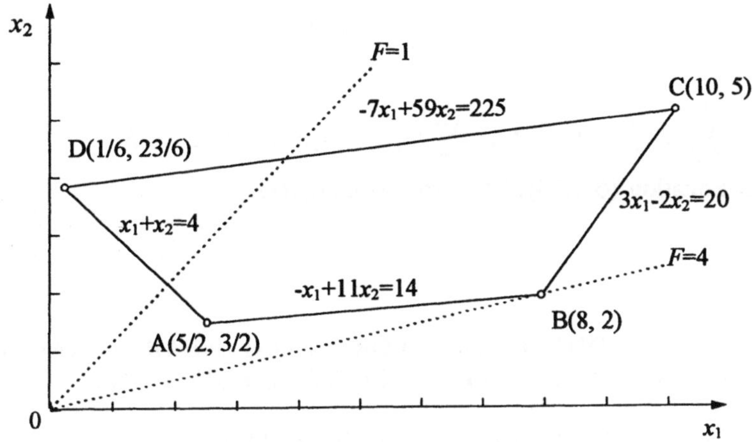

Figure 6.4.1

The optimal value of $F(x)$ is 4 and is obtained at the point $B(8, 2)$. Take a point (x_1^0, x_2^0) on the straight line segment AB *i.e.* $-x_1^0 + 11x_2^0 = 14$. This point is efficient for problem (6.4.12) associated to the given problem. Indeed, if (x_1^0, x_2^0) is not efficient, then there is $(x_1^*, x_2^*) \in S$ such that $x_1^* \geq x_1^0$ and $x_2^* < x_2^0$ or $x_1^* > x_1^0$ and $x_2^* \leq x_2^0$. From here, it follows that $-x_1^* + 11x_2^* < 14$, which contradicts the fact that $(x_1^*, x_2^*) \in S$. Hence, each point of the segment AB (it can be proved that also of the segment BC) is efficient for problem (6.4.12) associated to the given problem. But such points (except the point B) are not efficient for the given problem.

As a corollary to Theorem 6.4.4, one gets the following efficiency condition for the linear fractional programming problem with several objective functions.

Theorem 6.4.5. *Let* x^0 *be an efficient solution for problem (6.4.1) - (6.4.2). Then there are two vectors* $u^0 > 0$ *and* $v^0 > 0$, *such that*

$$\sum_{i=1}^{k} (u_i^0 c_i' - v_i^0 d_i')x^0 \geq \sum_{i=1}^{k} (u_i^0 c_i' - v_i^0 d_i')x, \quad \forall\ x \in S \tag{6.4.15}$$

Proof. Since x^0 is efficient solution for problem (6.4.1) - (6.4.2), according to Theorem 6.4.4, x^0 is also efficient solution for problem (6.4.12), and hence, according to Theorem 6.1.1, there are parameters $u^0 > 0$ and $v^0 > 0$ such that x^0 is optimal solution for problem

$$\max_{x \in S} \sum_{i=1}^{k} (u_i^0 c_i' - v_i^0 d_i')x$$

i.e.,

$$\sum_{i=1}^{k} (u_i^0 c_i' - v_i^0 d_i')x^0 \geq \sum_{i=1}^{k} (u_i^0 c_i' - v_i^0 d_i')x, \quad \forall\, x \in S$$

which represents Equation (6.4.15). The proof is complete.

6.5. OPTIMALITY CONDITIONS FOR NONLINEAR FRACTIONAL PROGRAMMING WITH SEVERAL OBJECTIVE FUNCTIONS

Consider the nonlinear fractional vector programming problem

$$\max F(x) = \{F_1(x) = f_1(x)/g_1(x), \ldots, F_k(x) = f_k(x)/g_k(x)\} \tag{6.5.1}$$

subject to

$$x \in S = \{x \in R^n \mid Ax \leq b, \ x \geq 0\} \tag{6.5.2}$$

where A is an $m \times n$ matrix, and b is a column vector with m components. Assume that

f_i, $i \in I$ are convex and differentiable

$$\text{and } f_i(x) \geq 0, \ \forall\, i \in I \text{ and } x \in S \tag{6.5.3}$$

g_i, $i \in I$ are concave and differentiable

$$\text{and } g_i(x) > 0, \ \forall\, i \in I \text{ and } x \in S \tag{6.5.4}$$

We construct a necessary condition for a solution to be efficient for problem (6.5.1) - (6.5.2). The condition corresponding to linear vector programming is both necessary and sufficient. Consider the vectors $\bar{u} > 0$ and $\bar{v} > 0$ having k components.

Lemma 6.5.1. *If $x^0 \in S$ is efficient solution for problem (6.5.1) - (6.5.2), then the following linear programming problem*

$$\max \sum_{i=1}^{k} \bar{u}_i y_i + \sum_{i=1}^{k} \bar{v}_i z_i \tag{6.5.5}$$

subject to

$$Ax \leq b; \ x \geq 0; \ x' \nabla f_i(x^0) - y_i = x^{0\prime} \nabla f_i(x^0), \ i \in I;$$

$$x' \nabla g_i(x^0) + z_i = x^{0\prime} \nabla g_i(x^0), \ i \in I; \ y, z \geq 0 \tag{6.5.6}$$

has an optimal solution (x^*, y^*, z^*) *with* $y^* = z^* = 0$.

Proof. By *reductio ad absurdum*, assume that problem (6.5.5.) - (6.5.6) admits an optimal solution (x^*, y^*, z^*), but $y^* \neq 0$. Assume, for example, that $y_k^* \neq 0$. Since (x^*, y^*, z^*) is feasible solution for problem (6.5.5) - (6.5.6), we have

$$x^{*\prime} \nabla f_k(x^0) > x^{0\prime} \nabla f_k(x^0) \tag{6.5.7}$$

$$x^{*\prime} \nabla f_i(x^*) \geq x^{0\prime} \nabla f_i(x^0), \quad i = 1,2, ..., k-1 \text{ and} \tag{6.5.8}$$

$$x^{*\prime} \nabla g_i(x^0) \leq x^{0\prime} \nabla g_i(x^0), \quad i \in I \tag{6.5.9}$$

From the convexity of the function f_k and from (6.5.7), it follows that

$$f_k(x^*) - f_k(x^0) \geq (x^* - x^0)' \nabla f_k(x^0) > 0$$

hence,

$$f_k(x^*) > f_k(x^0) \tag{6.5.10}$$

From the convexity of functions f_i, $i \in I$, $i \neq k$, and from (6.5.8), it follows that

$$f_i(x^*) - f_i(x^0) \geq (x^* - x^0)' \nabla f_i(x^*) \geq 0$$

i.e.,

$$f_i(x^*) \geq f_i(x^0), \quad i \in I, \quad i \neq k \tag{6.5.11}$$

Also, from the concavity of functions g_i, $i \in I$ and (6.5.9), it follows that

$$g_i(x^*) - g_i(x^0) \leq (x^* - x^0)' \nabla g_i(x^0) \leq 0$$

i.e.,

$$g_i(x^*) \leq g_i(x^0), \quad i \in I \tag{6.5.12}$$

From (6.5.10), (6.5.11), (6.5.12) and (6.5.3), it follows that $F_i(x^*) \geq F_i(x^0)$, $i \in I$, the inequality being strict for $i = k$. This contradicts the assumption that x^0 is efficient solution for problem (6.5.1) - (6.5.2). The proof follows in a similar manner if we assume that $z^* \neq 0$. In conclusion, problem (6.5.5) - (6.5.6) admits an optimal solution (x^*, y^*, z^*) with $y^* = z^* = 0$. The proof is complete.

Theorem 6.5.1. *If $x^0 \in S$ is efficient solution for problem (6.5.1) - (6.5.2), then the vectors $u^*, v^* > 0$ exist such that:*

$$u^{*\prime}[\nabla f(x^0)]' x^0 - v^{*\prime}[\nabla g(x^0)]' x^0 \geq u^{*\prime}[\nabla f(x^0)]' x - \\ - v^{*\prime}[\nabla g(x^0)]' x, \quad x \in S$$

where

$$\nabla f(x) = [\nabla f_1(x), ..., \nabla f_k(x)] \text{ is an } n \times k \text{ matrix.}$$

Proof. Since x^0 is efficient solution for problem (6.5.1) - (6.5.2), then, according to Lemma 6.5.1, the problem (6.5.5) - (6.5.6) has an optimal solution (x^*, y^*, z^*) with $y^* = z^* = 0$.

The dual of problem (6.5.5) - (6.5.6) is

$$\min \left(w'b - u'[\nabla f(x^0)]' x^0 + v'[\nabla g(x^0)]' x^0\right) \tag{6.5.13}$$

subject to

$$w'A - u'[\nabla f(x^0)]' + v'[\nabla g(x^0)]' \geq 0$$
$$u \geq \bar{u}^0 > 0, \quad v \geq \bar{v} > 0, \quad w \geq 0 \tag{6.5.14}$$

Corresponding to the optimal solution (x^*, y^*, z^*), we have an optimal solution (u^*, v^*, w^*) for the dual problem (6.5.13) - (6.5.14) and the optimal values coincide, *i.e.*

$$w^{*\prime}b - u^{*\prime}[\nabla f(x^0)]'x^0 + v^{*\prime}[\nabla g(x^0)]'x^0 = \sum_{i=1}^{k} \bar{u}_i y_i + \sum_{i=1}^{k} \bar{v}_i z_i = 0 \tag{6.5.15}$$

For u^* and v^* fixed, w^* is optimal solution of the problem:

$$\min\ w'b \tag{6.5.16}$$

subject to

$$w'A \geq u^{*\prime}[\nabla f(x^0)]' - v^{*\prime}[\nabla g(x^0)]', \quad w \geq 0 \tag{6.5.17}$$

The dual of the problem (6.5.16) - (6.5.17) is problem

$$\max\ u^{*\prime}[\nabla f(x^0)]'y - v^{*\prime}[\nabla g(x^0)]'y \tag{6.5.18}$$

subject to

$$Ay \leq b, \quad y \geq 0 \tag{6.5.19}$$

From (6.5.15), it follows that

$$w^{*\prime}b = u^{*\prime}[\nabla f(x^0)]'x^0 - v^{*\prime}[\nabla g(x^0)]'x^0$$

and since w^* is optimal solution of the problem (6.5.16) - (6.5.17), from duality theory, it follows that x^0 is optimal solution of the problem (6.5.15) - (6.5.20), *i.e.*,

$$u^{*\prime}[\nabla f(x^0)]'x^0 - v^{*\prime}[\nabla g(x^0)]'x^0 \geq u^{*\prime}[\nabla f(x^0)]'x - v^{*\prime}[\nabla g(x^0)]'x, \quad \forall\ x \in S$$

which is the condition mentioned in the statement of the theorem.

6.6. DETERMINATION OF EFFICIENT (PROPERLY EFFICIENT) SOLUTIONS FOR NONLINEAR FRACTIONAL VECTOR PROGRAMMING PROBLEMS

Consider the nonlinear fractional vector programming problem:

$$\max_{x \in S} F(x) = \{F_1(x) = f_1(x)/g_1(x), \ldots, F_k(x) = f_k(x)/g_k(x)\} \tag{6.6.1}$$

where $S \subseteq R^n$, $f_i, g_i : S \to R$, $\forall\ i \in I = \{1, 2, \ldots, k\}$. Regarding the functions g_i, $i \in I$, we make the assumption

$$\min \{ g_i(x) : i \in I \} > 0, \quad \forall\ x \in S \tag{6.6.2}$$

For an element $y \in S$, consider the functions $F_i^y : S \to R$ and $F^y : S \to R^k$ where

$$F_i^y(x) = f_i(x) \cdot g_i(y) - f_i(y) \cdot g_i(x), \quad \forall\ x \in S, \ i \in I \tag{6.6.3}$$

and

$$F^y(x) = (F_1^y(x), \ldots, F_k^y(x)) \tag{6.6.4}$$

For a given $y \in S$, to the nonlinear fractional vector programming problem associate an auxiliary nonlinear fractional programming problem:

$$PA(y): \max_{x \in S} F^y(x)$$

The following results show that, for problems (6.6.1) and $PA(y)$, the efficient solutions (weak efficient or properly efficient) coincide.

Theorem 6.6.1. *Under the assumption (6.6.2),* $x^0 \in S$ *is efficient solution for problem (6.6.1) if and only if* x^0 *is efficient solution for problem* $PA(x^0)$.

Proof. Let $x^0 \in S$ be efficient solution for problem (6.6.1) which means, according to Definition 6.1.3, that if:

$$f_i(x)/g_i(x) > f_i(x^0)/g_i(x^0) \text{ for a } i \in I \text{ and } x \in S, \tag{6.6.5}$$

then there is an index $j \in I_i$ such that

$$f_j(x)/g_j(x) < f_j(x^0)/g_j(x^0) \tag{6.6.6}$$

This is equivalent, using (6.6.2) and (6.6.3), with the fact that, if

$$F_i^{x^0}(x) > 0 = F_i^{x^0}(x^0) \text{ for a } i \in I \text{ and } x \in S, \tag{6.6.7}$$

then there is an index $j \in I_i$ for which

$$F_j^{x^0}(x) < 0 = F_j^{x^0}(x^0) \tag{6.6.8}$$

Hence, x^0 is efficient solution for problem $PA(x^0)$. The proof is complete.

A similar result takes place for the weak efficient solutions. To prove a similar result for the properly efficient solutions, we need the following assumption:

there is $M_1 > 0$ such that

$$\min \{g_i(x)/g_j(x) \mid i \in I, j \in I_i\} \leq M_1 \quad \forall\, x \in S \tag{6.6.9}$$

Such an assumption is verified if, for example, S is a compact set, the functions g_i, $i \in I$ are continuous on S, and (6.6.2) takes place. But if S is not compact and the functions g_i, $i \in I$ are not continuous, the assumption (6.6.9) is verified if

$$0 < \gamma \leq g_i(x) \leq \beta, \ \forall \ i \in I, \ \ x \in S \tag{6.6.10}$$

Condition (6.6.10) appears in Theorem 6.4.2 for the proof of a similar result in the case of linear fractional vector programming.

Theorem 6.6.2. *Under the assumptions (6.6.2) and (6.6.9),* $x^0 \in S$ *is properly efficient solution for problem (6.6.1) if and only if it is properly efficient solution for problem* $PA(x^0)$.

Proof. Necessity. Let $x^0 \in S$ be efficient solution for problem (6.6.1). Let's show that x^0 is properly efficient solution for problem $PA(x^0)$. Let $i \in I$ and $x \in S$ be such that

$$F_i^{x^0}(x) > F_i^{x^0}(x^0) = 0 \tag{6.6.11}$$

Using (6.6.2) and (6.6.3), it follows that

$$F_i(x) > F_i(x^0) \tag{6.6.12}$$

From (6.6.12) and from the fact that x^0 is properly efficient solution (hence, also efficient) for problem (6.6.1), it follows that there is a scalar $M > 0$ and $j \in I_i$, such that

$$F_j(x) < F_j(x^0) \tag{6.6.13}$$

and

$$(F_i(x) - F_i(x^0))/(F_j(x^0) - F_j(x)) \leq M \tag{6.6.14}$$

From Equations (6.6.2), (6.6.3), (6.6.9) and (6.6.14), it follows that

$$\frac{F_i^{x^0}(x) - F_i^{x^0}(x^0)}{F_j^{x^0}(x^0) - F_j^{x^0}(x)} \equiv \frac{F_i(x) - F_i(x^0)}{F_j(x^0) - F_j(x)} \cdot \frac{f_j(x) \cdot f_i(x^0)}{f_i(x) \cdot f_j(x^0)} \leq M \cdot M_1^2 \tag{6.6.15}$$

From (6.6.11) and (6.6.15), it follows that x^0 is a properly efficient solution for problem $PA(x^0)$.

Sufficiency. Assume that x^0 is properly efficient solution for problem $PA(x^0)$ hence, there is $M'' > 0$ such that, for any $i \in I$ and $x \in S$ for which

$$F_i^{x^0}(x) > F_i^{x^0}(x^0) = 0, \tag{6.6.16}$$

there is at least an index $j \in I_i$ such that

$$F_j^{x^0}(x) < F_j^{x^0}(x^0) = 0 \tag{6.6.17}$$

and

$$(F_i^{x^0}(x) - F_i^{x^0}(x^0))/(F_j^{x^0}(x^0) - F_j^{x^0}(x)) \leq M'' \tag{6.6.18}$$

Due to (6.6.2), Equation (6.6.15) is equivalent to (6.6.12), and (6.6.17) to (6.6.13). Even more, (6.6.15) and (6.6.18) imply that, for $M = M'' \cdot M_1^2$, the inequality (6.6.14) takes place. In conclusion, there is $M > 0$ such that, for any $i \in I$ and $x \in S$, for which (6.6.12) takes place, there is $j \in I_i$ such that (6.6.13) and (6.6.14) take place, which means that x^0 is properly efficient solution for problem (6.6.1). The proof is complete.

We now associate to the problem (6.6.1) - (6.6.2) another nonlinear vector programming problem, and give a necessary condition for x^0 to be efficient for problem (6.6.1) - (6.6.2).

$$\max_{x \in S} G(x) = \{f_1(x), \ldots, f_k(x), -g_1(x), \ldots, -g_k(x)\} \tag{6.6.19}$$

Additionally, to the assumption (6.6.2), we also add

$$f_i(x) \geq 0, \quad \forall\, i \in I \text{ and } \forall\, x \in S \tag{6.6.2'}$$

Theorem 6.6.3. *If $x^0 \in S$ is efficient solution to problem (6.6.1) - (6.6.2), then x^0 is efficient solution to problem (6.6.19).*

Proof. Let $x^0 \in S$ be efficient solution to problem (6.6.1) - (6.6.2). If x^0 is not efficient solution to problem (6.6.19), it follows that there is a $y \in S$ and a $j \in I$ such that

a) $f_i(y) \geq f_i(x^0), \quad \forall\, i \in I_j, \quad g_i(y) \leq g_i(x^0), \quad \forall\, i \in I$ and $f_j(y) > f_j(x^0)$

or

b) $f_i(y) \geq f_i(x^0), \quad \forall\, i \in I, \quad g_i(y) \leq g_i(x^0), \quad \forall\, i \in I_j$ and $g_i(y) < g_i(x^0)$

From either a) or b), and due to assumption (6.6.2'), it follows that

$$f_i(y)/g_i(y) \geq f_i(x^0)/g_i(x^0), \quad \forall\, i \in I_j \text{ and } f_j(y)/g_j(y) > f_j(x^0)/g_j(x^0)$$

which contradicts the fact that x^0 is efficient solution for problem (6.6.1) - (6.6.2). The proof is complete.

Theorem 6.6.4. *The feasible solution x^0 is efficient for problem (6.6.1) - (6.6.2) if and only if x^0 maximizes each f_i and minimizes each g_i on the set*

$$S_i = \{\, x \in S \mid f_j(x)g_j(x^0) \geq f_j(x^0)g_j(x), \quad \forall\, j \neq i \,\}$$

Proof. Consider that x^0 is efficient solution for problem (6.6.1) - (6.6.2) and assume that x^0 does not maximize each f_i and does not minimize each g_i on S_i. Consider that there is an index k and a $y \in S_i$, which maximizes f_k and minimizes g_k. Hence,

$$f_j(y)g_j(x^0) \geq f_j(x^0)g_j(y), \quad \forall\, j \neq k$$

and

$$f_k(y) > f_k(x^0), \quad g_k(y) < g_k(x^0)$$

From here, it follows that

$$f_j(y)/g_j(y) \geq f_j(x^0)/g_j(x^0), \quad \forall\, j \neq k$$

and

$$f_k(y)/g_k(y) > f_k(x^0)/g_k(x^0)$$

which contradicts the assumption that x^0 is efficient solution for problem (6.6.1) - (6.6.2). Conversely, assume that x^0 maximizes each f_i and minimizes each g_i on set S_i, but that x^0 is not efficient solution for problem (6.6.1) - (6.6.2). Hence, there is $y \in S$ such that

$$f_i(y)/g_i(y) \geq f_i(x^0)/g_i(x^0), \quad \forall\, i \neq h \tag{6.6.20}$$

and

$$f_h(y)/g_h(y) > f_h(x^0)/g_h(x^0) \tag{6.6.21}$$

From (6.6.20), it follows that $y \in S_h$. We distinguish three cases relative to f_h:

1) $f_h(y) > f_h(x^0)$, from where it follows that x^0 does not maximize the function f_h on S_h.

2) $f_h(y) = f_h(x^0)$, and considering (6.6.21), it follows that $g_h(x^0) > g_h(y)$ hence, x^0 does not minimize g_h on S_h.

3) $f_h(y) < f_h(x^0)$, and considering (6.6.21), it follows that $g_h(x^0) > g_h(y)$ hence, x^0 does not minimize g_h on S_h.

In all the three cases we obtained a contradiction. In a similar manner, one can consider three cases relative to g_h. Hence, x^0 is efficient solution to problem (6.6.1) - (6.6.2).

6.7. HISTORICAL NOTES. COMMENTARIES

The results presented in Section 6.2 and Section 6.3 are due to Choo and Atkins [313], [314]. The presentation of results in Section 6.4 followed Ţigan [991], and then Kaul and Gupta [545]. The optimality conditions presented in Section 6.5 are due to Bhatia and Gupta [180]. Theorems 6.6.1 and 6.6.2 of Section 6.6 for the equivalence of efficient (properly efficient) solutions for a nonlinear fractional vector programming problem and for a nonlinear non-fractional vector programming problem are due to Ţigan [992]. Theorems 6.6.3 and 6.6.4 from Section 6.6 are due to Bhatia and Datta [179].

Let's examine briefly other results obtained in fractional vector programming. Interactive solution methods for the vector programming problem with linear fractional objective functions were presented by Sannomiya, Nishikawa and Inazu [776] and by Thanassoulis [974]. In the particular case when only two linear fractional objective functions are present, Warburton [1033] proposed a parametric solution method. An interactive procedure for the pseudo-monotonic vector programming problem was presented by Weber [1037]. In the particular case of linear fractional functions, the problem coincides with that presented in Section 6.4. Kornbluth and Steuer [571], [572] presented a simplex-type algorithm for obtaining weak efficient solutions of a linear fractional programming problem with several objective functions. An important step in their algorithm is that of determining the greatest value (θ_{max}) of x_j for which the solutions associated to $x_j \leq \theta_{max}$ are weak efficient solutions and those for which $x_j > \theta_{max}$ are weak inefficient. Benson [169] revisited this point from the algorithm of Kornbluth and Steuer and presented a new method for the calculation of θ_{max} and gave numerical results from this method. Gupta [441] presented an algorithm for the determination of the set of all the efficient solutions in a vector programming problem with linear fractional objective functions and with zero-one variables. Zolkiewski [1085] (see also [1086]) presented an adaptation of the simplex method to the fractional case and found the set of efficient points for a linear vector programming problem. Nykowski and Zolkiewski [696], [697] proposed a solution procedure for the linear fractional vector programming problem. Slowinski [867], starting from a linear vector programming problem, in which the data were fuzzy numbers, arrived to a linear fractional vector programming problem which he solved through an interactive method. Goal programming with linear fractional objective functions was studied by Charnes and Cooper [286], Kornbluth and Steuer [573], and Stancu-Minasian and Ţigan [926] in the stochastic case.

The duality for the fractional vector programming problem was studied by Gupta [443], in the linear case, by Bhatia and Datta [179], in the nonlinear case using the notion of subgradient of a vector functions, and by Weir [1040]. Weir [1040] showed that to a properly efficient solution in the primal problem corresponds a properly efficient solution in the dual problem. Additionally, Bhatia and Datta [179] gave the necessary optimality condition according to Fritz John and to Kuhn-Tucker for nonlinear fractional vector programming problem (see problem (6.5.1) - (6.5.2)) in which the functions f_i and g_i are not necessarily differentiable.

CHAPTER 7

FRACTIONAL PROGRAMMING IN THE COMPLEX SPACE

In this chapter, we present several results which were obtained in fractional programming in complex space. The idea of studying the mathematical programming in complex space belongs to N. Levinson ("Linear programming in complex space", *J. Math. Anal. Appl.*, **14**, 44-62 (1966)) who extended to the complex space the basic theorems of linear programming. In particular, using a variant in the complex space of Farcas' Lemma from real space, he generalizes duality theorems from real linear programming. Since then, over 160 works of mathematical programming in the complex space (See Dorel I. Duca and I.M. Stancu -Minasian, "Mathematical programming in complex space : a comprehensive bibliography", *Rev. Anal. Numér. Théor. Approx.* **17**(2), 1988, 133-140) have appeared. Several works refer to fractional programming. A synthesis of these was given by Stancu-Minasian [908].

The first work on fractional programming in complex space appeared in 1970, when Swarup and Sharma [963] generalized to the complex space the results of Charnes and Cooper [281] from linear fractional programming. Later, complex fractional programming (linear and non-linear) was treated by Das and Swarup [344], Gulati and Chandra [436], Gulati [435], Saxena and Patkar [781], [782], Bector, Chandra and Gulati [145], [147], Datta and Bhatia [347], Saxena, Patkar and Parkash [784], Bector and Bhatt [137], Gupta [442], Kaul and Gupta [546], Bector and Bhatia [131], Parkash, Saxena and Patkar [705], Stancu-Minasian, Duca and Nishida [916].

7.1. NOTATIONS, DEFINITIONS, PRELIMINARY RESULTS

We use the following notations that appear in most works on mathematical programming in complex space: $C^n(R^n)$, the n-dimensional vector space of complex (real) numbers; $C^{m\times n}(R^{m\times n})$, the set of $m\times n$ complex (real) matrices; $R^n_+ = \{x \in R^n,$

$x_j \geq 0, j = 1, \ldots, n\}$, the non-negative orthant of R^n, $x \geq y(x, y \in R^n)$ means $x - y \in R^n_+$. If $A \in C^{m \times n}$, then $\bar{A}$, A^T, $A^H = \bar{A}^T$ represents conjugate matrix, transpose matrix, conjugate transpose matrix, respectively.

If $z, w \in C^n$, $\langle z, w \rangle = w^H z$ is inner product of z and w. If $z \in C^n$, then Re z is the real part of z, Im z is the imaginary part of z, arg z is the argument of z, $\|z\| = \langle z, z \rangle^{1/2}$ is the norm of z and $|z|$ is the modulus of z.

If $\varnothing \neq S \subset C^n$, then $S^* = \{y \in C^n \mid \forall z \in S \Rightarrow Re(y^H z) \geq 0\}$ constitutes the dual (polar) of S. Define variety $M = \{(z, w) \in C^{2n} \mid w = \bar{z}\}$. If $f : C^n \to C$ is a analytical function in a neighborhood of $z^0 \in C^n$, then $\nabla_z f(z^0) = \left[\dfrac{\partial f(z^0)}{\partial z_i}\right]$, $i = 1, \ldots, n$, is the gradient of function f in point z^0. Similarly, if the complex function $f(w^1, w^2)$ is analytical in the $2n$ variables (w^1, w^2) to $(z_0, \bar{z}_0) \in C^n$, define

$$\nabla_1 f(z_0, \bar{z}_0) = \nabla_z f(z_0, \bar{z}_0) = \left[\frac{\partial f(z_0, \bar{z}_0)}{\partial w_j^1}\right], \quad j = 1, \ldots, n$$

and

$$\nabla_2 f(z_0, \bar{z}_0) = \nabla_{\bar{z}} f(z_0, \bar{z}_0) = \left[\frac{\partial f(z_0, \bar{z}_0)}{\partial w_j^2}\right], \quad j = 1, \ldots, n.$$

If $f : C^n \to C^m$ is an analytical function in a neighborhood of $z^0 \in C^n$, define $D_z f(z^0) = \partial f_i(z^0)/\partial z_j$, $i = 1, \ldots, m; j = 1, \ldots, n$. If $f : C^n \times C^n \to C^m$, $f = (f_1, \ldots, f_m) = f(w^1, w^2)$ is analytical in the $2n$ variables (w^1, w^2) at $(z_0, \bar{z}_0) \in C^n \times C^n$, then define

$$D_1 f(z_0, \bar{z}_0) = D_1 f(z_0, \bar{z}_0) = \left[\frac{\partial f_i(z_0, \bar{z}_0)}{\partial w_j^1}\right], \quad i = 1, \ldots, m; \; j = 1, \ldots, n$$

and

$$D_2 f(z_0, \bar{z}_0) = D_{\bar{z}} f(z_0, \bar{z}_0) = \left[\frac{\partial f_i(z_0, \bar{z}_0)}{\partial w_j^2}\right], \quad i = 1, \ldots, m; \; j = 1, \ldots, n.$$

(i) a set $S \subseteq C^n$ is *convex* if, for all $0 \leq t \leq 1$, we have $tS + (1 - t)S \subseteq S$.

(ii) $S \subseteq C^n$ is *cone* if $0 \leq t$ implies $tS \subseteq S$.

(iii) $S \subseteq C^n$ is *cone convex* if $S + S \subseteq S$.

(iv) $S \subseteq C^n$ is *polyhedral cone* if there is $k \in N$ and $A \in C^{n \times k}$ such that $S = AR^k_+ = \{Ax \mid x \in R^k_+\}$ *i.e.*, S is generated by a finite number of vectors (the columns of A).

Equivalently, $S \subseteq C^n$ is polyhedral cone if it is the intersection of a finite number of closed half-spaces having the origin on the boundary, *i.e.*, there is a natural number p and p points $u_1, \ldots, u_p$ such that

$$S = \bigcap_{k=1}^{p} D(u_k) = \{z \in C^n \mid Re \langle z, u_k \rangle \geq 0\}$$

Let $S = \bigcap_{k=1}^{p} D(u_k)$ be a polyhedral cone in C^n or R^n and $z \in S$. Denote $S(z)$ the intersection of all half-spaces $D(u_k)$ which have z on the boundary, *i.e.*, $S(z) = \bigcap_{k \in B(z)} D(u_k)$ where $B(z) = \{k \in \{1, \ldots, p\} \mid Re \langle z, u_k \rangle = 0\}$. If $z \in$ int S (the interior of S), then $S(z) = C^n$.

Let $M \subseteq C^n$, $S \subseteq C^m$, M and S be closed and non-empty convex cones and $f : M \to C^m$.

(v) f is *convex* in $z^2 \in M$ with respect to S if $\forall\, z^1 \in M$, $z^1 \neq z^2$ and $\lambda \in (0, 1)$ we have $\lambda f(z^1) + (1-\lambda)f(z^2) - f[\lambda z^1 + (1-\lambda)z^2] \in S$ or, equivalently, if f is analytical: $f(z^1) - f(z^2) - [D_z f(z^2)]^T(z^1 - z^2) \in S$

(vi) f is *pseudoconvex* in $z^2 \in$ *int* M with respect to S if $\forall\, z^1 \in M$, $z^1 \neq z^2$,

$$[D_z f(z^2)]^T(z^1 - z^2) \in S \Rightarrow f(z^2) - f(z^1) \in S$$

(vii) is *quasiconvex* in $z^2 \in M$ with respect to S if $\forall\, z^1 \in M$, $z^1 \neq z^2$ and $\lambda \in (0, 1)$,

$$f(z^2) - f(z^1) \in S \Rightarrow f(z^2) - f[(1 - \lambda)z^2 + \lambda z^1] \in S,$$

The following results follow Adi Ben-Israel ("Linear equations and inequalities on finite dimensional, real or complex, vector spaces : A unified theory", *J. Math. Anal. Appl.*, **27** (1969), no.2, 367-389).

R1: A polyhedral cone in C^n is a closed convex cone.

R2: A non-empty set $S \subset C^n$ is a closed convex cone if and only if $S = S^{**}$.

R3: If S and T are closed convex cones, then $(S \cap T)^* = cl\ S^* + cl\ T^*$, where $cl\ S$ is the closure of S.

R4: (Farcas' Lemma in the complex space). Let $A \in C^{m \times n}$, $b \in C^m$ and let S be a polyheadral cone in C^n. The following statements are equivalent

i) The system $Az = b$, $z \in S$ is consistent,

ii) $A^H y \in S^* \Rightarrow Re \langle b, y \rangle \geq 0$.

R5: If S and T are convex cones in C^m and C^n respectively, then $(S \times T)^* = S^* \times T^*$.

R6: If $S = R_+$, then $S^* = R_+ + iR = \{u \in C : Re\ u \geq 0\}$.

R7: If $S = R^p$, then $S^* = iR^p = \{u \in C^p : Re\ u = 0\}$.

Let $f : C^n \to C$ and $g : C^n \to C^m$ be analytical in a neighborhood of z^0, S a polyhedral cone in C^m and let the problem

$$\min \{Re\ f(z) \mid g(z) \in S\}. \tag{7.1.1}$$

Consider a feasible point z^0 in problem (7.1.1). We say that, in z^0, *the Kuhn-Tucker constraint qualification* is satisfied, or that z^0 is *qualified*, if for all $z \in C^n$ for which relation $[D_z g(z^0)]z \in S(g(z^0))$ exist, a differentiable arc $a(\theta)$ and the real numbers $\varepsilon > 0$ and $k > 0$ exist such that $a(0) = z^0$, $g(a(\theta)) \in S$, $(0 \leq \theta \leq \varepsilon)$ and $a'(0) = kz$.

Theorem 7.1.1. *(Kuhn-Tucker conditions for problem (7.1.1)). Let $S \subseteq C^m$ be a polyhedral cone and let $f : C^n \to C$, $g : C^n \to C^m$ analytical in a neighborhood of the qualified point z^0. A necessary condition for z^0 to be a local minimum of problem (7.1.1) is that a vector $u \in [S(g(z^0))]^* \subset S^*$ exists such that the following conditions are satisfied:*

$$\overline{\nabla_z f(z^0)} = [D_z^H g(z^0)]u \text{ and } Re\ \langle g(z^0), u\rangle = 0.$$

Theorem 7.1.2. *Let $S \subseteq C^m$ be a polyhedral cone and let $f : C^{2n} \to C, g : C^{2n} \to C^m$ analytical functions in a neighborhood of the qualified point $(z^0, \overline{z^0})$. A necessary condition for $(z^0, \overline{z^0})$ to be a local minimum of problem*

$$\min \{Re\ f(z, \overline{z}) \mid g(z, \overline{z}) \in S\} \tag{7.1.2}$$

is that a vector $u \in [S(g(z^0, \overline{z^0}))]^* \subset S^*$ exists such that the following conditions are satisfied

$$\overline{\nabla_z f(z^0, \overline{z^0})} + \nabla_{\overline{z}} f(z^0, \overline{z^0}) = [D_z^H g(z^0, \overline{z^0})]u + \overline{[D_{\overline{z}}^T g(z^0, \overline{z^0})]u}$$
$$\text{and } Re\ \langle g(z^0, \overline{z^0}), u\rangle = 0$$

Let $g : C^{2n} \to C^m$ be analytical and let S, T be closed convex cones in C^m and R^m, respectively. Let K be a convex subset of the variety M and let $(z_0, \overline{z_0}) \in K$.

Then, the function g is :

(viii) *convex* (CX) with respect to S in z_0 on K if for all $(z, \overline{z}) \in K$ we have

$$g(z, \overline{z}) - g(z_0, \overline{z_0}) - [D_z\, g(z_0, \overline{z_0})]^T(z - z_0) - [D_{\overline{z}}\, g(z_0, \overline{z_0})]^T(\overline{z} - \overline{z_0}) \in S.$$

strict convex (SCX) with respect to S in z_0 on K if for all $(z, \bar{z}) \in K$, $z \neq z_0$ we have

$$g(z, \bar{z}) - g(z_0, \bar{z}_0) - [D_z\, g(z_0, \bar{z}_0)]^T(z - z_0) - [D_{\bar{z}}\, g(z_0, \bar{z}_0)]^T(\bar{z} - \bar{z}_0) \in int\ S.$$

quasiconvex (QX) with respect to S in z_0 on K if for all $(z, \bar{z}) \in K$,

$$g(z_0, \bar{z}_0) - g(z, \bar{z}) \in S \Rightarrow [D_z\, g(z_0, \bar{z}_0)]^T(z_0 - z) + [D_{\bar{z}}\, g(z_0, \bar{z}_0)]^T(\bar{z}_0 - \bar{z}) \in S.$$

pseudoconvex (PCV) with respect to S in z_0 on K if for all $(z, \bar{z}) \in K$

$$[D_z\, g(z_0, \bar{z}_0)]^T(z - z_0) + [D_{\bar{z}}\, g(z_0, \bar{z}_0)]^T(\bar{z} - \bar{z}_0) \in S \Rightarrow g(z, \bar{z}) - g(z_0, \bar{z}_0) \in S.$$

Similarly, other definitions hold for the real part of g.

E.g.,, the real part of g is *convex* with respect to T in z_0 on K if and only if for all $(z, \bar{z}) \in K$

$$\lambda\ Re\ g(z, \bar{z}) + (1-\lambda)Re\ g(z_0, \bar{z}_0) - Re\ g(\lambda z + (1-\lambda)z_0, \lambda\bar{z}_0 + (1-\lambda)\bar{z}_0) \in T$$

When $g(z, \bar{z})$ is analytical, the above inequality is equivalent to

$$Re\left\{g(z, \bar{z}) - g(z_0, \bar{z}_0) - [D_z\, g(z_0, \bar{z}_0)]^T(z - z_0) - [D_{\bar{z}}\, g(z_0, \bar{z}_0)]^T(\bar{z} - \bar{z}_0)\right\}$$

Also, the real part of g is *pseudoconvex* with respect to T in z_0 on K if for all $(z, \bar{z}) \in K$,

$$Re\left\{ [D_z\, g(z_0, \bar{z}_0)]^T(z - z_0) + [D_{\bar{z}}\, g(z_0, \bar{z}_0)]^T(\bar{z} - \bar{z}_0)\right\} \Rightarrow$$
$$\Rightarrow Re\ [g(z, \bar{z}) - g(z_0, \bar{z}_0)] \in T.$$

and *strict pseudoconvex* (SPCX) with respect to T in z_0 on K if for all $z(z \neq z_0)$ and $(z, \bar{z}) \in K$,

$$Re\left\{ [D_z\, g(z_0, \bar{z}_0)]^T(z - z_0) + [D_{\bar{z}}\, g(z_0, \bar{z}_0)]^T(\bar{z} - \bar{z}_0)\right\} \in T \Rightarrow$$
$$\Rightarrow Re\ [g(z, \bar{z}) - g(z_0, \bar{z}_0)] \in int\ T.$$

A function f is *concave, pseudoconcave, quasiconcave* in $z \in M$ with respect to S iff is convex, pseudo-convex, quasi-convex to $z \in M$ with respect to $-S = \{z \in C^m : -z \in S\}$.

7.2. LINEAR FRACTIONAL PROGRAMMING IN COMPLEX SPACE

Consider the following linear fractional programming problem in complex space:

$$\min f(z) = \frac{Re\ (c^H z + c_0)}{Re\ (d^H z + d_0)}, \tag{7.2.1}$$

subject to

$$|arg\ (Az - b)| \leq \beta, \tag{7.2.2}$$
$$|arg\ z| \leq \alpha,$$

where c, d, z are vectors of type $n \times 1$ with complex elements, A is a matrix of type $m \times n$ with complex elements, b is a vector of type $m \times 1$ with complex elements, α, β are real vectors of type $n \times 1$ respectively $m \times 1$ such that $0 \leq \alpha_i, \beta_i \leq \pi/2$, $i = 1, ..., n$; $j = 1, ..., m$; and c_0 and d_0 are complex numbers. Using the change of variable $y = tz$ we show that for the solution of problem (7.2.1) - (7.2.2) we need to solve at most two linear programming problems in the complex space. Denote by

$$S = \{ z \mid |arg\ (Az - b)| \leq \beta, \quad |arg\ z| \leq \alpha \}$$

the set of feasible solutions for the problem (7.2.1) - (7.2.2). Assume that S is non-empty and bounded *i.e.*, $\forall z \in S$ we have $\|z\| \leq M$ where M is a certain finite number. Consider the change of variable $y = tz$ where $t \geq 0$ is chosen such that

$$Re\ (d^H y + d_0 t) = \delta \neq 0.$$

The problem (7.2.1) - (7.2.2) becomes the following linear programming problem in the complex space:

$$\min g(y) = Re\ (c^H y + c_0 t) \tag{7.2.3}$$

subject to

$$|arg\ (Ay - bt)| \leq \beta,$$
$$Re\ (d^H y + d_0 t) = \delta, \tag{7.2.4}$$
$$|arg\ y| \leq \alpha, \quad |arg\ t| \leq 0.$$

Note that the complex number arguments are assumed to vary in the interval $(-\pi, \pi)$ and that the inequality $|arg\ z| \leq \alpha$ means $|arg\ z_k| \leq \alpha_k$, $k = 1, ..., n$. Before proving the equivalence theorem between the problems (7.2.1) - (7.2.2) and (7.2.3) - (7.2.4) we prove two lemmas.

Lemma 7.2.1. *If z_1 and z_2 are complex vectors of equal dimensions such that $|arg\ z_1| \leq \alpha$ and $|arg\ z_2| \leq \alpha$ $(0 \leq \alpha_i \leq \pi/2)$, then,*

$$|arg\ (z_1 + z_2)| \leq \alpha.$$

Proof. From the hypothesis, we have $|arg\ z_{1i}| \leq \alpha_i$ and $|arg\ z_{2i}| \leq \alpha_i$ thus, $|arg\ (z_{1i} + z_{2i})| \leq \alpha_i$ *i.e.*, $|arg(z_1 + z_2)| \leq \alpha$.

Lemma 7.2.2. *For every* (y_1, t) *which satisfies the constraints (7.2.4) we have* $t > 0$.

Proof. Assume that t can be zero *i.e.*, $(y_1, 0)$ satisfies the constraints (7.2.4). Let $z_1 \in S$. Since $|arg\ Ay_1| \leq \beta$, $|arg\ y_1| \leq \alpha$ we have $(z_1 + uy_1) \in S$ for all real $u > 0$. Indeed, based on Lemma 7.2.1 we have

$$| \ arg\ \{A(z_1 + uy_1) - b\} \ | = | \ arg\ \{(Az_1 - b) + uAy_1\} \ | \leq \beta$$

and

$$| \ arg\ (z_1 + uy_1)| \leq \alpha$$

Since u is a generic positive real number, it can take an indefinitely large value. But this means that S is unbounded which contradicts the hypothesis that S is bounded. Thus $t > 0$.

We now prove the equivalence theorem between the problems (7.2.1) - (7.2.2) and (7.2.3) - (7.2.4).

Theorem 7.2.1. *If*

a) $0 < sgn\ \delta = sgn\ Re\ (d^H z_1 + d_0)$ *for* z_1 *optimal solution of problem (7.2.1) - (7.2.2),*

b) (y_1, t_1) *is optimal solution of problem (7.2.3) - (7.2.4),*

then y_1/t_1 *is optimal solution of problem (7.2.1) - (7.2.2).*

Proof. By *reductio ad absurdum*, assume that y_1/t_1 is not optimal solution for the problem (7.2.1) - (7.2.2) thus, there is another optimal $z_1 \in S$, *i.e.*,

$$\frac{Re\ (c^H z_1 + c_0)}{Re\ (d^H z_1 + d_0)} < \frac{Re\ \{c^H(y_1/t_1) + c_0\}}{Re\ \{d^H(y_1/t_1) + d_0\}} \tag{7.2.5}$$

From condition a), we have

$$Re\ (d^H z_1 + d_0) = \theta\delta \text{ with } \theta > 0.$$

Consider $\hat{y} = \theta^{-1} z_1$, $\hat{t} = \theta^{-1}$, then it can be easily verified that $(\hat{y}, t)$ satisfies the constraints (7.2.4). We have

$$\frac{Re\ (c^H z_1 + c_0)}{Re\ (d^H z_1 + d_0)} = \frac{Re\ \theta^{-1}(c^H z_1 + c_0)}{Re\ \theta^{-1}(d^H z_1 + d_0)} =$$

$$= \frac{Re\ (c^H \hat{y} + c_0 \hat{t})}{Re\ (d^H \hat{y} + d_0 \hat{t})} = \frac{Re\ (c^H \hat{y} + c_0 \hat{t})}{\delta}, \tag{7.2.6}$$

$$\frac{Re\ \left[c^H(y_1/t_1) + c_0\right]}{Re\ \left[d^H(y_1/t_1) + d_0\right]} = \frac{Re\ \left[c^H y_1 + d_0 t_1\right]}{Re\ \left[d^H y_1 + d_0 t_1\right]} = \frac{Re\ \left[d^H y_1 + c_0 t_1\right]}{\delta}. \tag{7.2.7}$$

From relations (7.2.5), (7.2.6), (7.2.7), and from the fact that $sgn\ \delta > 0$, we have

$$Re\ (c^H\hat{y} + c_0\hat{t}) < Re\ (c^H y_1 + c_0 t_1)$$

which contradicts the fact that (y_1, t_1) is optimal solution of problem (7.2.3) - (7.2.4). Thus, the hypothesis that y_1/t_1 is not optimal solution of problem (7.2.1) - (7.2.2) is false.

If, for z_1 an optimal solution of problem (7.2.1) - (7.2.2), we have

$$sgn\ Re\ (d^H z_1 + d_0) < 0,$$

then we can change the sign of vectors (c^H, c_0), (d^H, d_0) without changing the value of function (7.2.1), and thus we have

$$sgn\ Re\ (d^H z_1 + d_0) > 0.$$

In conclusion, the solution of problem (7.2.1) - (7.2.2) is equivalent to the solution of two complex linear programming problems.

Problem I.

$$\min \Big\{ Re\ (c^H y + c_0 t) \mid \quad |arg(Ay - bt)| \leq \beta,$$
$$Re\ (d^H y + d_0 t) = 1,\ |arg\ y| \leq \alpha,\ |arg\ t| \leq 0 \Big\},$$

Problem II.

$$\min \Big\{ Re(-c^H y - c_0 t) \mid \quad |arg\ (Ay - bt)| \leq \beta,$$
$$Re\ (-d^H y - d_0 t) = 1,\ |arg\ y| \leq \alpha,\ |arg\ t| \leq 0 \Big\}.$$

If we know the sign of the denominator or of the numerator at the optimal solution, then we solve only one of the above problems.

Theorem 7.2.2. *If $Re\ (d^H z + d_0) = 0,\ \forall\ z \in S$, then both problems, I and II, are inconsistent.*

Proof. Indeed, if the condition from the statement of the theorem is satisfied, then we can not have $\pm\ Re\ t(d^H z + d_0) = \pm\ Re\ (d^H y + d_0 t) = 1$ and thus problems I and II are inconsistent.

If $Re\ (d^H z + d_0) = 0$ for $z \in S' \subset S$, then, any point for which $Re\ (d^H z + d_0) = 0$ is a limit point of the sequence $\{z^n\}$ with $Re\ (d^H z^n + d_0) = \varepsilon_n \neq 0$ and $\varepsilon_n \to 0$, ε_n is a real number. For $\{y^n, t_n\}$, we have $\pm Re\ t_n(d^H z^n + d_0) = Re\ (\pm d^H y^n + d_0 t_n) = t_n \varepsilon_n = 1$ *i.e.*, $t_n \to \infty$ as $\varepsilon_n \to 0$. Thus, for optimum f, we also include $\max f = +\infty$.

7.3. LINEAR FRACTIONAL PROGRAMMING IN COMPLEX SPACE WITH NONLINEAR CONSTRAINTS

Consider the following linear fractional programming problem in complex space having nonlinear constraints. Minimize the function:

$$f(z) = \frac{Re\ (c^H z + c_0)}{Re\ (d^H z + d_0)} \qquad (7.3.1)$$

subject to

$$g(z) \in S \qquad (7.3.2)$$

where $g : C^n \to C^m$ is an analytical and concave function satisfying the Kuhn-Tucker constraints qualification, S is a polyhedral convex cone in C^m. Assume that the denominator keeps the same sign on the set of feasible solutions.

Theorem 7.3.1. *Any local minimum of function f is also a global minimum.*

Proof. Accept the contrary, *i.e.*, there is z_1 local minimum different from z_0 global minimum. Since $z_0 \neq z_1$, we have

$$f(z_0) < f(z_1) \qquad (7.3.3)$$

Let $z = \lambda z_1 + (1 - \lambda)z_0$ with $0 \leq \lambda \leq 1$, such that $g(z) \in S$. Using $Re\ (d^H z_0 + d_0) \neq 0$, we calculate

$$f(z) = \frac{\lambda\ Re(c^H z_1 + c_0)/Re(d^H z_0 + d_0) + (1 - \lambda)f(z_0)}{\lambda Re(d^H z_1 + d_0)/Re(d^H z_0 + d_0) + (1 - \lambda)}.$$

Using relation (7.3.3) we get

$$f(z) < \frac{\lambda Re(c^H z_1 + c_0)/Re(d^H z_1 + d_0) + (1 - \lambda)f(z_1)}{\lambda Re(d^H z_1 + d_0)/Re(d^H z_0 + d_0) + (1 - \lambda)} =$$

$$= f(z_1)\frac{\lambda Re(d^H z_1 + d_0)/Re(d^H z_0 + d_0) + (1 - \lambda)}{\lambda Re(d^H z_1 + d_0)/Re(d^H z_0 + d_0) + (1 - \lambda)} = f(z_1).$$

Thus,

$$f(z) < f(z_1) \qquad (7.3.4)$$

On the other hand, we can choose λ such that z is sufficiently close to the local minimum, z_1, and thus

$$f(z) \geq f(z_1). \qquad (7.3.5)$$

Relations (7.3.4) and (7.3.5) are contradictory, thus, the hypothesis that $z_1 \neq z_0$ is false.

Now, we give the necessary and sufficient condition for the existence of the optimum to problem (7.3.1) - (7.3.2).

Theorem 7.3.2. *Let $S \subset C^m$ be a polyhedral cone and let $f : C^n \to C$ and $g : C^n \to C^m$ be analytical functions in a neighborhood of the qualified point z_0. The necessary and sufficient condition for point z_0 to be a minimum point for the problem (7.3.1) - (7.3.2) is that a vector $u \in [S(g(z_0))]^* \subset S^*$ exists such that*

$$\overline{\nabla_z f(z_0)} = [D_z^H g(z_0)]u \tag{7.3.6}$$

and

$$Re \ \langle g(z_0), u\rangle = 0 \tag{7.3.7}$$

Proof. We prove the necessity. Let

$$Z = \left\{z \in C^n \ \middle| \ [D_z g(z_0)]z \in S(g(z_0)), \ \ Re\left\langle \overline{\nabla_z f(x_0)}, z\right\rangle < 0\right\}.$$

If $Z = \varnothing$, then

$$[D_z g(z_0)]z \in S(g(z_0)) \Rightarrow Re \left\langle \overline{\nabla_z f(z_0)}, z\right\rangle \geq 0 \tag{7.3.8}$$

The cone $S(g(z_0))$ is polyhedral and thus, according to R1, Section 7.1, it is convex and closed, and according to R2, Section 7.1, in (7.3.8), $S(g(z_0))$ can be replaced by $([S(g(z_0))]^*)^*$. Applying Farcas' Lemma (R4, Section 7.1) we have that (7.3.8) is equivalent to the existence of a $u \in [S(g(z_0))]^*$ such that

$$\overline{\nabla_z f(z_0)} = [D_z^H g(z_0)]u$$

Hence, to demonstrate that relation (7.3.6) exists, it is sufficient to show that $Z = \varnothing$. For this, let z be such that $[D_z g(z_0)]z \in S(g(z_0))$. But, z_0 is a qualified point, and thus there is a feasible arc $a(\theta)$ with $a(0) = z_0$, $a'(\theta) = kz (k > 0)$. By the hypothesis, z_0 is a point of local minimum of $Re\, f(z)$ along $a(\theta)$, thus, $\frac{d}{d\theta} Re[f(a(\theta))]\,|_{\theta=0} \geq 0$ or $Re \frac{d}{d\theta} f(a(\theta))\,|_{\theta=0} \geq 0$. Hence,

$$Re \ [\nabla_z^T f(z_0) a'(0)] \geq 0$$

or, using $a'(\theta) = kz$, $Re \ \langle \overline{\nabla f(z)}, z\rangle \geq 0$, which shows that $Z = \varnothing$. Let's prove relation (7.3.7). We have,

$$[S(g(z_0))]^* = \left(\bigcap_{k \in B(g(z_0))} H_{u_k}\right)^* = cl \sum_{k \in B(g(z_0))} (H_{u_k})^* =$$

$$= cl \sum_{k \in B(g(z_0))} \{\alpha_k u_k : \alpha_k \geq 0\} = \sum_{k \in B(g(z_0))} \{\alpha_k u_k : \alpha_k \geq 0\}$$

The second equality stems from R3, Section 7.1, and the last results from the fact that $\sum_{k \in B(g(z_0))} \{\alpha_k u_k; \alpha_k \geq 0\}$ is a polyhedral cone, thus closed. Consequently, if $u \in [S(g(z_0))]^*$, then,

$$u = \sum_{i \in B(g(z_0))} \beta_i u_i \quad (\beta_i \geq 0) \tag{7.3.9}$$

From the definition of $B(g(z_0))$, we have $Re\ \langle u_i, g(z_0)\rangle = 0$ for $i \in B(g(z_0))$. Thus (7.3.7) results from (7.3.9). The proof of necessity is complete.

For proof of sufficiency, consider a point z which verifies the constraints (7.3.2) and assume that there are z_0 and u satisfying relations (7.3.6) and (7.3.7). We have, $Re\ \langle u, g(z)\rangle \geq 0$ since $g(z) \in S$ and $u \in S^*$.

Through a simple calculation we get

$$\nabla f(z_0) = \frac{c}{d^H z_0 + d_0} - f(z_0)\left[\frac{d}{(d^H z_0 + d_0)}\right], \quad (z - z_0)^H\, \nabla f(z_0) =$$

$$= \frac{c^H z - c^H z_0}{d^H z_0 + d_0} - f(z_0)\frac{d^H z - d^H z_0}{d^H z_0 + d_0} =$$

$$= \frac{(c^H z + c_0) - (c^H z_0 + c_0)}{d^H z_0 + d_0} - f(z_0)\frac{(d^H z + d_0) - (d^H z_0 + d_0)}{d^H z_0 + d_0} =$$

$$= \left[\frac{d^H z + d_0}{d^0 z_0 + d_0}\right][f(z) - f(z_0)]$$

thus,

$$f(z) - f(z_0) = \left[\frac{d^H z_0 + d_0}{d^H z + d_0}\right](z - z_0)^H\, \nabla f(z_0)$$

Also, from (7.3.6) by multiplication by $(z - z_0)^H$, we get

$$(z - z_0)^H\, \overline{\nabla_z f(z_0)} = (z - z_0)^H \left[D_z^H g(z_0)\right] u$$

From the concavity of function g, we have

$$-g(z) + g(z_0) + D_z^T g(z_0)(z - z_0) \in S$$

and from $u \in S^*$, we have

$$Re\ \langle u, -g(z) + g(z_0) + D_z^T g(z_0)(z - z_0)\rangle \geq 0$$

or

$$Re\ \langle (z - z_0)^H, (D_z g(z_0))u\rangle \geq Re\ \langle u, g(z_0) - g(z)\rangle.$$

Since $u \in S^*$, we have $Re\ \langle u, g(z_0) - g(z)\rangle \geq 0$, *i.e.*,

$$(z - z_0)^H\, \overline{\nabla_z f(z_0)} \geq 0$$

From the continuity of function $Re\ (d^H z + d_0)$, for z sufficiently close to z_0, we have $\dfrac{Re\ (d^H z_0 + d_0)}{Re\ (d^H z + d_0)} > 0$. Thus, $f(z) - f(z_0) \geq 0$, *i.e.*, $f(z) \geq f(z_0)$, which means that z_0 is a point of local minimum. By Theorem 7.3.1, z_0 is also global minimum.

For the solution of problem (7.3.1) - (7.3.2) one can utilize the change of variable $y = tz\ \ (t \geq 0)$, and one can prove a theorem similar to Theorem 7.2.1.

7.4. NONLINEAR FRACTIONAL PROGRAMMING IN COMPLEX SPACE WITH NONLINEAR CONSTRAINTS

Consider the nonlinear fractional programming problem in complex space over arbitrary polyhedral cones

$$\max \frac{Re\, f(z, \bar{z})}{Re\, g(z, \bar{z})} \tag{7.4.1}$$

subject to

$$-h(z, \bar{z}) \in S \tag{7.4.2}$$

where $f, g : W \subset C^n \times C^n \to C$ are analytic functions with concave and convex real part with respect to R_+ on the manifold W, $h : W \to C^m$ is analytic function, convex with respect to S on the manifold W, S is a closed convex cone in C^m. Let $S_P = \{(z, \bar{z}) \mid -h(z, \bar{z}) \in S\}$ be the domain of feasible solutions for problem (7.4.1) - (7.4.2). We assume that $Re\, f(z, \bar{z}) \geq 0$ and $Re\, g(z, \bar{z}) > 0$ for all $(z, \bar{z}) \in S_P$. Let $\phi : R \to R$ be a monotonically strictly increasing function, with $\phi(t) > 0$ for $t > 0$. For $t \in R_+$ and $(y, \bar{y}) \in C^{2n}$ define the functions

$$\begin{aligned} Re\, F(y, \bar{y}; t) &= Re\, f\left[\frac{y}{t}, \frac{\bar{y}}{t}\right]\phi(t) \\ Re\, G(y, \bar{y}; t) &= Re\, g\left[\frac{y}{t}, \frac{\bar{y}}{t}\right]\phi(t) \end{aligned} \tag{7.4.3}$$

Let

$$\begin{aligned} Re\, F(y, \bar{y}; 0) &= \lim_{t \to 0} Re\, F(y, \bar{y}; t) \\ Re\, G(y, \bar{y}; 0) &= \lim_{t \to 0} Re\, G(y, \bar{y}; t) \end{aligned} \tag{7.4.4}$$

for the case when the limits exist and make the hypothesis that $G(\mathbf{0}; 0) = 0$. For the solution of problem (7.4.1) - (7.4.2) use the change of variable

$$y = tz \ ; \quad \bar{y} = t\bar{z}, \tag{7.4.5}$$

where $t > 0$ is a scalar chosen such that, for a given function ϕ, and for $\gamma \in R$, $\gamma \neq 0$, we have

$$Re\, G(y, \bar{y}; t) = \gamma \tag{7.4.6}$$

Multiplying the denominator and the numerator of (7.4.1) by $\phi(t)$, and using (7.4.3), problem (7.4.1) - (7.4.2) becomes

$$Max\ Re\, F(y, \bar{y}; t) \tag{7.4.7}$$

subject to

$$\begin{aligned} -h\left[\frac{y}{t}, \frac{\bar{y}}{t}\right] &\in S \\ Re\, G(y, \bar{y}; t) &= \gamma \\ t &> 0 \end{aligned} \tag{7.4.8}$$

Theorem 7.4.1. *If*

(i) $0 < sgn\,\gamma = Re\ g(z_1, \bar{z}_1)$ *for the optimal solution* $(z_1, \bar{z}_1)$ *of problem (7.4.1) - (7.4.2), and*

(ii) $(y_1, \bar{y}_1; t_1)$ *is an optimal solution of problem (7.4.7) - (7.4.8),*

then $(y_1/t_1, \bar{y}_1/t_1)$ *is an optimal solution of problem (7.4.1) - (7.4.2).*

Proof. Assume, by *reductio ad absurdum,* that $(y_1/t_1, \bar{y}_1/t_1)$ is not optimal solution of problem (7.4.1) - (7.4.2). Thus, there is $(z_1, \bar{z}_1) \in S_P$ $((z_1, \bar{z}_1) \neq (y_1/t_1, \bar{y}_1/t_1))$ optimal solution of problem (7.4.1) - (7.4.2) *i.e.,*

$$\frac{Re\ f(z_1, \bar{z}_1)}{Re\ g(z_1, \bar{z}_1)} > \frac{Re\ f(y_1/t_1, \bar{y}_1/t_1)}{Re\ g(y_1/t_1, \bar{y}_1/t_1)} \tag{7.4.9}$$

From condition (i), we have $Re\ g(z_1, \bar{z}_1) = \theta\gamma \quad (\theta > 0)$

If we consider

$$y = \phi^{-1}\left[\frac{1}{\theta}\right] z_1\ ;\ \bar{y} = \phi^{-1}\left[\frac{1}{\theta}\right]\bar{z}_1,\quad t = \phi^{-1}\left[\frac{1}{\theta}\right], \tag{7.4.10}$$

it can be easily verified that $(y, \bar{y}; t)$ is feasible solution for problem (7.4.7) - (7.4.8). We have

$$\frac{Re\ f(z_1, \bar{z}_1)}{Re\ g(z_1, \bar{z}_1)} = \frac{\phi(t)\ Re\ f(z_1, \bar{z}_1)}{\phi(t)\ Re\ g(z_1, \bar{z}_1)} = \frac{Re\ F(y, \bar{y}; t)}{Re\ G(y, \bar{y}; t)} = \frac{Re\ F(y, \bar{y}; t)}{\gamma} \tag{7.4.11}$$

Also,

$$\frac{Re\ f(y_1/t_1, \bar{y}_1/t_1)}{Re\ g(y_1/t_1, \bar{y}_1/t_1)} = \frac{\phi(t_1)\ Re\ f(y_1/t_1, \bar{y}_1/t_1)}{\phi(t_1)\ Re\ g(y_1/t_1, \bar{y}_1/t_1)} = \tag{7.4.12}$$

$$= \frac{Re\ F(y_1, \bar{y}_1; t_1)}{Re\ G(y_1, \bar{y}_1; t_1)} = \frac{Re\ F(y_1, \bar{y}_1; t_1)}{\gamma}.$$

From relations (7.4.9), (7.4.11) and (7.4.12), we have

$$Re\ F(y, \bar{y}; t) > Re\ F(y_1, \bar{y}_1; t_1)$$

which contradicts the hypothesis that $(y_1, \bar{y}_1; t_1)$ is an optimal solution of problem (7.4.7)-(7.4.8). Thus, $(y_1/t_1, \bar{y}_1/t_1)$ is optimal solution of problem (7.4.1) - (7.4.2).

Note that the domain of feasible solutions for problem (7.4.7) - (7.4.8) is not convex even if $Re\ G\ (y, \bar{y}; t)$ is convex. If we consider $\phi(t) = t$, then problem (7.4.7) - (7.4.8) can be replaced by the following equivalent problem of convex programming

$$\max Re\ F(y, \bar{y}; t)$$

subject to

$$-h\left[\frac{y}{t}, \frac{\bar{y}}{t}\right] \in S$$

$$Re\ G(y, \bar{y}; t) \leq \gamma$$
$$t > 0$$

7.5. NONLINEAR AND NON-DIFFERENTIABLE FRACTIONAL PROGRAMMING IN THE COMPLEX SPACE

Consider the following nonlinear and non-differentiable fractional programming problem in the complex space:

Minimize the function

$$f(z) = \frac{Re\left[c^H z + \left(z^H H z\right)^{1/2} + c_0\right]}{Re\left[d^H z + d_0\right]}, \tag{7.5.1}$$

subject to

$$Az - b \in S, \quad z \in T, \tag{7.5.2}$$

where $A \in C^{m \times n}$, $c, d \in C^n$, c_0, d_0 are constants, $H \in C^{n \times n}$ is a positive semi-definite Hermitian matrix, $S \subset C^m$, $T \subset C^n$ are polyhedral convex cones. Denote by $S_P = \{ z \mid Az - b \in S,\ z \in T \}$ the set of feasible solutions for problem (7.5.1) - (7.5.2) which we assume, as usually, to be non-empty and bounded. For the solution of problem (7.5.1) - (7.5.2), utilize the change of variable $y = tz$, where $t \geq 0$ is chosen such that $Re\left[d^H y + d_0 t\right] = \delta$, ($\delta$ is a real number $\neq 0$) to obtain the following equivalent problem

$$\min h(y, t) = Re\left[c^H y + \left(y^H H y\right)^{1/2} + c_0 t\right], \tag{7.5.3}$$

subject to

$$Ay - bt \in S,$$
$$Re\left[d^H y + d_0 t\right] = \delta, \tag{7.5.4}$$
$$y \in T.$$

The following two lemmas and the connecting theorem between problems (7.5.1) - (7.5.2) and (7.5.3) - (7.5.4) are similar to Lemmas 7.2.1, 7.2.2 and Theorem 7.2.1.

Lemma 7.5.1. *If* $z_1, z_2 \in S_P$, *then* $z_1 + z_2 \in S_P$.

Proof. Indeed, if $z_1, z_2 \in S_P$, then for every component of these vectors, we have $z_{i1} \in S_P$, $z_{i2} \in S_P$, thus, $z_{i1} + z_{i2} \in S_P$ since S and T are polyhedral convex cones and consequently $z_1 + z_2 \in S_P$.

Lemma 7.5.2. *For every* (y_1, t) *which satisfies the constraints (7.5.4), we have* $t > 0$.

Proof. By *reductio ad absurdum*, assume that $(y_1, 0)$ satisfies the constraints (7.5.4) and let $z_1 \in S_P$. We have $Ay_1 \in S$ and $y_1 \in T$. For $\mu > 0$ real, we have, using Lemma 7.5.1, that

$$A(z_1 + \mu y_1) - b = (Az_1 - b) + \mu A y_1 \in S \text{ and } z_1 + \mu y_1 \in T$$

which gives $z_1 + \mu y_1 \in S_P$. Since μ can take a value whatever great, we conclude that S_P is unbounded. Contradiction. Thus $t > 0$.

Theorem 7.5.1. *If:*

a) $0 < sgn\ \delta = sgn\ Re\left[d^H z_1 + d_0\right]$ *for* z_1 *optimal solution of problem (7.5.1) - (7.5.2), and*

b) (y_1, t_1) *is optimal solution of problem (7.5.3) - (7.5.4),*

then y_1/t_1 *is optimal solution of problem (7.5.1) - (7.5.2).*

Proof. By *reductio ad absurdum*, assume that y_1/t_1 is not optimal solution for problem (7.5.1) - (7.5.2) thus, there is another optimum $z_1 \in S_P$, *i.e.*,

$$\frac{Re\ [c^H z_1 + (z_1^H H z_1)^{1/2} + c_0]}{Re\ [d^H z_1 + d_0]} \tag{7.5.5}$$

$$< \frac{Re\ [c^H (y_1/t_1) + \{(y_1/t_1)^H H (y_1/t_1)\}^{1/2} + c_0]}{Re\ [d^H (y_1/t_1) + d_0]}$$

$$= \frac{Re\ [c^H y_1 + (y_1^H H y_1)^{1/2} + c_0 t_1]}{Re\ [d^H y_1 + d_0 t_1]}$$

But, $Re\ [d^H z_1 + d_0] = \theta\delta$, $\theta > 0$, real. Choosing $\hat{y} = \theta^{-1} z_1$, $\hat{t} = \theta^{-1}$ it can be easily verified that point $(\hat{y}, \hat{t})$ satisfies (7.5.4). We have:

$$\frac{Re\ [c^H z_1 + (z_1^H H z_1)^{1/2} + c_0]}{Re\ [d^H z_1 + d_0]} = \frac{Re\ \theta^{-1}\ [c^H z_1 + (z_1^H H z_1)^{1/2} + c_0]}{Re\ \theta^{-1}\ [d^H z_1 + d_0]} = \tag{7.5.6}$$

$$= \frac{Re\ [c^H \hat{y} + (\hat{y}^H H \hat{y})^{1/2} + c_0 \hat{t}]}{Re\ [d^H \hat{y} + d_0 \hat{t}]} = \frac{Re\ [c^H \hat{y} + (\hat{y}^H H \hat{y})^{1/2} + c_0 \hat{t}]}{\delta}$$

and

$$\frac{Re\ [c^H (y_1/t_1) + \{(y_1/t_1)^H H (y_1/t_1)\}^{1/2} + c_0]}{Re\ [d^H (y_1/t_1) + d_0]} = \tag{7.5.7}$$

$$= \frac{Re\ [c^H y_1 + (y_1^H H y_1)^{1/2} + c_0 t_1]}{Re\ [d^H y_1 + d_0 t_1]} = \frac{Re\left[c^H y_1 + (y_1^H H y_1)^{1/2} + c_0 t_1\right]}{\delta}$$

From relations (7.5.5), (7.5.6), (7.5.7), and from $\delta \neq 0$, we have

$$Re\ [c^H \hat{y} + (\hat{y}^H H \hat{y})^{1/2} + c_0 \hat{t}] < Re\ [c^H y_1 + (y_1^H H y_1)^{1/2} + c_0 t_1].$$

This relation contradicts the hypothesis that (y_1, t_1) is an optimal solution for problem (7.5.3) - (7.5.4). In conclusion, y_1/t_1 is an optimal solution of problem (7.5.1) - (7.5.2).

If $sgn\ Re\ [d^H z_1 + d_0] < 0$ for an optimal solution of problem (7.5.1) - (7.5.2), then multiplying by (-1) both the denominator and the numerator of f, we revert to previous case and the result of Theorem 7.5.1. is perfectly applicable.

After the change of variable $y = tz$, the equivalent problem is

$$\min Re\left[-c^H y - (y^H H y)^{1/2} - c_0 t\right], \tag{7.5.8}$$

subject to

$$\begin{gathered} Ay - bt \in S, \\ Re\left[-d^H y - d_0 t\right] = 1, \\ y \in T. \end{gathered} \tag{7.5.9}$$

Theorem 7.5.2. *If*

$$Re\left[d^H z + d_0\right] = 0, \quad (\forall)\ z \in S_P,$$

then problems (7.5.3) - (7.5.4) and (7.5.8) - (7.5.9) are inconsistent.

The proof is the same as for Theorem 7.2.2.
By Theorem 7.5.1, the solution of problem (7.5.1) - (7.5.2) has been reduced to the solution of one of problems (7.5.3) - (7.5.4) or (7.5.8) - (7.5.9). But the objective functions of these problems are not differentiable, which means that for their solution we can not use convex programming techniques.

7.6. DUALITY FOR LINEAR FRACTIONAL PROGRAMMING IN COMPLEX SPACE

In this section we show that the following two problems are dual.

Primal problem.

(P)

$$\min f(z) = \frac{Re\ (c^H z + c_0)}{Re\ (d^H z + d_0)},$$

subject to

$$Az - b \in S,\ z \in T.$$

Dual problem.

(**D**) $$\max g(v, w) = v,$$

subject to

$$(-A^H w + c - vd) \in T^*$$
$$Re\,(b^H w + c_0 - vd_0) = 0$$
$$w \in S^*,\ v \in R$$

where $c,\ d,\ z \in C^n$, $c_0,\ d_0 \in C$, $b \in C^m$, $A \in C^{m \times n}$, $S \subset C^m$, $T \subset C^n$ are polyhedral convex cones.

Make the hypothesis that the denominator of f is positive for all z from the domain under consideration.

Theorem 7.6.1. *(Weak duality theorem). If z is feasible for the problem* **P** *and (v, w) is feasible for the problem* **D**, *then*

$$f(z) \geq g(v, w).$$

The proof is immediate and is omitted.

Corollary 7.6.1. *If for z and (v, w), feasible solutions for* (**P**) *and* (**D**), *respectively, we have $f(z) = g(v, w)$, then z is optimal solution for problem* **P** *and (v, w) for problem* **D**.

Theorem 7.6.2. *(Direct duality theorem). If z_0 is optimal solution for problem* **P**, *then there are $v_0 \in R$ and $w_0 \in S^*$ such that (v_0, w_0) is optimal for problem* **D** *and the optimal values are equal.*

Proof. Since z_0 is optimal solution for problem **P**, according to Theorem 7.1.1. there are $w \in S^*$ and $x \in T^*$ such that

$$\frac{(c \cdot Re\,(d^H z_0 + d_0) - d \cdot Re\,(c^H z_0 + c_0))}{(Re\,(d^H z_0 + d_0))^2} = A^H w + x \tag{7.6.1}$$

and

$$Re\,(z_0^H A^H w - b^H w + z_0^H x) = 0 \tag{7.6.2}$$

Choose

$$v_0 = \frac{Re\,(c^H z_0 + c_0)}{Re\,(d^H z_0 + d_0)},\ w_0 = w\, Re\,(d^H z_0 + d_0),\ x_0 = x\, Re\,(d^H z_0 + d_0).$$

Evidently, $w_0 \in S^*$, $x_0 \in T^*$. Point (v_0, w_0) is feasible for problem **D**. Indeed, from (7.6.1) and (7.6.2), we have

$$-A^H w_0 + c - v_0 d = x_0 \in T^* \tag{7.6.3}$$

$$Re\,(z_0^H A^H w_0 - b^H w_0 + z_0^H x_0) = 0 \tag{7.6.4}$$

and from (7.6.3), we have

$$Re\,(-z_0^H A^H w_0 + c^H z_0 - v_0 d^H z_0) = Re\,(z_0^H x_0).$$

This last relation, together with relation (7.6.4), and from the way we defined v_0, implies

$$Re\ (b^H w_0 + c_0 - v_0 d_0) = 0.$$

From the way we defined v_0, we have $f(z_0) = g(v_0, w_0)$, and thus, by Corollary 7.6.1, we conclude that (v_0, w_0) is optimal solution for problem **D**.

To prove the converse duality theorem make the additional hypotheses:

(i) The set $K = \{ z : Az - b \in S,\ z \in T \}$ contains at least two elements.

(ii) $Az \in S,\ z \in T \Rightarrow z = 0$

We also state the following lemma with obvious proof:

Lemma 7.6.1. *Under conditions (i), (ii) for all* $(z, \lambda) \in G = \{(z, \lambda) \mid Az - b\lambda \in S,\ z \in T,\ \lambda \in R\}$ *we have* $\lambda \geq 0$.

Theorem 7.6.2. *(Converse duality theorem). If* (v_0, w_0) *is an optimal solution for problem* **D**, *then there is* z_0 *optimal solution for problem* **P** *and the two optimal values are equal.*
Proof. Since (v_0, w_0) is optimal solution for problem **D**, then, according to Theorem 7.1.1, there is (x, y, λ) such that we have

$$Ay - b\lambda = x \in S, \tag{7.6.5}$$

$$y \in T,\ \ \lambda \in R, \tag{7.6.6}$$

$$Re\ (d^H y + \lambda d_0) = 1, \tag{7.6.7}$$

$$Re\ [(-w_0^H A + c^H - v_0 d^H)y + w_0^H x + (b^H w_0 + c_0 - v_0 d_0)\lambda] = 0. \tag{7.6.8}$$

We have $\lambda > 0$ since, from Lemma 7.6.1, we have $\lambda \geq 0$. We can not have $\lambda = 0$ since relations (7.6.5) - (7.6.7) would contradict hypothesis **ii)**. Choosing $z_0 = y/\lambda$, relations (7.6.5) - (7.6.6) imply that z_0 is feasible solution for problem **P**. From relation (7.6.5), we have

$$Re\ (w_0^H Ay - w_0^H b\lambda) = Re\ (w_0^H x).$$

From here, and from relation (7.6.8), we have

$$g(v_0, w_0) = v_0 = \frac{Re\ (c^H y + \lambda c_0)}{Re\ (d^H y + \lambda d_0)} = \frac{Re\ (c^H z_0 + c_0)}{Re\ (d^H z_0 + d_0)} = f(z_0).$$

Corollary 7.6.1 implies that z_0 is optimal solution for problem **P**.

7.7. MULTIPLE OBJECTIVE LINEAR FRACTIONAL PROGRAMMING PROBLEM IN COMPLEX SPACE

In this section, we consider multiple objective linear fractional programming problem in complex space. Characterization of efficiency is given for this class of problems. The principal tools employed are the change of variables, which reduces a multiple objective linear fractional programming problem in complex space to a multiple objective linear programming in complex space. Also, a theorem of the alternative for complex nonhomogeneous linear inequality is given.

Stancu-Minasian, Duca and Nishida [916], considered a multiple objective linear fractional program in complex space of the form;

$$\text{Minimize} \left[\frac{Re\,(\langle c^1, z\rangle + c_1^0)}{Re\,(\langle d^1, z\rangle + d_1^0)}, \ldots, \frac{Re\,(\langle c^p, z\rangle + c_p^0)}{Re\,(\langle d^p, z\rangle + d_p^0)} \right] \tag{7.7.1}$$

subject to

$$Az - b \in S, \quad z \in T, \tag{7.7.2}$$

where

(i) $c^1, \ldots, c^p, \ d^1, \ldots, d^p, z$ are $n \times 1$ column vectors in complex space;

(ii) $c_1^0, \ldots, c_p^0, \ d_1^0, \ldots, d_p^0$ are complex numbers;

(iii) S is a convex cone in C^m;

(iv) T is a convex cone in C^n;

(v) A is an $m \times n$ complex matrix;

(vi) b is an $m \times 1$ column vector in complex space.

For problem (7.7.1) - (7.7.2), we assume that the feasible set

$$X = \{ z \in C^n : Az - b \in S, z \in T \},$$

is regular, i.e. non-empty and bounded. Further, it is assumed that the denominators $Re\,(\langle d^k, z\rangle + d_k^0)$, $k = 1, \ldots, p$ do not change the sign on the feasible set X. The term "minimize" being used in problem (7.7.1) - (7.7.2) is for finding all efficient solutions in a minimization sense in terms of the following definition:

Definition 7.7.1. A point $z^0 \in C^n$ is said to be efficient solution for problem (7.7.1) - (7.7.2) if $z^0 \in X$ and if there exists no $z \in X$ such that

$$\frac{Re\,(\langle c^k, z\rangle + c_k^0)}{Re\,(\langle d^k, z\rangle + d_k^0)} \leq \frac{Re\,(\langle c^k, z^0\rangle + c_k^0)}{Re\,(\langle d^k, z^0\rangle + d_k^0)}, \quad k = 1, \ldots, p \tag{7.7.3}$$

and such that for at least one $j \in \{1, \ldots, p\}$ one has

$$\frac{Re\,(\langle c^j, z\rangle + c_j^0)}{Re\,(\langle d^j, z\rangle + d_j^0)} < \frac{Re\,(\langle c^j, z^0\rangle + c_j^0)}{Re\,(\langle d^j, z^0\rangle + d_j^0)} \tag{7.7.4}$$

Let us introduce the change of variables $w = tz$, where t is a non-negative real scalar quantity chosen such that

$$Re\left(\langle d^k, w\rangle + d_k^0 t\right) = \delta_k \quad \text{for all } k \in \{1, ..., p\} \tag{7.7.5}$$

Here δ_k, for all $k \in \{1, ..., p\}$, are specified real numbers. Introducing the change of variables $w = tz$ in problem (7.7.1) - (7.7.2) and taking (7.7.5) into account, we obtain the following multiple objective linear programming problem in complex space:

$$\text{Minimize } Re\left\{\frac{1}{\delta_1}\left(\langle c^1, w\rangle + c_1^0 t\right), ..., \frac{1}{\delta_p}\left(\langle c^p, w\rangle + c_p^0 t\right)\right\} \tag{7.7.6}$$

subject to

$$\left.\begin{aligned} Aw - bt &\in S \\ Re\left(\langle d^k, w\rangle + d_k^0 t\right) &= \delta_k, \quad k = 1, ..., p \\ w &\in T \\ t &\in R_+ \end{aligned}\right\} \tag{7.7.7}$$

because S and T are cones.

If we denote

$$D = \begin{bmatrix} d^{1^H} \\ ... \\ d^{p^H} \end{bmatrix} \in C^{p \times n}, \quad d^0 = \begin{bmatrix} d_1^0 \\ ... \\ d_p^0 \end{bmatrix} \in C^p, \quad \delta = \begin{bmatrix} \delta_1 \\ ... \\ \delta_p \end{bmatrix} \in R^p,$$

then the constraints (7.7.7) may be written in the form:

$$\left.\begin{aligned} Aw - bt &\in S \\ Re\left(Dw + d^0 t\right) &= \delta \\ w &\in T \\ t &\in R_+ \end{aligned}\right\} \tag{7.7.8}$$

or

$$\left.\begin{aligned} (A\ \ -b)\begin{bmatrix} w \\ t \end{bmatrix} &\in S \\ Re\ (D\ d^0)\begin{bmatrix} w \\ t \end{bmatrix} &= \delta \\ \begin{bmatrix} w \\ t \end{bmatrix} &\in T \times R_+. \end{aligned}\right\} \tag{7.7.9}$$

Let Y be the feasible set of problem (7.7.6), (7.7.9), i.e.

$$Y = \left\{ \begin{bmatrix} w \\ t \end{bmatrix} \in C^{n+1} : (A\ \ -b)\begin{bmatrix} w \\ t \end{bmatrix} \in S,\ Re\ (D\ d^0)\begin{bmatrix} w \\ t \end{bmatrix} = \delta,\ \begin{bmatrix} w \\ t \end{bmatrix} \in T \times R_+ \right\}$$

It is possible that Y to be empty for certain values of δ. In what follows, we give some conditions for that Y to be non-empty. For this we will give a theorem of the alternative for complex nonhomogeneous linear inequalities.

Theorem 7.7.1. *Let $A \in C^{m\times n}$, $B \in C^{p\times n}$, $b \in C^m$, $\delta \in R^p$ and let $S \subseteq C^m$, $T \subseteq C^n$ be polyhedral cones. Then either*

$$\left.\begin{aligned} Az - b &\in S \\ Re\ Bz &= \delta \\ z &\in T \end{aligned}\right\} \text{ has a solution } z \in C^n \tag{7.7.10}$$

or

$$\left.\begin{aligned} A^H u + B^H v &\in T^* \\ -u &\in S^* \\ v &\in R^p \\ Re\ \langle b, u\rangle + \langle \delta, v\rangle &< 0 \end{aligned}\right\} \text{ has a solution } u \in C^m,\ v \in C^p \tag{7.7.11}$$

but never both.

Proof. Systems (7.7.10) and (7.7.11) cannot have both solutions, for then

$$\begin{aligned} 0 &\le Re\ (\langle Az - b, -u\rangle + \langle z, A^H u + B^H v\rangle) = \\ &= Re\ (-\langle Az, u\rangle + \langle b, u\rangle + \langle Az, u\rangle + \langle Bz, v\rangle) = \\ &= Re\ \langle b, u\rangle + \langle \delta, v\rangle < 0, \end{aligned}$$

by (7.7.10), (7.7.11), the properties of the inner product and the definitions of T^* and S^*.

Suppose now that (7.7.10) and (7.7.11) are inconsistent. By inconsistency of (7.7.11) we have

$$\begin{bmatrix} A & -I & 0 \\ B & 0 & I \end{bmatrix}^H \begin{bmatrix} u \\ v \end{bmatrix} \in T^* \times S^* \times R^p \Rightarrow Re \left\langle \begin{bmatrix} b \\ \delta \end{bmatrix}, \begin{bmatrix} u \\ v \end{bmatrix} \right\rangle \geq 0$$

Then, by Farcas' lemma (R4, Section 7.1), the system

$$\begin{cases} \begin{bmatrix} A & -I & 0 \\ B & 0 & I \end{bmatrix} \begin{bmatrix} z \\ w \\ r \end{bmatrix} = \begin{bmatrix} b \\ \delta \end{bmatrix} \\ \begin{bmatrix} z \\ w \\ r \end{bmatrix} \in \left(T^* \times S^* \times R^p\right)^* \end{cases}$$

have a solution $z \in C^n$, $w \in C^m$, $r \in C^p$. Since $(T^* \times S^* \times R^p)^* = T \times S \times (iR^p)$, it results that the system

$$\left. \begin{aligned} Az - w &= b \\ Bz - r &= \delta \\ z &\in T \\ w &\in S \\ r &\in iR^p \end{aligned} \right\}$$

has a solution $z \in C^n$, $w \in C^m$, $r \in C^p$. Then there exist $z \in C^n$, $w \in C^m$, $r \in C^p$ such that $Az - b = w \in S$, $Bz - \delta = r \in iR^p$, $z \in T$, and hence system (7.7.10) is consistent, which contradicts the hypothesis. The proof is complete.

Theorem 7.7.2. *Let* $A \in C^{m \times n}$, $D \in C^{p \times n}$, $b \in C^m$, $d^0 \in C^p$, $\delta \in R^p$ *and let* $S \subseteq C^m$, $T \subseteq C^n$ *polyhedral cones. The feasible set* Y *of problem (7.7.6), (7.7.9) is non-empty if and only if the system*

$$\left. \begin{aligned} A^H u + D^H v &\in T^* \\ Re \langle b, u \rangle &\leq Re \langle d^0, v \rangle \\ -u &\in S^* \\ v &\in R^p \\ \langle \delta, v \rangle &< 0, \end{aligned} \right\} \qquad (7.7.12)$$

has no solution $u \in C^m$, $v \in C^p$.

Proof. The set Y is non-empty if and only if system (7.7.9) has a solution $w \in C^n$, $t \in R$. By Theorem (7.7.1), system (7.7.9) has a solution $w \in C^n$, $t \in R$ if and only if the system

$$\left.\begin{array}{c} (A \ -b)^H u + (D \ \ d^0)^H v \in (T \times R_+)^* \\ -u \in S^* \\ v \in R^p \\ \langle \delta, v \rangle < 0, \end{array}\right\} \qquad (7.7.13)$$

has no solution $u \in C^m$, $v \in C^p$. Since $(T \times R_+)^* = (R_+^* + iR)$, it results that system (7.6.13) has no solution if and only if the system

$$\left.\begin{array}{c} A^H u + D^H v \in T^* \\ -b^H u + d^{0H} v \in R_+ + iR \\ -u \in S^* \\ v \in R^p \\ \langle \delta, v \rangle < 0, \end{array}\right\}$$

has no solution $u \in C^m$, $v \in C^p$. Hence, the proof of theorem follows.

The following four theorems give necessary conditions for the feasible set Y to be non-empty for all $\delta \in R^p$.

Theorem 7.7.3. *Let* $A \in C^{m \times n}$, $D \in C^{p \times n}$, $b \in C^m$, $d^0 \in C^p$. *Let* $S \subseteq C^m$ *and* $T \subseteq C^n$ *be convex cones such that*

$$\left.\begin{array}{l} Az - b \in \text{int } S \\ z \in \text{int } T \end{array}\right\}$$

has a solution $z \in C^n$. *If the feasible set* Y *of problem (7.7.6) - (7.7.7) is non-empty for all* $\delta \in R^p$, *then the system*

$$\left\{\begin{array}{l} Aw - bt \in \text{int } S \\ Re\,(Dw + d^0 t) = 0 \\ w \in \text{int } T \\ t > 0, \end{array}\right. \qquad (7.7.14)$$

has a solution $w \in C^n$, $t \in R$.

Proof. Let $z^1 \in \text{int } T$ be such that $Az^1 - b \in \text{int } S$. Since, for all $\delta \in R^p$, the set Y is non-empty, it follows that there exist $w^1 \in C^n$ and $t^1 \in R$ such that

$$\begin{cases} Aw^1 - bt^1 \in S \\ Re\,(Dw^1 + d^0t^1) = -Re\,(Dz^1 + d^0) \\ w^1 \in T \\ t^1 \in R_+ \end{cases}$$

Let $w = z^1 + w^1$ and let $t = 1 + t^1$. Then

$$Aw - bt = (Az^1 - b) + (Aw^1 - bt^1) \in \text{int } S + S = \text{int } S,$$
$$Re\,(Dw + d^0t) = Re\,(Dz^1 + d^0) + Re\,(Dw^1 + d^0t^1) = 0,$$
$$t = 1 + t^1 > 0,$$

hence, $w = z^1 + w^1 \in C^n$, $t = 1 + t^1 \in R$ is a solution of system (7.7.14).

Theorem 7.7.4. *Let $A \in C^{m \times n}$, $D \in C^{p \times n}$, $b \in C^m$, $d^0 \in C^p$. Let $S \subseteq C^m$ and $T \subseteq C^n$ be convex cones such that the system.*

$$\begin{cases} Az - b \in S \\ z \in \text{int } T, \end{cases}$$

has a solution $z \in C^n$. If the feasible set Y of problem (7.7.6) - (7.7.7) is non-empty for all $\delta \in R^p$, then the system

$$\begin{cases} Aw - bt \in S \\ Re\,(Dw + d^0t) = 0 \\ w \in \text{int } T \\ t > 0, \end{cases}$$

has solution $w \in C^n$, $t \in R$.

The proof is similar to that given for Theorem 7.7.3.

Theorem 7.7.5. *Let $A \in C^{m \times n}$, $D \in C^{p \times n}$, $b \in C^m$, $d^0 \in C^p$. Let $S \subseteq C^m$ and $T \subseteq C^n$ be convex cones such that the system*

$$\begin{cases} Az - b \in \text{int } S \\ z \in T, \end{cases}$$

has a solution $z \in C^n$. If the feasible set Y of problem (7.7.6)-(7.7.7) is non-empty for all $\delta \in R^p$, then the system

$$\begin{cases} Aw - bt \in \operatorname{int} S \\ Re\left(Dw + d^0 t\right) = 0 \\ w \in T \\ t > 0, \end{cases}$$

has a solution $w \in C^n$, $t \in R$.

The proof is similar to that given for Theorem 7.7.3.

Theorem 7.7.6. *Let* $A \in C^{m \times n}$, $D \in C^{p \times n}$, $b \in C^m$, $d^0 \in C^p$. *Let* $S \subseteq C^m$ *and* $T \subseteq C^n$ *be convex cones such that the system*

$$\begin{cases} Az - b \in S \\ \\ z \in T, \end{cases}$$

has a solution $z \in C^n$. *If the feasible set* Y *of problem (7.7.6)-(7.7.7) is non-empty for all* $\delta \in R^p$, *then the system*

$$\begin{cases} Aw - bt \in S \\ Re\left(Dw + d^0 t\right) = 0 \\ w \in T \\ t > 0, \end{cases}$$

has a solution $w \in C^n$, $t \in R$.

The proof is similar to that given for Theorem 7.7.3.

Theorem 7.7.7. *Let* $\delta_1, ..., \delta_p \in R$. *If* $\begin{bmatrix} w \\ t \end{bmatrix}$ *is a feasible solution of problem (7.7.6)-(7.7.7), then* $t > 0$.

Proof. Suppose, by contradiction, that $\begin{bmatrix} w \\ 0 \end{bmatrix}$ is a feasible solution of problem (7.7.6)-(7.7.7). Let z be a feasible solution of problem (7.7.1) - (7.7.2). Then, for all real $r > 0$, we have

$$A(z + rw) - b = (Az - b) + r(Aw) \in S + rS \subseteq S + S \subseteq S$$

and

$$z + rw \in T + rT \subseteq T + T \subseteq T,$$

because S and T are convex cones. Therefore, $z + rw \in X$, for all $r > 0$. But this implies that X is unbounded, contradicting our regularity assumption. Hence $t > 0$.

The following theorem establishes a connection between the efficient solutions of problems (7.7.1)-(7.7.2) and (7.7.6.)-(7.7.7).

Theorem 7.7.8. *Let* $\delta_1, ..., \delta_p \in R$*, be such that for all* $z \in X$ *there exist* $w \in C^n$ *and* $t \in R$ *such that*

$$\begin{bmatrix} w \\ t \end{bmatrix} \in Y$$

and

$$\frac{Re\,(\langle c^k, w\rangle + c_k^0 t)}{Re\,(\langle d^k, w\rangle + d_k^0 t)} \leq \frac{Re\,(\langle c^k, z\rangle + c_k^0)}{Re\,(\langle d^k, z\rangle + d_k^0)}, \text{ for all } k \in \{1, ..., p\}.$$

If $\begin{bmatrix} w^0 \\ t^0 \end{bmatrix}$ *is an efficient solution of problem (7.7.6)-(7.7.7), then* $(1/t^0)w^0$ *is an efficient solution of problem (7.7.1)-(7.7.2).*

Proof. Suppose that $\begin{bmatrix} w^0 \\ t^0 \end{bmatrix}$ is an efficient solution of problem (7.7.1)-(7.7.2). Then, by Theorem 7.7.7, we have $t^0 > 0$ and hence $(1/t^0)w^0$ is finite. Moreover, since $\begin{bmatrix} w^0 \\ t^0 \end{bmatrix}$ is an efficient solution for problem (7.7.6)-(7.7.7), we have

$$Aw^0 - bt^0 \in S \text{ and } w^0 \in T$$

and hence

$$A\left[(1/t^0)w^0\right] - b \in S \text{ and } (1/t^0)w^0 \in T,$$

because S and T are cones and $t^0 > 0$. Then $(1/t^0)w^0$ is a feasible solution for (7.7.1)-(7.7.2), i.e. $(1/t^0)w^0 \in X$. Assume, by contradiction, that $(1/t^0)w^0$ is not an efficient solution for problem (7.7.1)-(7.7.2). Then there exists $z \in X$ such that

$$\frac{Re\,(\langle c^k, z\rangle + c_k^0)}{Re\,(\langle d^k, z\rangle + d_k^0)} \leq \frac{Re\,(\langle c^k, (1/t^0)w^0\rangle + c_k^0)}{Re\,(\langle d^k, (1/t^0)w^0\rangle + d_k^0)} \tag{7.7.15}$$

for all $k \in \{1, ..., p\}$ and

$$\frac{Re\,(\langle c^j, z\rangle + c_j^0)}{Re\,(\langle d^j, z\rangle + d_j^0)} < \frac{Re\,(\langle c^j, (1/t^0)w^0\rangle + c_j^0)}{Re\,(\langle d^j, (1/t^0)w^0\rangle + d_j^0)} \tag{7.7.16}$$

for at least one $j \in \{1, ..., p\}$.

Since $z \in X$, by hypothesis, it follows that there exists $\begin{bmatrix} w \\ t \end{bmatrix}$ such that

$$\frac{Re\ (\langle c^k, w\rangle + c_k^0 t)}{Re\ (\langle d^k, w\rangle + d_k^0 t)} \le \frac{Re\ (\langle c^k, z\rangle + c_k^0)}{Re\ (\langle d^k, z\rangle + d_k^0)}, \tag{7.7.17}$$

for all $k \in \{1, ..., p\}$. From (7.7.15), (7.7.16) and (7.7.17) we deduce that

$$\frac{Re\ (\langle c^k, w\rangle + c_k^0 t)}{Re\ (\langle d^k, w\rangle + d_k^0 t)} \le \frac{Re\ (\langle c^k, w^0\rangle + c_k^0 t^0)}{Re\ (\langle d^k, w^0\rangle + d_k^0 t^0)}$$

for all $k \in \{1, ..., p\}$, and

$$\frac{Re\ (\langle c^j, w\rangle + c_j^0 t)}{Re\ (\langle d^j, w\rangle + d_j^0 t)} < \frac{Re\ (\langle c^j, w^0\rangle + c_j^0 t^0)}{Re\ (\langle d^j, w^0\rangle + d_j^0 t^0)}$$

for at least one $j \in \{1, ..., p\}$, contradicting the fact the $\begin{bmatrix} w^0 \\ t^0 \end{bmatrix}$ is an efficient solution of problem (7.7.6)-(7.7.7). The proof is complete.

7.8. HISTORICAL NOTES. COMMENTARIES

The results presented in Section 7.2 are due to Swarup and Sharma [963] and generalize to the complex space the results of Charnes and Cooper [281] from linear fractional programming in real space (Chapter 3, Section 3.4). Saxena, Patkar and Parkash [784] obtained the same results for the more general problem when the constraints (7.2.2) are replaced by nonlinear constraints of form (7.5.2) (see also, Saxena and Patkar [781]). The results presented in Section 7.2 can be easily extended to the solution of quadratic fractional programming problem in complex space [963]:

$$\min \frac{Re\ (z^H E z + c^H z + c_0)}{Re\ (z^H F z + d^H z + d_0)}$$

with the constraints (7.2.2), where $E(F)$ is a positive semi-definite (negative semi-definite) Hermitian matrix of type $m \times n$. In a similar manner, one can show that, under the hypothesis that the denominator keeps, say, the positive sign on the whole set of the feasible solutions, through the change of variable $y = tz$, the quadratic programming problem is made equivalent to the following complex nonlinear programming problem:

$$\min \{Re(y^H E y + c^H y t + c_0 t^2) \mid |\ arg\ (Ay - bt)\ | \le \beta,$$
$$Re(y^H F y + d^H y t + d_0 t^2) = 1, \quad |\ arg\ y\ | \le \alpha, \quad |\ arg\ t\ | \le 0\}$$

The aspects presented in Section 7.3 represent an extension to the complex space of the results of Swarup [944] [957] from the real space. Theorems 7.3.1 and 7.3.2 of Section 7.3 are due to Das and Swarup [344]. The results of Section 7.4 are due to Datta and Bhatia [347] and represent an extension to complex space of the results presented in Section 4.2, Chapter 4. The development presented in Section 7.5 was constructed based on Saxena and Patkar [782]. The duality theorems of Section 7.6 were formulated by Gulati and Chandra [436] (see also Saxena [777]). Saxena and Patkar [781] generalized a series of results from real and complex. Thus, if in problems **P** and **D** presented in Section 7.6, we consider $\boldsymbol{d} = \boldsymbol{0}$, $c_0 = 0$, $d_0 = 1$, we get the pair of dual problems in the complex space considered by R.A. Abrams and A. Ben-Israel ("Complex mathematical programming", Report No. 69-11, Northwest Univ., Evanston, Illinois, Nov. 1969)

(P) $\min \{ Re\ (c^H z) \mid Az - b \in S,\ z \in T \}$ and

(D) $\max \{ Re\ (b^H w) \mid -A^H w + c \in T^*,\ w \in S^* \}$,

If, in problems **P** and **D**, we consider $S = R^m_-$, $T = R^n_+$, $c_0 = d_0 = 0$, the other matrices being real, we get the pair of real linear fractional dual problems studied by Chadha [233], with the only difference that the constraint $Re\ (b^H w + c_0 - v d_0) = 0$ is considered in the form $b^T w \leq 0$. Bector [127] observed by that it should have been $b^T w = 0$.

The dual problems under consideration in Section 7.6 were also studied by Saxena, Patkar and Parkash [784], with the only difference that the constraint$Re\ (b^H w + c_0 - v d_0) \geq 0$ appeared in the dual problem.

For nonlinear and non-differentiable fractional programming problem in the complex space (7.5.1)-(7.5.2), Saxena and Patkar [782] constructed the dual problem and proved duality theorems of form

$$\max g(u, v, w) = u$$

subject to

$$-A^H v - du + Hw + c \in T^*$$
$$Re[b^H v - d_0 u + c_0] \geq 0$$
$$w^H H w \leq 1;\ v \in S^*$$

The duality results generalize also a series of known results. Thus, if $\boldsymbol{H} = \boldsymbol{0}$, $c_0 = 0$, $\boldsymbol{d} = \boldsymbol{0}$, $d_0 = 1$, then we recover the dual problems in complex space studied by Abrams and Ben-Israel [*op. cit*].

If, additionally, we have $S = S^* = R^m_+$, $T = T^* = R^n_+$, then we recover the dual problems from real space.

If $\boldsymbol{H} = \boldsymbol{0}$, then we recover the duality results in complex space for linear fractional programming problems presented in Section 7.6.

Also, if $S = S^* = R_+^m$, $T = T^* = R_+^n$, then we get the dual of a real nonlinear fractional programming problem, studied by Aggarwal,and Saxena [23]. Bector, Chandra and Gulati [145] studied a complex nonlinear fractional problem. Using a complex version of Dinkelbach theorem [360], they reduced the problem to a convex parametric problem in complex space for which the results of R. A. Abrams ("Nonlinear programming in complex space. Sufficient conditions and duality", *J. Math. Anal. Appl.* **38** (3),1972, 619-632) become applicable. In another paper, Bector, Chandra and Gulati [147] give a Langrangian approach to duality of the general complex nonlinear fractional programming problem over cones. These results give an alternative and more natural presentation of their results in [145] and extend the results of Bector [123] to complex fractional programming problems over cones. The problem considered by they is the following nonlinear fractional program as primal problem:

(P1) $$\text{Minimize } \phi(z, \bar{z}) = \frac{Re\, f(z, \bar{z})}{Re\, h(z, \bar{z})}$$

subject to $(z, \bar{z}) \in D$
where

i) $f : C^{2n} \to C$, $h : C^{2n} \to C$ and $g : C^{2n} \to C^m$ are analytic functions on the manifold $M = \{(w_1, w_2) \in C^{2n} : w_2 = \bar{w}_1\}$ and $S \subset C^m$ is a polyhedral cone.

ii) $D = \{(z, \bar{z}) : (z, \bar{z}) \in K,\ g(z, \bar{z}) \in S\}$ is compact and contains at least two points.

iii) K is a convex subset of M.

iv) $Re\, f(z, \bar{z}) \geq 0$ and $Re\, h(z, \bar{z}) > 0$ for all $(z, \bar{z}) \in K$

v) f (respectively h) has convex (respectively concave) real part with respect to R_+ on K and g is concave with respect to S on K.

vi) **(P1)** satisfied Kuhn-Tucker constraint qualification.
Let the Langrangian of problem **(P1)** be defined as

$$L(z, \bar{z}, u) = \frac{Re\, f(z, \bar{z}) - Re\, \langle u, g(z, \bar{z})\rangle}{Re\, h(z, \bar{z})}$$

Bector, Chandra and Gulati [147] proved the following:

<u>Saddle point theorem</u>: *The point $(z_0, \bar{z}_0) \in K$ is optimal for problem (P1) if and only if there exists a $u_0 \in S^*$ such that*

$$L(z, \bar{z}, u_0) \geq L(z_0, \bar{z}_0, u_0) \geq L(z_0, \bar{z}_0, u)$$

for all $(z, \bar{z}) \in K$ and all $u \in S^*$.

Also, they introduced the following nonlinear fractional programs as dual problem to the primal problem **(P1)**.

(D1) Maximize $L(w, \bar{w}, u) = \dfrac{Re\, f(w, \bar{w}) - Re\, \langle u, g(w, \bar{w})\rangle}{Re\, h(w, \bar{w})}$

subject to

$$\nabla_w \left[\frac{f(w, \bar{w})}{Re\, h(w, \bar{w})}\right] + \overline{\nabla_{\bar{w}} \left[\frac{f(w, \bar{w})}{Re\, h(w, \bar{w})}\right]}$$

$$- u^H D_w \left[\frac{g(w, \bar{w})}{Re\, h(w, \bar{w})}\right] - \overline{u^T D_{\bar{w}} \left[\frac{g(w, \bar{w})}{Re\, h(w, \bar{w})}\right]} = 0$$

$$(w, \bar{w}) \in K, \quad u \in S^*$$

and proved weak duality theorem, direct duality theorem, Mangasarian's (strict) converse duality theorem, and Hanson's converse duality theorem.

Problem **(D1)** is equivalent to problem:

(D2) Maximize $\dfrac{Re\, f(w, \bar{w}) - Re\, \langle u, g(w, \bar{w})\rangle}{Re\, h(w, \bar{w})} \quad (=v)$

subject to

$$\nabla_w f(w, \bar{w}) - \overline{\nabla_{\bar{w}} f(w, \bar{w})} - v\, \nabla_w h(w, \bar{w}) - v\, \overline{\nabla_{\bar{w}} h(w, \bar{w})}$$

$$= u^H D_w\, g(w, \bar{w}) + u^T\, \overline{D_{\bar{w}}\, g(w, \bar{w})}$$

$$Re\, f(w, \bar{w}) - v\, Re\, h(w, \bar{w}) - Re\, \langle u, g(w, \bar{w})\rangle = 0$$

$$(w, \bar{w}) \in K, \quad u \in S^*, \quad v \in R$$

which is the same problem as obtained in [145] except that v is not restricted to be non-negative.

The results obtained by Bector, Chandra, and Gulati [147] contain a number of well known results:

1) If $K = M$ and $Re\, h(z, \bar{z}) = 1$ for all $(z, \bar{z}) \in M$, then we obtain the results of Abrams (*op. cit.*).

2) If $K = M$, $f(z, \bar{z}) = c^H z$, $g(z, \bar{z}) = d^H z$ and the constraints are $Az - b \in S$, $z \in T$, then problems **(P1)** and **(D2)** reduce to the problems studies by Gulati and Chandra [436] and presented in Section 7.6.

From problems **(P1)** and **(D1)** we also obtain the dual of linear, quadratic and nonlinear programming in complex space. Similar results can be found in Bector and Bhatia [131] and Bector and Bhatt [137].

Duality theorems for the fractional programming problems in complex space were given by Gulati [435]. Also, Jain [517] presented duality theorems for the following nonlinear and non-differentiable problem in complex space.

$$\min \left\{ \Phi(z, \bar{z}) = \frac{Re\left[f(z, \bar{z}) + (z^H B z)^{1/2}\right]}{Re\left[g(z, \bar{z}) - (z^H D z)^{1/2}\right]} \mid Az - b \in S,\ z \in T \right\}$$

where $f, g : C^{2n} \to C$ are analytical functions, f convex and g concave, $B, D \in C^{n \times n}$ are positive semi-definite Hermitian matrices. Additionally, Parkash, Saxena and Patkar [705] considered the same problem but with the constraints of form $h(z, \bar{z}) \in S$, where $h : C^{2n} \to C^m$ is an analytical function and $S \subset C^m$ is a polyhedral cone. They obtained the existence conditions for the solution, and the duality theorems.

In the last decade, certain results from mathematical programming in the real space with several objective functions were extended to the complex space. Thus, we presented in Section 7.7 the results obtained by Stancu-Minasian, Duca and Nishida [916] for the multiple objective linear fractional programming problem in complex space. Other aspects were treated by Gupta [442] and Kaul and Gupta [546]. Let

$$h(z) = \left\{ \frac{Re\ c_1^H z}{Re\ d_1^H z}, \ldots, \frac{Re\ c_p^H z}{Re\ d_p^H z} \right\}$$

and

$$S = \left\{ z \in C^n : b - Az \in P,\ z \in Q \right\}$$

Kaul and Gupta [546] considered the problem

(P2) $$\max \left\{ h(z) \mid z \in S \right\}$$

and obtained the necessary condition for a point $z^0 \in S$ to be efficient solution. To this purpose, one can associate to problem **(P2)** the following problem

(P3) $$\underset{z \in S}{Max}\ Re \left\{ c_1^H z, \ldots, c_p^H z, \ldots, -d_1^H z, \ldots, -d_p^H z \right\}$$

and showed that, if $z^0 \in S$ is efficient solution for problem **(P2)**, then it is efficient for **(P3)** also. This result represents a generalization to the complex space of Theorem 6.4.4, Chapter 6. Gupta [442] considered problem

(P4) $$\min \left\{ h(z) \mid z \in S \right\}$$

to which he associated the following dual problem

(D4) $$\max \phi(u) = (u_1, \ldots, u_p)$$

subject to

$$A^H v_i - u_i d_i + c_i \in Q^*, \quad i = 1, \ldots, p$$

$$Re\ (b^H v_i) = 0, \quad v_i \in P^*, \quad u_i \in R, \quad i = 1, \ldots, p$$

CHAPTER 8

SPECIAL LINEAR FRACTIONAL PROGRAMMING PROBLEMS

In this chapter, we consider several special problems which appear in linear fractional programming. Thus, in Section 8.1, we consider a problem of linear fractional programming in which the variables of the objective function appear in absolute-value, and the constraints lack the condition for non-negativity. We show that, under certain conditions, it is possible to apply the Simplex algorithm developed for the standard linear fractional problem. In Section 8.2, we consider the separable linear fractional programming problem, and we reduce it to the solution of a sequence of quadratic programming problems. In Section 8.3, we study the linear fractional programming problem with disjunctive constraints for which we present a solution method and we construct the Dual problem.

In the classical problems de mathematical programming, in general, and in the fractional programming problems, in particular, it is accepted that the coefficients which appear in problem are known exactly. This condition is not satisfied in most real problems. Usually, the problem data are marked by uncertainty and imprecision. The introduction of uncertainty and imprecision in the modeling process is an important gain in tackling real problems.

Initially, a quantitative assessment of uncertainty was attempted using the concepts of the theory of probabilities. This implied that notion of uncertainty can be superposed to the notion of randomness. Probabilistic analysis, specific to phenomena which can be reproduced and which have appeared for sufficiently many times, assumes that certain parameters can be represented by random variables. Such a treatment needs sufficient statistical data such that information about the distribution functions of the random variables which appear in the mathematical model can be extracted. The introduction of random variables also leads to several computational difficulties. Modeling of uncertainty for data difficult to predict can be also done using fuzzy sets. The case in which the coefficients of the variables are fuzzy sets corresponds to the situation in which the coefficients take preferred values in a certain set.

The above mentioned cases do not cover all the real situations. A new chapter of mathematical programming has lately emerged: the inexact mathematical programming in which the constraints of the problem are of inclusion type, and in which, about the coefficients, one only knows that they belong to an usually convex set. This Chapter will cover all the above mentioned aspect of imprecision as applied to fractional programming. The stochastic case will be treated in Sections 8.4 - 8.8. The case of fractional programming with inclusion constraints will be analyzed in Section 8.9, and the case of fuzzy fractional programming will be treated in Section 8.10.

8.1. LINEAR FRACTIONAL PROGRAMMING WITH ABSOLUTE-VALUE FUNCTIONALS

This section will consider a linear fractional programming problem in which the variables appear in the objective function in absolute-value, and in which the constraints lack the non-negativity condition. Consider the following problem

$$max \frac{\sum_{i=1}^{n} c_i|x_i| + c_0}{\sum_{i=1}^{n} d_i|x_i| + d_0}, \tag{8.1.1}$$

subject to

$$\sum_{i=1}^{n} a_{ji} x_i = b_j , \quad j = 1 , \dots , m \tag{8.1.2}$$

$$x_i \text{ without sign constraint.}$$

The notations are usual($|x_i|$ is the absolute-value of x_i). Also, and in this case, consider that the domain of feasible solutions is a regular set.

Since the Simplex method was developed for problems having the non-negativity condition for the variables, it follows that for solving the problem (8.1.1) - (8.1.2) one cannot, in general, apply the Simplex method. We show that, with certain constraints onto the coefficients c_i and d_i from the objective function and upon the values of the denominator and of the numerator, one can still apply the Simplex method (since only in these conditions one can realize a certain condition for entering the base). Otherwise, the Simplex method must be seriously modified. Solution of problem (8.1.1) - (8.1.2) is achieved by transforming it into a usual linear fractional programming problem with a restrictive condition for entering the base.

Each variable x_i is written as the difference between two positive variables, $x_i = x_i^+ - x_i^-$, $x_i^+, x_i^- \geq 0$, for example one can choose $x_i^+ = max(0, x_i)$, $x_i^- = max(0, -x_i)$. If $x_i^+ x_i^- = 0$, then $|x_i| = |x_i^+ - x_i^-| = x_i^+ + x_i^-$. In this way, the problem (8.1.1) - (8.1.2) becomes

$$max \frac{\sum_{i=1}^{n} c_i(x_i^+ + x_i^-) + c_0}{\sum_{i=1}^{n} d_i(x_i^+ + x_i^-) + d_0}, \tag{8.1.3}$$

subject to

$$\sum_{i=1}^{n} a_{ji}(x_i^+ - x_i^-) = b_j, \quad j = 1, \ldots, m; \tag{8.1.4}$$

$$x_i^+, x_i^- \geq 0$$

and the condition for entering the base is

$$x_i^+ x_i^- = 0, \quad i = 1, \ldots, n. \tag{8.1.5}$$

If we denote $\hat{A} = (A, -A) = (a_1^+, a_2^+, \ldots, a_n^+, \; a_1^-, a_2^-, \ldots, a_n^-)$, $\hat{x} = (x_1^+, x_2^+, \ldots, x_n^+, \; x_1^-, x_2^-, \ldots, x_n^-)$, then problem (8.1.3) - (8.1.4) is written as

$$max \frac{\sum_{i=1}^{n} c_i x_i^+ + \sum_{i=1}^{n} c_i x_i^- + c_0}{\sum_{i=1}^{n} d_i x_i^+ + \sum_{i=1}^{n} d_i x_i^- + d_0} = \frac{N}{D}. \tag{8.1.6}$$

subject to

$$\hat{A}\hat{x} = \boldsymbol{b}, \quad \hat{x} \geq \boldsymbol{0} \text{ and constraint (8.1.5).} \tag{8.1.7}$$

Let $\boldsymbol{x}_B$ an initial basic solution for problem (8.1.6) - (8.1.7) corresponding to base $\boldsymbol{B} = (b_1, b_2, \ldots, b_m)$ from matrix $\hat{A}$. We have $\boldsymbol{x}_B = \boldsymbol{B}^{-1}\boldsymbol{b}$ and $\boldsymbol{x}_B \geq \boldsymbol{0}$. Denote by $Z^{(1)} = \boldsymbol{c}_B'\boldsymbol{x}_B + c_0$, $Z^{(2)} = \boldsymbol{d}_B'\boldsymbol{x}_B + d_0$ the values of the numerator and denominator of function (8.1.6) for basic solution $\boldsymbol{x}_B$. The coefficients $\boldsymbol{c}_B$ and $\boldsymbol{d}_B$ are those coefficients from the numerator and denominator corresponding to base $\boldsymbol{B}$. Assume that, for the basic solution $\boldsymbol{x}_B$, the following elements $\boldsymbol{u}_j^+ = \boldsymbol{B}^{-1}\boldsymbol{a}_j^+$; $\boldsymbol{u}_j^- = \boldsymbol{B}^{-1}\boldsymbol{a}_j^-$; $z_j^{+(1)} = \boldsymbol{c}_B'\boldsymbol{u}_j^+$; $z_j^{+(2)} = \boldsymbol{d}_B'\boldsymbol{u}_j^+$; $z_j^{-(1)} = \boldsymbol{c}_B'\boldsymbol{u}_j^-$; $z_j^{-(2)} = \boldsymbol{d}_B'\boldsymbol{u}_j^-$, are known for each column, $\boldsymbol{a}_j$, of $\hat{A}$, but not of $\boldsymbol{B}$. Let

$$\Delta_j^+ = Z^{(2)}(c_j - z_j^{+(1)}) - Z^{(1)}(d_j - z_j^{+(2)}), \text{ and}$$

$$\Delta_j^- = Z^{(2)}(c_j - z_j^{-(1)}) - Z^{(1)}(d_j - z_j^{-(2)}).$$

According to Section 3.1, Chapter 3, if $\Delta_j > 0$, and at least a $u_{ij} > 0$ $(i = 1, 2, \ldots, m)$, then it is possible to obtain a new basic feasible solution substituting one of the columns of $\boldsymbol{B}$ with $\boldsymbol{a}_j$, and the value of the objective function to increase. Swarup and Bedi [701] showed that only in two cases one can apply the Simplex algorithm for solution of problem (8.1.6) - (8.1.7).

Case 1. $c_i < 0,\ d_i < 0 \quad (i = 1, 2, \dots, n)$, $N < 0$ and $D > 0$ for all feasible solutions.

This is the same as showing that, in this case, when applying the Simplex algorithm, condition (8.1.5) is satisfied. I.e., if we assume that $x_i^+ > 0$, i.e., x_i^+ is in base, then, necessarily $x_i^- = 0$, i.e., x_i^- is outside the base. Since $x_i^+ > 0$, $\boldsymbol{a}_i^+$ is in base and we have $z_i^{+(1)} - c_i = 0$ $(z_i^{+(1)} = c_i < 0)$ and $z_i^{+(2)} - d_i = 0$ $(z_i^{+(2)} = d_i < 0)$ i.e., $\Delta_i^+ = 0$. To calculate Δ_i^-, we have

$$z_i^{-(1)} = \boldsymbol{c}_B{}'\boldsymbol{u}_i^- = \boldsymbol{c}_B{}'\boldsymbol{B}^{-1}\boldsymbol{a}_j^- = -\boldsymbol{c}_B{}'\boldsymbol{B}^{-1}\boldsymbol{a}_j^+ = -\boldsymbol{c}_B{}'\boldsymbol{u}_i^+ = -z_i^{+(1)} > 0$$

since $\boldsymbol{a}_i^+ = -\boldsymbol{a}_i^-$. Similarly, $z_i^{-(2)} = -z_i^{+(2)} > 0$. Thus,

$$z_i^{-(1)} - c_i = -z_i^{+(1)} - c_i = -(z_i^{+(1)} + c_i) > 0 \text{ and}$$

$$z_i^{-(2)} - d_i = -z_i^{+(2)} - d_i = -(z_i^{+(2)} + d_i) > 0\ .$$

It follows that

$$\Delta_i^- = Z^{(2)}(c_i - z_i^{-(1)}) - Z^{(1)}(d_i - z_i^{-(2)}) < 0\ .$$

This means that in Case 1 if $\boldsymbol{a}_i^+$ is in base (thus, $x_i^+ > 0$), then $\boldsymbol{a}_i^-$ cannot be candidate to enter the base (thus, $x_i^- = 0$). The condition (8.1.5) is satisfied and one can apply the Simplex algorithm to obtain a finite optimum. From Case 1, by multiplying both N and D with (-1), one can obtain another applicability case for the Simplex algorithm.

Case 1'. $c_i > 0$, $d_i > 0\ \forall\ i = 1, \dots, n$, $N > 0$ and $D < 0$, for all feasible solutions.

Case 2. $c_i < 0$ and $d_i > 0$, $\forall\ i = 1, 2, \dots, n$, $N > 0$ and $D > 0$, for all feasible solutions. Indeed, consider $x_i^+ > 0$ thus, column $\boldsymbol{a}_i^+$ is basic i.e., $z_i^{+(1)} - c_i = 0$ (thus, $z_i^{+(1)} = c_i < 0$) and $z_i^{+(2)} - d_i = 0$ (i.e., $z_i^{+(2)} = d_i > 0$). Similarly, $z_i^{-(1)} = -z_i^{+(1)} > 0$ and $z_i^{-(2)} = -z_i^{+(2)} < 0$. Thus, $z_i^{-(1)} - c_i = -z_i^{+(1)} - c_i = -(z_i^{+(1)} + c_i) > 0$ and $z_i^{-(2)} - d_i = -z_i^{+(2)} - d_i = -(z_i^{+(2)} + d_i) < 0$. Hence, $\Delta_i^- = Z^{(2)}(c_i - z_i^{-(1)}) - Z^{(1)}(d_i - z_i^{-(2)}) < 0$ which means that $\boldsymbol{a}_i^-$ cannot be a candidate to enter the base and Simplex method will converge to a finite optimum.
By multiplying by -1 both N and D, one gets another applicability case for the Simplex algorithm.

Case 2'. $c_i > 0$, $d_i < 0$, $\forall\, i = 1, 2, \ldots, n$, $N < 0$ and $D < 0$ for all feasible solutions.

Table 8.1.1

	For all $i = 1, \ldots, n$		For all feasible solutions	
1	$c_i > 0$,	$d_i > 0$	$N < 0$,	$D > 0$
2	$c_i < 0$,	$d_i < 0$	$N > 0$,	$D < 0$
3	$c_i < 0$,	$d_i > 0$	$N < 0$,	$D < 0$
4	$c_i > 0$,	$d_i < 0$	$N > 0$,	$D > 0$
5	$c_i > 0$,	$d_i > 0$	$N > 0$,	$D > 0$
6	$c_i > 0$,	$d_i > 0$	$N < 0$,	$D < 0$
7	$c_i < 0$,	$d_i < 0$	$N > 0$,	$D > 0$
8	$c_i < 0$,	$d_i < 0$	$N < 0$,	$D < 0$
9	$c_i > 0$,	$d_i < 0$	$N > 0$,	$D < 0$
10	$c_i > 0$,	$d_i < 0$	$N < 0$,	$D > 0$
11	$c_i < 0$,	$d_i > 0$	$N > 0$,	$D < 0$
12	$c_i < 0$,	$d_i > 0$	$N < 0$,	$D > 0$

Note that the 4 constraints presented above are only sufficient conditions to obtain an optimum solution of problem (8.1.3)-(8.1.4) through the Simplex algorithm, but not also necessary. It is possible to have a problem of linear fractional programming with objective function containing the absolute-value variables, which admits a solution with the Simplex algorithm, but which does not have any of cases 1, 1', 2, 2', but one of the cases given in Table 8.1.1 (see also Example 3).

Example 8.1.1. (Case 1) max $\{(-|x| - 2)/(-2|x| + 15) \mid -3 \le x \le -1,\ x$ without sign restriction$\}$. Through the change of variable $x = x^+ - x^-$, x^+ , $x^- \ge 0$, the problem becomes max $\{ (-(x^+ + x^-) - 2)/(-2(x^+ + x^-) + 15) \mid x^+ - x^- \le -1$, $x^+ - x^- \ge -3$, x^+, $x^- \ge 0$ $x^+ \cdot x^- = 0 \}$. Using the Simplex algorithm, one gets the optimum solution $x = -1$, for which the value of objective function is $Z = -3/13$.

Example 8.1.2. (Case 2) max $\{(-|x| + 4)/(3|x| + 2) \mid -3 \le x \le -1\ ,\ x$ without sign restriction$\}$. The optimum solution is $x = -1$ and the optimum value of the objective function is $Z = 3/5$.

Example 8.1.3. max $\{(2|x| + 7)/(|x| + 3) \mid -3 \le x \le 1\ ,\ x$ without sign restriction$\}$. Though conditions 1, 1', 2 and 2' are not satisfied, one can still apply the Simplex algorithm. The optimum solution is $x = 0$ and the value of the objective function is $Z = 7/3$.

8.2. SEPARABLE LINEAR FRACTIONAL PROGRAMMING

Consider a special type of fractional programming problems, *viz.* when the objective function and the constraints are functions separable. The function F is separable if

$$F(x_1\ ,\ \dots\ ,\ x_n) = \sum_{j=1}^{n} F_j(x_j)$$

i.e., can be represented as a sum of functions, each of them depending only on one independent variable.

Consider the non-linear and separable programming problem

$$\max\ (\min)\ F(x) = \sum_{j=1}^{n} F_j(x_j)\ , \tag{8.2.1}$$

subject to

$$h_i(x) = \sum_{j=1}^{n} h_{ij}(x_j) \le b_i\ ,\quad i = 1\ ,\ \dots\ ,\ m\ ; \tag{8.2.2}$$

$$x_j \ge 0\ ,\quad j = 1\ ,\ \dots\ ,\ n\ ,$$

in which the functions h_i are assumed separable and continuous. The basic idea in the solution of problem (8.2.1)-(8.2.2) is to consider linear approximations (piecewise) of the non-linear functions and to solve the resulting linear programming problems. Their solutions constitutes approximations of the solution to the initial problem. These approximate solutions converge towards the solution of the initial problem. If the non-linear function f_j depends on only one variable x_j, the variation interval of variable x_j is split into subintervals using a finite number of points x_{kj}, $k = 1\ ,\ \dots\ ,\ p$. Introducing the non-negative variables $\lambda_{1j}\ ,\ \dots\ ,\ \lambda_{pj}$, one gets the polygon approximating the non-linear function f_j as

$$\tilde{f}_j = \sum_{k=1}^{p} f_j(x_{kj})\lambda_{kj} \text{ with conditions} \tag{8.2.3}$$

$$\sum_{k=1}^{p} \lambda_{kj} = 1 , \tag{8.2.4}$$

$$\sum_{k=1}^{p} \lambda_{kj} x_{kj} = x_j \tag{8.2.5}$$

and not more than two λ_{kj} can be positive and these must be adjacent.

If, in problem (8.2.1)-(8.2.2), we substitute each non-linear function of x_j, $j = 1 , \dots , n$ through polygonal approximations in the way described by relations (8.2.3)-(8.2.5), the problem (8.2.1)-(8.2.2) becomes an approximate problem in variables λ_{kj}:

$$\max \sum_{j=1}^{n} \sum_{k=1}^{p} F_j(x_{kj}) \lambda_{kj} , \tag{8.2.6}$$

subject to

$$\begin{gathered} \sum_{j=1}^{n} \sum_{k=1}^{p} h_{ij}(x_{kj}) \lambda_{kj} \leq b_i , \quad i = 1 , \dots , m ; \\ \sum_{k=1}^{p} \lambda_{kj} = 1 , \quad \lambda_{kj} \geq 0 \quad j = 1 , \dots , n ; \quad k = 1 , \dots , p . \end{gathered} \tag{8.2.7}$$

If λ_{kj}^* is an optimal solution of problem (8.2.6)-(8.2.7), then

$$x_j^* = \sum_{k=1}^{p} \lambda_{kj}^* x_{kj} , \quad j = 1 , \dots , n ,$$

will be an approximate solution of problem (8.2.1)-(8.2.2). The literature provides numerous works about separable programming. Charnes and Lemke[1] proposed an approximate technique for the minimization of a non-linear convex and separable function with linear constraints. Miller[2] generalized this technique to solve problems having non-linear and separable constraints. Recently, these methods have been applied for the solution of separable fractional programming problems.

Consider the following separable linear fractional programming problem

$$\min F(x_1 , \dots x_n) = \sum_{j=1}^{n} c_j/(x_j + s_j), \quad c_j > 0 , \quad s_j > 0, \tag{8.2.8}$$

$$Ax \leq b , \quad x \geq 0 , \tag{8.2.9}$$

[1]Charnes, A., Lemke, C. E., "Minimization of nonlinear separable convex functionals". *Naval Res. Logist. Quart.*, 1(4)(1954), 301-312.

[2]Miller, C. E., "The simplex method for local separable programming", *Recent Advances in Mathematical Programming* (R.L. Graves and P. Wolfe Eds.) McGraw-Hill Book Company, London, 1963.

where $A = (a_{ij})$ is a matrix $m \times n$, x and $\mathbf{0}$ are vectors with n components and b is a vector with m components. Denote by $D = \{ x | Ax \le b \, , \; x \ge \mathbf{0} \}$ the domain of feasible solutions for problem (8.2.8)-(8.2.9) which is assumed to be a convex polyhedron, closed and non-empty in R^n. To prove the theorem on which the calculation procedure for the optimal solution of problem (8.2.8)-(8.2.9) is based, reconsider Theorem 4.4.1, Chapter 4, in a particular case.

Theorem 8.2.1. *Consider the mathematical programming problem*

$$\min_{x \in D} f(x) \, , \tag{8.2.10}$$

where f is a convex function and D is a convex set. Then x^0 is an optimal solution for problem (8.2.10) *if and only if x^0 is an optimal solution of problem $\min_{x \in D} x^t \nabla f(x^0)$, where $\nabla f(x^0)$ is the gradient of function f calculated in point $x = x^0$.*

Also, recall the following result from the convex programming theory:

Theorem 8.2.2. *If there is an optimal solution for problem* (8.2.10), *where f is a strict convex function and D a convex set, closed and non-empty, then its optimal solution is unique.*

Theorem 8.2.3. *If*

$$F(x) = \sum_{j=1}^{n} c_j/(x_j + s_j) \quad (c_j > 0 \, , \; s_j > 0)$$

and

$$\Phi^0(x) = \sum_{j=1}^{n} \lambda_j x_j^2 - \sum_{j=1}^{n} p_j x_j$$

where

$$\lambda_j = -\frac{1}{s_j} \frac{\partial F}{\partial x_j^0} = \frac{c_j}{s_j (x_j^0 + s_j)^2} > 0 \, , \quad p_j = \frac{-(2x_j^0 + s_j)}{s_j} \frac{\partial F}{\partial x_j^0}$$

and

$x^0 = (x_1^0 \, , \, \ldots \, , \, x_n^0)$ is a generic point in D, then

a) $$F(x^0) = \sum_{j=1}^{n} \frac{c_j}{s_j} + \Phi^0(x^0) \, ,$$

b) $$F(x) \le \sum_{j=1}^{n} \frac{c_j}{s_j} + \Phi^0(x) \quad \forall \, x \ge \mathbf{0} \, ,$$

c) $$\frac{\partial \Phi^0(x)}{\partial x^0} = \frac{\partial F(x)}{\partial x^0} \, .$$

Proof. One can write the function F as $F(\mathbf{x}) = \sum_{j=1}^{n} c_j f(x_j)$, where $f(x_j) = 1/(x_j + s_j)$, or as

$$F(\mathbf{x}) = \sum_{j=1}^{n} \frac{c_j}{s_j} - \sum_{j=1}^{n} \frac{c_j x_j}{s_j(x_j + s_j)} .$$

Since the function $f(x_j)$ is concave and differentiable, we have $f(x_j) \geq f(x_j^0) + (x_j - x_j^0) \dfrac{\partial f}{\partial x_j^0}$. Multiplying by $(-c_j/s_j)\, x_j \leq 0$ and summing, we get

$$- \sum_{j=1}^{n} \frac{c_j}{s_j} x_j f(x_j) \leq - \sum_{j=1}^{n} \frac{c_j}{s_j} x_j (x_j - x_j^0) \frac{\partial f}{\partial x_j^0} - \sum_{j=1}^{n} \frac{c_j}{s_j} x_j f(x_j^0) .$$

Thus

$$F(\mathbf{x}) = \sum_{j=1}^{n} \frac{c_j}{s_j} - \sum_{j=1}^{n} \frac{c_j x_j}{s_j(x_j + s_j)} \leq$$

$$\leq \sum_{j=1}^{n} \frac{c_j}{s_j} - \sum_{j=1}^{n} \frac{c_j}{s_j} x_j (x_j - x_j^0) \frac{\partial f}{\partial x_j^0} - \sum_{j=1}^{n} \frac{c_j}{s_j} x_j f(x_j^0) =$$

$$= \sum_{j=1}^{n} \frac{c_j}{s_j} - \sum_{j=1}^{n} \frac{c_j}{s_j} \frac{\partial f}{\partial x_j^0} x_j^2 - \sum_{j=1}^{n} \frac{c_j}{s_j} \left[f(x_j^0) - x_j^0 \frac{\partial f}{\partial x_j^0} \right] x_j =$$

$$= \sum_{j=1}^{n} \frac{c_j}{s_j} + \sum_{j=1}^{n} \lambda_j x_j^2 - \sum_{j=1}^{n} p_j x_j = \sum_{j=1}^{n} \frac{c_j}{s_j} + \Phi^0(\mathbf{x}) , \tag{8.2.11}$$

where we denoted $\lambda_j = - \dfrac{c_j}{s_j} \dfrac{\partial f}{\partial x_j^0} = - \dfrac{1}{s_j} \dfrac{\partial F}{\partial x_j^0}$; $\Phi^0(\mathbf{x}) = \sum_{j=1}^{n} \lambda_j x_j^2 - \sum_{j=1}^{n} p_j x_j$ and

$$p_j = \frac{c_j}{s_j} \left[f(x_j^0) - x_j^0 \frac{\partial f}{\partial x_j^0} \right] = \frac{c_j}{s_j} \frac{2x_j^0 + s_j}{(x_j^0 + s_j)^2} = \frac{-(2x_j^0 + s_j)}{s_j} \frac{\partial F}{\partial x_j^0}.$$

Relation (8.2.11) means relation (b). We have

$$F(\mathbf{x}^0) = \sum_{j=1}^{n} \frac{c_j}{s_j} - \sum_{j=1}^{n} \frac{c_j x_j^0}{s_j(x_j^0 + s_j)} . \tag{8.2.12}$$

Also,

$$\Phi(\mathbf{x}^0) = \sum_{j=1}^{n} \lambda_j (x_j^0)^2 - \sum_{j=1}^{n} p_j x_j^0 = \sum_{j=1}^{n} \frac{c_j}{s_j} \frac{(x_j^0)^2}{s_j(x_j^0 + s_j)^2} - \sum_{j=1}^{n} \frac{c_j}{s_j} \frac{(2x_j^0 + s_j) x_j^0}{(x_j^0 + s_j)^2} =$$

$$= - \sum_{j=1}^{n} \frac{c_j}{s_j} \frac{x_j^0 (x_j^0 + s_j)}{(x_j^0 + s_j)^2} = - \sum_{j=1}^{n} \frac{c_j x_j^0}{s_j(x_j^0 + s_j)} . \tag{8.2.13}$$

Relation (a) follows from (8.2.12) and (8.2.13). Also, by simple differentiation, one also gets relation (c).

Theorem 8.2.3 allows us, instead of problem (8.2.8)-(8.2.9), to consider the problem

$$\min_{x \in D} \Phi^0(x) \tag{8.2.14}$$

Relation (b) of Theorem 8.2.3 implies the following result:

Theorem 8.2.4. *The existence of an optimum solution for problem* (8.2.8)-(8.2.9) *implies the existence of an optimum solution for problem* (8.2.14).

Since the functions $F(x)$ and $\Phi^0(x)$ are strict convex, Theorem 8.2.2 implies:

Corollary 8.2.1. *The optimal solutions of problems* (8.2.8)-(8.2.9) *and* (8.2.14), *if they exist, are unique.*

We can now give a procedure for the solution of problem (8.2.8)-(8.2.9) which consists in the minimization of a sequence of convex quadratic functions.

Start from a general feasible solution $x^0 = (x_1^0 , \dots , x_n^0) \in D$, we construct the function $\Phi^0(x)$. Using Corollary 8.2.1, it follows that problem (8.2.14) has an unique solution; let this be x^1. Two cases are apparent:

a) If $x^1 = x^0$, then, from Theorem 8.2.1, it follows that x^0 is also an optimal solution of the problem:

$$\min_{x \in D} x \frac{\partial \Phi^0(x)}{\partial x^0} . \tag{8.2.15}$$

But the result of Theorem 8.2.3 implies

$$x \frac{\partial \Phi^0(x)}{\partial x^0} = x \frac{\partial F(x)}{\partial x^0}$$

thus, x^0 is also optimal solution for the problem

$$\min_{x \in D} x \frac{\partial F(x)}{\partial x^0} \tag{8.2.16}$$

and, applying again Theorem 8.2.1, it follows that x^0 is an optimal solution of problem (8.2.8)-(8.2.9), and the algorithm stops.

b) If $x^1 \neq x^0$, and since x^1 is an unique optimal solution, we have $\Phi^0(x^1) < \Phi^0(x^0)$. Apply Theorem 8.2.3, points (a) and (b), and obtain:

$$F(x^1) \leq \sum_{j=1}^{n} \frac{c_j}{s_j} + \Phi^0 (x^1) < \sum_{j=1}^{n} \frac{c_j}{s_j} + \Phi^0(x^0) = F(x^0) .$$

Similarly, define the function $\Phi^1(x)$ with x^1 instead of x^0, and repeat the previous procedure. Thus we solve problem

$$\min_{x \in D} \Phi^1(x) \tag{8.2.17}$$

having the solution x^2. Excluding the case when $x^2 = x^1$, i.e., when the algorithm actually stops, we have $x^2 \neq x^1$, and by a similar path we obtain $F(x^2) < F(x^1) < F(x^0)$. Repeating the reasoning, we obtain the sequence $\{F(x^i)\}$ $i = 0, 1, \ldots$ which is monotone strict decreasing and bounded from below by $F(x^*)$, where x^* is the optimal solution of problem (8.2.8)-(8.2.9). Thus, $F(x^n) \to F(x^0)$. The iterative procedure stops either when $x^i = x^{i-1}$, or when a further substantial reduction of the function value $F(x)$, can no longer be achieved, i.e., $F(x^{i-1}) - F(x^i) < \delta$, where δ depends on the required approximation accuracy.

Observation 8.2.1. Since the function F is convex, the problem (8.2.8)-(8.2.9) represents the minimization problem of a function that is convex, separable, and with linear constraints. Consequently, we can use the method of Charnes and Lemke (*op. cit.*) to get a solution.

Observation 8.2.2. The possibility exists to calculate the first terms of sequence $\{x^i\}$ much easier, i.e.: Consider the free minimum of the function $\Phi^0(x)$ as given by the equations

$$\frac{\partial \Phi^0(x)}{\partial x_j} = 2\lambda_j x_j - p_j = 0\,, \quad j = 1, \ldots, n, \text{ i.e.,}$$

$$x_j = \frac{p_j}{2\lambda_j} = \frac{(2x_j^0 + s_j)}{2} = x_j^0 + \frac{s_j}{2}\,.$$

Consider $\bar{s} = (s_1, \ldots, s_n)$. If $x^0 + \bar{s}/2 \in D$, this is an optimal solution for problem (8.2.14). Let $x^{(k)} = x^0 + k\bar{s}$, where k is a positive number and let $k_0 = \max\{k \mid x^{(k)} \in D,\ k \geq 0\}$. If we consider $t = k_0 - \frac{1}{2}[2k_0]$, then $x^{(k)}$ and x^t are separated by $[2k_0]$ terms in the sequence $\{x^i\}$ where $[x]$ represents the integer part of x, and we have $F(x^{k_0}) < F(x^t) < F(x^0)$. Hence, by this procedure, after we selected point x^0, we can jump a series of terms and find the term x^{k_0}.

Observation 8.2.3. A similar procedure was used by Anand and Swarup [58] for the solution of the problem $\max\{\sum_{j=1}^{n}(x_j + \alpha_j)/(x_j + \beta_j)\}$ $(\alpha_j, \beta_j \geq 0)$ on the domain defined de (8.2.9). Wadhwa [1025] use this procedure to obtain the solution of the separable non-linear fractional problem $\min\{\sum_{j=1}^{n}(k_j + m_j x_j^2)/(x_j + c_j)\}$, $c_j > 0$, $k_j > 0$, $m_j \geq 0$, on the domain (8.2.9).

8.3. LINEAR FRACTIONAL PROGRAMMING WITH DISJUNCTIVE CONSTRAINTS

In the mathematical programming problems, the constraints are of the general form "and". In practice, one can also meet situations when the constraints are disjunctive i.e., of type "or". Thus, Németi and Radó[3] considered a scheduling and production launching problem, i.e. to find the minimum fabrication time for a product that proceeds through a sequence of machine-tools in a predetermined order. The mathematical model for such a problem leads to a fractional programming problem with disjunctive constraints. Reference [417] considers a scheduling problem for transportation system which leads to a problem of fractional programming with disjunctive constraints.

In general, in a disjunctive programming problem, a series of constraints (linear or not), united through the connectives ($\vee$, -(*non*) , $\rightarrow$), appear. For a discussion of the various variants of disjunctive fractional programming problems one can consult reference [719]. Now, we will indicate a solution method for the problem

$$\min \frac{cx + c_0}{dx + d_0}, \tag{8.3.1}$$

subject to

$$A_1x \le b_1 \vee A_2x \le b_2 \vee \ldots \vee A_px \le b_p \, ; \tag{8.3.2}$$
$$x \ge 0,$$

where A_i are matrices of the form $m_i \times n$, b_i are m_i dimensional column vectors, c, d and x are row or column vectors with n components, and c_0 and d_0 real constants. The problem consists in the determination of a vector x which minimizes the function given in (8.3.1), and which verifies at least a constraints system $A_ix \le b_i$. The domain of feasible solutions is no longer convex, and, consequently, the methods presented in Chapter 3 are no longer directly applicable. On this domain, assume $dx + d_0 > 0$.

The basic idea is that of joining together the p constraints systems using certain indicator variable δ_i. The indicator δ_i takes the value 1 when x must verify the constraint $A_ix \le b_i$, and zero otherwise. If we denote by M_i an upper bound of the expression $A_ix - b_i$, then one can substitute the constraint $A_ix \le b_i$ with the constraint $A_ix + M_i\delta_i \le M_i + b_i$ or, equivalently, by $A_ix + M_i\delta_i \le b_i'$. With these observations, one can substitute constraint (8.3.2) with the following constraints:

$$A_ix + M_i\delta_i \le b_i' , \quad i = 1 , \ldots , p , \tag{8.3.3}$$

$$\sum_{i=1}^{p} \delta_i \ge 1 , \quad \delta_i \in \{0, 1\}, \quad x \ge 0. \tag{8.3.4}$$

---

[3]L. Németi and F. Radó, "Ein Wartezeitproblem in der programmieurung der Production", *Mathematica*, 5(1963), 69-95.

Theorem 8.3.1. *If* (x^0, δ^0) *is an optimal solution of problem* (8.3.1), (8.3.3)-(8.3.4), *then* x^0 *is an optimal solution of problem* (8.3.1)-(8.3.2).

Proof. Let (x^0, δ^0) be the optimal solution of problem (8.3.1), (8.3.3)-(8.3.4). The constraints (8.3.4) ensure that at least a system from the constraints $A_i x \le b_i$ is satisfied, and since (8.3.3) represents an extended form of (8.3.2), it follows that x^0 is a feasible solution of problem (8.3.1) - (8.3.2). Since the problems have same objective function, it follows that x^0 is an optimal solution of problem (8.3.1)-(8.3.2).

Problem (8.3.1), (8.3.3)-(8.3.4) constitutes an integer and mixed integer linear fractional programming problem. This type of problems will be discussed in the next chapter.

Example 8.3.1. $\min F(x) = \dfrac{2x_1 + x_2 + 1}{3x_1 + 5x_2 + 3}$ subject to $x \in S_1 \cup S_2 \cup S_3$, where

$$S_1 = \{(x_1, x_2) \mid x_2 \le 2,\ x_1 + x_2 \le 3,\ x_1, x_2 \ge 0\},$$
$$S_2 = \{(x_1, x_2) \mid -x_1 + 2x_2 \le 1,\ 2x_1 - x_2 \le 4,\ x_1, x_2 \ge 0\},$$
$$S_3 = \{(x_1, x_2) \mid x_1 \le 4,\ 3x_2 \le 2,\ x_1, x_2 \ge 0\}.$$

Figure 8.3.1 shows clearly that, though each S_i is convex, their reunion ABCDEFGO is not convex.

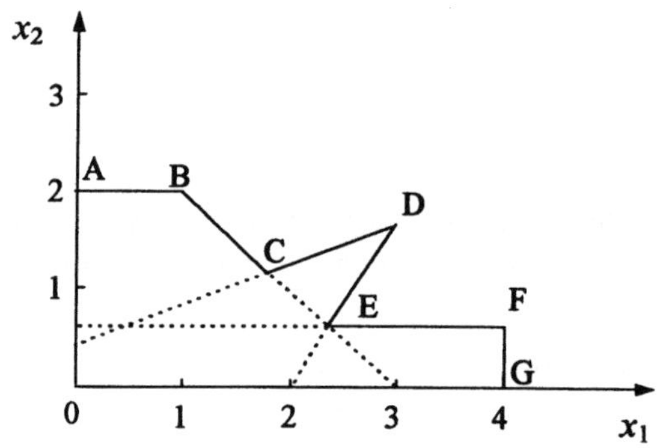

Figure 8.3.1.

Introducing the variables δ_i, the constraints become

$$x_2 + \delta_1 \le 3,$$
$$x_1 + x_2 + 4\delta_1 \le 7,$$
$$-x_1 + 2x_2 + 5\delta_2 \le 6,$$
$$2x_1 - x_2 + 4\delta_2 \le 8,$$

$$x_1 \le 4 ,$$
$$3x_2 + 2\delta_3 \le 3 ,$$
$$\delta_1 + \delta_2 + \delta_3 \ge 1 ,$$
$$x_1 , x_2 \ge 0 ,$$
$$\delta_i \in \{0, 1\} , \quad i = 1, 2, 3.$$

The new problem has the solution $x_1 = 0$, $x_2 = 2$ and $\delta_1 = 1$. The minimal value of F is 3/13.

Now, we formulate the Dual problem for a disjunctive linear fractional programming problem of the form

$$\min F(x) = \frac{cx + c_0}{dx + d_0}, \tag{8.3.5}$$

subject to

$$\bigvee_{h\in Q} \{ A^h x \ge b^h ; \ x \ge 0 \} \tag{8.3.6}$$

where A^h is a matrix $m\times n$ for all $h \in Q$, c, d are $1 \times n$ vectors, b^h is a vector $m\times 1$ for all $h \in Q$, x is a $n\times 1$ vector , c_0 , d_0 are scalars and Q is a finite set de indices. Also, assume that $dx + d_0 > 0$ for the domains determined by each inequalities system from the parentheses (these systems do not contain other disjunctions). For $dx + d_0 \equiv 1$, one gets the problem studied by E. Balas ("A note on duality in disjunctive programming", *J. Optim. Theory Appl.*, **21** (1977), 523-528).

The dual of problem (8.3.5)-(8.3.6) is

$$\max G(z) = z , \tag{8.3.7}$$

subject to

$$\bigwedge_{h \in Q} \{u^h A^h + dz \le c , \ -u^h b^h + d_0 z \le c_0 ; \ u^h \ge 0\} . \tag{8.3.8}$$

The dual problem is feasible if a z exists, and if for any $h \in Q$ exists a u^h such that $(u^h , z) = (u_1^h , ... , u_m^h , z)$ satisfies the system of constraints given in the parentheses.

Let $X_h = \{x : A^h x \ge b ; \ x \ge 0\}$, $\bar{X}_h = \{x : A^h x \ge 0 ; \ x \ge 0\}$.
$U_h = \{(u^h , z) : u^h A^h + dz \le c , \ -u^h b^h + d_0 z \le c_0 : u^h \ge 0\}$,
$\bar{U}_h = \{(u^h , z) : u^h A^h + dz \le 0, \ -u^h b^h + d_0 z \le 0 ; \ u^h \ge 0\}$,
$Q^* = \{h \in Q : X_h \ne \varnothing\}$, $Q^{**} = \{h \in Q : U_h = \varnothing\}$.

Assume that the following regularity condition is satisfied
(RC1) $\quad (Q^* \ne \varnothing, \ Q\backslash Q^{**} \ne \varnothing) \Rightarrow Q^*\backslash Q^{**} \ne \varnothing$;
i.e., if primal problem is feasible and the dual problem is not feasible, then $h \in Q$ exists such that $X_h \ne \varnothing$ and $U_h = \varnothing$. We show that problem (8.3.7)-(8.3.8) is the dual

of problem (8.3.5) -(8.3.6). Using the change of variables

(TR) $y = tx\ ,\quad t = \dfrac{1}{dx + d_0} > 0,$

the primal problem becomes:

$$\min H(y, t) = cy + c_0 t\ ,$$

subject to

(I) $\bigvee_{h\in Q} \{A^h y - b^h t \geq 0\ ,\ dy + d_0 t \geq 1\ ,\ -dy - d_0 t \geq -1\ ,\ y \geq 0\ ,\ t > 0\}$

or

$$\min H(y, t) = (c, c_0)\begin{bmatrix} y \\ t \end{bmatrix}$$

subject to

(II)
$$\bigvee_{h\in Q}\left\{\begin{array}{l}\begin{bmatrix} A^h & -b^h \\ d & d_0 \\ -d & -d_0 \end{bmatrix}\begin{bmatrix} y \\ t \end{bmatrix} \geq \begin{bmatrix} 0 \\ 1 \\ -1 \end{bmatrix} \\ \\ y \geq 0 \qquad t > 0 \end{array}\right.$$

Using the results of Balas (*op. cit.*), the dual of problem (II) is

$$\max\ (u_1^h\ , \dots , u_m^h\ , v\ , w)\begin{bmatrix} 0 \\ 1 \\ -1 \end{bmatrix}$$

subject to

(III)
$$\bigwedge_{h\in Q}\left\{\begin{array}{l}(u_1^h\ , \dots , u_m^h\ , v, w)\begin{bmatrix} A^h & b^h \\ d & d_0 \\ -d & -d_0 \end{bmatrix} \leq (c, c_0)\ , \\ \\ (u_1^h\ , \dots , u_m^h\ , v\ , w) \geq 0 \end{array}\right.$$

i.e.,

(IV)
$$\max\ (v - w)\ ,$$

subject to

$$\bigwedge_{h\in Q}\{u^h A^h + d(v - w) \leq c,\ \ -u^h b^h + d_0(v - w) \leq c_0,\ \ u^h \geq 0,\ \ v, w \geq 0\}$$

Performing the substitution $z = v - w$ (z without sign constraint) this problem becomes

$$\max G(z) = z$$

subject to

(V) $\bigwedge_{h\in Q}\{\ u^h A^h + dz \leq c\ ,\ \ -u^h b^h + d_0 z \leq c_0\ ,\ \ u^h \geq 0\ \}$

which is Dual problem (8.3.7)-(8.3.8).

Since the problems (8.3.5)-(8.3.6) and (I) (or (II)) are equivalent and problem (V) (or III) or (IV) are the dual of (I), it follows that problem (8.3.7)-(8.3.8) is the dual of problem (8.3.5)-(8.3.6). Hence, the Dual problem is an ordinary linear programming problem, and consequently, can be easily solved.

To prove the duality theorem, the regularity condition (RC1) must be expressed in terms of problems (I) and (V). Let

$Y_h = \{(y, t) : A^h y - b^h t \geq 0, \; dy + d_0 t = 1 \; ; \; y \geq 0, \; t > 0\}$

$\overline{Y}_h = \{ (y, t) : A^h y - b^h t \geq 0 \; , \; dy + d_0 t = 0 \; ; y \geq 0, \; t > 0 \}$ and

$Q^{***} = \{ h \in Q : Y_h \neq \emptyset \}$.

Consider the following regularity condition:

(RC2) $\quad (Q^{***} \neq \emptyset \; , \; Q \backslash Q^{**} \neq \emptyset \;) \Rightarrow Q^{***} \backslash Q^{**} \neq \emptyset \; ,$

i.e., if (I) is feasible and the dual problem is infeasible, then exists an $h \in Q$ such that $Y_h \neq \emptyset$ and $U_h = \emptyset$.

The two regularity conditions (RC1) and (RC2) are equivalent as it follows from the following lemma.

Lemma 8.3.1. *If the condition* (RC1) *is satisfied, then* (RC2) *is also satisfied and conversely.*

Proof. The proof follows from the fact that the change of variables (TR) of the domain of feasible solutions of problem (8.3.5)-(8.3.6) in the domain of problem (I) is a one-to-one mapping. Thus, (8.3.5)-(8.3.6) is feasible (infeasible) if and only if (I) is feasible (infeasible).

Now we can state the Duality theorem:

Theorem 8.3.2. *Assume that the problems* (8.3.5)-(8.3.6) *and* (8.3.7)-(8.3.8) *satisfy the regularity condition* (RC1). *Then, one of the following situations take place*:

a) *Both problems are feasible; each has an optimal solution and the optimal values are equal.*

b) *One of the problems is infeasible; the other one either is infeasible or does not have a finite optimum.*

The proof follows from Lemma 8.3.1 and from the duality theorem corresponding to the linear case (see Balas, *op. cit.*).

Example 8.3.2. Reconsider Example 8.3.1. The dual problem is:

$$\max \; G(z) = z \; ,$$

subject to

$$-u_2^1 + 3z \le 2 ,$$
$$-u_1^1 - u_2^1 + 5z \le 1 ,$$
$$2u_1^1 + 3u_2^1 + 3z \le 1 ,$$
$$u_1^2 - 2u_2^2 + 3z \le 2 ,$$
$$-2u_1^2 + u_2^2 + 5z \le 1 ,$$
$$u_1^2 + 4u_2^2 + 3z \le 1 ,$$
$$-u_1^3 + 3z \le 2 ,$$
$$-3u_2^3 + 5z \le 1 ,$$
$$4u_1^3 + 2u_2^3 + 3z \le 1 ,$$
$$u_i^k \ge 0 , \quad i = 1, 2 ; \quad k = 1, 2, 3.$$

The optimum solution is $u_1^1 = \frac{2}{13}$, $u_1^2 = \frac{1}{13}$, $u_2^3 = \frac{2}{39}$, and $z = \frac{3}{13}$. Note that the optimal values of the two problems are equal. The phenomenon of "duality gap" does not exists.

Example 8.3.3. min $F(x) = \dfrac{3x_1 - 4x_2 + 1}{x_1 + x_2 + 3}$,

subject to

$$\left\{ \begin{matrix} -x_1 + x_2 \ge 0 , -x_1 - x_2 \ge -4 \\ x_1 \ge 0 , \; x_2 \ge 0 \end{matrix} \right\} \vee \left\{ \begin{matrix} -x_1 + x_2 \ge 0 , \; x_1 - x_2 \ge 2 , \\ x_1 \ge 0 , \; x_2 \ge 0 \end{matrix} \right\}$$

The optimum solution is $x^0 = (0, 4)$ with $F(x^0) = -\frac{15}{7}$. Dual problem is

$$\max \; G(z) = z ,$$

subject to

$$-u_1^1 - u_2^1 + z \le 3 ,$$
$$u_1^1 - u_2^1 + z \le -4 ,$$
$$4u_2^1 + 3z \le 1 ,$$
$$-u_1^2 + u_2^2 + z \le 3 ,$$
$$u_1^2 - u_2^2 + z \le -4 ,$$
$$-2u_2^2 + 3z \le 1 ,$$
$$u_1^k \ge 0 , \quad i, k = 1, 2.$$

The dual problem is infeasible. This is due to the fact that the regularity condition (RC1) is violated since $Q^* = \{1\}$, $Q \backslash Q^{**} = \{2\}$,

$$X_2 = \varnothing \quad \text{and} \quad U_2 = \varnothing \quad \text{but} \quad Q^* \backslash Q^{**} = \varnothing .$$

8.4. STOCHASTIC FRACTIONAL PROGRAMMING PROBLEM

Consider the following stochastic fractional programming problem

$$\gamma(\omega) = \underset{x \in S}{\text{optimum}} \frac{z_1(x, \omega)}{z_2(x, \omega)} \tag{8.4.1}$$

where $S = \{x \mid Ax = b;\ x \geq 0\}$.

It is considered that z_1 and z_2 are linear functions of the same random variable $t(\omega)$ (the simple randomization case) having a distribution function $T(z)$ continuous and strict increasing i.e. $z_1(x, \omega) = (c + t(\omega)c_1)'x$ and $z_2(x, \omega) = (d + t(\omega)d_1)'x$.

For problem (8.4.1), we make the following assumptions:

a) The set S is non-empty and bounded,

b) The denominator of the objective function preserves the same sign (let us assume it to be positive) on S, consequently $P\left\{\omega \mid (d + t(\omega)d_1)'x > 0\right\} = 1$,

c) All basic solutions are nondegenerate.

Stancu-Minasian [894] gives a representation of the probability distribution of the optimal value $\gamma(\omega)$ by means of the characteristic values of the corresponding parametric programs, taking $t(\omega) = \lambda$ as a parameter ranging in a finite interval $[\delta_1, \delta_2]$.

It is well know [1014] that for parametric fractional problem, the interval $[\delta_1, \delta_2]$ may be divided into a finite number of so-called "critical regions" which are defined by the various combinations of variables forming optimal bases.
Let λ_j $(1 \leq j \leq p-1)$ be the characteristic value $(\delta_1 \leq \lambda_1 < \lambda_2 < \ldots < \lambda_{p-1} \leq \delta_2)$ and let $x^j (1 \leq j \leq p)$ be the optimal solutions corresponding to the intervals $(\lambda_{j-1}, \lambda_j)$ $j = 1, \ldots, p$. Denote $\delta_1 = \lambda_0$ and $\delta_2 = \lambda_p$. Since the random variable $t(\omega)$ is supposed to have a continuous distribution function, the intersection of two such critical regions has zero probability.

Theorem 8.4.1. [894]. *Let $F(z)$ be the distribution function of the random variable $\gamma(\omega)$ and let* $u_j(z) = \dfrac{(zd' - c')x^j}{(c_1' - zd_1')x^j}$. *Then* $F(z) = \sum_{j=1}^{p} H_j(z)$ *where*

$$H_j(z) = \begin{cases} T(\lambda_j)-T(\lambda_{j-1}) & \text{if } \begin{cases} (c_1' - zd_1')x^j>0 \text{ and } u_j(z)\geq\lambda_j \\ (c_1' - zd_1')x^j<0 \text{ and } u_j(z)\leq\lambda_{j-1} \\ (c_1' - zd_1')x^j= 0 \text{ and } c'x^j<zd'x^j \end{cases} \\ 0 & \text{if } \begin{cases} (c_1' - zd_1')x^j>0 \text{ and } u_j(z)\leq\lambda_{j-1} \\ (c_1' - zd_1')x^j<0 \text{ and } u_j(z)\geq\lambda_j \\ (c_1' - zd_1')x^j= 0 \text{ and } c'x^j\geq zd'x^j \end{cases} \\ \begin{matrix} T(u_j(z))-T(\lambda_{j-1}) \\ T(\lambda_j)-T(u_j(z)) \end{matrix} & \text{if } \lambda_{j-1}<u_j(z)<\lambda_j \text{ and } \begin{cases} (c_1' - zd_1')x^j>0 \\ (c_1' - zd_1')x^j<0 \end{cases} \end{cases} \quad (1\leq j\leq p)$$

In Chapter 1, Example 5, we defined the minimum-risk model (or aspiration criterion model). Akin to the minimum-risk problem is the problem of Kataoka [910] (or fractile criterion model) which consists in maximization of the α - fractile of the cumulative distribution of the objective function with the value of α preassigned by the decision maker

$$\max \{f = F_\alpha(c'x) \mid x \in S \text{ and } P\{\omega \mid c'(\omega)x \geq f\} \geq \alpha\}$$

Extensions of the fractile criterion model and aspiration criterion model to the case of stochastic linear fractional programming are given by Stancu-Minasian [896], [899], [910] and Gupta and Swarup [457] (see, also Cambini [214]). Callahan and Bector [212] consider Kataoka's problem for a product and for a quotient of general functions.

Theorem 8.4.2. [896]. *The minimum-risk problem for the ratio of two linear functions whose numerator has random coefficients (normal non-degenerated, with mean value $\bar{c}$ and covariance matrix V), i.e. the problem*

$$\max_{x \in S} P\left\{\omega \mid \frac{c'(\omega)x}{d'x} \leq k\right\} \tag{8.4.2}$$

is equivalent to the following nonlinear fractional programming problem

$$\max_{x \in S} \frac{kd'x - \bar{c}'x}{\sqrt{x'Vx}} \tag{8.4.3}$$

Theorem 8.4.3. [896]. *Kataoka's problem for the ratio of two linear functions whose numerator has random coefficients (normal non-degenerated, with mean value $\bar{c}$ and covariance matrix V), i.e. the problem*

$$\max\left\{k \mid P\left\{\omega \mid \frac{c'(\omega)x}{d'x} \leq k\right\} = \alpha,\ x \in S\right\} \tag{8.4.4}$$

is equivalent to the following nonlinear fractional programming problem

$$\max_{x \in S} \frac{\bar{c}'x - q\,(x'Vx)^{1/2}}{d'x} \tag{8.4.5}$$

where $\phi^{-1}(\alpha) = -q$ *and* $\phi(\cdot)$ *is Laplace's function.*

It should be noted that in some conditions, problem (8.4.5) has a finite optimal solution and that the objective function is explicit quasiconcave.

However, the approach in these models has been to assume that the random variables involved are normally distributed, and that the denominator of the fractional objective function is fixed. Gupta *et al.* [451] studied a stochastic linear fractional programming problem, in which the parameters of both the numerator and the denominator are assumed to be mutually independent Cauchy variates

$$\max \{ f \mid P\{\frac{(c'x - \bar{c}'x)}{(d'x - \bar{d}'x)} \le f\} = p \text{ and } x \in S \} \tag{8.4.6}$$

where the vectors c and d are assumed to be random with their components c_j and d_j $j = 1, ..., n$, all being independent Cauchy variates with parameters $(\bar{c}_j, q_j)$ and $(\bar{d}_j, r_j)$, respectively. Gupta and Jain [449] consider Kataoka's problem for the linear fractional objective functions under chance constraints with both the right hand and the left hand coefficients being random.

8.5. VECTORIAL MINIMUM-RISK PROBLEM

This section is concerned with the vectorial minimum-risk problem with linear and linear fractional objective functions. It is show that in the case when the objective functions are simply randomized, these stochastic programming problems are equivalent with deterministic linear fractional programming problems with multiple objective functions.

Stancu-Minasian considered in several papers [897], [898], [899], [902], [910], [928] a more general case of the minimum-risk problem, i.e., the case when the probabilities that the values of the r objective functions exceed some levels of performance u_k are maximized.

$$V\max_{x \in S} \{ f_k(x) = P[\omega \in \Omega : c_k'(\omega)x \ge u_k];\ k \in I = \{1, ..., r\} \} \tag{8.5.1}$$

where S is an arbitrary set in R^n. In particular, we can consider that S is a polyhedral subset of R^n.

Stancu-Minasian called such problem a multiple minimum-risk problem with levels $u_1, ..., u_r$. For solving problem (8.5.1), Stancu-Minasian proposed a method which consists in the sequential solving of some minimum-risk problems (containing also quadratic constraints). Related to this method , the multiple minimum-risk solution concept is introduced and a relaxation-type method for obtaining this is given.

We assume now that

$$c_k(\omega) = c_k^1 + c_k^2 t_k(\omega), \quad k \in I \tag{8.5.2}$$

where $c_k^i \in R^n$ $(i = 1, 2)$ are given constant vectors and $t_k(\omega)$ is a random variable on the probability space $\{\Omega_k, K_k, P_k\}$. It should be mentioned that generality is unimpaired if the random variables t_k are assumed to be defined on the same probability space $\{\Omega, K, P\}$. In what follows, we consider that "Vmax" means that the efficient points are regarded as optimal solutions to (8.5.1) according to Definition 6.1.1 Chapter 6. Thus, we can formulate:

Definition 8.5.1. A point $x^0 \in S$ is said to be a *vectorial minimum-risk solution* for problem (8.5.1) if it is an efficient solution for the vector-valued problem

$$\operatorname*{Vmax}_{x \in S} \{f_k(x) : k \in I\}$$

that is (*cf. e.g.*, [922]) if $f_i(x) > f_i(x^0)$ for some $x \in S$ and some $i \in I$ implies that there exists at least one $j \in I - \{ i \}$ such that $f_j(x) < f_j(x^0)$.

We shall show that, in the case when the objective functions are simply randomized, problem (8.5.1) is equivalent with a deterministic linear fractional programming problem with multiple objective functions. Further, these results will be extended to the fractional case.

Concerning to problem (8.5.1), we adopt the following assumptions:

(H1) S is a non-empty and compact set,

(H2) $c_k^2 x > 0$ for all $x \in S$ and for all $k \in I$,

(H3) the distribution functions T_k of the random variables $t_k(\omega)$ are continuous and strict increasing.

First, we give

Lemma 8.5.1. *If assumptions (H1) - (H3) are satisfied, then for all* $k \in I$, *we have*

$$f_k(x) = 1 - T_k\left[\frac{u_k - c_k^1 x}{c_k^2 x}\right] \quad \textit{for all } x \in S. \tag{8.5.3}$$

Proof. The proof is similar to that of Theorem 4.1, Ref. [910] (see also [171]). Thus, from (8.5.2) and (H2), we have

$$f_k(x) = P\{\omega \in \Omega : (c_k^1 + t_k(\omega)c_k^2)x \geq u_k\} =$$

$$= P\left\{ \omega \in \Omega : t_k(\omega) \geq \frac{u_k - c_k^1 x}{c_k^2 x} \right\}$$

Further, according to the fact that T_k is a continuous and strict increasing function, we have $f_k(x) = 1 - T_k \left[\frac{u_k - c_k^1 x}{c_k^2 x} \right]$ i.e. (8.5.3). The proof is now complete.

Remark 8.5.1. Since $f_k = (1 - T_k) \circ g_k$ for all $k \in I$, and since $1 - T_k$ is strict decreasing and g_k, where

$$g_k(x) = \frac{u_k - c_k^1 x}{c_k^2 x}, \text{ for all } x \in S \tag{8.5.4}$$

is quasi-monotonic, it results that, when S is a convex set, f_k is a quasi-monotonic function. Worth mentioning is that Weber [1037] considered pseudo-monotonic multi-objective programming, which does not ordinarily include problem (8.5.1) as a particular case.

We can now state and prove a theorem which relates the multi-objective minimum-risk problem (8.5.1) to a linear fractional programming problem with several objective functions.

Theorem 8.5.1. *Let assumptions (H1) - (H3) be satisfied. Then a point $x' \in S$ is a vectorial minimum-risk solution for problem (8.5.1) if and only if x' is an efficient solution for the following linear fractional programming problem with several objective functions.*

(PF)
$$\operatorname*{Vmax}_{x \in S} \left\{ \frac{c_k^1 x - u_k}{c_k^2 x} : k \in I \right\}$$

Proof. Indeed, by Definition 8.5.1, $x' \in S$ is a vectorial minimum-risk solution for problem (8.5.1) if for any $x'' \in S$ and $i \in I$ for which $f_i(x'') > f_i(x')$, there exists a $j \in I - \{i\}$ such that $f_j(x'') < f_j(x')$. But then, in virtue of Lemma 8.5.1 and by (8.5.4), it results that for all $x'' \in S$ and $i \in I$ for which $T_i(g_i(x'')) < T_i(g_i(x'))$ there exists a $j \in I - \{i\}$ such that $T_j(g_j(x'')) > T_j(g_j(x'))$. Since T_i and T_j are strict increasing, these inequalities are equivalent with $g_i(x'') < g_i(x')$ and $g_j(x'') > g_j(x')$, respectively. Consequently, this yields that for all $x'' \in S$ and $i \in I$ for which $\frac{c_i^1 x'' - u_i}{c_i^2 x''} > \frac{c_i^1 x' - u_i}{c_i^2 x'}$ there exists a $j \in I - \{i\}$ such that $\frac{c_j^1 x'' - u_j}{c_j^2 x''} < \frac{c_j^1 x' - u_j}{c_j^2 x'}$.
This means that x' is an efficient solution for problem (PF). The converse part of the

theorem can be shown using similar arguments.

Remark 8.5.2. By Theorem 8.5.1, we can conclude that the vectorial minimum-risk problem can be solved with a method used for solving the multi-objective linear fractional programming problem (PF). For instance, an interactive procedure to solve (PF) was given by Weber [1037].

Similarly, we can approach the case of the vectorial minimum-risk problem with linear fractional objective functions, which includes the previously discussed linear case as a particular case.

Thus, we consider the vectorial linear fractional problem

$$\text{(P1)} \qquad \operatorname*{Vmax}_{x \in S} \left\{ Z_k(x) = \frac{c_k x + \alpha_k}{d_k x + \beta_k} : k \in I \right\}$$

where c_k, d_k $(k \in I)$ are given constant vectors in R^n, α_k, β_k $(k \in I)$ are scalar constants. We also assume that, for all $k \in I$, $d_k x + \beta_k$ does not change in sign (say positive) on the set S.

Let now c_k and d_k assume random values for all $k \in I$: $c_k(\omega) = c_k^1 + c_k^2 t_k(\omega)$, $d_k(\omega) = d_k^1 + d_k^2 t_k(\omega)$, where c_k^i, $d_k^i \in R^n$ $(i = 1, 2)$ are given constant vectors and $t_k(\omega)$ is a random variable on the probability space $\{\Omega, K, P\}$.

As in problem (8.5.1), we can state the vectorial minimum-risk problem corresponding to levels $u_1, u_2, ..., u_r$

$$\operatorname*{Vmax}_{x \in S} \left\{ f_k(x) = P \left\{ \omega \in \Omega : \frac{c_k(\omega)x + \alpha_k}{d_k(\omega)x + \beta_k} \geq u_k \right\} : k \in I \right\} \tag{8.5.5}$$

Regarding problem (8.5.5), additional to (H1) and (H3), we make the following assumptions:

(H4) $P\{ \omega \in \Omega : (d_k^1 + d_k^2 t_k(\omega))'x + \beta_k > 0 \} = 1$, for all $x \in S$ and for all $k \in I$;

(H5) $(c_k^2 - d_k^2 u_k)x > 0$ for all $x \in S$ and for all $k \in I$.

The corresponding results, like those of Lemma 8.5.1 and Theorem 8.5.1, also hold for problem (8.5.5):

Lemma 8.5.2. *If assumptions (H1) and (H4) are satisfied, then, for all* $k \in I$, *we have*

$$f_k(x) = 1 - T_k\left[\frac{\left(d_k^1 u_k - c_k^1\right)x + u_k\beta_k - \alpha_k}{\left(c_k^2 - d_k^2 u_k\right)x}\right] \quad \text{for all } x \in S.$$

Theorem 8.5.2. *Let assumptions (H1), (H4) and (H5) be satisfied. Then a point* $x' \in S$ *is a vectorial minimum-risk solution for problem (8.5.5) if and only if* x' *is efficient for the following deterministic linear fractional programming problem with several objective functions :*

$$(PF1) \qquad \underset{x \in S}{\mathrm{Vmax}}\left\{\frac{\left(c_k^1 - d_k^1 u_k\right)x + \alpha_k - u_k\beta_k}{\left(c_k^2 - d_k^2 u_k\right)x} : k \in I\right\}$$

The proofs are similar to those of Lemma 8.5.1 and Theorem 8.5.1. and are omitted.

Remark 8.5.3. The feasible set S in the formulation of previous problems is not necessarily a convex set. For instance, we can consider a discrete feasible set S, as in the pseudo-boolean programming case and the results of Theorems 8.5.1 and 8.5.2 remain valid. In this case, the minimum-risk solutions for problems (8.5.1) and (8.5.5) can be obtained by solving some appropriate vectorial fractional programming problems with zero-one variables, using, for instance, the algorithm described by Bina Gupta [441].

8.6. STOCHASTIC PROGRAMMING WITH MULTIPLE FRACTILE CRITERIA

This section deals with a more general case of Kataoka's problem. i.e., the determination of the largest values of r fractional objective functions such that the probabilities of obtaining such results exceed (or are equal to) some given thresholds. Probabilistic constraints are introduced instead of linear conditions. Under suitable assumptions, this model is shown to be equivalent to a deterministic multiple non-linear fractional programming problem with non-linear constraints (P_{II}). Via a variable transformation and parametrization of the denominators in P_{II} by a vector $\eta = (\eta_i)$, a multiple objective non-linear programming problem (P_{III}) is obtained. A theorem on the connection between efficient solutions to (P_{II}) and the efficient solutions to (P_{III}) is given. In what follows, we shall need the *Cantelli inequality* : If X is a random variable such that $E(X) = \mu$ and $E(X - \mu)^2 = \sigma^2$ ($E(X)$ is the means value of X), then

$$P(X - \mu \leq \lambda) \begin{cases} \leq \dfrac{\sigma^2}{\sigma^2 + \lambda^2} & \text{if } \lambda < 0 \\ \geq 1 - \dfrac{\sigma^2}{\sigma^2 + \lambda^2} & \text{if } \lambda \geq 0 \end{cases}$$

We consider Kataoka's problem in the case of multiple criteria [914]

Problem P_I

$$\text{" maximize " } F = (f_1, ..., f_r) \tag{8.6.1}$$

subject to

$$P\left[\frac{g^1x}{h^1x + h_0^1} \geq f_1\right] = p_1, ..., P\left[\frac{g^rx}{h^rx + h_0^r} \geq f_r\right] = p_r \tag{8.6.2}$$

$$P\left(\sum_j a_{1j}x_j \leq b_1\right) \geq q_1^*, ..., P\left(\sum_j a_{mj}x_j \leq b_m\right) \geq q_m^* \tag{8.6.3}$$

where $x_j \geq 0$, $(j = 1, ..., n)$ are decision variables; $g^i = \left(g_1^i, ..., g_n^i\right)$, $h^i = \left(h_1^i, ..., h_n^i\right)$, $(i = 1, ..., r)$, $a_s = \left(a_{s1}, a_{s2}, ..., a_{sn}\right)$, $(s = 1, ..., m)$ are random vectors; $b_1, ..., b_m$ are random variables; $p_1, ..., p_r, q_1^*, ..., q_m^*$ are preassigned probability levels of the chance constraints and $h_0^1, ..., h_0^r$ are real constants.

Suppose that

(i) $1 \geq p_i > 1/2$, $1 > q_s^* > 1/2$ $(i = 1, ..., r;\ s = 1, ..., m)$

(ii) For each $i = 1, ..., r$, the random vector $g^i = \left(g_j^i\right)$, $(j = 1, ..., n)$ has a known normal distribution with mean vector $G^i = \left(G_1^i, G_2^i, ..., G_n^i\right)$ and covariance matrix $\hat{D}^i = \left(d_{jk}^i\right)$, where $d_{jk}^i = E\left(g_j^i - G_j^i\right)\left(g_k^i - G_k^i\right)$, $(j, k = 1, ..., n)$

(iii) $h^1, ..., h^r$ are constant vectors.

Let $R_i = g^ix/(h^ix + h_0^i)$ $(i = 1, ..., r)$. Since R_i is a linear combination of normal variates, R_i is normally distributed with mean $\bar{R}_i = G^ix/(h^ix + h_0^i)$ and variance $S_i^2 = x'\hat{D}^ix/(h^ix + h_0^i)^2$. Hence

$$P\left(R_i \geq f_i\right) = P\left[\frac{R_i - \bar{R}_i}{S_i} \geq \frac{f_i - \bar{R}_i}{S_i}\right] = 1 - P\left[\frac{R_i - \bar{R}_i}{S_i} < \frac{f_i - \bar{R}_i}{S_i}\right]$$

Thus, constraints (8.6.2) are equivalent to $1 - p_i = \phi\left[\dfrac{f_i - \bar{R}_i}{S_i}\right]$ or $f_i = \bar{R}_i - q_i S_i = \dfrac{G^i x - q_i\,(x'\hat{D}^i x)^{1/2}}{h^i x + h_0^i}$, where $-q_i = \phi^{-1}(1 - p_i)$ with ϕ being the standard Laplace function. If $p_i > 1/2$ (assumption (i)), then $1 - p_i < 0.5$ and $\phi^{-1}(1 - p_i) = -q_i \leq 0$. Let us find a deterministic equivalent to constraints (8.6.3).

Suppose that one of the following statements is true :

(iv) For each $i = 1, ..., m$, the random vector $(a_{i1}, a_{i2}, ..., a_{in}, -b_i)$ has a known normal distribution with mean vector $(A_{i1}, A_{i2}, ..., A_{in}, -B_i)$ and known covariance matrix $\hat{E}_i$.

(v) For each $i = 1, ..., m$, the random vector $(a_{i1}, a_{i2}, ..., a_{in}, -b_i)$ has an unknown distribution with mean vector $(A_{i1}, A_{i2}, ..., A_{in}, -B_i)$ and known covariance matrix $\hat{E}_i$.

Write

$$\hat{E}_i = \begin{bmatrix} e_{11}^i & \dots & e_{1n}^i & w_{i1} \\ e_{21}^i & \dots & e_{2n}^i & w_{i2} \\ \dots & \dots & \dots & \dots \\ e_{n1}^i & \dots & e_{nn}^i & w_{in} \\ w_{i1} & \dots & w_{in} & T_i^2 \end{bmatrix}$$

with $e_{jk}^i = E(a_{ij} - A_{ij})(a_{ik} - A_{ik})$, $w_{ij} = E(a_{ij} - A_{ij})(-b_i + B_i)$, $j, k = 1, ..., n$, $T_i^2 = E(-b_i + B_i)^2$.

If **(iv)** holds, let $U_i = \sum_j a_{ij}x_j - b_i x_{n+1}$, with $x_{n+1} = 1$. Since U_i are linear combinations of normal variates, U_i are normally distributed with mean $\bar{U}_i = A_i X - B_i$ and variances $V_i^2 = X'\hat{E}_i X$, where $X' = (x_1, ..., x_{n+1})$, $A_i = (A_{i1}, ..., A_{in}, 0)$. Constraints (8.6.3) are equivalent to $\bar{U}_i + t_i V_i \leq 0$, with $t_i = \phi^{-1}(q_i^*)$, or to $A_i X + t_i (X'\hat{E}_i X)^{1/2} \leq B_i$, $i = 1, ..., m$. Note that, since $q_i^* > 1/2$, we have $t_i \geq 0$.

If **(v)** holds, by the Cantelli inequality for $t_i \geq 0$ we have $P[(U_i - \bar{U}_i)/V_i \leq t] \geq t_i^2/(1 + t_i^2)$, or $P[U_i \leq \bar{U}_i + t_i V_i] \geq t_i^2/(1 + t_i^2)$, $(i = 1, ..., m)$. Taking $t_i^2 = q_i^*(1 - q_i^*)$, we find that for any X satisfying $\bar{U}_i + t_i V_i \leq 0$ $(i = 1, ..., m)$ constraints (8.6.3) necessarily hold. Obviously, this is the same condition as that obtained under **(iv)**.

Thus, problem P_I is replaced by a deterministic multiple non-linear fractional programming problem with non-linear inequality constraints, namely Problem P_{II}:

$$\text{"maximize" } F = \left[\frac{\bar{G}^1 X - q_1(X'\bar{D}^1 X)^{1/2}}{H^1 X + h_0^1}, \ldots, \frac{\bar{G}^r X - q_r(X'\bar{D}^r X)^{1/2}}{H^r X + h_0^r} \right]$$

subject to $X \in D^+ = \{X \in R^{n+1} \mid A_i X + (X'\bar{E}_i X)^{1/2} \leq B_i, (i = 1, \ldots, m)$ $A_{m+1}X = 1,\ X \geq 0\}$, where $\bar{G}^i = (G_1^i, \ldots, G_n^i, 0)$, $H^i = (h_1^i, h_2^i, \ldots, h_n^i, 0)$, $(i = 1, \ldots, r)$, $A_{m+1} = (0, \ldots, 0, 1)$ are $(n+1)$ - vectors, $\bar{E}_i = t_i^2 \hat{E}_i$ and $\bar{D}^i = \begin{bmatrix} \hat{D}_i & 0 \\ 0 & 0 \end{bmatrix}$ is an $(n+1) \times (n+1)$ symmetric positive semi-definite matrix.

Now, by Theorem 2.6 of reference [910], $(X'\bar{D}^i X)^{1/2}$ and $(X'\bar{E}_i X)^{1/2}$ are convex functions. Hence, D^+ is convex. By Theorem 4.12 of reference [910], the function $F_i = \dfrac{\bar{G}^i X - q_i(X'\bar{D}^i X)^{1/2}}{H^i X + h_0^i}$ is explicit quasiconcave on the convex domain of feasible solutions. Therefore, P_{II} is a problem of maximizing a vectorial explicit quasiconcave function over a convex region. The term "maximize" being used above in Problem P_{II} means finding all efficient solutions.

Assume that :

(vi) the set D^+ is regular (i.e. non-empty and bounded)

(vii) $H^i X + h_0^i > 0$ for all $X \in D^+$ and for all i.

According to Definition 6.1.1 we have :

Definition 8.6.1. A point X^0 is said to be efficient for Problem P_{II} if $X^0 \in D^+$, and if there exists no $X \in D^+$ such that for all $i = 1, \ldots, r$ one has

$$\frac{\bar{G}^i X^0 - q_i(X^{0\prime}\bar{D}_i X^0)^{1/2}}{H^i X^0 + h_0^i} \leq \frac{\bar{G}^i X - q_i(X'\bar{D}^i X)^{1/2}}{H^i X + h_0^i} \tag{8.6.4}$$

and for at least one $j \in \{1, \ldots, r\}$ one has

$$\frac{\bar{G}^j X^0 - q_j(X^{0\prime}\bar{D}^j X^0)^{1/2}}{H^j X^0 + h_0^j} < \frac{\bar{G}^j X - q_j(X'\bar{D}^j X)^{1/2}}{H^j X + h_0^j} \tag{8.6.5}$$

Let us introduce the new variable $y = tX$, where t is a non-negative real scalar quantity such that

$$H^i y + h_0^i t = \eta_i, \quad \eta_i \in R, \quad i = 1, \ldots, r \tag{8.6.6}$$

Here η_i $(i = 1, ..., r)$ are given real numbers.
Using the transformation given above, Problem P_{II} amounts to a problem of convex vectorial mathematical programming.
Hence, we can state Problem P_{III} :

$$\text{" maximize " } F^*(y, t) = \left(\bar{G}^1 y - q_1(y'\bar{D}^1 y)^{1/2}, ..., \bar{G}^r y - q_r(y'\bar{D}^r y)^{1/2}\right)$$

subject to

$$(y, t) \in D^{++} = \left\{(y, t) \mid A_i y + \left(y'\bar{E}_i y\right)^{1/2} \leq B_i t \ (i=1, ..., m); \right.$$
$$\left. A_{m+1} y = t; H^i y + h_0^i t = \eta_i \ (i = 1, ..., r); y \geq 0, t \geq 0\right\}$$

Let $D^{+++} = \left\{(y, t) \mid A_{m+1} y = t; H^i y + h_0^i t = \eta_i, i = 1, ..., r; y \geq 0, t \geq 0\right\}.$
It is possible that D^{++} be empty for certain values of η_i. Conditions for D^{+++} (and hence D^{++}) to be non-empty for all given η_i, have been given in Kall's theorem [see S. Vajda, *Probabilistic Programming*, Academic Press, New-York, 1972, p.5].

Lemma 8.6.1. *We have* $t > 0$ *for any feasible solution of* P_{III}.

The proof is similar to that of Lemma 3.4.1 and will be omitted. The following theorem establishes a connection between the efficient solutions of Problems P_{II} and P_{III}.

Theorem 8.6.1. *If*

(a) $0 < \text{sgn } \eta_i = \text{sgn}(H^i X^+ + h_0^i)$ *for all* $i = 1, ..., r$ *where* X^+ *is an optimal solution to Problem* P_{II};

(b) (y_1, t_1) *is an efficient solution to Problem* P_{III},

Then y_1/t_1 *is efficient solution for Problem* P_{II}.

Proof. By Lemma 8.6.1. we have $t_1 > 0$. Since (y_1, t_1) is an efficient solution to Problem P_{III}, we have

$$\bar{G}^i y_1 - q_i(y_1'\bar{D}^i y_1)^{1/2} \geq \bar{G}^i y - q_i(y'\bar{D}^i y)^{1/2} \tag{8.6.7}$$

for all i and

$$\bar{G}^i y_1 - q_i(y_1'\bar{D}^i y_1)^{1/2} > \bar{G}^i y - q_i(y'\bar{D}^i y)^{1/2} \tag{8.6.8}$$

for at least one i, whatever be $(y, t) \in D^{++}$. If y_1/t_1 were not an efficient solution to Problem P_{II}, then there would exist an $X^0 \in D^+$ with $X^0 \neq y_1/t_1$ such that

$$\frac{\bar{G}^i X^0 - q_i(X^{0\prime}\bar{D}^i X^0)^{1/2}}{H^i X^0 + h_0^i} \geq \frac{\bar{G}^i(y_1/t_1) - q_i[(y_1/t_1)'\bar{D}^i(y_1/t_1)]^{1/2}}{H^i(y_1/t_1) + h_0^i}, \text{ for all } i, \tag{8.6.9}$$

and

$$\frac{\bar{G}^i X^0 - q_i(X^{0\prime}\bar{D}^i X^0)^{1/2}}{H^i X^0 + h_0^i} > \frac{\bar{G}^i(y_1/t_1) - q_i[(y_1/t_1)'\bar{D}^i(y_1/t_1)]^{1/2}}{H^i(y_1/t_1) + h_0^i}, \text{ for at least one } i. \tag{8.6.10}$$

By (a), $H^iX + h_0^i = \theta\eta_i$ for $i = 1, ..., r$ and for some real $\theta > 0$. Let $t_0 = \theta^{-1} = \eta_i/(H^iX + h_0^i)$ and $y_0 = t_0X^0$. One can easily show that (y_0, t_0) is a feasible solution to Problem P_{III}. Also, by Lemma 8.6.1, $t_0 > 0$. From (8.6.9) and (8.6.10) we have

$$\frac{\bar{G}^i(y_0/t_0) - q_i\left[(y_0/t_0)'\bar{D}^i(y_0/t_0)\right]^{1/2}}{H^i(y_0/t_0) + h_0^i} \geq \frac{\bar{G}^i(y_1/t_1) - q_i\left[(y_1/t_1)'\bar{D}^i(y_1/t_1)\right]^{1/2}}{H^i(y_1/t_1) + h_0^i}$$

for all i and

$$\frac{\bar{G}^i(y_0/t_0) - q_i\left[(y_0/t_0)'\bar{D}^i(y_0/t_0)\right]^{1/2}}{H^i(y_0/t_0) + h_0^i} > \frac{\bar{G}^i(y_1/t_1) - q_i\left[(y_1/t_1)'\bar{D}^i(y_1/t_1)\right]^{1/2}}{H^i(y_1/t_1) + h_0^i}$$

for at least one i. This means that

$$\frac{\bar{G}^iy_0 - q_i(y_0'\bar{D}^iy_0)^{1/2}}{H^iy_0 + h_0^it_0} \geq \frac{\bar{G}^iy_1 - q_i(y_1'\bar{D}^iy_1)^{1/2}}{H^iy_1 + h_0^it_0} \tag{8.6.11}$$

for all i, and

$$\frac{\bar{G}^iy_0 - q_i(y_0'\bar{D}^iy_0)^{1/2}}{H^iy_0 + h_0^it_0} > \frac{\bar{G}^iy_1 - q_i(y_1'\bar{D}^iy_1)^{1/2}}{H^iy_1 + h_0^it_0} \tag{8.6.12}$$

for at least one i. Since (y_0, t_0) and (y_1, t_1) are feasible solutions to P_{III}, it follows from (8.6.11) and (8.6.12) that

$$\bar{G}^iy_0 - q_i(y_0'\bar{D}^iy_0)^{1/2} \geq \bar{G}^iy_1 - q_i(y_1'\bar{D}^iy_1)^{1/2}$$

for all i, and

$$\bar{G}^iy_0 - q_i(y_0'\bar{D}^iy_0)^{1/2} > \bar{G}^iy_1 - q_i(y_1'\bar{D}^iy_1)^{1/2}$$

for at least one i, thus, contradicting the fact that (y_1, t_1) is an efficient solution to Problem P_{III}. The proof is complete.

Therefore, an efficient solution to P_{II} may be obtained by solving P_{III}.

8.7. A STOCHASTIC APPROACH TO SOME LINEAR FRACTIONAL GOAL PROGRAMMING PROBLEMS

This section extends goal programming with linear fractional criteria and linear constraints to include the case when the objective functions or target values are random variables. Three stochastic problems are considered and it is shown that these problems can be reduced, under certain hypotheses, to deterministic linear fractional min-max problems with linear or convex constraints.

The classical goal programming problem is formulated as shown below.

Problem GP :

$$\text{Minimize } g(x, y, z)$$

subject to

$$F_i(x) - y_i + z_i = G_i \quad \text{for } i \in I = \{1, 2, ..., r\} \tag{8.7.1}$$

$$x \in S, \quad y = (y_1, y_2, ..., y_r) \geq \mathbf{0}, \quad z = (z_1, z_2, ..., z_r) \geq \mathbf{0} \tag{8.7.2}$$

where:

(i) $F_i: S \rightarrow R$, $(i \in I)$ are the objective functions;

(ii) $G_i \in R$, $(i \in I)$ are the levels (goals, target values) to be reached by the objective functions;

(iii) y and z are unknown deviational vectors, whose components measure the deviation upwards and downwards, respectively, of $F_i(x)$ values from the corresponding goal values G_i $(i \in I)$;

(iv) $S \subseteq R^n$ is the set of feasible solutions to the problem;

(v) $g: R_+^{n+2r} \rightarrow R_+$ represents a global deviation function depending on the deviational vectors y and z.

Ordinarily, the objective function g assumes the form

$$g(x, y, z) = \sum_{i \in I} (\alpha_i y_i + \beta_i z_i) \tag{8.7.3}$$

where α_i and β_i are intragoal positive weights and specify the relative penalties to be applied for over- or under-achievements in G_i. However, here we shall consider another form for $g(x, y, z)$ (see, (8.7.12)).

The goal programming problem has been investigated and generalized by several authors who gave many efficient solution methods. Thus, for instance, Agrawal, Swarup and Garg [48] considered that the objective functions F_i $(i \in I)$ are linear fractional and proposed an algorithm for solving the goal programming problem based on Swarup's procedure [942] for solving linear fractional functional programming problems. Kornbluth and Steuer [573] considered the case when F_i $(i \in I)$ are linear fractional and either **(a)** each linear fractional criterion is assigned to its own priority level or **(b)** all the criteria are at the same level.

In Case **(a)**, a sequence of linear fractional programming problems has to be solved, whereas in Case **(b)**, one has to solve only one linearly constrained programming problem with a nonlinear objective function which is the sum of the linear fractional objective functions.

Stancu-Minasian [906] considered that all the criteria are at the same priority level and a goal objective function of Chebyshev type must be minimized (see, *e.g.* (8.7.11)).

Peteanu and Ţigan [728] examined the case when all the objective functions are linear fractional and have the same priority level and, additionally, the goals G_i $(i \in I)$ are extended to the interval goals $[G_i', G_i'']$ $(i \in I)$. It is shown that the interval goal programming problem can be reduced to a min-max linear fractional problem with linear constraints.

Let us replace Problem GP by the following equivalent Problem GP1:

$$\text{Minimize } g(x, y, z)$$

subject to

$$F_i(x) - y_i \leq G_i, \quad i \in I \tag{8.7.4}$$

$$F_i(x) + z_i \geq G_i, \quad i \in I \tag{8.7.5}$$

$$x \in S, \ \ y_i \geq 0, \ \ z_i \geq 0, \ \ i \in I \tag{8.7.6}$$

It can be easily shown that problems GP and GP1 have the same optimal solution set when function g is of the form (8.7.3) with positive weights. Moreover, it can be proved that GP and GP1 have the same optimal solution set when $g(x, y, z)$ is a strict increasing with respect to y and z.

a) Random target values

In what follows, we shall assume that the objective functions F_i in Problem GP1 are well determined, and that the target values G_i are independent random variables with known distributions $T_i(\cdot)$, $(i \in I)$. In this situation, we consider that constraints (8.7.4) - (8.7.5) are satisfied in a proportion of cases or, in other words, with certain given probabilities, rather than always satisfied. We consider the following stochastic goal programming problem:

Problem SGP:

$$\text{Minimize } g(x, y, z)$$

subject to

$$P\{\omega \mid F_i(x) \geq G_i(\omega) - z_i\} \geq p_i, \quad i \in I \tag{8.7.7}$$

$$P\{\omega \mid F_i(x) < G_i(\omega) + y_i\} \geq q_i, \quad i \in I \tag{8.7.8}$$

$$x \in S, \ \ y_i \geq 0, \ \ z_i \geq 0, \ \ i \in I \tag{8.7.9}$$

where p_i and $q_i \in [0, 1]$ are lower bounds of these probabilities.

Let us find the deterministic equivalent of Problem SGP. Let

$$T_i^{-1}(\theta) = \inf\{\eta: T_i(\eta) \geq \theta\}, \ \ 0 \leq \theta \leq 1, \ \ i \in I$$

and

$$\hat{T}_i^{-1}(\theta) = \sup\{\eta: T_i(\eta) \leq \theta\}, \ \ 0 \leq \theta \leq 1, \ \ i \in I$$

We rewrite (8.7.7) as :

$$P\left\{\omega \mid F_i(x) \geq G_i(\omega) - z_i\right\} = P\left\{\omega \mid G_i(\omega) \leq F_i(x) + z_i\right\} =$$
$$= T_i(F_i(x) + z_i) \geq p_i, \quad i \in I$$

or, equivalently,

$$F_i(x) + z_i \geq T_i^{-1}(p_i), \quad i \in I \tag{8.7.7'}$$

Similarly, (8.7.8) is re-written as

$$F_i(x) - y_i \leq \hat{T}_i^{-1}(1 - q_i), \quad i \in I \tag{8.7.8'}$$

Thus, we have

Theorem 8.7.1. *If $G_i(i \in I)$ are independent random variables with distributions $T_i(\cdot)(i \in I)$, then Problem SGP is equivalent to the following deterministic problem:*

Problem PGD:

$$\textit{Minimize } g(x, y, z)$$

subject to (8.7.7'), (8.7.8') and (8.7.6).

Further, we need supplementary conditions on F_i and g and we shall show that the stochastic goal problem SGP reduces to a linear fractional min-max problem with linear constraints. Thus, we assume that:

(H1) the objective functions are linear fractional:

$$F_i(x) = \frac{c_i x + c_{i0}}{d_i x + d_{i0}}, \text{ for all } x \in S, \tag{8.7.10}$$

where $c_i \in R^n$, $d_i \in R^n$, $c_{i0}, d_{i0} \in R$.

(H2) For each $i \in I$:

$$d_i x + d_{i0} > 0, \text{ for all } x \in S. \tag{8.7.11}$$

(H3) the global deviation function g is of Chebyshev type, i.e.

$$g(x, y, z) = \max_{i \in I} (\alpha_i y_i + \beta_i z_i), \quad (\alpha_i, \beta_i > 0, \ i \in I) \tag{8.7.12}$$

Such functions were considered, for example, in [728] and [906].

(H4) the set S is polyhedral non-empty and bounded subset of R^n, defined by $S = \{x \in R^n \mid Ax = b, \ x \geq 0\}$ where A is a given $m \times n$ real matrix and $b \in R^m$.

Using these statements, problem PGD becomes:

Problem PGD1: Find

$$\min \max_{i \in I} (\alpha_i y_i + \beta_i z_i)$$

subject to

$$\frac{c_t x + c_{i0}}{d_t x + d_{i0}} + z_i \geq T_i^{-1}(p_i), \quad i \in I \tag{8.7.13}$$

$$\frac{c_t x + c_{i0}}{d_t x + d_{i0}} - y_i \leq \hat{T}_i^{-1}(1 - q_i), \quad i \in I \tag{8.7.14}$$

$$x \in S, \quad y_i \geq 0, \quad z_i \geq 0, \quad i \in I \tag{8.7.15}$$

Using assumption (H2), we transform constraints (8.7.13) and (8.7.14) into the following constraints :

$$(c_i - T_i^{-1}(p_i)d_i)x + z_i(d_t x + d_{i0}) \geq -c_{i0} + d_{i0}T_i^{-1}(p_i), \ i \in I \tag{8.7.13'}$$

$$(c_i - \hat{T}_i^{-1}(1 - q_i)d_i)x - y_i(d_t x + d_{i0}) \leq -c_{i0} + d_{i0}\hat{T}_i^{-1}(1 - q_i), \ i \in I \tag{8.7.14'}$$

Therefore, problem PGD1 becomes:

Problem PGD2: Find

$$\min \max_{i \in I} (\alpha_i y_i + \beta_t z_i)$$

subject to constraints (8.7.13'), (8.7.14') and (8.7.15).

The constraints of Problem PGD2 are nonlinear, but we can linearize them by the Kornbluth-Steuer transformation [566]

$$u_i = z_i(d_t x + d_{i0}); \quad v_i = y_i(d_t x + d_{i0}), \quad i \in I. \tag{8.7.16}$$

Employing this change of variable, problem PGD2 becomes a deterministic linear fractional min-max problem with linear constraints:

Problem PM : Find

$$\min \max_{i \in I} \frac{\alpha_i v_i + \beta_t u_i}{d_t x + d_{i0}}$$

subject to

$$(c_i - T_i^{-1}(p_i)d_i)x + u_i \geq -c_{i0} + d_{i0}T_i^{-1}(p_i), \quad i \in I$$

$$(c_i - \hat{T}_i^{-1}(1 - q_i)d_i)x - v_i \leq -c_{i0} + d_{i0}\hat{T}_i^{-1}(1 - q_i), \quad i \in I$$

$$x \in S, \quad u = (u_1, u_2, ..., u_r) \geq \mathbf{0}, \quad v = (v_1, v_2, ..., v_r) \geq \mathbf{0}$$

Under assumptions (H1) - (H4), the following theorem is immediate:

Theorem 8.7.2. *If G_i are independent random variables and if (x^*, u^*, v^*) is an optimal solution of Problem PM, then (x^*, y^*, z^*) with*

$$y_i^* = \frac{v_i^*}{d_t x^* + d_{i0}}, \quad z_i^* = \frac{u_i^*}{d_t x^* + d_{i0}}, \quad i \in I$$

is an optimal solution to Problem SGP and conversely.

Proof. The proof follows from the fact that Problem SGP is equivalent to Problem PGD (Theorem 8.7.1), Problem PGD is equivalent to Problem PGD1 or Problem PGD2 by relations (8.7.10) - (8.7.12) and Problem PGD2 reduces to Problem PM by the variable transformation (8.7.16).

b) Random objective functions

Now, we assume that the target values G_i $(i \in I)$ in Problem GP1 are well determined, but the objective functions F_i are random. We adopt the following assumption:

(H5) the objective functions F_i $(i \in I)$ are of the form:

$$F_i(x, \omega) = \frac{c_i(\omega)x + c_{i0}(\omega)}{d_i x + d_{i0}} \tag{8.7.17}$$

where the components of the vectors $(c_i, c_{i0}) \in R^{n+1}$ $(i \in I)$ are normal independent random variables with expectations $\bar{c}_{ij}$ and variances σ_{ij}^2, $(i \in I,\ j = 0, 1, ..., n)$.

We also assume that assumptions (H2) - (H4) are true. The stochastic goal programming problem is:

Problem GPF: Find

$$\min \max_{i \in I} (\alpha_i y_i + \beta_i z_i)$$

subject to

$$P\{\omega \mid F_i(x, \omega) \geq G_i - z_i\} \geq p_i, \quad i \in I \tag{8.7.18}$$

$$P\{\omega \mid F_i(x, \omega) < G_i + y_i\} \geq q_i, \quad i \in I \tag{8.7.19}$$

$$x \in S, \quad y_i \geq 0, \quad z_i \geq 0, \quad i \in I \tag{8.7.20}$$

Recalling (8.7.11) and (8.7.17), constraints (8.7.18) and (8.7.19) can be written as

$$P\{\omega \mid c_i(\omega)x + c_{i0}(\omega) \geq G_i(d_i x + d_{i0}) - z_i(d_i x + d_{i0})\} \geq p_i, \quad i \in I$$

$$P\{\omega \mid c_i(\omega)x + c_{i0}(\omega) < G_i(d_i x + d_{i0}) + y_i(d_i x + d_{i0})\} \geq q_i, \quad i \in I,$$

or, equivalently, as

$$\frac{G_i(d_i x + d_{i0}) - z_i(d_i x + d_{i0}) - \bar{c}_i x - \bar{c}_{i0}}{\left(\sum_{j=1}^{n} \sigma_{ij}^2 x_j^2 + \sigma_{i0}^2\right)^{1/2}} \leq \phi^{-1}(1 - p_i), \quad i \in I \tag{8.7.21}$$

$$\frac{G_i(d_i x + d_{i0}) + y_i(d_i x + d_{i0}) - \bar{c}_i x - \bar{c}_{i0}}{\left(\sum_{j=1}^{n} \sigma_{ij}^2 x_j^2 + \sigma_{i0}^2\right)^{1/2}} \geq \phi^{-1}(q_i), \quad i \in I, \tag{8.7.22}$$

where $\phi(\cdot)$ is the probability distribution function of the standard normal variable $N(0,1)$. Thus, problem GPF becomes:

Problem GPF1: Find

$$\min \max_{i \in I} (\alpha_i y_i + \beta_i z_i)$$

subject to constraints (8.7.21), (8.7.22) and (8.7.20).

Employing the change of variable (8.7.16), problem GPF1 becomes a deterministic linear fractional min-max problem with nonlinear constraints:

Problem GPF2: Find

$$\min \max_{i \in I} \frac{\alpha_i v_i + \beta_i u_i}{d_i x + d_{i0}}$$

subject to

$$\frac{G_i(d_i x + d_{i0}) - u_i - \bar{c}_i x - \bar{c}_{i0}}{(\sum_{j=1}^{n} \sigma_{ij}^2 x_j^2 + \sigma_{i0}^2)^{1/2}} \leq \phi^{-1}(1 - p_i), \quad i \in I$$

$$\frac{G_i(d_i x + d_{i0}) + v_i - \bar{c}_i x - \bar{c}_{i0}}{(\sum_{j=1}^{n} \sigma_{ij}^2 x_j^2 + \sigma_{i0}^2)^{1/2}} \geq \phi^{-1}(q_i), \quad i \in I$$

$$x \in S, \quad u_i \geq 0, \quad v_i \geq 0, \quad i \in I.$$

It is reasonable to choose $p_i > 0.5$ and $q_i > 0.5$ such that the constraints of Problem GPF2 define a convex set. Hence, we have

Theorem 8.7.3. *If the objective functions F_i $(i \in I)$ are random of the form (8.7.17), if assumptions (H2) - (H4) are true, and if (x^*, u^*, v^*) is an optimal solution of Problem GPF2, then (x^*, y^*, z^*), with*

$$y_i^* = \frac{v_i^*}{d_i x^* + d_{i0}}, \quad z_i^* = \frac{u_i^*}{d_i x^* + d_{i0}}, \quad i \in I$$

is an optimal solution to Problem GPF and conversely.

The proof is similar to that of Theorem 8.7.2 and so, it will be omitted.

c) Simple randomization of objective functions

Now, we adopt the following assumptions for Problem GPF:

(H6) the objective functions F_i $(i \in I)$, given by (8.7.17), are simply randomized, i.e.,

$$c_i(\omega) = c_i' + t_i(\omega)c_i'', \quad c_{i0}(\omega) = c_{i0}' + t_i(\omega)c_{i0}'', \quad i \in I \qquad (8.7.23)$$

where $c_i^{\prime}, c_i^{\prime\prime} \in R^n$ $(i \in I)$ are constant vectors, $c_{i0}^{\prime}, c_{i0}^{\prime\prime} \in R$, $(i \in I)$ are scalar constants and $t_i(\cdot)$, $(i \in I)$ are random variables on a probability space (Ω, K, P) with continuous and strict increasing distribution functions $T_i(\cdot)$, $(i \in I)$,

(H7) the global function g assumes the form (8.7.12),

(H8) $d_i x + d_{i0} > 0$, and $c_i^{\prime\prime} x + c_{i0}^{\prime\prime} > 0$ for all $i \in I$ and for all $x \in S$.

Hence, problem GPF becomes:

Problem GPSR: Find

$$\min \max_{i \in I} (\alpha_i y_i + \beta_i z_i)$$

subject to

$$P\left\{ \omega \mid \frac{(c_i^{\prime} + t_i(\omega)c_i^{\prime\prime})x + (c_{i0}^{\prime} + t_i(\omega)c_{i0}^{\prime\prime})}{d_i x + d_{i0}} \geq G_i - z_i \right\} \geq p_i, \quad i \in I \tag{8.7.24}$$

$$P\left\{ \omega \mid \frac{(c_i^{\prime} + t_i(\omega)c_i^{\prime\prime})x + (c_{i0}^{\prime} + t_i(\omega)c_{i0}^{\prime\prime})}{d_i x + d_{i0}} < G_i + y_i \right\} \geq q_i, \quad i \in I \tag{8.7.25}$$

$$x \in S, \quad y_i \geq 0, \quad z_i \geq 0, \quad i \in I \tag{8.7.26}$$

Let us find the deterministic equivalent of Problem GPSR. Under assumptions (H6), (H7), and (H8), we transform constraints (8.7.24) and (8.7.25) into the following constraints:

$$P\left\{ \omega \mid t_i(\omega) \geq \frac{G_i(d_i x + d_{i0}) - z_i(d_i x + d_{i0}) - c_i^{\prime} x - c_{i0}^{\prime}}{c_i^{\prime\prime} x + c_{i0}^{\prime\prime}} \right\} \geq p_i, \quad i \in I$$

$$P\left\{ \omega \mid t_i(\omega) < \frac{G_i(d_i x + d_{i0}) + y_i(d_i x + d_{i0}) - c_i^{\prime} x - c_{i0}^{\prime}}{c_i^{\prime\prime} x + c_{i0}^{\prime\prime}} \right\} \geq q_i, \quad i \in I$$

or

$$\frac{G_i(d_i x + d_{i0}) - z_i(d_i x + d_{i0}) - c_i^{\prime} x - c_{i0}^{\prime}}{c_i^{\prime\prime} x + c_{i0}^{\prime\prime}} \leq T_i^{-1}(1 - p_i), \quad i \in I \tag{8.7.27}$$

$$\frac{G_i(d_i x + d_{i0}) + y_i(d_i x + d_{i0}) - c_i^{\prime} x - c_{i0}^{\prime}}{c_i^{\prime\prime} x + c_{i0}^{\prime\prime}} \geq T_i^{-1}(q_i), \quad i \in I \tag{8.7.28}$$

According to assumption (H8), constraints (8.7.27) and (8.7.28) become

$$(G_i d_i - c_i^{\prime} - T_i^{-1}(1 - p_i)c_i^{\prime\prime})x - z_i(d_i x + d_{i0}) \leq c_{i0}^{\prime} + c_{i0}^{\prime\prime} T_i^{-1}(1 - p_i) - d_{i0} G_i, \quad i \in I \tag{8.7.29}$$

$$(G_i d_i - c_i^{\prime} - T_i^{-1}(q_i)c_i^{\prime\prime})x + y_i(d_i x + d_{i0}) \geq c_{i0}^{\prime} + c_{i0}^{\prime\prime} T_i^{-1}(q_i) - d_{i0} G_i, \quad i \in I \tag{8.7.30}$$

Therefore, problem GPSR can be formulated as follows:

Problem GPSR1: Find

$$\min \max_{i \in I} (\alpha_i y_i + \beta_i z_i)$$

subject to constraints (8.7.26), (8.7.29) and (8.7.30).

By the change of variable, Problem GPSR1 becomes:

Problem GPSR2: Find

$$\min \max_{i \in I} \frac{\alpha_i v_i + \beta_i u_i}{d_i x + d_{i0}}$$

subject to

$$(G_i d_i - c_i' - T_i^{-1}(1 - p_i) c_i'') x - u_i \leq c_{i0}' + c_{i0}'' T_i^{-1}(1 - p_i) - d_{i0} G_i, \quad i \in I$$
$$(G_i d_i - c_i' - T_i^{-1}(q_i) c_i'') x + v_i \geq c_{i0}' + c_{i0}'' T_i^{-1}(q_i) - d_{i0} G_i, \quad i \in I$$
$$x \in S, \quad u_i \geq 0, \quad v_i \geq 0, \quad i \in I$$

We have

Theorem 8.7.4. *If assumptions (H4) - (H9) are true, and if (x^*, u^*, v^*) is an optimal solution of the min-max problem GPSR2, then (x^*, y^*, z^*), where*

$$y_i^* = \frac{v_i^*}{d_i x^* + d_{i0}}, \quad z_i^* = \frac{u_i^*}{d_i x^* + d_{i0}}, \quad i \in I$$

is an optimal solution of the stochastic goal programming problem GPSR, and conversely.

8.8. STOCHASTIC FRACTIONAL MAX-MIN PROBLEM

a) The linear fractional case

Consider the following linear fractional max-min problem:

$$\max_x \min_y F(x, y) = \frac{cx + dy + f}{ex + gy + h} \tag{8.8.1}$$

subject to :

$$Ax + By \leq b, \quad x \geq 0, \quad y \geq 0 \tag{8.8.2}$$

where $x \in R^p$, $y \in R^q$ are variable vectors, whereas $c, e \in R^p$, $d, g \in R^q$, $b \in R^m$ and $A \in R^{m \times p}$, $B \in R^{m \times q}$, $f, h \in R$ are given vectors and matrices, respectively.

Denote by:

$$S = \{(x, y) \in R^p \times R^q : Ax + By \leq b, \ x \geq 0, y \geq 0\} \tag{8.8.3}$$

the feasible set of the problem (8.8.1) - (8.8.2), and

$$Q(x) = \{ y \geq 0 : By \leq b - Ax \} \tag{8.8.4}$$

$$X = \{ x \geq 0 : Q(x) \neq \varnothing \} \tag{8.8.5}$$

Following reference [323], for a max-min problem having the objective function $F : S \to R$, we consider the following definition:

Definition 8.8.1. A point $(x', y') \in S$ is an optimal solution for the max-min problem: $\max_x \min_y F(x, y)$, subject to $(x, y) \in S$ if:

(i) $F(x', y') = \min_{y \in Q(x')} F(x', y)$

(ii) $F(x', y') \geq \min_{y \in Q(x)} F(x, y)$, for all $x \in X$.

We consider the case when c and d are random vectors on a given probability space (Ω, K, P) and the minimum-risk approach is used to obtain the solution of this stochastic problem.

The minimum-risk approach to the problem (8.8.1) - (8.8.2) corresponding to the level w consists in the following programming problem :

SFP. Find

$$V(w) = \max_{x \in X} P \left\{ \omega : \min_{y \in Q(x)} \frac{c(\omega)x + d(\omega)y + f}{ex + gy + h} \geq w \right\} \tag{8.8.6}$$

Denote

$$H(x, y, \omega) = \frac{c(\omega)x + d(\omega)y + f}{ex + gy + h}, \quad \forall\ (x, y) \in S, \quad \omega \in \Omega.$$

According to Definition 8.8.1, we have:

Definition 8.8.2. A point (x', y') is an optimal solution for the minimum-risk problem SFP if:

(i) $P\{\omega : H(x', y', \omega) \geq w\} = P\{\omega : \min_{y \in Q(x')} H(x', y, \omega) \geq w\}$

and

(ii) $P\{\omega : H(x', y', \omega) \geq w\} \geq P\{\omega : \min_{y \in Q(x)} H(x, y, \omega) \geq w\} \quad \forall\ x \in X$

We make the following assumptions on the problem SFP:

(H1) S is a non-empty and compact set;

(H2) $ex + gy + h > 0$, for all $(x, y) \in S$.

Next, we show that, by an appropriate change of variable, the linear fractional stochastic max-min problem SFP can be reduced to a linear stochastic max-min problem. If we take

$$t = \frac{1}{ex + gy + h}, \quad u = tx, \quad v = ty, \tag{8.8.7}$$

then we can associate to problem SFP the following linear stochastic max-min problem:

SLP. Find

$$V_1(w) = \max_{u} P\{\omega : \min_{v,\, t} (c(\omega)u + d(\omega)v + ft) \geq w\}$$

subject to:

$$Au + Bv - bt \leq 0 \tag{8.8.8}$$

$$eu + gv + ft = 1 \tag{8.8.9}$$

$$u \geq 0, \quad v \geq 0, \quad t \geq 0 \tag{8.8.10}$$

Let U be the feasible set for problem SLP. Using similar arguments as in [995], we can prove:

Theorem 8.8.1. *If the assumption H1 holds, then* $(u, v, 0)$ *is not a feasible solution for problem SLP.*

Proof. It must be proved that every (u, v, t) verifying (8.8.9) - (8.8.10) has $t > 0$. In the contrary case, there exists a pair (u, v) such that $Au + Bv \leq 0, \; u \geq 0, \; v \geq 0$. Then by (8.8.9), it follows that (u, v) is not the null vector. But, if the pair (x, y) verifies (8.8.2), then the pair $(x + zu, \; y + zv)$ verifies (8.8.2) for every $z \geq 0$. This fact contradicts the boundness of the feasible set S supposed by H1. The proof is complete.

Theorem 8.8.2. *If the assumptions H1 and H2 hold and if* (u'', v'', t'') *is an minimum-risk solution for problem SLP, then* $(u''/t'', v''/t'')$ *is a minimum-risk solution for problem SFP.*

Proof. By Theorem 8.8.1 it follows that $t'' > 0$. Then from (8.8.8) we get $A\frac{u''}{t''} + B\frac{v''}{t''} \leq b$, which means $(u''/t'', v''/t'') \in S$.

Since

$$\left\{\omega : \min_{(v,\, t) \in Q'(u)} (c(\omega)u + d(\omega)v + ft) \geq w\right\}$$

$$= \left\{\omega : \min_{y \in Q(u/t)} \frac{c(\omega)u/t + d(\omega)y + f}{eu/t + gy + h} \geq w\right\}$$

$(Q'(u) = \{(v, t) : (u, v, t) \in U\})$ for all $t > 0$ and $(u, v, t) \in U$, it follows that the point (x'', y'') where $x'' = u''/t''$, $y'' = v''/t''$ verifies the conditions from Definition 8.8.1. Hence, it is an optimal solution for SFP. The proof is complete.

Now, we consider the simple randomization case:

$$c(\omega) = c_1 + t_1(\omega)c_2, \quad d(\omega) = d_1 + t_1(\omega)d_2 \tag{8.8.11}$$

where $c_i \in R^p$, $d_i \in R^q$ $(i = 1, 2)$ are constant given vectors whereas $t_1(\omega)$ is a random variable on a probability space (Ω, K, P) with a continuous and strict increasing distribution function T.

We also make the assumption:
(H3) $c_2x + d_2y > 0$, for all $(x, y) \in S$.

Then, as in reference [1008] for the stochastic linear max-min problem, one can establish a relationship between the SFP and a deterministic fractional max-min problem which does not depend on the distribution function T.

Theorem 8.8.3. *If assumptions H1, H2, H3 hold, and if the distribution function T is continuous and strict increasing, then the optimal solutions of problem SFP does not depend on T and can be determined by solving the following max-min problem :*

$$\max_x \min_y \frac{(c_1 - ew)x + (d_1 - gw)y - wh + f}{c_2x + d_2y}, \tag{8.8.12}$$

$$\text{subject to } (x, y) \in S.$$

Proof. Using assumption H2 yields:

$$P\{\omega : \min_{y \in Q(x)} H(x, y, \omega) \geq w\} =$$

$$P\{\omega : \min_{y \in Q(x)} (c(\omega) - ew)x + (d(\omega) - gw)y \geq wh - f\}$$

Hence, the SFP problem is equivalent with a linear stochastic max-min problem having the objective function $F(x, y) = (c(\omega) - ew)x + (d(\omega) - gw)y$. Using Theorem 1 of reference [1008], we get the conclusion of the present theorem, and the proof is complete.

b) The bilinear fractional case.

b1) The minimum-risk problem

Consider the following bilinear fractional max-min problem

P. Find

$$\max_{x \in X} \min_{y \in Y} \frac{xAy + ax + by + c}{xBy + dx + ey + f} \tag{8.8.13}$$

where

$$X = \{ x \in R^n : Cx \leq g, \;\; x \geq 0 \} \tag{8.8.14}$$

$$Y = \{ y \in R^m : Dy \leq h, \;\; y \geq 0 \} \tag{8.8.15}$$

and x and y are variable vectors, whereas $A, B \in R^{n \times m}$, $C \in R^{p \times n}$, $D \in R^{q \times m}$, $a, d \in R^n$, $b, e \in R^m$, $g \in R^p$, $h \in R^q$, $c, f \in R$ are given matrices, vectors and real numbers, respectively.

We adopt the following assumption for problem P:

(H4) $xBy + dx + ey + f > 0$, for all $(x, y) \in X \times Y$.

Analog to Definition 8.8.1., we have

Definition 8.8.3. Given a function $F : S \times Q \to R$, (S subset of $R^{n'}$ and Q subset of $R^{m'}$), we say that $(x', y') \in S \times Q$ is an optimal solution for the max-min problem:

$$\max_{x \in S} \min_{y \in Q} F(x, y)$$

if (x', y') verifies the following conditions:

$$F(x', y') = \min \{ F(x', y) : y \in Q \} \tag{8.8.16}$$

$$F(x', y') \geq \min \{ F(x, y) : y \in Q \}, \text{ for all } x \in S \tag{8.8.17}$$

In what follows, we shall assume that the matrices A, B and the vectors a, b, d, e are random with simple randomization, i.e. of the form:

$$A(\omega) = A' + t(\omega)A'', \quad B(\omega) = B' + t(\omega)B'', \tag{8.8.18}$$

$$a(\omega) = a' + t(\omega)a'', \quad b(\omega) = b' + t(\omega)b'', \tag{8.8.19}$$

$$d(\omega) = d' + t(\omega)d'', \quad e(\omega) = e' + t(\omega)e'', \tag{8.8.20}$$

where $A', A'', B', B'' \in R^{n \times m}$ are given matrices, $a', a'', d', d'' \in R^n$, $b', b'', e', e'' \in R^m$ are given constant vectors, whereas $t(\omega)$ is a random variable on a probability space (Ω, K, P) with a continuous and strict increasing distribution function $T(z)$.

The minimum-risk problem corresponding to level z, associated to problem P is:

PR. Find

$$v(z) = \max P \{ \omega : Z(x, \omega) > z \} \tag{8.8.21}$$

where

$$Z(x, \omega) = \min \{ H(x, y, \omega) : y \in Y \} \tag{8.8.22a}$$

and

$$H(x, y, \omega) = \frac{xA(\omega)y + a(\omega)x + b(\omega)y + c}{xB(\omega)y + d(\omega)x + e(\omega)y + f} \tag{8.8.22b}$$

Concerning problem PR, we make an assumption similar to (H4):

(H5) $xB(\omega)y + d(\omega)x + e(\omega)y + f > 0, \quad \forall\ (x, y) \in X \times Y$ and $\forall\ \omega \in \Omega$.

Likewise, Definition 8.8.3 implies:

Definition 8.8.4. A pair $(x', y') \in X \times Y$ is called minimum-risk solution for problem PR if the following conditions hold:

$$P\{\omega : H(x', y', \omega) > z\} = P\{\omega : \min\{H(x', y, \omega) : y \in Y\} > z\} \tag{8.8.23}$$

$$P\{\omega : H(x', y', \omega) > z\} \geq P\{\omega : \min\{H(x, y, \omega) : y \in Y\} > z\} \tag{8.8.24}$$

for all $x \in X$.

Denote:

$$V(x, y) = x(zB' - A')y + (zd' - a')x + (ze' - b')y + zf - c$$

and

$$W(x, y) = x(A'' - zB'')y + (a'' - zd'')x + (b'' - ze'')y,$$

for all $(x, y) \in X \times Y$.

Further, suppose that:

$$W(x, y) > 0, \text{ for all } (x, y) \in X \times Y \tag{8.8.25}$$

X and Y are bounded and non-empty sets. (8.8.26)

Thus, we show that, under assumptions (8.8.25) and (8.8.26), the minimum-risk problem PR, can be solved by a deterministic problem, which does not depend on the distribution function of the random variable $t(\omega)$. This fact generalizes a results obtained in reference [920] for bilinear programming.

Theorem 8.8.4. *If the conditions (8.8.25) and (8.8.26) hold, and if the distribution function T is continuous and strict increasing, then the minimum-risk solution of problem PR does not depend on T and it can be obtained by solving the following min-max bilinear problem:*

PA.

$$\min_{x \in X} \max_{y \in Y} \frac{V(x, y)}{W(x, y)}$$

Proof. From (8.8.18) - (8.8.20) and (8.8.21) - (8.8.22), we get:

$$G(x, z) = P\{\omega : \min\{H(x, y, \omega) : y \in Y\} > z\} =$$
$$= P\{\omega : H(x, y, \omega) > z, \text{ for all } y \in Y\}.$$

Hence, according to (8.8.25), we have

$$\begin{aligned} G(x, z) &= P\{\omega : t(\omega) > V(x, y)/W(x, y), \ \forall\ y \in Y\} = \\ &= P\{\omega : t(\omega) > \max\{V(x, y)/W(x, y) : y \in Y\}\} = \\ &= 1 - P\{\omega : t(\omega) < \max\{V(x, y)/W(x, y) : y \in Y\}\} \end{aligned}$$

Then, the problem PR is equivalent to:
$\max\{G(x, \omega) : x \in X\} = \max\{1 - T(\max\{V(x, y)/W(x, y) : y \in Y\}) : x \in X\}$.

Therefore, by the assumption that T is continuous and strict increasing, we get:

$$v(z) = \max\{G(x, z) : x \in X\} = 1 - T\left[\min_{x \in X}\ \max_{y \in Y} \frac{V(x, y)}{W(x, y)}\right]$$

i.e., a minimum-risk solution for problem PR can be obtained by solving problem PA.

Consider now a particular case of problem PR, when only the objective denominator is random, while the objective nominator has a deterministic special form. In this max-min problem, denoted by PRF, the function H is expressed as

$$H(x, y, \omega) = \frac{xA(\omega)y + a(\omega)x + b(\omega)y + c}{(dx + f)(ey + r)} \tag{8.8.27}$$

for all $(x, y) \in X \times Y$ and $\omega \in \Omega$.
In (8.8.27), $d \in R^n$, $e \in R^m$, $f, r \in R$ are given real vectors and numbers respectively, and the other notation have the same meaning as in the statement of problem PR. From problem PRF, performing the variable transformations,

$$u = \theta x, \quad v = sy \tag{8.8.28}$$

we get the following minimum-risk problem:

PRL. Find

$$V'(z) = \max_{(u, \theta) \in X'} P\{\omega : Z'(u, \theta, \omega) > z\} \tag{8.8.29}$$

where:

$$X' = \{(u, \theta) \in R^{n+1} : Cu - g\theta \le 0,\ du + f\theta = 1,\ u \ge 0, \theta \ge 0\} \tag{8.8.30}$$

$$Z'(u, \theta, \omega) = \min\{H'(u, \theta, v, s, \omega) : (v, s) \in Y'\} \tag{8.8.31}$$

$$H'(u, \theta, v, s, \omega) = uA(\omega)v + a(\omega)su + b(\omega)\theta v + c\theta s \tag{8.8.32}$$

$$Y' = \{(v, s) \in R^{m+1} : Dv - hs \le 0,\ ev + rs = 1,\ v \ge 0, s \ge 0\} \tag{8.8.33}$$

Suppose that:

$$dx + f > 0 \text{ and } ey + r > 0, \text{ for all } x \in X,\ y \in Y \tag{8.8.34}$$

Theorem 8.8.5. *Assume that* (8.8.26) *and* (8.8.34) *hold. Then, we have:*
(i) *if* $(u, \theta) \in X'$ *and* $(v, s) \in Y'$, *then* θ *and* s *are strict positive numbers;*
(ii) *if* (u', θ', v', s') *is minimum-risk solution for PRL, then* $(u'/\theta', v'/s')$ *is a minimum-risk solution for PRF.*

Proof. Assertion (i) can be easily proved using assumption (8.8.26). Part (ii) of the theorem results from performing the variable transformation (8.8.28) in problem PRF.

Theorem 8.8.5. provides an approach to transforming the fractional max-min problem PRF into a bilinear stochastic max-min problem (problem PRL), which, by Theorem 8.8.4., is equivalent with a deterministic bilinear fractional max-min problem. In what follows, we will present a numerical procedure for solving the deterministic bilinear fractional problem PA, and, by Theorems 8.8.4 and 8.8.5., the stochastic problem PR (or PRF). With this aim in mind, we employ a parametrical procedure which represents a specific particularization to problem PA, of a general iterative method given in references [1003] and [1005] (see also reference [920]).

The parametrical procedure.

Step 1. Choose $x_0 \in X$ and take $k = 0$

Step 2. Solve the linear fractional program

$$t_k = \min_{y \in Y} \frac{V(x_k, y)}{W(x_k, y)} \tag{8.8.35}$$

and let $y_k \in Y$ be an optimal solution for (8.8.35)

Step 3. Solve the linear program:
PL(t_k). Find

$$\beta(t_k) = \max_{x, \mu} (M_k x + h\mu + zf - c)$$

subject to

$$\mu D \geq xL_k + N_k \tag{8.8.36}$$

$$\mu \geq 0, \quad x \in X \tag{8.8.37}$$

where $\mu \in R^q$ is the vector of dual variables and

$$M_k = (zd' - a') - t_k(a'' - zd'')$$

$$N_k = (ze' - b') - t_k(b'' - ze'')$$

$$L_k = (zB' - A') - t_k(A'' - zB'')$$

Let (x_{k+1}, μ_{k+1}) be an optimal solution of program PL(t_k).

Step 4.

i) If $\beta(t_k) = 0$, then (x_k, y_k) is an optimal solution for PA and the procedure stops. The point (x_k, y_k) is also an optimal solution for PR.

ii) If $\beta(t_k) > 0$, then go to Step 2, with k replaced by k+1.

A proof for the convergence of this procedure was presented, for instance, by Ţigan [1005] (or reference [1003]), in the frame of the parametrical method theory for general fractional max-min problems.

Given that the parametrical procedure generally needs an infinite number of iterations, in order to obtain a finite iterative method , one may use in Step 4 an approximate stop criterion of the form:

Step 4'.

i) If $\beta(t_k) \leq r'$, then the algorithm stops and (x_k, y_k) is an approximate solution for PA (or PR).

ii) If $\beta(t_k) > r'$, then go to Step 2 with k replaced by k+1.

Here, r' is a positive number which represents a measure of the desired approximation. This algorithmic approach needs to solve, at each iteration, by Simplex techniques, only two programming problems: a linear fractional program in Step 2, and an ordinary linear programming problem in Step 3. At iteration k, the solving of the linear fractional program (8.8.35) in Step 2 may begin, except the first iteration, with y_{k-1} as the initial solution (the optimal solution of (8.8.35) obtained in the previous iteration). In Step 3, for solving the linear program PL(t_k), a primal-dual simplex algorithm may be suitably used, in order to employ as initial solution, at iteration k, the optimal solution obtained in the previous iteration. This approach seems to be especially efficient when the number of constraints which define the set X is large with respect to the number q of constraints (8.8.36).

b2) Kataoka's model.

The following generalized Kataoka's problem is related to the minimum-risk problem PR, and can be associated with problem P:

PK. Find

$$\max z$$

subject to

$$P\{\omega : \min_{y \in Y} H(x, y, \omega) \geq z\} \geq \alpha \qquad (8.8.38)$$

$$x \in X, \quad z \in R \qquad (8.8.39)$$

where $\alpha \in [0, 1]$ is a given lower level for the probability appearing in (8.8.38).

Next, let assume that in the fractional objective function H (see (8.8.22b)) only the numerator is random, that is, in (8.8.18), (8.8.19) and (8.8.20), one has:

$$B'' = 0, \quad d'' = 0, \quad e'' = 0, \quad B = B', \quad d = d', \quad e = e' \tag{8.8.40}$$

Moreover, we suppose that

$$D''(x, y) = xA''y + a''x + b''y > 0, \quad \forall\, (x, y) \in X \times Y \tag{8.8.41}$$

and for any $(x, y) \in X \times Y$, we denote

$$D'(x, y) = xA'y + a'x + b'y + c, \; N(x, y) = xBy + dx + ey + f \tag{8.8.42}$$

Theorem 8.8.6. *If assumption (H4), (8.8.26), (8.8.40) and (8.8.41) hold, and if the distribution function T is continuous and strict increasing, then any optimal solution of problem PK can be obtained by solving the following bilinear fractional max-min problem:*

PKA. Find

$$\max_{x \in X} \; \min_{y \in Y} \; \frac{D'(x, y) + T^{-1}(1 - \alpha)D''(x, y)}{N(x, y)}$$

Proof. Indeed, by (H4) and (8.8.41), we have

$$P\{\, \omega : \min_{y \in Y} H(x, y, \omega) \geq z \,\} = P\{\, \omega : H(x, y, \omega) \geq z, \;\; \forall\, y \in Y \,\} =$$

$$= P\{\, \omega : xA(\omega)y + a(\omega)x + b(\omega)y + c \geq zN(x, y), \;\; \forall\, y \in Y \,\} =$$

$$= P\left\{\, \omega : t(\omega) \geq \frac{zN(x, y) - D'(x, y)}{D''(x, y)}, \;\;\; \forall\, y \in Y \right\} =$$

$$= P\left\{\, \omega : t(\omega) \geq \max_{y \in Y} \frac{zN(x, y) - D'(x, y)}{D''(x, y)} \right\} =$$

$$= 1 - T\left[\, \max_{y \in Y} \frac{zN(x, y) - D'(x, y)}{D''(x, y)} \right] \geq \alpha$$

Then, the inequality (8.8.38) may be rewritten as:

$$\max_{y \in Y} \frac{zN(x, y) - D'(x, y)}{D''(x, y)} \leq T^{-1}(1 - \alpha)$$

which, by (H4) and (8.8.41), is equivalent to

$$z \leq \min_{y \in Y} \frac{D'(x, y) + T^{-1}(1 - \alpha)D''(x, y)}{N(x, y)}$$

The last inequality implies that any optimal solution of problem PK may be obtained solving max-min bilinear fractional PKA.

8.9. INEXACT LINEAR FRACTIONAL PROGRAMMING WITH SET-INCLUSIVE CONSTRAINTS

Consider the following linear fractional programming problem:

$$\sup \frac{c'x}{d'x} \tag{8.9.1}$$

subject to

$$Ax \le b \,, \quad x \ge 0 \,. \tag{8.9.2}$$

If we denote the columns of A with $a_j = (a_{1j}, a_{2j}, \dots, a_{mj})'$, $(j = 1, \dots, n)$, the constraints (8.9.2) are written as

$$x_1 a_1 + x_2 a_2 + \dots + x_n a_n \le b \,, \quad x \ge 0 \,. \tag{8.9.3}$$

Consider now problem (8.9.1), (8.9.3) in a different mode, i.e., by substituting constraints (8.9.3) with inclusion constraints. More precisely, consider that the vectors $a_1, \dots, a_n$ are variable, each varying in the compact, convex and non-empty sets $K_j \subseteq R^m$. Also, accept that the vector b belongs to the non-empty sets K, closed and convex from R^m. We will substitute the constraints (8.9.3) with a more general constraint in which the left hand side has a finite sum of convex sets that must be included in a given convex set. We arrive at the following optimization problem:

$$\sup \frac{c'x}{d'x} \tag{8.9.4}$$

subject to

$$x_1 K_1 + x_2 K_2 + \dots + x_n K_n \subseteq K \,, \quad x \ge 0 \,. \tag{8.9.5}$$

Note that

$$A + B = \{ z \mid z = x + y, \ x \in A, \ y \in B \},$$
$$aA = \{ z \mid z = ax, \ x \in A \}, \ a \in R.$$

We call problem (8.9.4)-(8.9.5) linear fractional programming problem with inexact data with set-inclusive constraints. The notion of mathematical programming problem with set-inclusive constraints was introduced by Soyster[4]. By analogy with the usual case, we call the sets K_j activity sets, and K the resource set.

---

[4]A.L. Soyster, "Convex programming with set-inclusive constraints and application to inexact linear programming", *Oper. Res.*, 21(5)(1973), 1154-1157.

Definition 8.9.1. The vector $x = (x_1 , \ldots , x_n)'$ is feasible solution for problem (8.9.4)-(8.9.5) if and only if for each possible set of vectors $a_j \in K_j$, $j = 1 , \ldots , n$, we have $x_1a_1 + x_2a_2 + \ldots + x_na_n \in K(x_j \geq 0)$.

If K contains the null vector, then (8.9.5) has the feasible solution $x_j = 0$ for all j. If K has a particular structure, then the programming with set-inclusive constraints can be considered a development of the idea of generalized programming.

Let $b \in R^m$, $K(b) = \{ y \in R^m \mid y \leq b \}$ be an obviously convex set, and define the *generalized linear fractional programming problem*

$$sup \frac{c'x}{d'x} \tag{8.9.6}$$

subject to

$$x_1a_1 + \ldots + x_na_n \leq b, \quad x_j \geq 0, \quad a_j \in K_j \tag{8.9.7}$$

For problem (8.9.4)-(8.9.5), we must determine the vectors x which maximize the fractional function whichever the vectors $a_j \in K_j$, while for problem (8.9.6)-(8.9.7) we must determine a certain system de vectors $a_1 , a_2 , \ldots , a_n$, and a certain x, which maximize the function in (8.9.6).
In other words, in problem (8.9.6)-(8.9.7), the vectors a_j are decision variables like x_j. From this point of view, Soyster's approach represents a conservative strategy for the inexact programming, while generalized programming is a nonconservative strategy.

The particularization of the resource set $K = K(b)$ allows us to obtain an optimal solution for problem (8.9.4)-(8.9.5), *via* an auxiliary linear fractional programming problem.
Let $A' = (a_1', \ldots, a_n')$ be the technological matrix of this program, where the activity vectors $a_j'(j \in J = \{1, \ldots, n\})$ are obtained *via* the support functionals of the convex sets $K_j(j \in J)$. The components of the activity vectors a_j' are given by: $a_{ij}' = sup \{e_ia_j : a_j \in K_j\}$,$i \in I = \{1, 2, \ldots, m\}$, $j \in J$, where e_i is a unit row vector with 1 in the i-th component. The auxiliary linear fractional program will be:

$$LFP(A'). \quad \max \left\{ \frac{c'x}{d'x} : A'x \leq b, \ x \geq 0 \right\}$$

a) Inexact multi-objective programs.

In what follows, we consider the multi-objective programming with set coefficients and we shall present several types of efficiency. As objective function, we consider the vector function

$f : H \times X \rightarrow R^p$, $f(C, x) = (f_1(C, x), ..., f_p(C, x))$ where $x = (x_1, ..., x_n) \in X \subseteq R^n$ is the vector of decision variables, while C is a $p \times s$ matrix (containing the objective coefficients), which belongs to a given convex compact subset H of the euclidean space $R^{p\times s}$. Denote $I = \{1,2, ..., p\}$ and $J = \{1,2, ..., s\}$. Several kinds of set coefficients for the objective function f_i, $(i \in I)$ can be considered. In reference [929], it was assumed that each column $k_j = (c_{1j}, ..., c_{pj})'$ of the matrix C varies independently into a given variation domain K_j, $(j \in J)$, which is a convex compact subset of R^p. Therefore, in this particular case, we have:

$$H = K_1 \times K_2 \times ... \times K_s \tag{8.9.8}$$

Given a matrix $C \in H$, we denote by $c_i(C)$ and $k_j(C)$, $(i \in I,\ j \in J)$ the rows and columns of C, respectively. Likewise, for any $i \in I$ and any $j \in J$, we define the sets:

$$M_i(H) = \{c_i(C) : C \in H\} \text{ and } K_j(H) = \{k_j(C) : C \in H\} \tag{8.9.9}$$

Next, suppose that:

(H1) H is a convex compact non-empty subset of $R^{p\times s}$.
(H2) the objective functions $f_i(i \in I)$ are continuous on $H \times X$.

These assumptions imply that, for any $x' \in X$ and for any $i \in I$, the set $f_i(H, x') = \{f_i(C, x') : C \in H\}$, is a closed bounded interval. Therefore, when the coefficients of the objective f_i are perturbed in a prescribed set, the values of f_i at a given point $x \in X$ cover a certain interval. Hence, it is natural for certain notions of efficient solutions for multi-objective programming with inexact objectives f_i, $(i \in I)$ to be related with a semi-order relation defined on the compact, interval set:

$$I(R) = \{[a, b] : a, b \in R,\ a \leq b\}$$

of the bounded closed intervals in R. Also, a semi-order relation on $I(R)$ induces a corresponding semi-order relation on $I(R)^p$, which can be employed to introduce a certain type of efficiency for the multi-objective problem. Thus, in order to formulate a conservative inexact multi-objective program, we consider the following order relation on $I(R)$:

$$[a, b] \leq_c [a', b'] \Leftrightarrow a \leq a' \tag{8.9.10}$$

and the strict relation:

$$[a, b] <_c [a', b'] \Leftrightarrow a < a' \tag{8.9.11}$$

Definition 8.9.2. **i)** A point $x' \in X$ is said to be a *c-efficient point* (in maximum sense) for f on X, if there exists no point $x \in X$, such that for any $i \in I$, $f_i(H, x) \geq_c f_i(H, x')$ and for at least an index i the strict inequality holds.
ii) A point $x' \in X$ is said to be *weakly c-efficient* for f on X, if there exists no point $x \in X$, such that for any $i \in I$, $f_i(H, x) >_c f_i(H, x')$.

We can formally state the conservative multi-objective problem with inexact data as:

PMC. Find $Vmax_c \{f(H, x) : x \in X\}$,
which consists in finding of all c-efficient points of f on X and where $f(H, x) = \{f(C, x) : C \in H\}$ for any $x \in X$.

We associate to problem PMC the following ordinary multi-objective program with set-inclusive constraints.

PSI. Find $\quad Vmax\ y = (y_1, ..., y_p)$,
subject to

$$f(H, x) \subseteq K(y) = \{z \in R^p : z \geq y\} \tag{8.9.12}$$

$$x \in X, \quad y \in R^p \tag{8.9.13}$$

Concerning the multi-objective programs PMC and PSI, the following property holds:

Theorem 8.9.1. *Let assumptions (H1) and (H2) be valid. Then a point $x' \in X$ is a c-efficient solution for PMC if and only if there exists $y' \in R^p$, such that (x', y') is an ordinary efficient solution for PSI.*

Proof. From (8.9.10) and (8.9.11), it follows that problem PMC can be replaced by the following ordinary equivalent multi-objective problem:

PMC1. Find $Vmax \{G(x) : x \in X\}$
where $G : X \to R^p$ is a vector function whose components $G_i : X \to R$ $(i \in I)$ for any $x \in X$, are defined by:

$$G_i(x) = inf \{f_i(C, x) : C \in H\}.$$

It is easy to see that problem PMC1 has same efficient values as problem:

PMC2. Find $Vmax\ y$
subject to (8.9.13) and:

$$f(C, x) \geq y, \text{ for all } C \in H. \tag{8.9.14}$$

Constraints (8.9.14) are equivalent to (8.9.12), hence problem PMC2 is equivalent to PSI.

The proof is now complete.

In order to obtain a problem equivalent (without set-inclusive constraints) to problem PSI, we suppose that H is a Cartesian product of the form (8.9.8). On the sets K_j, ($j \in J$), we make the following assumption:

(H3) For any $j \in J$, the set K_j is a given non-empty convex compact subset of R^p.
(H4) For every $i \in I$, the function f_i depends only on row $c_i(C)$, i.e., $f_i(C, x) = f_i(c_i(C), x)$, for any point $x \in X$ and any matrix $C \in H$.
(H5) For any point $x \in X$ and any $i \in I$, function $f_i(\cdot, x)$ is a non-decreasing function on the set M_i (defined by (8.9.9)) of R^s.
Now, associate to problem PSI the following auxiliary linear multi-objective problem with nonlinear constraints:
PMCD. Find $Vmax\ \{y : y \in R^p, f(C', x) \geq y, x \in X\}$,
where $C' = (c'_{ij})$ is a $p \times s$ matrix whose elements are the support functional of K_j.

Theorem 8.9.2. *If assumptions (H3) - (H5) hold, then problems PSI and PMCD have the same feasible set and the same efficient solutions.*

Proof. By using the support functional technique employed by Soyster (*op. cit.*) the theorem can be obtained as a consequence of Theorem 2.3 from reference [929], applied to problem PSI.

Theorems 8.9.1 and 8.9.2 provide sufficient conditions for the multi-objective problem PMC with inexact data in the objectives to be solved *via* the auxiliary multi-objective problem PMCD. In the particular case when f_i ($i \in I$) are linear functions and X is a convex polyhedron, problem PMCD reduces to an ordinary linear multi-objective program.

Theorems 8.9.1 and 8.9.2 imply the following theorem:

Theorem 8.9.3. *Let assumptions (H2) - (H5) be valid. Then $x' \in X$ is a c-efficient solution for PMC if and only if there exists $y' \in R^p$ such that (x', y') is an ordinary efficient solution for PMCD.*

Now, we consider a non-conservative notion of efficiency for the vectorial function f.

Definition 8.9.3. A point $x' \in X$ is said to be an *nc-efficient point* for f on X if there exists $C' \in H$ such that (x', C') is an ordinary (maximum) efficient point of f on the set $H \times X$.

Thus, the non-conservative multi-objective programming problem with inexact objectives can be stated as:

PMNC. Find $Vmax\ \{f(C, x) : C \in H,\ \ x \in X\}$

We can associate to problem PMNC the following multi-objective problem with inexactness in the constraints:

PMNC1. Find $Vmax\ \{y : f(C, x) \geq y,\ \ y \in R^p,\ C \in H,\ x \in X\}$

The following connection between problems PMNC and PMNC1 is obvious:

Theorem 8.9.4. *The point $x' \in X$ is an nc-efficient solution for PMNC if and only if there exist $y' \in R^p$ and $C' \in H$ such that (x', C', y') is an ordinary (maximum) efficient point for PMNC1.*

When f_i, $(i \in I)$ are linear (or linear fractional) functions and X is a polyhedral convex set, it can be shown that problem PMNC1 can be reduced to a linear (or linear fractional) multi-objective problem with generalized linear constraints.

For these last problems, we suggest in the following subsection an interactive approach to obtain (weakly) nc-efficient solutions *via* interval goal programming.
Now, we consider some particular cases of the inexact multi-objective programming problems, namely the cases when the criteria are linear fractional. We assume that X is a polyhedral non-empty and bounded subset of R^n, defined by: $X = \{x \in R^n : Ax = b,\ x \geq 0\}$ where A is a given $m \times n$ real matrix and $b \in R^m$.

We also, suppose that uncertainty appears only in the numerator of the linear fractional criteria f_i $(i \in I)$, i.e.:

$$f_i(C, x) = U_i(C, x)/W_i(D, x) \tag{8.9.15}$$

where:

$$U_i(C, x) = c_{i1}x_1 + \dots + c_{in}x_n + c_{i0} \tag{8.9.16}$$

$$W_i(D, x) = d_{i1}x_1 + \dots + d_{in}x_n + d_{i0} \tag{8.9.17}$$

Here, C is a variable $p \times n$ matrix, whose columns k_j, $(j \in J = \{1,2, ..., n\})$ belong to the perturbation sets K_j $(K_j \subseteq R^p)$, D is a fixed $p \times n$ matrix and $k_0 = (c_{10}, ..., c_{p0})$, $d_0 = (d_{10}, ..., d_{p0})$ have fixed real components. We set $H = K_1 \times \dots \times K_n$

We assume that:

$$W_i(D, x) > 0, \text{ for all } x \in X \text{ and for every } i \in I, \tag{8.9.18}$$

$$K_j \text{ is a convex compact non-empty set in } R^p (j \in J) \tag{8.9.19}$$

By Theorem 8.9.3., problem PMC with inexact fractional criteria of the form (8.9.15) can be reduced to a particular multi-objective problem PMCD, as follows:

LP. Find $\quad Vmax\ y$
subject to

$$Ax = b,\ x \geq 0 \tag{8.9.20}$$

$$U_i(C', x)/W_i(D, x) \geq y_i,\quad y_i \in R, \text{ for any } i \in I,$$

where the elements of the matrix C' are the support functional of K_j.

Using assumption (8.9.18), we can transform problem LP with fractional constraints into the following equivalent problem:

LP1. Find $\quad Vmax\ y$
subject to

$$(8.9.20) \text{ and } U_i(C', x) \geq y_i W_i(D, x),\ y_i \in R,\ i \in I.$$

Using the Kornbluth-Steuer transformation [573]

$$u_i = y_i W_i(D, x),\quad i \in I \tag{8.9.21}$$

problem LP1 becomes:

FP1. Find $\quad \text{Vmax } \{u_i/W_i(D, x) : i \in I\},$
subject to

$$(8.9.20) \text{ and } U_i(C', x) \geq u_i,\ u_i \in R,\ i \in I \tag{8.9.22}$$

Now, from Theorem 8.9.3, we get the following result:

Theorem 8.9.5. *If f_i $(i \in I)$ are linear fractional functions (of the form (8.9.15) - (8.9.17)) which verify condition (8.9.18) and the perturbation set K_j verifies (8.9.19) for any $j \in J$, then x' is a c-efficient point for the linear fractional problem PMC if and only if there exists $u' \in R^p$ such that (x', u') is an ordinary efficient point for FP1 (which is a usual linear fractional multi-objective program with linear constraints).*

Weakly efficient solutions for FP1, and, consequently, weakly c-efficient solutions for PMC with linear fractional inexact criteria, may be obtained *via* max-min linear fractional programs of the form:

PLF. Find $\max \{ \min \{u_i/W_i(D, x) : i \in I \}\}$, subject to (8.9.20), (8.9.22).

A similar study can be also made in the case of the nonconservative approach of the inexact multi-criterion problem with fractional objectives.

Thus, by Theorem 8.9.4, problem PMNC with generalized fractional criteria of the form 8.9.15, can be rewritten as:

FGP. Find $Vmax\ y$ subject to (8.9.20) and

$$U_i(C, x)/W_i(D, x) \geq y_i,\ y_i \in R,\ i \in I,\ C \in H \tag{8.9.23}$$

But, under condition (8.9.18), using the Kornbluth-Steuer variable change (8.9.21), problem FGP becomes:

FGP1. Find $Vmax\ \{u_i/W_i(D, x) : i \in I\}$, subject to (8.9.20), (8.9.23) and $U_i(C, x) \geq u_i,\ u_i \in R,\ i \in I$.

Then, by Theorem 8.9.4, we get the following result:

Theorem 8.9.6. *Let f_i, ($i \in I$) be linear fractional functions, verifying condition (8.9.18), and let H be a perturbation set which verifies condition (8.9.19). Then x' is an nc-efficient point for the multi-objective linear fractional problem PMNC if and only if there exists $u' \in R^p$ such that (x', u') is an ordinary efficient point for FGP1.*

Problem FGP1 is a linear fractional multi-objective program with generalized linear constraints.

b). Goal programming with inexact data.

In this subsection, we present some extensions of the goal programming in order to include criterion functions with inexact data. These extensions establish a natural link between the goal programming, multi-objective programming, inexact programming with set-inclusive constraints and generalized linear and linear fractional programming.

The goal programming problem was presented in Section 8.7. Here we recall this problem in the form:

GP. Find the lexicographic minimum

$$Lex\ \min\ \{g^j(y, z) : j \in J = \{1, 2, ..., k\}\}$$

subject to:

$$F_i(x) - y_i + z_i = G_i, \text{ for all } i \in I = \{1, ..., r\}$$

$$x \in S \subset R^n,\ y = (y_1, ..., y_r) \geq 0,\ z = (z_1, ..., z_r) \geq 0$$

where, additionally to the notations of Section 8.7, we have $g^j : R_+^{2r} \to R_+$, representing a deviation function associated to the j-th priority level ($j \in J$):

Consider the functions $f : \mathfrak{C} \times S \to R^r$, $g : \mathfrak{Z} \times S \to R^r$

$$f(C, x) = c_0 + c_1x_1 + ... + c_nx_n,\ c_j \in K_j \subset R^r, j \in M = \{1, 2, ..., n\} \tag{8.9.24}$$

$$g(D, x) = d_0 + d_1x_1 + ... + d_nx_n,\ d_j \in H_j \subset R^r, j \in M \tag{8.9.25}$$

where $\mathfrak{C} = K_1 \times K_2 \times \ldots \times K_n$, $\mathfrak{Z} = H_1 \times H_2 \times \ldots \times H_n$, $C \in \mathfrak{C}$ and $D \in \mathfrak{Z}$ are $r \times n$ real matrices having the columns c_j and d_j $(j \in M)$ respectively, and c_0, d_0 are given real vectors in R^r.

We assume that:

$$g(D, x) > 0, \text{ for all } D \in \mathfrak{Z},\ x \in S \text{ and } i \in I \tag{8.9.26}$$

$$K_j \text{ and } H_j,\ (j \in M) \text{ are convex non-empty and compact subsets of } R^r. \tag{8.9.27}$$

Then, we can define the fractional functions $F_i : P \times S \rightarrow R$

$$F_i(C, D, x) = \frac{c_{i0} + c_{i1}x_1 + \ldots + c_{in}x_n}{d_{i0} + d_{i1}x_1 + \ldots + d_{in}x_n},\ i \in I$$

where $P = \mathfrak{C} \times \mathfrak{Z}$ and c_{ij} and d_{ij} are the i-th components of the vectors c_j and d_j, $(i \in I, j = 0,1, \ldots, n)$ respectively.

By the assumptions (8.9.26) and (8.9.27) and by the continuity of functions $F_i (i \in I)$, for a given point $x' \in S$, the set $F_i(P, x') = \{ F(C, D, x') : C \in \mathfrak{C},\ D \in \mathfrak{Z} \}$ is a bounded interval.

Now, we formulate the interval goal problem with inexact data, as follows:

IGP. Find $Lex \min \left\{ g^j(y, z) : j \in J \right\}$
subject to

$$F_i(P, x) \subset \left[G' - y_i,\ G'' + z_i\right],\ i \in I, \tag{8.9.28}$$
$$x \in S,\ y \geq 0,\ z \geq 0,$$

where $[G_i', G_i'']$, $(i \in I)$ are interval goals and y_i, z_i, $(i \in I)$ are deviational variables, which measure the deviation upward and downward, respectively, of the interval $F_i(P, x')$ from the goal interval $[G_i', G_i'']$.

We remark that the goal programming problem IGP is a lexicographic optimization problem with set-inclusive constraints.

In what follows, we assume that all the criterion functions have the same priority level and the associate deviational objective function g has the form:

$$g(y, z) = \max_{i \in I} (\max \{\alpha_i y_i,\ \beta_i z_i\}) \tag{8.9.29}$$

where for each $i \in I$, α_i and β_i are given intragoal real non-negative weights.

Then the linear fractional goal programming problem (IGP) with inexact data can be rewritten as the following min-max problem with inexact nonlinear constraints:

TP. Find $\min \max_{i \in I} (\max \{\alpha_i y_i, \beta_i z_i\})$
subject to (8.9.28) and:

$$G_i' + y_i \leq \frac{f_i(C, x)}{g_i(D, x)}, \text{ for all } C \in \mathfrak{C}, \quad D \in \mathfrak{Z}, \quad i \in I \tag{8.9.30}$$

$$G_i'' - z_i \geq \frac{f_i(C, x)}{g_i(D, x)}, \text{ for all } C \in \mathfrak{C}, \quad D \in \mathfrak{Z}, \quad i \in I \tag{8.9.31}$$

Let $C' = [c_{ij}']$, $C'' = [c_{ij}'']$, $D' = [d_{ij}']$, $D'' = [d_{ij}'']$, be $r \times n$ matrices, whose elements are defined by:

$$c_{ij}' = \min \{e_i c_j : c_j \in K_j\}, \; c_{ij}'' = \max \{e_i c_j : c_j \in K_j\}$$
$$d_{ij}' = \max \{e_i d_j : d_j \in H_j\}, \; d_{ij}'' = \min \{e_i d_j : d_j \in H_j\}$$

where

$$e_i = (e_{i1}, e_{i2}, ..., e_{ir}) \in R^r \text{ and } e_{it} = \begin{cases} 0, \text{ if } & i \neq t \\ 1, \text{ if } & i = t \end{cases}$$

We remark that c_{ij}', c_{ij}'', d_{ij}', d_{ij}'' are well defined, since by (8.9.27) the sets K_j and H_j are supposed compact and non-empty. Now, we assume that:

$$f_i(C, x) \geq 0, \text{ for all } C \in \mathfrak{C}, \quad x \in S \text{ and } i \in I \tag{8.9.32}$$

Then it is reasonable to suppose that:

$$G_i' \geq 0 \; (G_i'' \geq 0), \; i \in I \tag{8.9.33}$$

and, besides the constraints (8.9.30), to consider:

$$z_i \leq G_i', \; i \in I \tag{8.9.34}$$

It is clear that if $z_i > G_i'$, then by (8.9.26) and (8.9.32) the constraints (8.9.30) are evidently verified.
Thus, instead of the problem (TP) we consider the problem:
TP1. Find $\min \max_{i \in I} (\max \{\alpha_i y_i, \beta_i z_i\})$
subject to (8.9.28), (8.9.30), (8.9.31) and (8.9.34).

Theorem 8.9.7. *If (x', y', z') is an optimal solution for problem TP1 then it is an optimal solution for problem TP.*
Proof. The theorem results from the fact that (x', y', z') is a feasible solution of problem TP, and if (x'', y'', z'') is a feasible solution of problem TP, such that $z_i'' > G_i'$, for some $i \in I$, then $g(y', z') \leq g(y'', z'')$.

Let associate to problem TP1, the following min-max problem:

FL. Find $\min \max_{i \in I} (\max \{\alpha_i y_i, \beta_i z_i\})$
subject to (8.9.28), (8.9.34) and

$$f_i(C', x) \geq (G_i' - z_i) g_i(D', x), \; i \in I \tag{8.9.35}$$

$$f_i(C'', x) \leq (G'' + y_i) g_i(D'', z), \; i \in I \tag{8.9.36}$$

Theorem 8.9.8. *If the assumptions (8.9.26), (8.9.27) and (8.9.32) hold, then the problems TP1 and FL have the same feasible (optimal) solution sets.*

Proof. First, note that, by (8.9.26), the constraints (8.9.30) and (8.9.31) are equivalent with the following constraints:

$$f_i(C, x) \geq (G_i' - z_i) g_i(D, x) \text{ for all } \mathbf{C} \in \mathfrak{C}, \; D \in \mathfrak{Z}, \; i \in I \tag{8.9.37}$$

$$f_i(C, x) \leq (G_i'' + y_i) g_i(D, x) \text{ for all } C \in \mathfrak{C}, \; D \in \mathfrak{Z}, \; i \in I \tag{3.9.38}$$

Using a similar argument as in the case of the mathematical programming with separable set-inclusive constraints (see *e.g.*, reference [925]) since, from (8.9.33) and (8.9.34), $G_i' + z_i \geq 0$ and $G_i'' - y_i \geq 0$, it results that the constraints (8.9.37) and (8.9.38) are equivalent with (8.9.35) and (8.9.36). Therefore, the problems TP1 and FL have the same feasible set, which proves the theorem.

Using the change of variables:

$$u_i = y_i \, g_i(D', x), \; v_i = z_i \, g_i(D'', x) \quad i \in I,$$

the problem FL becomes:

LFP. Find $\min \max_{i \in I} \left[\max \left\{ \dfrac{\alpha_i u_i}{g_i(D', x)}, \dfrac{\beta_i v_i}{g_i(D'', x)} \right\} \right]$

subject to

$$f_i(C', x) \geq G_i' g_i(D', x) - v_i, \quad i \in I$$

$$f_i(C'', x) \leq G_i'' g_i(D'', x) + u_i, \quad i \in I$$

$$Ax = b, \; x \geq 0, \; u \geq 0, \; v \geq 0$$

From (8.9.24) and (8.9.25), it follows that LFP is a max-min problem with linear fractional criteria and linear constraints.

Theorem 8.9.9. *If $F_i (i \in I)$ are linear fractional functions and the assumptions (8.9.26), (8.9.27) and (8.9.32) hold, then if (x', u', v') is an optimal solution for problem LFP then (x', y', z') is an optimal solution for the problem TP, where $y_i' = u_i'/g_i(D', x')$ and $z_i' = v_i'/g_i(D'', x')$, for all $i \in I$.*

Proof. The theorem results from Theorems 8.9.7 and 8.9.8 and from the fact that, if (x', u', v') is an optimal solution for problem LFP, then (x', y', z') is an optimal solution for problem FL.

Thus, from Theorem 8.9.9, we conclude that the goal programming problem TP with inexact criteria can be solved by means of the min-max problem LFP with linear fractional criteria and linear constraints.

c) Generalized pseudo-fractional max-min problems.

Consider some nonlinear max-min problems with generalized linear constraints and pseudo-fractional objective functions.

GPF. Find $V = \max\limits_{x,C,A} \min\limits_{y} F(C, x, y) = \dfrac{xCy + dx + ey + w}{\max\limits_{1 \le k \le r} (h_k y + s_k)} + f(x)$

subject to

$$x_1 a_1 + \dots + x_n a_n \le a, \quad x \ge 0 \tag{8.9.39}$$

$$\begin{bmatrix} c_j \\ a_j \end{bmatrix} \in K_j, \; j \in J = \{1, 2, \dots, n\} \tag{8.9.40}$$

$$By \le b, \quad y \ge 0 \tag{8.9.41}$$

Here $x = (x_1, \dots, x_n) \in R^n$ and $y = (y_1, \dots, y_m) \in R^m$, the components of vectors $c_j \in R^m$ and $a_j \in R^p$ are unknown coefficients of the objective and constraints. They belong to the non-empty convex sets $K_j \subset R^{m+p}$, $(j \in J)$, and can be also considered as decision variables. In the statement of the problem GPF, the vectors $d \in R^n$, $e \in R^m$, $h_k \in R^m$, $(k \in I = \{1, \dots, r\})$, $a \in R^p$, $b \in R^q$, the $q \times m$ real matrix B, the real numbers w and s_k $(k \in I)$ and the function $f : R^n \to R$ are given. Let $\mathfrak{C}$ and $\mathring{A}$ be the sets of all matrices $C = (c_1 c_2 \dots c_n)$ and $A = (a_1 a_2 \dots a_n)$ respectively, whose columns c_j and a_j, $(j \in J)$ satisfy the constraints (8.9.40) and such that

$$S(A) = \{x \in R^n \mid x_1 a_1 + \dots + x_n a_n \le a, \; x \ge 0\} \ne \varnothing .$$

We will show that, if the function f is linear, then the generalized max-min problem GPF can be reduced to some ordinary generalized linear programming problems.

Let X and Y denote:

$$X = \{ x \in R^n \mid \exists \; A \in \mathring{A} \text{ such that } x_1 a_1 + \dots + x_n a_n \le a, \; x \ge 0 \}$$

$Y = \{ y \in R^m \mid By \leq b, y \geq 0 \}.$

Definition 8.9.4. A pair $(x', y') \in X \times Y$ is an optimal solution for the problem GPF if there exist $A' \in \mathring{A}$ and $C' \in \mathfrak{C}$ such that $x' \in S(A')$ and

$$F(C', x', y') = V \tag{8.9.42}$$

$$\min_{y \in Y} F(C', x', y) = F(C', x', y') \tag{8.9.43}$$

In the following we suppose that:

(H1) X and Y are non-empty bounded sets;

(H2) the denominator of the objective function is positive on the set Y, i.e.

$$G(y) = \max_{k \in I} (h_k y + s_k) > 0, \quad \forall \; y \in Y \tag{8.9.44}$$

We make a transformation of the problem GPF into a generalized max-min bilinear problem (when f is a linear function) having a nonlinear constraint, by means of the change of variable:

$$u = ty, \quad t \geq 0 \tag{8.9.45}$$

Thus, we get the problem:

GPL. Find $V' = \max\limits_{x, C, A} \; \min\limits_{u, t} H(C, x, u, t) = (xCu + tdx + tw + eu + f(x))$

subject to (8.9.39), (8.9.40) and

$$Bu - bt \leq 0 \tag{8.9.46}$$

$$\max_{k \in I} (h_k u + s_k t) \geq 1 \tag{8.9.47}$$

$$u \geq 0, \; t \geq 0 \tag{8.9.48}$$

Denote by U the solution set of the constraints (8.9.46) - (8.9.48). The following lemma is obviously:

Lemma 8.9.1. *Let Y be a bounded set. If $(u', t') \in U$ then $t' > 0$.*

We have the following relation between the optimal solutions of the GPF and GPL problems.

Theorem 8.9.10. *Let Y be a bounded non-empty set and suppose that condition (8.9.44) holds. If (x', u', t') is an optimal solution of the problem GPL, then $\left(x', \dfrac{u'}{t'}\right)$ is an optimal solution of the problem GPF and $V = V'$.*

Proof. Since (x', u', t') is an optimal solution of the problem GPL, there exist $C' \in \mathfrak{C}$ and $A' \in \mathring{A}$ such that $x' \in S(A')$ and

$$H(C', x', u', t') = V' \tag{8.9.49}$$

$$H(C', x', u', t') = \min_{(u,t) \in U} H(C', x', u, t) \tag{8.9.50}$$

From Lemma 8.9.1., it follows that $t' > 0$. Then it is sufficient to show that $y' = u'/t'$ is an optimal solution of the problem:

GP. $\min_{y \in Y} F(C', x', y)$.

Suppose that y' is not an optimal solution of GP, i.e., there exists $y'' \in Y$ such that:

$$F(C', x', y'') < F(C', x', y') \tag{8.9.51}$$

Let's denote:

$$u'' = y''/G(y'') \text{ and } t'' = 1/G(y''). \tag{8.9.52}$$

But (x', u'', t'') is a feasible solution of GPL, such that, from (8.9.47), (8.9.51), (8.9.52) we have

$$F(C', x', y'') = H(C', x', u'', t'') < F(C', x', y') =$$
$$= \frac{x'C'y' + dx' + ey' + w}{G(y')} + f(x') =$$
$$= \frac{x'C'u' + t'dx' + eu' + wt'}{\min_{k \in I} (h_k u' + t's_k)} + f(x') \leq H(C', x', u', t')$$

which contradicts the optimality of (x', u', t').

The equality $V = V'$ results from the fact that $F(C, x, y) = H(C, x, u, t)$ for every $t > 0$ and $y = u/t$.

In order to solve the problem GLP, we associate to this problem the following r max-min problem with linear separate constraints:

GPL(k). Find $V_k = \max_{x, C, A} \min_{u, t} H(C, x, u, t)$

subject to the constraints (8.9.39), (8.9.40), (8.9.46), (8.9.48), and

$$h_k u + s_k t \geq 1 \tag{8.9.53}$$

For every $k \in I$, we denote by U_k the solution set of the constraints (8.9.46), (8.9.48) and (8.9.53). If the solution set U_K is empty we take $V_k = -\infty$.

Theorem 8.9.11. *If the assumptions (H1) and (H2) hold then*

$$V = \max_{k \in I} V_k$$

and for every $k \in I$, *such that* $V = V_k$, *the optimal solution* (x^k, u^k, t^k) *of problem GPL(k) is an optimal solution of problem GPL.*

Proof. The proof results from the fact that $U = \bigcup_{k=1}^{r} U_k$.

Now, from the duality property of linear programming, we can associate to each problem GPL(k), ($k \in I$) for which $U_k \neq \varnothing$, the following generalized programming problem:

PL(k). Find $V_k' = \max_{x, C, A, z, v_k} (f(x) + v_k)$

subject to (8.9.39), (8.9.40), and

$$-zB + v_k h_k - Cx \leq e \tag{8.9.54}$$

$$zb + v_k s_k - dx \leq w \tag{8.9.55}$$

$$z \geq \mathbf{0}, \quad v_k \geq 0 \tag{8.9.56}$$

where $z = (z_1, ..., z_q) \in R^q$ and v_k are dual variables.

Lemma 8.9.2. *If* $U_k \neq \varnothing$, *and if* (x', z', v_k') *is an optimal solution for PL(k), then there exists* $(u', t') \in U_k$ *such that* (x', u', t') *is an optimal solution for GPL(k), and* $V_k = V_k'$.

Proof. This Lemma is an immediate consequence of the equality

$$Q(x, C, A) = Q'(x, C, A), \quad \forall\, (x, C, A) \in X \times \mathfrak{C} \times \text{Å},$$

between the optimal values of the following dual linear programming problem:

P_k. Find $Q(x, C, A) = \min_{(u, t) \in U_k} (xCu + tdx + tw + eu + f(x))$

D_k. Find $Q'(x, C, A) = \max_{z, v_k} (v_k + f(x))$

subject to (8.9.54) - (8.9.56).

The following result follows from Theorems 8.9.10 and 8.9.11 and Lemma 8.9.2.

Theorem 8.9.12. *Let suppose that (H1) and (H2) hold and that* (x', z', v_k') *is an optimal solution of the problem PL(k) for which* $V_k = \max_{i \in I} V_i$. *Then, there exists* $(u', t') \in U_k$ *such that* $\left(x', \dfrac{u'}{t'}\right)$ *is an optimal solution of the problem GPF.*

Hence, from Theorem 8.9.12, it follows that problem GPF can be reduced to the solving of the generalized programming problems PL(k). Moreover, if the function f is linear and if K_j, $(j \in J)$ are polyhedral non-empty sets, then PL(k) are generalized linear programming problems.

8.10. MULTI-OBJECTIVE FUZZY LINEAR FRACTIONAL PROGRAMMING

Generalization of the linear fractional programming to the fuzzy case has been made for case of multi-objective function. Consider the following fuzzy multi-objective linear fractional programming (FMOLFP) [767].

$$\min z(x, \tilde{c}, \tilde{d}) = (z_1(x, \tilde{c}_1, \tilde{d}_1), \ldots, z_k(x, \tilde{c}_k, \tilde{d}_k)) \tag{8.10.1}$$

subject to

$$x \in X(\tilde{a}, \tilde{b}) = \{ x \in R^n \mid \tilde{a}_j x \leq \tilde{b}_j, j = 1, \ldots, m; x \geq 0 \} \tag{8.10.2}$$

where

$$z_i(x, \tilde{c}_i, \tilde{d}_i) = p_i(x, \tilde{c}_i) / q_i(x, \tilde{d}_i) =$$

$$= (\tilde{c}_{i1}x_1 + \ldots + \tilde{c}_{in}x_n + \tilde{c}_{i0}) / (\tilde{d}_{i1}x_1 + \ldots + \tilde{d}_{in}x_n + \tilde{d}_{i0})$$

$$\tilde{c}_i = (\tilde{c}_{i1}, \ldots, \tilde{c}_{in}, \tilde{c}_{i0}), \quad \tilde{d}_i = (\tilde{d}_{i1}, \ldots, \tilde{d}_{in}, \tilde{d}_{i0}), \quad \tilde{a}_j = (\tilde{a}_{j1}, \ldots, \tilde{a}_{jn})$$

and $\tilde{b}_j$ represent fuzzy parameters. Fuzzy parameters are characterized as the fuzzy numbers introduced by D. Dubois and H. Prade [*Fuzzy Sets and Systems: Theory and Applications*, Academic Press, 1980]. A real fuzzy number $\tilde{p}$ is a convex continuous fuzzy subset of real line with membership function $\mu_{\tilde{p}}(p) : R \to [0, 1]$ (continuous mapping) (Figure 8.10.1)

$$\mu_{\tilde{p}}(p) = \begin{cases} 0 & p \in (-\infty, p_1] \\ \text{strictly increasing} & p \in [p_1, p_2] \\ 1 & p \in [p_2, p_3] \\ \text{strictly decreasing} & p \in [p_3, p_4] \\ 0 & p \in [p_4, \infty] \end{cases}$$

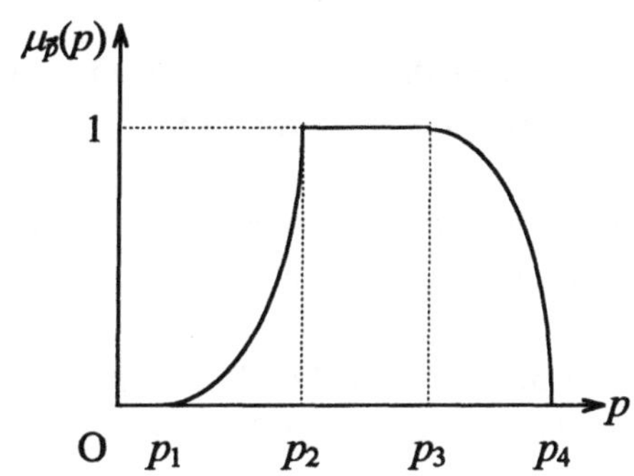

Figure 8.10.1

Let $\mu_{\tilde{c}_i}(c_i)$, $\mu_{\tilde{d}_i}(d_i)$, $\mu_{\tilde{a}_j}(a_j)$ and $\mu_{\tilde{b}_j}(b_j)$ be the membership functions of the fuzzy number vectors $\tilde{c}_i$, $\tilde{d}_i$, $\tilde{a}_j$ and $\tilde{b}_j$.

Let also,

$$c = (c_1, ..., c_k), \quad \tilde{c} = (\tilde{c}_1, ..., \tilde{c}_k), \quad d = (d_1, ..., d_k), \quad \tilde{d} = (\tilde{d}_1, ..., \tilde{d}_k)$$
$$a = (a_1, ..., a_m), \quad \tilde{a} = (\tilde{a}_1, ..., \tilde{a}_m), \quad b = (b_1, ..., b_m), \quad \tilde{b} = (\tilde{b}_1, ..., \tilde{b}_m)$$
$$\mu_{\tilde{c}}(c) = (\mu_{\tilde{c}_1}(c_1), ..., \mu_{\tilde{c}_k}(c_k)), \quad \mu_{\tilde{d}}(d) = (\mu_{\tilde{d}_1}(d_1), ..., \mu_{\tilde{d}_k}(d_k))$$
$$\mu_{\tilde{a}}(a) = (\mu_{\tilde{a}_1}(a_1), ..., \mu_{\tilde{a}_m}(a_m)), \quad \mu_{\tilde{b}}(b) = (\mu_{\tilde{b}_1}(b_1), ..., \mu_{\tilde{b}_m}(b_m))$$

Definition 8.10.1. The α-level set of the fuzzy numbers $\tilde{a}$, $\tilde{b}$, $\tilde{c}$ and $\tilde{d}$ is defined as the ordinary set $L_\alpha(\tilde{a}, \tilde{b}, \tilde{c}, \tilde{d})$ for which the degree of their membership functions exceeds the level α:

$$L_\alpha(\tilde{a}, \tilde{b}, \tilde{c}, \tilde{d}) = \{(a, b, c, d) \mid \mu_{\tilde{a}}(a) \geq \alpha, \quad \mu_{\tilde{b}}(b) \geq \alpha, \quad \mu_{\tilde{c}}(c) \geq \alpha, \quad \mu_{\tilde{d}}(d) \geq \alpha\}$$

For a fixed value of α, problem (8.10.1)-(8.10.2) is a nonfuzzy α-multi-objective linear fractional programming (α-MOLFP) problem

$$\min z(x, c, d) = (z_1(x, c_1, d_1), ..., z_k(x, c_k, d_k)) \tag{8.10.3}$$

subject to

$$x \in X(a, b) = \{ x \in R^n \mid a_j x \leq b_j \ , \ j = 1, ..., m, \quad x \geq \mathbf{0} \}$$
$$(a, b, c, d) \in L_\alpha(\tilde{a}, \tilde{b}, \tilde{c}, \tilde{d}) \tag{8.10.4}$$

where

$$z_i(x, c_i, d_i) = p_i(x, c_i) \; / \; q_i(x, d_i) =$$
$$= (c_{i1}x_1 + ... + c_{in}x_n + c_{i0}) \; / \; (d_{i1}x_1 + ... + d_{in}x_n + d_{i0})$$

In problem (8.10.3) - (8.10.4) the parameters (a,b,c,d) are decision variables. Assume that $q_i(x, d_i) > 0$ for all **x**.

Now, we present the concept of α-Pareto optimality which extend the ordinary Pareto optimality (Definition 6.1.1) on the basis of the α-level sets of the fuzzy numbers.

Definition 8.10.2 [767]. The point $x^* \in X(a, b)$ is said to be an α-Pareto optimal solution to the α-MOLFP, if and only if there does not exist another $x \in X(a, b)$, $(a, b, c, d) \in L_\alpha(\tilde{a}, \tilde{b}, \tilde{c}, \tilde{d})$ such that $z_i(x, c_i, d_i) \leq z_i(x^*, c_i^*, d_i^*)$, $i = 1, ..., k$ with strict inequality holding for at least one i, where the corresponding values of parameters (a^*, b^*, c^*, d^*) are called α-**level optimal parameters**.

Let $X^p(\alpha)$ denotes the set of α-Pareto optimal solutions for the fixed level α. It is easy to verify that if $\alpha^1 > \alpha^2$, then for any $x^1 \in X^p(\alpha^1)$, exists $x^2 \in X^p(\alpha^2)$ such that $z(x^1, c, d) \geq z(x^2, c, d)$. In order to find an α-Pareto optimal solution, Sakawa and Yano [767] propose an interactive algorithm. The decision-maker is asked to give the degree α and the reference levels $\bar{z}_i$ of the objective functions. Then, the α-Pareto optimal solution can be obtained by solving the following min-max problem

$$\min \max_{1 \leq i \leq k} (z_i(x, c_i, d_i) - \bar{z}_i) \tag{8.10.5}$$

subject to (8.10.4).

This problem is equivalent to

$$\min_{x,v,a,b,c,d} v \tag{8.10.6}$$

subject to (8.10.4) and

$$z_i(x, c_i, d_i) - \bar{z}_i \leq v, \quad i = 1, ..., k \tag{8.10.7}$$

Due to $q_i(x, d_i) > 0$, $(x \in X)$, the constraint (8.10.7) is equivalent to

$$p_i(x, c_i) - \bar{z}_i q_i(x, d_i) \leq v q_i(x, d_i), \quad i = 1, ..., k \tag{8.10.8}$$

We remark that the constraints (8.10.4) and (8.10.8) are nonlinear. To deal with such nonlinearlities, consider the following set-valued function $S_i(\cdot)$ and $T_i(\cdot)$.

$$S_i(x, v, c_i, d_i) = \{ (x, v) \mid p_i(x, c_i) - \bar{z}_i q_i(x, d_i) \leq v q_i(x, d_i) \} \tag{8.10.9}$$

$$T_j(x, a_j, b_j) = \{ x \mid a_j x \leq b_j \} \tag{8.10.10}$$

For $x \geq \mathbf{0}$, these set-valued functions have the following properties

a) $S_i(x, v, c_i^1, d_i) \supset S_i(x, v, c_i^2, d_i)$ if $c_i^1 \leq c_i^2$

b) $S_i(x, v, c_i, d_i^1) \subset S_i(x, v, c_i, d_i^2)$ if $d_i^1 \leq d_i^2$

c) $T_j(x, a_j^1, b_j) \supset T_j(x, a_j^2, b_j)$ if $a_j^1 \leq a_j^2$

d) $T_j(x, a_j, b_j^1) \subset T_j(x, a_j, b_j^2)$ if $b_j^1 \leq b_j^2$

Also, from the properties of the α-level for the fuzzy numbers $\tilde{c}_i$, $\tilde{d}_i$, $\tilde{a}_j$ and $\tilde{b}_j$, it follows that the feasible regions for c_i, d_i, a_j and b_j can be considered the intervals $[c_{i\alpha}^L, c_{i\alpha}^R]$, $[d_{i\alpha}^L, d_{i\alpha}^R]$, $[a_{j\alpha}^L, a_{j\alpha}^R]$ and $[b_{j\alpha}^L, c_{j\alpha}^R]$. Thus, the problem (8.10.4), (8.10.6) and (8.10.7) is equivalent to

$$\min_{v,\, x \geq 0} v \tag{8.10.11}$$

subject to

$$p_i(x, c_{i\alpha}^L) \leq (\bar{z}_i + v) q_i(x, d_{i\alpha}^R), \quad i = 1, ..., k \tag{8.10.12}$$

$$a_{j\alpha}^L x \leq b_{j\alpha}^R, \quad j = 1, ..., m \tag{8.10.13}$$

If the value of v is fixed, the problem (8.10.11)-(8.10.12) reduces to a set of linear inequalities. Thus, the finding of the optimal solution v^* is equivalent to determining the minimum value of v such that there exists an admissible set satisfying the constraints (8.10.12) - (8.10.13).

Let

$$V_{\min} = \min_{1 \le i \le k} \left\{ \min_{x \in X(a_{j\alpha}^L, b_{j\alpha}^R)} (p_i(x, c_{i\alpha}^L) / q_i(x, d_{i\alpha}^R)) - \bar{z}_i \right\} \tag{8.10.14}$$

$$V_{\max} = \max_{1 \le i \le k} \left\{ \min_{x \in X(a_{j\alpha}^L, b_{j\alpha}^R)} (p_i(x, c_{i\alpha}^L) / q_i(x, d_{i\alpha}^R) - \bar{z}_i \right\} \tag{8.10.15}$$

Due to v satisfying $V_{\min} \le v \le V_{\max}$, Sakawa and Yano [767] proposed a method for solving problem (8.10.11) - (8.10.13) by the combined use of the bisection method and of the linear programming method.

Step 1: Set $v = V_{\min}$. Does an admissible set satisfying the constraints (8.10.12) - (8.10.13) exists?

If YES, then set $v^* = V_{\min}$

If NO, set $v^L = V_{\min}$, $v^R = V_{\max}$ and go to the next step.

Step 2: For the value $v = (v^L + v^R)/2$, update the value of v^L or v^R using the bisection method as follows:

$v^R = v$ if admissible set exists for v,

$v^L = v$ if no admissible set exists for v.

The decision maker selects an appropriate starting objective from among the objectives $z_i(x)$. Let it be $z_1(x)$. The following linear fractional programming problem is solved

$$\min_{x \ge 0} \; p_1(x, c_{i\alpha}^L) / q_1(x, d_{i\alpha}^R) \tag{8.10.16}$$

subject to (8.10.12) (with v^* instead of v) and (8.10.13).
The link between the optimal solutions to (8.10.11) - (8.10.13) and the α-Pareto optimal concept of the α-MOLFP is established through the following theorem:

Theorem 8.10.1. *If x^* is an unique optimal solution to problem (8.10.11) - (8.10.13), then x^* is an α-Pareto optimal solution to the α-MOLFP.*

Proof. Assume that x^* is an unique optimal solution to problem (8.10.11) - (8.10.13), but x^* is not an α-Pareto optimal solution to the α-MOLFP. Then there must exist $x \in X(a, b)$ and $(a, b, c, d) \in L_\alpha(\tilde{a}, \tilde{b}, \tilde{c}, \tilde{d})$ such that

$$z(x, c, d) \le z(x^*, c_\alpha^L, d_\alpha^R)$$

On the other hand,

$$\max_{1 \le i \le k} (z_i(x, c_{i\alpha}^L, d_{i\alpha}^R) - \bar{z}_i) \le \max_{1 \le i \le k} (z_i(x, c_i, d_i) - \bar{z}_i) \le$$
$$\le \max_{1 \le i \le k} (z_i(x^*, c_{i\alpha}^L, d_{i\alpha}^R) - \bar{z}_i)$$

which contradicts the fact that x^* is an unique optimal solution to (8.10.11) -(8.10.13). Hence, x^* must be an α-Pareto optimal solution to the α-MOLFP.

Theorem 8.10.2. *If x^* is an α-Pareto optimal solution and (a^*, b^*, c^*, d^*) is an α-level optimal parameter to the α-MOLFP, then x^* is an optimal solution to (8.10.11) - (8.10.13) for some $\bar{z} = (\bar{z}_1, ..., \bar{z}_k)$.*

Proof. Assume by *reductio ad absurdum* that x^* is not an optimal solution to (8.10.11) - (8.10.13) for any $\bar{z}$ such that

$$z_1(x^*, c_1^*, d_1^*) - \bar{z}_1 = ... = z_n(x^*, c_k^*, d_k^*) - \bar{z}_k = v^k.$$

Then there exist $x \in X(a, b)$ and $(a, b, c, d) \in L_\alpha(\tilde{a}, \tilde{b}, \tilde{c}, \tilde{d})$ such that

$$z_i(x^*, c_i^*, d_i^*) - \bar{z}_i > z_i(x, c_i, d_i) - \bar{z}_i \ge z_i(x, c_{i\alpha}^L, d_{i\alpha}^R) - \bar{z}_i$$

i.e.

$$z_i(x, c_{i\alpha}^L, d_{i\alpha}^R) < z_i(x^*, c_i^*, d_i^*)$$

which contradicts the fact that x^* is an α-Pareto optimal solution. The proof is complete.

From Theorem 8.10.1 it follows that x^* is an α-Pareto optimal if x^* is an unique solution to problem (8.10.11) - (8.10.13) which is not easy to check numerically. Due to this fact, Sakawa and Yano [767] proposed an optimality test for a current optimal solution x^*. We solve the linear programming problem

$$\bar{w} = \max \sum_{i=1}^{k} \varepsilon_i \tag{8.10.17}$$

subject to (8.10.13), $x \ge 0$ and

$$p_i(x, c_{i\alpha}^L) + \varepsilon_i = z_i(x^*, c_{i\alpha}^L, d_{i\alpha}^R) q_i(x, d_{i\alpha}^R),\ \varepsilon_i \ge 0,\ i = 1, ..., k$$

Denote by $\bar{x}$ and $\bar{\varepsilon}$ the optimal solution of this problem. If $\bar{w} = 0$, then x^* is an α-Pareto optimal solution. If $\bar{w} > 0$ then at least one $\varepsilon_i > 0$. Perform the following operations:

Step 1. For any j such that $\varepsilon_j > 0$ solve the problem

$$\min_{x \ge 0} z_j(x, c_{j\alpha}^L, d_{j\alpha}^R) \tag{8.10.18}$$

subject to (8.10.13) and

$$z_i(x, c_{i\alpha}^L, d_{i\alpha}^R) = z_i(\bar{x}, c_{i\alpha}^L, d_{i\alpha}^R),\quad i \in \{ j \mid \bar{\varepsilon}_j = 0 \}$$
$$z_i(x, c_{i\alpha}^L, d_{i\alpha}^R) \le z_i(\bar{x}, c_{i\alpha}^L, d_{i\alpha}^R),\quad i \in \{j \mid \bar{\varepsilon}_j > 0\}$$

Step 2. Test the α-Pareto optimality for the solution to (8.10.18)

Step 3. If $\bar{w} = 0$, stop. In the contrary case, return to step 1. Repeating this process at least $k - 1$ iterations, an α-Pareto optimal solution can be obtained. Having the α-Pareto optimal solution corresponding to the degree α and the reference levels $\bar{z}_i$, the decision maker has to answer to whether or not he/she is satisfied with the current solution. If the decision maker is satisfied with the current solution, stop. Otherwise, the decision maker must update his/her reference levels and/or the degree α. To this purpose, the decision maker is supplied with trade-off rates between the objective functions and the degree α. Finally a satisfactory solution for the decision maker is derived from an α-Pareto optimal solution set.

8.11. HISTORICAL NOTES. COMMENTARIES.

The presentation of Section 8.1 followed reference [961]. Bansal [105] reconsidered the study of problem (8.1.1)-(8.1.2) and showed that the optimum solution, assumed to be non-zero, is found on the boundary of the feasible solutions domain. Bansal applied the Simplex algorithm to obtain a local optimal solution which is successively improved using cuts until the optimal solution is obtained. Saxena, Patkar, and Parkash [785] considered a similar problem

$$\max \{Z = \sum_{i=1}^{n} c_i'|x_i| + c_0 + (\sum_{i=1}^{n} d_i'|x_i| + d_0)/(\sum_{i=1}^{n} e_i'|x_i| + e_0) \mid$$

$$\sum_{i=1}^{n} a_{ji}x_i = b_j\,,\ j = 1, \dots, m\,;\ x_i \text{ without sign restriction } \}.$$

and showed that one can apply the Simplex algorithm only in two cases. The model (8.2.8)-(8.2.9) from Section 8.2 was studied by Jagannathan [511]. Other models of separable fractional programming were also studied in literature. Gogia [414] considered the problem

$$\max \{F(x) = (\sum_{j=1}^{n} f_j(x_j))/(\sum_{j=1}^{n} g_j(x_j)) \mid \sum_{j=1}^{n} h_{ij}(x_j) \le b_i,$$
$$i = 1, \dots, m,\quad x_j \ge 0,\quad j = 1, \dots, n\}.$$

The functions f_j, g_j and h_{ij}, assumed to be continuous, are approximated through piece-wise linear functions following the technique described by Miller (*op.cit.*). The initial problem becomes a linear fractional programming problem whose optimal solution constitutes an approximate solution of the initial problem.

Arora and Aggarwal [74] considered the maximization problem of a sum of linear concave (convex) fractional and separable functions

$$\max \{ V_s = \sum_{j=1}^{s} v_j/(c_j'X_j + \alpha_j) \mid \sum_{j=1}^{n} B_jX_j = \bar{b},\ B_jX_j = b_j\,,\ X_j \ge 0, j = 1, \dots, s \}$$

Černov and Lange [306] gave an approximate method for the solution of transport problems with fractional and separable objective function:

$$\min \{F(x) = \sum_{i=1}^{m}\sum_{j=1}^{n} f_{ij}(x_{ij})/\sum_{i=1}^{m}\sum_{j=1}^{n} g_{ij}(x_{ij}) \mid \sum_{j=1}^{n} x_{ij} = a_i \; ; \sum_{i=1}^{m} x_{ij} = b_j, \; \alpha_{ij} \le x_{ij} \le \beta_{ij}\}$$

The functions f_{ij} and g_{ij} are linearized, and the new problem has the form

$$\min \{ \; (\sum\sum c'_{\mu\nu} y_{\mu\nu} + c'_0)/(\sum\sum c''_{\mu\nu} y_{\mu\nu} + c''_0) \mid \sum_{\nu} y_{\mu\nu} = A_\mu \; ,$$

$$\sum_{\mu} y_{\mu\nu} = B_\nu \; , \; y_{\mu\nu} \ge 0 \; \}$$

and $y_{\mu\nu}$ must satisfy certain supplementary conditions.
Černov [302] considered a separable fractional programming problem similar to that studied by Gogia [414], with the difference that the objective function is of minimum type, and the constraints also contain limits on the variables. Thus, the problem has the form

$$\min F(x) = \sum_{j=1}^{n} f_j(x_j) \; / \sum_{j=1}^{n} g_j(x_j) \; ,$$

$$x \in \{ \; x \mid \sum_{j=1}^{n} h_{ij}(x_j) \le b_i \; , \; i = 1 \; , \dots , n \; ; \; \alpha_j \le x_j \le \beta_j \; , \; j = 1 \; , \dots , n \; \}$$

The solution method is the same, i.e., all the separable functions are approximated with linear functions which contain a greater number of variables, and in this way the problem is reduced to a conventional linear fractional programming problem.

The results presented in Section 8.3 were obtained by Patkar and Stancu-Minasian [719], [721]. Reference [720] presents an iterative algorithm for the solution of the linear fractional programming problem with a certain type of disjunctive constraints. The results presented in Section 8.5, 8.7, 8.8 were obtained by Stancu-Minasian and Ţigan [920], [922], [924], [926], [928], [931] and the results from Section 8.6 were obtained by Stancu-Minasian [914]. In references [1010], [1011], [1012], Stancu-Minasian and Ţigan (see also reference [932]) considered the max-min bicriterion fractional problem with separable polyhedral constraints.

$$\text{P(max-min)}: \; V = \max_{x \in X} \; \min_{y \in Y} \; h(F(x), \; Q(x, \; y))$$

where $X \subset R^n$ and $Y \subset R^m$ are two given non-empty compact convex polyhedral sets, $h : D \to R \;\; (D \subset R^2)$ is a real-valued function, nondecreasing and continuous on the set:

$$D = \left\{ \; (F(x), \; Q(x, \; y)) \; : \; x \in X, \; y \in Y \right\},$$

and $F : X \to R$, $Q : X \times Y \to R$ are fractional functions. In references [1010], [1011] the functions F and Q have the form $F(x) = \dfrac{cx + p}{fx + q}$, $x \in X$ and

$Q(x, \; y) = \dfrac{xCy + ey + gx}{dx + r}$, $x \in X$, $y \in Y$, whereas in reference [1012],

$Q(x, y) = \dfrac{xCy + ey + gx}{dy + r}$, and in reference [932], $Q(x, y) = \dfrac{xCy + ey + gx + \gamma}{(\alpha x + \beta)(dy + r)}$. It is shown that the problem P(max-min) can be transformed into a bicriterion fractional maximization problem for which is given an algorithm based on the Warburton's parametrical procedure [1033].

The particular case of problem P(max-min), when $h(F(x), Q(x,y)) = \max\ (F(x), Q(x,y))$ was studied by Belenkii [163]. References [323], [917] and [1004] deal with nonlinear fractional or quasimonotonic max-min problems with linked constraints.
Further extensions to max-min problems under separate linear constraints having nonlinear objective functions have been studied by Cimoca and Ţigan [320] and Ţigan [1001].

The results presented in Section 8.9 a) - c) are due to Stancu-Minasian and Ţigan [925], [927], [929], [930], [931], [1009]. In reference [550], Kaur presents a method for solving the inexact linear fractional problem

$$\sup \left\{ \inf_{\substack{c \in C \\ d \in D}} \frac{c'x}{d'x} \mid x_1K_1 + \dots + x_nK_n \subseteq K,\ x_j \geq 0 \right\},$$

where C is a closed convex set in R^n, D and K_j are compact convex sets in R^n, and K is a closed convex set in R^m defined by $K = \{ y \in R^m \mid By \leq b \}$, where B is a non-negative $r \times m$ matrix, $d'x > 0$ for every feasible solution x and for all $d \in D$ and $c'x \geq 0$ for every feasible solution x and for all $c \in C$.

Singh [854] considered the problem

$$\sup \{ \inf_{(a,\, c) \in A \times C} \frac{c'x + a}{d'x + e} \mid x_1K_1 + \dots + x_nK_n \subseteq K(b),\ x \geq 0 \}.$$

which through the change of variable $z_0 = \dfrac{1}{d'x + e}$, $z = z_0x$ ($\forall\ x \in R^n$) transforms into a non-fractional problem to which one can apply the results existing in the literature, including duality results. Kaul, Kaur and Lyall [547] considered the uncertainty of coefficients d and e.

$$\sup \{ \inf_{\substack{(a,\, c) \in A \times C \\ (d,\, e) \in D \times E}} \frac{c'x + a}{d'x + e} \mid x_1K_1 + \dots + x_nK_n \subseteq K_0\ ,\ x_j \geq 0 \}.$$

The Dual of this problem is constructed, and the duality results are given. The same problem is considered by Xu [1059] with the difference that $x \in X = \{ x \in R^n \mid Ax \leq b,\ x \geq 0 \}$. Two dual problems are presented together with strong duality theorems and a saddle-point optimality criterion. Kaur and Bhatia [552] constructed the Dual of the generalized linear fractional programming problem (8.9.6)-(8.9.7) and showed that this Dual has the form of a linear programming problem with inexact data. Generalized linear fractional programming problem was considered by

Chadha [234], Kaul and Bhatia [541], Patkar *et al.* [714], Lyall and Kaur [617] and Wadhwa [126].

The results presented in Section 8.10 are due to Sakawa and Yano [767]. Other works are also given in the literature in which fuzzy approaches were used for solving multiple objective linear fractional programming (MOLFP).
In a previous paper [771], Sakawa and Yumine presented a fuzzy approach for solving MOLFP by combined use of the bisection method and the linear programming method. Five types of membership functions (linear, exponential, hyperbolic, hyperbolic inverse and piece-wise linear functions) were proposed to quantify the fuzzy or imprecise goals of the decision-maker. In reference [769] Sakawa and Yano presented a new interactive fuzzy decision method for solving MOLFP problem by assuming that the decision maker has fuzzy goals for each of the objective functions. The concept of M-Pareto optimal solution is introduced and an interactive decision making technique is proposed to choose satisfactory solutions for the decision maker among a set of M-Pareto optimal solutions. In reference [1066], Yano and Sakawa treat two types of fuzzy goals (called fuzzy equal and fuzzy min or fuzzy max) and define M-α-Pareto optimal solution. They present a new interactive fuzzy decision making method for obtaining the solution among an M-α-Pareto optimal solution set. In reference [770], Sakawa and Yano introduce I-α-Pareto optimal solutions as new solution concepts taking account of two types of uncertainties and propose an interactive decision making technique to choose satisfactory solutions for the decision maker among a set of I-α-Pareto optimal solutions. For other questions, see also references [768], [772], [773], [774]. Sakawa *et al.* [774] extend the ordinary Pareto optimality concepts based on the concepts of possibility and necessity for fuzzy numbers. Using four indices for ranking two fuzzy numbers, they define four types of Pareto optimality for MOLFP having fuzzy parameters and examine the relationships among them. Luhandjula [614] uses a linguistic variable approach to solve MOLFP. Imprecise aspirations of the decision-maker are represented by structured linguistic variables. The aggregation of membership functions is done with a compensatory operator which does not guarantee the efficiency of the optimal solution. Dutta *et al.* [373] modified the linguistic approach of Luhandjula [614] such as to obtain efficient solutions of MOLFP. In reference [371], Dutta *et al.* performed sensitivity analysis of fuzzy linear fractional programming problem (FLFPP) where some of the constraints are deterministic (soft) and some are fuzzy, and also the objective functions are fuzzy. The problem (FLFPP) is converted into an equivalent fuzzy linear programming problem (FLP). The sensitivity analysis of the FLFPP is obtained by the sensitivity analysis of the problem (FLP).
Slowinski [867], [868] solved a fuzzy MOLF problem using an interactive technique involving a linear programming procedure in the calculation phase. Lee and Tcha [603] proposed a fuzzy set theoretic approach, based on the observation that a fuzzified MOLFP problem can be reduced to a sequence of linear inequality problems. Lee *et al.* [604] established a duality for the fuzzy MOLFP problem *via* Farkas' lemma, based on the works of Jagannathan and Schaible [515], [516], and which satisfies the nice strong duality relations. Exploiting the structural symmetry between the primal and dual problems, they proposed a parallel algorithm which runs in a dual-parallel processor system.

CHAPTER 9

INTEGER AND MIXED INTEGER LINEAR FRACTIONAL PROGRAMMING

Some of the problems mentioned in Chapter 1 required that either part of the variables, or all of them take integer values. This chapter will study such problems. In particular, we will address the bivalent programming in which part of the variables or all of them can take only values 0 or 1 (Section 9.1). Distinction will be made between integer fractional programming where all the variables must be integers, and mixed integer fractional programming, where only part of the variables need be integers. To solve these problems we will try to adapt some of the methods presented in Chapter 3 for fractional programming with continuous variables, e.g., methods based on the Simplex algorithm, parametric methods, and the method of the change of variable. Additionally, we will also present branch and bound methods, and methods using the cutting planes.

9.1. LINEAR FRACTIONAL PROGRAMMING WITH BIVALENT VARIABLES

Let's consider first the case without constraints. Consider the problem of minimizing the function

$$F(x_1 , \dots , x_n) = \frac{\sum_{i=1}^{n} a_i x_i + a_0}{\sum_{i=1}^{n} b_i x_i + b_0} , \tag{9.1.1}$$

$$x_i \in \{0, 1\} \tag{9.1.2}$$

where $a_i \geq 0$, $b_i \geq 0$, $a_i + b_i > 0$, $i = 1 , \dots , n$. If $b_i = 0$, then the condition $a_i + b_i > 0$ implies $a_i > 0$ and in the optimal solution of problem (9.1.1)-(9.1.2) we get $x_i = 0$. Hence, without restricting the generality, we can assume that $b_i > 0$. A vector $X^* = (x_1^* , \dots , x_n^*)$ is a local minimum of the function F, if for any point $X = (x_1 , \dots , x_n)$ neighboring X^* (*i.e.* $\sum_{i=1}^{n} |x_i - x_i^*| = 1$), one gets $F(X^*) \leq F(X)$. Any point of local minimum of the function F is also point of global minimum [465]. The algorithm presented next for solving the problem (9.1.1)-(9.1.2) is extremely simple and consists in exploring once the ratios $\{a_i/b_i\}$. Fie $N = \{1, 2, \ldots, n\}$ and $I \subseteq N$. For a certain solution X denote by a_{00}, b_{00} the value of the numerator and denominator of F, and put $RI = \dfrac{a_{00}}{b_{00}}$. The steps of the algorithm are the following :

1) Consider $I = \varnothing$ and $j := 0$

2) Consider $j := j + 1$. If $j = n + 1$, go to Step 7, else continue with Step 3.

3) If

$$RI(= a_j/b_j) > \frac{a_0 + \sum_{i\in I} a_i}{b_0 + \sum_{i\in I} b_i} = \left[\frac{a_{00}}{b_{00}} = RI\right],$$

go to Step 2, else continue with next step.

4) Consider

$$I := I \cup \{j\}, \quad RI = \frac{a_{00} + a_j}{b_{00} + b_j}$$

and go to Step 5.

5) Consider

$$K = \{\, l \in I \mid RI < a_l/b_l \,\}$$

If $K=\varnothing$, go to Step 2, and if $K\neq\varnothing$, continue with Step 6.

6) Consider

$$RI = \frac{a_{00} - \sum_{l\in K} a_l}{b_{00} - \sum_{l\in K} b_l}, \quad I := I - K$$

and go to Step 5.

7) The optimal solution is

$$X^* = (x_1^*, \ldots, x_n^*) \text{ with } x_i^* = \begin{cases} 1, & i \in I \\ 0, & i \in N - I \end{cases}$$

and the value of the objective function is RI.

We shall prove that the vector X^* obtained through the above algorithm is optimal solution of problem (9.1.1)-(9.1.2). In the proof we use the following obvious result:

Lemma 9.1.1. *Consider the fractions a_1/b_1 and a_2/b_2 with a_i, $b_i > 0$, $i = 1, 2$. If $a_1b_2 - a_2b_1 \neq 0$, then*

$$\min\,[a_1/b_1\,,\ a_2/b_2] < (a_1 + a_2)/(b_1 + b_2) < \max\,[a_1/b_1\,,\ a_2/b_2].$$

Note that, at the end of the algorithm, one get the set of indices I representing those indices of fractions a_j/b_j that satisfy

$$\frac{a_j}{b_j} \le \frac{\sum_{i\in I} a_i + a_0}{\sum_{i\in I} b_i + b_0}, \quad j \in I \tag{9.1.3}$$

and

$$\frac{a_j}{b_j} > \frac{\sum_{i\in I} a_i + a_0}{\sum_{i\in I} b_i + b_0}, \quad j \in N - I \tag{9.1.4}$$

To show that X^* obtained through this algorithm is optimal solution of problem (9.1.1)-(9.1.2), it is sufficient to show that for any other point X^{**} neighboring X^*, we have $F(X^{**}) > F(X^*)$. Assume that X^{**} is obtained from X^* by replacing x_i^* with $\bar{x}_i^*$. Using Lemma 9.1.1, we get

$$\frac{\sum_{i\in I} a_i + a_j + a_0}{\sum_{i\in I} b_i + b_j + b_0} > \frac{\sum_{i\in I} a_i + a_0}{\sum_{i\in I} b_i + b_0} = F(X^*), \quad j \in N - I \tag{9.1.5}$$

and

$$F(X^*) = \frac{\sum_{i\in I} a_i - a_j + a_j + a_0}{\sum_{i\in I} b_i - b_j + b_j + b_0} \le \max\left[\frac{a_j}{b_j}, \frac{\sum_{i\in I} a_i - a_j + a_0}{\sum_{i\in I} b_i - b_j + b_0}\right], \quad j \in I. \tag{9.1.6}$$

From (9.1.3) and (9.1.6) one gets

$$F(X^*) \le \frac{\sum_{i\in I} a_i - a_j + a_0}{\sum_{i\in I} b_i - b_j + b_0}, \quad j \in I$$

Hence, X^* is a local minimum for the problem (9.1.1)-(9.1.2). Also, X^* is a global minimum of F.

Observation 1. The solution of problem (9.1.1)-(9.1.2) is not necessary unique. The above algorithm furnishes the optimal solution with the largest number of non-zero variables.

Observation 2. Let's order the fractions a_i/b_i in increasing order $a_{i_1}/b_{i_1} \le a_{i_2}/b_{i_2} \le \dots \le a_{i_n}/b_{i_n}$ and let's consider an integer k, $(1 \le k \le n)$, such that

$$\frac{\sum_{j=1}^{k} a_{i_j} + a_0}{\sum_{j=1}^{k} b_{i_j} + b_0} \le a_{i_s}/b_{i_s}, \quad s > k, \qquad \frac{\sum_{j=1}^{k} a_{i_j} + a_0}{\sum_{j=1}^{k} b_{i_j} + b_0} \ge a_{i_s}/b_{i_s}, \quad s \le k.$$

Then the vector

$$X^* = (x_1^*, \ldots, x_n^*) \quad with \quad x_{i_j}^* = \begin{cases} 1 & j \le k \\ 0 & j > k \end{cases}$$

represents the optimal solution of problem (9.1.1)-(9.1.2).

Example 9.1.1. Determine

$$X^* = (x_1^*, \ldots, x_{10}^*), \quad x_i^* \in \{0, 1\}, \quad i = 1, \ldots, 10,$$

which minimizes the function

$$F(X) = \frac{1 + 2x_1 + 3x_2 + 4x_3 + 5x_4 + 6x_5 + 3x_6 + 8x_7 + 9x_8 + 10x_9 + 11x_{10}}{2 + 11x_1 + 10x_2 + 9x_3 + 8x_4 + 7x_5 + 16x_6 + 5x_7 + 44x_8 + 3x_9 + 2x_{10}}.$$

The sequence of algorithm steps (shown in parentheses) is the following:

(1) $I = \varnothing$, $j := 0$,

(2) $j := 1$;

(3) Since $2/11 = a_1/b_1 < a_0/b_0 = 1/2$ go to Step 4.

(4) $I = \{1\}$, $RI := 3/13 = 0.23076$,(5) $K = \varnothing$ and go to Step 2.

(2) - (3) Perform steps (2) and (3) four times to eliminate the fractions 3/10; 3/9; 5/8 and 6/7 greater than 3/13. Continue with Step 2.

(2) $j := 6$, (3) Since $a_6/b_6 = 3/16 < 3/13$ go to Step 4.

(4) $I = \{1, 6\}$ $RI := 6/29 = 0.2068$, (5) $K = \varnothing$ and go to Step 2.

(2) $j := 7$, (3) Since $a_7/b_7 = 8/5 > 8/29$ go to Step 2.

(2) $j := 8$, (3) Since $a_8/b_8 = 9/44 < 6/29$ go to Step 4.

(4) $I = \{1, 6, 8\}$, $RI = 15/73 = 0.2054$, (5) $K = \varnothing$ and go to Step 2.

(2) - (3) Perform steps (2) and (3) two times to eliminate the last two fractions, since $10/3 > 15/73$ and $11/2 > 15/73$. Thus, $I = \{1, 6, 8\}$ and $RI = 17/73 = 0.2054$. The optimal solution is

$$X^* = (1\ 0\ 0\ 0\ 0\ 1\ 0\ 1\ 0\ 0).$$

Consider now the case of bivalent fractional programming with constraints

$$\min F(x_1, \ldots, x_n) = \frac{\sum_{i=1}^{n} a_i x_i + a_0}{\sum_{i=1}^{n} b_i x_i + b_0}, \tag{9.1.7}$$

subject to

$$x_i \in \{0, 1\}, \tag{9.1.8}$$

$$a_i \ge 0, \quad b_i > 0, \quad i = 1, \ldots, n, \tag{9.1.9}$$

$$f_i(x_1, \ldots, x_n) \ge F_i, \quad i = 1, \ldots, k, \tag{9.1.10}$$

where each of the pseudo-Boolean functions f_i is non-decreasing. Recall that a pseudo-Boolean function $f : B_2^n = \underbrace{B_2 \times \ldots \times B_2}_{n \text{ times}} \to R$ is non-decreasing if $(x_1, x_2, \ldots, x_n) \le (x_1', x_2', \ldots, x_n')$ implies $f(x_1, x_2, \ldots, x_n) \le f(x_1', x_2', \ldots, x_n')$.

In the space B_2^n, $(x_1, x_2, \dots, x_n) \le (x_1', x_2', \dots x_n')$ if and only if $x_i \le x_i' (i = 1, \dots, n)$ and, in the Boolean algebra $B_2 = \{0, 1\}$, $x \le y$, if and only if $xy = x (x, y \in B_2)$. Also,

$(x_1, x_2, \dots, x_n) < (x_1', x_2', \dots, x_n')$ if and only if $x_i \le x_i'$ for each i and for at least an index j we have $x_j < x_j'$.

Lemma 9.1.2. *The problem* (9.1.7)-(9.1.10) *is feasible (i.e., it has at least a feasible solution) if and only if the vector* (1, 1, ..., 1) *satisfies the constraints* (9.1.10).

Proof. If (1, ..., 1) satisfies the constraints (9.1.10), then, evidently, the problem (9.1.7)-(9.1.10) is feasible. Conversely, if the problem (9.1.7)-(9.1.10) is feasible, then exist a vector $(x_1, \dots, x_n)$ which satisfies the constraints (9.1.10). Since $(1, 1, \dots, 1) \ge (x_1, \dots, x_n)$ and f_i are non-decreasing, it follows that (1, 1, ..., 1) satisfies the constraints (9.1.10).

We now consider that, for the problem (9.1.7)-(9.1.10), (1, ..., 1) satisfies the constraints (9.1.10).

Theorem 9.1.1. *If* $X^* = (x_1^*, \dots, x_n^*)$ *is optimal solution of problem* (9.1.7)-(9.1.10) *and if* I *is the set of integers with the property that* $x_i^* = 1$, *for* $i \in I \subseteq N$ *and* $x_i^* = 0$ *for* $i \in N - I$, *then*

$$F(X^*) = \frac{\sum_{i \in I} a_i + a_0}{\sum_{i \in I} b_i + b_0} \le \frac{a_j}{b_j}, \quad j \in N - I .$$

Proof. By *reductio ad absurdum*, assume that an index $j \in N-I$ exists such that

$$F(X^*) = \frac{\sum_{i \in I} a_i + a_0}{\sum_{i \in I} b_i + b_0} > \frac{a_j}{b_j} .$$

Then using Lemma 9.1.1 yields

$$\frac{\sum_{i \in I} a_i + a_j + a_0}{\sum_{i \in I} b_i + b_j + b_0} < \max \left[\frac{\sum_{i \in I} a_i + a_0}{\sum_{i \in I} b_i + b_0}, \frac{a_j}{b_j} \right] = F(X^*). \tag{9.1.11}$$

Since the vector $X^{**} = (x_1^*, \dots, \bar{x}_j^* (=1), \dots, x_n^*) \ge X^*$ then X^{**} satisfies the constraints (9.1.10), and (9.1.11) implies that X^{**} minimizes the function F. Contradiction.

This theorem yields the following two corollaries.

Corollary 9.1.1. *If* X^0 *minimizes the function* (9.1.7) *without constraints and has the smallest number of non-zero components and* X^* *is the solution of problem* (9.1.7)-(9.1.10), *then* $X^0 \le X^*$.

Corollary 9.1.2. *If* X^1 *minimizes the function* (9.1.7) *without constraints and has the greatest number of non-zero components, and does not satisfy the constraints* (9.1.10), *then* $X^1 < X^*$.

This implies that if we found the optimal solution X^1 of function (9.1.7) without constraints, then the optimal solution X^* of the problem with constraints has just as many components of 1 as X^1, and will have additional others due to condition $X^1 < X^*$. Having the optimal solution X^1 for the function F not verifying the constraints (9.1.10), we go to problem (9.1.7)-(9.1.10) and replace with 1 the variables which have the value 1 in X^1 and obtain a problem of smaller dimension:

$$\min F(x) = \frac{\sum_{i=1}^{m} a_i x_i + a_0}{\sum_{i=1}^{m} b_i x_i + b_0} \tag{9.1.12}$$

subject to

$$a_0/b_0 < a_i/b_i\ ,\quad (i = 1\ ,\ \dots\ ,\ m)\ ;\quad a_0 \ge 0\ ,\quad b_0 > 0\ , \tag{9.1.13}$$

$$g_1(x_1\ ,\ \dots\ ,\ x_m) > G_1\ ,\ \dots\ ,\ g_k(x_1\ ,\ \dots\ ,\ x_m) > G_k\ . \tag{9.1.14}$$

Theorem 9.1.2. *If* $X^* = (x_1^*\ ,\ \dots\ ,\ x_n^*)$ *is the solution of problem* (9.1.12)-(9.1.14) *with the smallest number de non-zero variables, and if* $I \subseteq M = \{1,\ 2,\ \dots,\ m\}$ *is such that* $x_i^* = 1$ *if* $i \in I$ *and* $x_i^* = 0$ *if* $i \in M - I$, *then*

$$F(X^*) = \frac{\sum_{i \in I} a_i + a_0}{\sum_{i \in I} b_i + b_0} \le \frac{a_j}{b_j}\ ,\quad i \in M - I\ . \tag{9.1.15}$$

If for $j \in I$ *we have*

$$a_j/b_j = \max_{i \in I}\ (a_i/b_i)\ , \tag{9.1.16}$$

then $(x_1^*,\ \dots,\ \bar{x}_j^*,\ \dots,\ x_m)$ *does not satisfy the constraints* (9.1.14).

Proof. We prove only (9.1.16) since (9.1.15) is proved in Theorem 9.1.1. From Lemma 9.1.1. and from $a_0/b_0 < a_i/b_i$, we have

$$F(X^*) \le a_j/b_j \text{ and hence } F(x_i^*,\ \dots,\ \bar{x}_j^*,\ \dots,\ x_m^*) \le F(x_1^*,\ \dots,\ x_m^*).$$

If $(x_1^*,\ \dots,\ \bar{x}_j^*,\ \dots,\ x_m^*)$ satisfied the constraints (9.1.14), then either we get a better solution than X^* or at least a solution with a smaller number of non-zero variables.

Contradiction.

We give next an algorithm for solving problem (9.1.12)-(9.1.14). The algorithm belongs to the class of branch-and-bound methods. In stage k a vector $X^k = (x_1^k, x_2^k, ..., x_m^k)$ is generated that minimizes the function (9.1.12), verifies the constraints (9.1.14) and has at most $k - 1$ non-zero components. Also generated are a set of vectors C_k such that $X = (x_1, ..., x_m) \in C_k$ if and only if a) $\sum_{i=1}^{m} x_i = k - 1$, b) X does not satisfy (9.1.14), c) $F(X) \leq a_j/b_j$ for all indices j such that $x_j = 0$, d) $F(X) < F(X^k)$. The transition from stage k to stage $k + 1$ is done considering that $X^{k+1} = X^k$, and for each vector $X' \in C_k$ one forms the class $L(X')$ of all the vectors $X = (x_1, ..., x_m)$ with the properties:

a') $X' < X$, b') $\sum_{i=1}^{m} x_i = k$, c') $F(X) \leq a_j/b_j$ for all indices j such that $x_j = 0$,

d') $F(X) < F(X^{k+1})$. Define

$$C_{k+1} = \bigcup_{X' \in C_k} \{ X \mid X \in L(X') \text{ and } X \text{ does not satisfy (9.1.14)}\}$$

$$C^{*}_{k+1} = \bigcup_{X' \in C_k} \{ X | X \in L(X') \text{ and } X \text{ satisfies (9.1.14)}\}$$

If $\min_{X \in C^{*}_{k+1}} F(X) < F(X^{k+1})$, consider $X^{k+1} = X^*$, X^* being a vector from C^{*}_{k+1} such that

$$F(X^*) = \min_{X \in C^{*}_{k+1}} F(X)$$

If $C_{k+1} = \varnothing$, then the algorithm stops. The optimal solution is $X^* = X^{k+1}$. If $C_{k+1} \neq \varnothing$, go to Step k+2. Initially, consider $X^1 = (1, 1, ..., 1)$ and $C_1 = (0, 0, ..., 0)$.

Theorem 9.1.3. *If the algorithm previously described stops to Step* $k + 1$, *then* X^{k+1} *is optimal solution of problem* (9.1.12)-(9.1.14). *This solution has the smallest number of non-zero variables.*

Proof. Consider the optimal solution of problem (9.1.12)-(9.1.14) and which has the smallest number de non-zero components, say r. Let j be an integer with the property that $x_j^* = 0$ and $a_j/b_j \geq a_i/b_i$ for any i such that $x_i^* = 0$. Theorem 9.1.2 yields that $(x_1^*, ..., \bar{x}_j^*, ..., x_m^*)$ does not satisfy the constraints (9.1.14).
The vector $(x_1^*, ..., \bar{x}_j^*, ..., x_m^*)$ has $r - 1$ non-zero components and has the property that $F(x_i^*, ..., \bar{x}_j^*, ..., x_m^*) \leq F(X^*) \leq F(X^r)$, where X^r is any vector with at most $r - 1$ non-zero components and satisfies (9.1.14). Also, $F(x_i^*, ..., \bar{x}_j^*, ..., x_n^*) \leq a_k/b_k$ for any

$k \neq j$ such that $x_k^* = 0$. It follows that the vector $(x_i^*, \ldots. \bar{x}_j^*, \ldots, x_m^*)$ is in class C_r. In the next stage, X^{r+1} will take the value X^*. If the algorithm stops at stage $k + 1$, we accept that $X^{r+1} = X^*$. The proof is complete.

In practical applications, it is advisable to start the algorithm at stage $s + 1$, where s is the number of non-zero variables of the optimal solution X^0 of the unconstrained problem, and to take $C_{s+1} = X^0$ and $X^{s+1} = (1, \ldots, 1)$.

Example 9.1.2. Reconsider the previous example:

$$\min F(x) = \frac{1 + 2x_1 + 3x_2 + 4x_3 + 5x_4 + 6x_5 + 3x_6 + 8x_7 + 9x_8 + 10x_9 + 11x_{10}}{2 + 11x_1 + 10x_2 + 9x_3 + 8x_4 + 7x_5 + 16x_6 + 5x_7 + 44x_8 + 3x_9 + 2x_{10}}$$

subject to

$$1 + 3x_1 + 2x_2 + x_3 + 5x_4 + 4x_5 + 7x_6 + 2x_7 + 3x_8 + 7x_9 + 5x_{10} > 15 \quad (9.1.17)$$

The solution of the unconstrained problem previously determined is

$$X^0 = (1\ 0\ 0\ 0\ 0\ 1\ 0\ 1\ 0\ 0)\ .$$

It can be easily shown that X^0 does not satisfy constraint (9.1.17). The algorithm starts at stage 4.

Stage 4. $X^4 = (1111111111)$; $F(X^4) = 62/117 = 0.5299145$, $C_4 = \{X^0\} = (1\ 0\ 0\ 0\ 0\ 1\ 0\ 1\ 0\ 0)$.

Stage 5. The following vectors are candidates to enter in class $L(X^0)$ (the values of function F for these vectors are given in parentheses).

1) (1 1 0 0 0 1 0 1 0 0) (F = 0.2168674)
2) (1 0 1 0 0 1 0 1 0 0) (F = 0.2317053)
3) (1 0 0 1 0 1 0 1 0 0) (F = 0.2469135)
4) (1 0 0 0 1 1 0 1 0 0) (F = 0.2625000)
5) (1 0 0 0 0 1 1 1 0 0) (F = 0.2948717)
6) (1 0 0 0 0 1 0 1 1 0) (F = 0.3289473)
7) (1 0 0 0 0 1 0 1 0 1) (F = 0.3466666)

Applying the condition (d') yields that all 7 vectors remain in class $L(X^0)$. Then, applying the condition (c'), one eliminates vectors 6, 7. Hence

$$L(X^0) = \begin{cases} (1\ 1\ 0\ 0\ 0\ 1\ 0\ 1\ 0\ 0) \\ (1\ 0\ 1\ 0\ 0\ 1\ 0\ 1\ 0\ 0) \\ (1\ 0\ 0\ 1\ 0\ 1\ 0\ 1\ 0\ 0) \\ (1\ 0\ 0\ 0\ 1\ 1\ 0\ 1\ 0\ 0) \\ (1\ 0\ 0\ 0\ 0\ 1\ 1\ 1\ 0\ 0) \end{cases}$$

$$C_5 = (1\ 0\ 1\ 0\ 0\ 1\ 0\ 1\ 0\ 0) = Y_1 \quad (2)$$

$$C_5^* = \begin{cases} (1\ 1\ 0\ 0\ 0\ 1\ 0\ 1\ 0\ 0) = Y_1' & (1) \\ (1\ 0\ 0\ 1\ 0\ 1\ 0\ 1\ 0\ 0) = Y_2' & (3) \\ (1\ 0\ 0\ 0\ 1\ 1\ 0\ 1\ 0\ 0) = Y_3' & (4) \\ (1\ 0\ 0\ 0\ 0\ 1\ 1\ 1\ 0\ 0) = Y_4' & (5) \end{cases}$$

$$\min_{X \in C_5^*} = \min\,[0.216;\ 0.2469;\ 0.2625;\ 0.2948] = 0.216$$

$$X_5 = (1\ 1\ 0\ 0\ 0\ 1\ 0\ 1\ 0\ 0)\ ;\quad F(X^5) = 0.216$$

Stage 6. The following vectors are candidates to enter in class $L(Y_1)$

(1 1 1 0 0 1 0 1 0 0) (F = 0.2391304)
(1 0 1 1 0 1 0 1 0 0) (F = 0.2666666)
(1 0 1 0 1 1 0 1 0 0) (F = 0.2808988)
(1 0 1 0 0 1 1 1 0 0) (F = 0.3103448)
(1 0 1 0 0 1 0 1 1 0) (F = 0.3411764)
(1 0 1 0 0 1 0 1 0 1) (F = 0.3571428).

This yields that, for all these vectors, the value of the objective function is greater than the previously found value for F which was 0.216. Hence $L(C_5) = \varnothing$, $C_6 = \varnothing$. The optimal solution is $X^* = (1\ 1\ 0\ 0\ 0\ 1\ 0\ 1\ 0\ 0)$ and $F(X^*) = 0.216$.

The parametric method presented in Section 3.5, Chapter 3, and in Section 4.5, Chapter 4, can be adapted in the case of bivalent linear fractional programming. For problem (9.1.7)-(9.1.10) denote by

$$S = \{\ X = (x_1, ..., x_n)\ |x_i \in \{0, 1\},\ f_j(x_1, ..., x_n) \geq F_j\ ,\quad i = 1, ..., n\ ;\quad j = 1, ..., k\ \}$$

the domain of feasible solutions. It is obvious that S is a finite set. One constructs the auxiliary problem which depends on parameter λ:

$$\min_{X \in S} g(X, \lambda) = \sum_{i=1}^{n} (a_i - \lambda b_i)x_i + (a_0 - \lambda b_0).$$

Solution of problem (9.1.7)-(9.1.10) reduces to the solution of a sequence of problems of pseudo-boolean programming with initial constraints. The algorithm consists in the following steps :

1) Determine a feasible solution $X^0 \in S$; let $\lambda = F(X^0)$.

2) Determine X^+ which minimizes the function $g(X, \lambda)$ for $X \in S$.

3) If $g(X^+, \lambda) = 0$, then X^+ is the optimal solution of problem (9.1.7)-(9.1.10). Otherwise, consider $\lambda = f(X^+)$ and restart the algorithm from Step 2.

9.2. CUTTING PLANES METHOD FOR INTEGER LINEAR FRACTIONAL PROGRAMMING PROBLEM

Consider the following problem of integer linear fractional programming

$$\max \frac{N(x)}{D(x)} = \frac{\sum_{j=1}^{n} c_j x_j + c_0}{\sum_{j=1}^{n} d_j x_j + d_0} \tag{9.2.1}$$

subject to

$$\sum_{j=1}^{n} a_{ij} x_j \leq b_i , \quad i = 1, ..., m, \tag{9.2.2}$$

$$x_j \geq 0, \text{ integers}, \; j = 1, ..., n \tag{9.2.3}$$

Denote by S the domain of feasible solutions for the problem (9.2.1)-(9.2.3). Then, assume that: 1) S is a regular set, 2) the coefficients c_j, d_j, b_i and a_{ij} are integers, 3) $D(x) > 0$ for any $x \in S$.

The principle of the method for solving the problem of integer linear fractional programming using the cutting planes method is the following : Solve first the problem (9.2.1)-(9.2.3) neglecting the condition that the values of x are integers. If the obtained solution has all components integers, then it is also the solution for problem (9.2.1)-(9.2.3). Otherwise, add an additional constraint (cut) to the Simplex table which is satisfied by any integer feasible solution of initial problem. Apply the Dual Simplex algorithm. If the components of the solution are integers, the problem is solved. Otherwise, generate a new additional constraint, etc. If, during the application of the Dual Simplex algorithm, one reaches the conclusion that the problem does not have solution, then the initial problem does not have integer solutions.

Swarup [943] (see also Hartmann [474] and Agrawal [35]) proposes to solve first the problem (9.2.1)-(9.2.3), without condition x_j integer, through Primal Simplex algorithm (see Section 3.1, Chapter 3) and then, if the solution is not integer, to add a Gomory cut[1].

Following reference [508], we are going to associate to problem (9.2.1)-(9.2.3) an auxiliary linear programming problem $P(\lambda)$, and to use Young's simplified Primal algorithm[2] to get an integer solution.

$$P(\lambda) : \max_{x \in S} N(x) - \lambda D(x) = \sum_{j=1}^{n} (c_j - \lambda d_j) x_j + c_0 - \lambda d_0.$$

[1]Gomory R. E., *All integer programming algorithm*, I.B.M. Res. Centre, Res.Rep. R.C., 189, Mimeo, 1960.

[2]Young, R., *A simplified (all-integer) integer programming algorithm*. Oper. Res., **16** (1968), 750-782.

Similar to theorems 3.5.2. (Section 3.5, Chapter 3) and 4.5.1 (Section 4.5, Chapter 4) we have:

Theorem 9.2.1. *If* $z(\lambda)$ *is the optimal value of problem* $P(\lambda)$, *and* λ^* *is the optimal value of problem* (9.2.1)-(9.2.3), *then*

a) $\lambda < \lambda^* \Leftrightarrow z(\lambda) > 0$, b) $\lambda = \lambda^* \Leftrightarrow z(\lambda) = 0$, c) $\lambda > \lambda^* \Leftrightarrow z(\lambda) < 0$.

Theorem 9.2.2. $z(\lambda)$ *is a strict decreasing convex function of* λ.

Additionally, we use and the following theorem which has an obvious proof:

Theorem 9.2.3. *Let* $x, \bar{x} \in S$ *and* $\lambda = N(x)/D(x)$. *Then* $N(\bar{x}) - \lambda D(\bar{x}) > 0$, *if and only if*

$$N(\bar{x})/D(\bar{x}) > N(x)/D(x) .$$

For a given integer linear programming problem, Young's simplified Primal algorithm can be applied by bringing the problem to a secondary form. The initial table is written as

$$\begin{aligned} \max x_0 &= c_0 + \sum_{j=1}^{n} (-c_j)(-x_j) , \\ x_{n+i} &= b_i + \sum_{j=1}^{n} a_{ij}(-x_j) \; ; \quad i = 1, ..., m \\ x_k &\geq 0, \quad k = 1, 2, ..., n + m \end{aligned} \tag{9.2.4}$$

where we assume that $b_i \geq 0$ $(i = 1, ..., m)$ such that the solution resulting from (9.2.4) is feasible.

Let's see how to proceed from a given Simplex table. Accept that $t_j (j = 1, ..., n)$ are non-basic variables, and $u_i (i = 1, ..., m')$ are basic variables. In m' we included possible cuts which add to the initial constraints. With these clarifications, the current table will take the form

$$\begin{aligned} \max x_0 &= \alpha_{00} + \sum_{j=1}^{n} \alpha_{0j}(-t_j), \\ u_i &= \alpha_{i0} + \sum_{j=1}^{n} \alpha_{ij}(-t_j) \quad i = 1, ..., m', \\ t_j &\geq 0, \; j = 1, ..., n, \; u_i \geq 0, \; i = 1, ..., m'. \end{aligned} \tag{9.2.5}$$

For (9.2.5), assume that the condition of feasible solution is satisfied i.e., $\alpha_{i0} \geq 0$, $i = 1, ..., m$. If, additionally, the solution given in Table (9.2.5) is dual feasible, i.e., $\alpha_{0j} \geq 0$, $j = 1, ..., n$, then this solution is optimal and the algorithm stops. Otherwise, generate a cut following a rule to be described later, and perform pivoting on this line. For the new table, the following three situations are possible:

a) The solution is Dual feasible. Then, Young's simplified Primal algorithm is finished. The last Simplex table that we obtained gives the optimal solution.

b) The solution is not Dual feasible but following the pivoting operation the coefficient α_{00} is transformed in α'_{00} such that $\alpha'_{00} > \alpha_{00}$ (this case is called *transition cycle*). Apply the Dual Simplex algorithm with a pivoting operation.

If Table (9.2.5) has a L-row (this line intervenes in Case c)), then this line is erased before the pivoting operation.

c) The solution is not Dual feasible and $\alpha'_{00} = \alpha_{00}$ in the following table (this case is called *stationary cycle*). Determine the pivoting element using a special line called L-row. If the current table does not have L-row, generate a L-row and perform a pivoting operation.

Applying a number of times the above cycles, Young's Primal algorithm guarantees that we reach Case a) and thus an optimal solution of problem (9.2.4) is obtained.

We apply the Young's Primal algorithm to problem $P(\lambda)$. The initial table is written in the form

$$\max x_0 = c_0 - \lambda d_0 + \sum_{j=1}^{n} \{(-c_j) - \lambda(-d_j)\}(-x_j),$$

$$z_1 = c_0 + \sum_{j=1}^{n} (-c_j)(-x_j), \quad z_2 = d_0 + \sum_{j=1}^{n} (-d_j)(-x_j), \tag{9.2.6}$$

$$x_{n+i} = b_i + \sum_{j=1}^{n} a_{ij}(-x_j), \quad i = 1, ..., m, \quad x_j \geq 0, \text{ integers}, \; j = 1, ..., n+m$$

for which assume that $b_i \geq 0$ $(i = 1, ..., m)$. Notice that Table (9.2.6) differs from Table (9.2.4) through the fact that the objective function is made parametrical through λ and contains the additional lines z_1 and z_2. These lines are used to calculate the new line corresponding to the objective function (line x_0) when the value of λ is modified. To describe the transition from a table to the other, assume that we have the following Simplex table:

$$\max x_0 = \alpha_{00}(\lambda) + \sum_{j=1}^{n} \alpha_{0j}(\lambda)(-t_j),$$

$$z_1 = \beta_{00} + \sum_{j=1}^{n} \beta_{0j}(-t_j), \quad z_2 = \gamma_{00} + \sum_{j=1}^{n} \gamma_{0j}(-t_j), \tag{9.2.7}$$

$$u_i = \alpha_{i0} + \sum_{j=1}^{n} \alpha_{ij}(-t_j), \quad i = 1, ..., m',$$

$$t_j \geq 0, \; j = 1, ..., n, \quad u_i \geq 0, \; i = 1, ..., m'.$$

for which we assume that the solution is feasible, i.e., $\alpha_{i0} \geq 0$, $i = 1, ..., m'$ and all the coefficients α_{ij}, β_{0j}, γ_{0j} are integers. The line x_0 follows from the two lines z_1 and z_2:

$$\alpha_{0j}(\lambda) = \beta_{0j} - \lambda\gamma_{0j}, \quad j = 1, ..., n. \tag{9.2.8}$$

Initially consider $\lambda = c_0/d_0$. It follows that $\alpha_{00}(\lambda) = 0$.

If, in the initial table, the solution is not dual feasible, then Young's Primal algorithm will lead, in a finite number of pivoting operations, to the case when the coefficient $\alpha_{00}(\lambda)$ is strict increasing (i.e., $\alpha_{00}(\lambda) > 0$) or the solution is dual feasible (with $\alpha_{00}(\lambda) = 0$). If $\alpha_{00}(\lambda) > 0$ (which corresponds to the case $\alpha'_{00} > \alpha_{00}$), then the value of λ is taken as

$\lambda = \beta_{00}/\gamma_{00}$, and the line x_0 is recalculated with (9.2.8), which yields $\alpha_{00}(\lambda) = 0$. If we obtained a table dual feasible, then the algorithm stops. The last table gives the optimal solution.

The algorithm for solving the integer linear fractional programming problem (9.2.1)-(9.2.3) consists of the following steps:

Step 1. Consider the initial solution $x_j^I = 0$ $(j = 1, ..., n)$. Take $\lambda = \dfrac{N(x^I)}{D(x^I)}$ and construct the initial Simplex table.

Step 2. If $\alpha_{0j}(\lambda) \geq 0$ $(j = 1, ..., n)$, then go to Step 9. Otherwise, perform the following steps, in which we add a cut and perform a pivoting operation.

Step 3. For each j such that $\alpha_{0j}(\lambda) < 0$, calculate $\theta_j = \min \{\alpha_{i0}/\alpha_{ij} | \alpha_{ij} \geq 0$, i is different from the lines x_0, z_1, z_2 and L }. If $J \neq \varnothing$ (which corresponds to the case $\alpha'_{00} > \alpha_{00}$ in the following table) go to Step 4. If $J = \varnothing$ (which corresponds to the case $\alpha'_{00} = \alpha_{00}$ in the following table) go to Step 5.

Step 4. If the current table has an L-row, erase this line. Select the column of the pivot $\alpha_{j_0}(j_0 \in J)$, and go to Step 7.

Step 5. If the current table has an L-row, go to Step 6. Otherwise before going to Step 6, add an L-row to the current table, i.e., $x_L = \alpha_{L_0} + \sum_{j=1}^{n}(-t_j)$, where $\alpha_{L_0} = \sum_{j=1}^{n} d_j$, and d_j is an upper margin of $t_j (0 \leq t_j \leq d_j$, $\forall$ $x \in S)$.

Step 6. If $\alpha_{Lj}(j = 1, ..., n)$ are the elements of the L-row, then, for each column $\alpha_j (j = 1, ..., n)$ such that $\alpha_{Lj} > 0$, calculate the column:

$$R_j = (\alpha_{0j}/\alpha_{Lj}, \ \alpha_{lj}/\alpha_{Lj}, ..., \alpha_{mj}/\alpha_{Lj})'.$$

The pivot column is selected α_{j0}. This realizes the lexicographic[3] minimum of the vectors R_j. Go to Step 7.

Step 7. Calculate i_0 such that

$$\alpha_{i_0 0}/\alpha_{i_0 j_0} = \min \{ \alpha_{i0}/\alpha_{ij_0} | \alpha_{ij_0} \geq 0, \ i = 1, ..., m' \},$$

and generate a Gomory cut which is attached to the current table:

$$s = [\alpha_{i_0 0}/\mu] + \sum_{j=1}^{n}[\alpha_{i_0 j}/\mu](-t_j), \quad \text{where } \mu = \alpha_{i_0 j_0} (> 0).$$

where [x] represents the integer part of x.

[3] A vector is lexicographic positive (negative) if the first non-zero component is positive (negative). For the given vectors $\boldsymbol{a}, \boldsymbol{b} \in R^n$, $\boldsymbol{a}$ is lexicographic greater than $\boldsymbol{b}$, denoted $\boldsymbol{a} \succeq \boldsymbol{b}$), if $\boldsymbol{a} - \boldsymbol{b}$ is lexicographic positive. For the given vectors $\boldsymbol{a}_1, \boldsymbol{a}_2, ..., \boldsymbol{a}_n$, the vector $\boldsymbol{a}_k$ is lexicographic minimal ($\boldsymbol{a}_k = \min lex (\boldsymbol{a}_1, ..., \boldsymbol{a}_n)$), if $\boldsymbol{a}_k \preceq \boldsymbol{a}_j$ for any $j = 1, ..., n$.

A pivoting operation is performed with the pivoting element $[\alpha_{i,j_0}/\mu](=1)$ of the newly introduced s-line.

Step 8. If $\alpha_{00}(\lambda) > 0$, then consider $\lambda = \beta_{00}/\gamma_{00}$, and update the line x_0 with the help of (9.2.8), and go to Step 2. If $\alpha_{00}(\lambda) = 0$, go directly to Step 2.

Step 9. The optimal solution has been obtained. The current value of λ represents the optimal value of problem (9.2.1)-(9.2.3). The optimal solution is

$$u_i = \alpha_{i0}, \quad i = 1, ..., m', \quad t_j = 0, \quad j = 1, ..., n.$$

We shall prove that the algorithm previously described finishes in a finite number of iterations, and that, in the end, the optimal solution of problem (9.2.1)-(9.2.3) is obtained. Since S is a regular set, the number of feasible solutions is finite. The problem $P(\lambda)$ has the same feasible domain S. Even more, each Simplex table offers a feasible solution. When we update the value of λ, the new λ satisfies the relation $\lambda = \beta_{00}/\gamma_{00} = N(\bar{x})/D(\bar{x})$, where $\bar{x}$ is the feasible solution presented in the Simplex table. According to Theorem 9.2.3, this new value of λ is strictly greater than the preceding value. Also, it is known that[4] for each λ, either the condition $\alpha_{00}(\lambda) = 0$ is satisfied (in which case the value of λ is updated), or a dual feasible solution is obtained following a finite number of iterations. This implies finiteness of the algorithm. When Step 9 is reached, the Simplex table will furnish the optimal solution for the problem $P(\lambda)$. Since we also have $\alpha_{00}(\lambda) = 0$, then Theorem 9.2.3 implies that this solution is also optimal for problem (9.2.1)-(9.2.3).

Example 9.2.1. max $N(x)/D(x) = (6x_1 + 8x_2 + 3)/(2x_1 + 3x_2 + 2)$
subject to

$$x_1 + 2x_2 \le 4,$$
$$7x_1 + 3x_2 \le 10,$$
$$x_1, x_2 \ge 0, \text{ integers}.$$

Introduce the slack variables x_3 and x_4, and write the constraints in the form

$$x_3 = 4 + (-x_1) + 2(-x_2) \; ;$$
$$x_4 = 10 + 7(-x_1) + 3(-x_2) \; .$$

This implies :

$$0 \le x_1 \le \min \{[4/1], [10/7]\} = 1 \quad (= d_1),$$
$$0 \le x_2 \le \min \{[4/2], [10/3]\} = 2 \; (= d_2), \tag{9.2.9}$$

Step 1. Consider the initial solution $x^I = (x_1^I, x_2^I, x_3^I, x_4^I) = (0, 0, 4, 10)$. Consider

[4]Salkin, H., P. Shroff and S. Mehta, *Primal all-integer programming applied to tableaux with rational coefficients*. Technical Memorandum, No 298, Department of Operations Research, Case Western Reserve University, April 1974.

$\lambda := N(x^1)/D(x^1) = 3/2$ and construct the initial Simplex table (Table 9.2.1).
Step 2. Since $\alpha_{01}(\lambda) = -3$, $\alpha_{02}(\lambda) = -7/2$ the solution from Table 9.2.1 is not Dual feasible.
Step 3. Calculate $\theta_1 = \min(4/1, 10/7) = 10/7 > 1$, $\theta_2 = \min(4/2, 10/3) = 4/2 > 1$, and hence $J = \{1, 2\}$.
Step 4. The column x_2 is chosen for pivoting. Note the fact that just as well we could choose the column x_1.
Step 7. Calculate min (4/2, 10/3) = 2 corresponding to line x_3. $\mu = \alpha_{i_r j_o} = 2$. Generate a Gomory cut $s_1 = [4/2] + [1/2](-x_1) + [2/2](-x_2)$ or,

$$s_1 = 2 + 0(-x_1) + (-x_2) \tag{9.2.10}$$

which represents the last line of Table 9.2.1. Perform a pivoting operation (the pivot is marked by a rectangular border in Table 9.2.1) and thus obtain Table 9.2.2.

From (9.2.9) and (9.2.10) it follows $0 \leq s_1 \leq 2$.
Step 8. Since $\alpha_{00}(\lambda) = 7 > 0$ consider $\lambda = 19/8$, and update the line x_0 with the help of the formula (9.2.8) and obtain Table 9.2.3.
Step 2. Since $\alpha_{01}(\lambda) = -5/4$, the solution given by Table 9.2.3 is not Dual feasible.
Step 3. Calculate $\theta_1 = \min(0/1, 4/7) = 0 < 1$, hence $J = \varnothing$.
Step 5. Table 9.2.3 does not have a L-row. Add a L-row

$$x_L = 3 + 1(-x_1) + 1(-s_1)$$

Step 6. Consider the vectors $R_1 = (-5/4, 1/1, 7/1, 0/1)$ and $R_2 = ((7/8)/1, -2/1, -3/1, 1/1)$.

The lexicographically smallest vector is R_1, which corresponds to the column x_1. Hence the column x_1 is chosen for pivoting.

Step 7. Calculate min (0/7, 4/7) = 0 which corresponds to line x_3. We have $\mu_{31} = 1$. Generate a Gomory cut

$$s_2 = [0/1] + [1/1](-x_1) + [-2/1](-s_1)$$

i.e.,

$$s_2 = 0 + (-x_1) + (-2)(-s_1).$$

Note that the generated cut coincides with the line x_3. A pivoting operation is performed and Table 9.2.4 is obtained.

Step 8. Since $\alpha_{00}(\lambda) = 0$, go to Step 2.
Step 2. Since $\alpha_{02}(\lambda) = -13/8$, the solution in Table 9.2.4 is not dual feasible.
Step 3. Calculate $\theta_2 = \min(4/11, 2/1) = 4/11 < 1$, hence $J = \varnothing$.
Step 5. Since Table 9.2.4 has a L-row, go to Step 6.

Step 6. The column s_1 is chosen for pivoting.
Step 7. Calculate min (4/11, 2/10) = 4/11 which corresponds to line x_4, hence $\mu_{i_1j_0} = 11$. Generate a Gomory cut

$$s_2 = [4/11] + [-7/11](-x_1) + [11/11](-s_1)$$

or

$$s_2 = 0 + (-1)(-x_1) + 1(-s_1).$$

This cut is considered as a new line of Table 9.2.4. A pivoting operation is performed and Table 9.2.5 is obtained. In the new table, the line s_1 is erased.

Step 8. Since $\alpha_{00}(\lambda) = 0$, go to Step 2.
Step 2. Since $\alpha_{01}(\lambda) = -3/8$, the solution of Table 9.2.5 is not Dual feasible.
Step 3. Calculate $\theta_1 = \min (4/4, 2/1) = 1$, hence $J = \varnothing$.
Step 4. Erase the line x_L from Table 9.2.5. Select x_3 as pivot column.
Step 7. Calculate min (4/4, 2/1) = 1 which corresponds to line x_4, hence $\mu_{i_1j_0} = 4$. Generate a Gomory cut

$$s_4 = [4/4] + [4/4](-x_3) + [-11/4](-s_3).$$

which is considered as a new line in Table 9.2.5. A pivoting operation is performed, and Table 9.2.6 is obtained.

Step 8. Since $\alpha_{00}(\lambda) = 3/8 > 0$, consider $\lambda = 17/7$, and update the line x_0 with formula (9.2.8). Table 9.2.7 is obtained.
Step 2. Since

$$\alpha_{01}(\lambda) = 3/7 > 0, \quad \alpha_{02}(\lambda) = 2/7 > 0,$$

the solution of Table 9.2.7 is Dual feasible. Go to Step 9.
Step 9. The optimal solution has been obtained, $\lambda = 17/7$ is the value of the objective function, and the solution is $x_1^0 = x_2^0 = x_3^0 = 1$, $x_4^0 = 0$.

Table 9.2.1

		$-x_1$	$-x_2$
x_0 =	0	-3	-7/2
z_1 =	3	-6	-8
z_2 =	2	-2	-3
x_3 =	4	1	2
x_4 =	10	7	3
s_1 =	2	0	[1]

Table 9.2.2

		$-x_1$	$-s_1$
x_0 =	7	−3	7/2
z_1 =	19	−6	8
z_2 =	8	−2	3
x_3 =	0	1	−2
x_4 =	4	7	−3
x_2 =	2	0	1

Table 9.2.3

		$-x_1$	$-s_1$
x_0 =	0	−5/4	7/8
z_1 =	19	−6	8
z_2 =	8	−2	3
x_3 =	0	$\boxed{1}$	−2
x_4 =	4	7	−3
x_2 =	2	0	1
x_L =	3	1	1

Table 9.2.4

		$-x_3$	$-s_1$
x_0 =	0	5/4	−13/8
z_1 =	19	6	−4
z_2 =	8	2	−1
x_1 =	0	1	−2
x_4 =	4	−7	11
x_2 =	2	0	1
x_L =	3	−1	3
s_3 =	0	−1	$\boxed{1}$

Table 9.2.5

		$-x_3$	$-s_3$
x_0 =	0	-3/8	13/8
z_1 =	19	2	4
z_2 =	8	1	1
x_1 =	0	-1	2
x_4 =	4	4	-11
x_2 =	2	1	-1
x_L =	3	2	-3
s_4 =	1	1	-3

Table 9.2.6

		$-s_4$	$-s_3$
x_0 =	3/8	3/8	4/8
z_1 =	17	-2	10
z_2 =	7	-1	4
x_1 =	1	1	-1
x_4 =	0	-4	1
x_2 =	1	-1	2
x_3 =	1	1	-3

Table 9.2.7

		$-s_4$	$-s_3$
x_0 =	0	3/7	2/7
z_1 =	17	-2	10
z_2 =	7	-1	4
x_1 =	1	1	-1
x_4 =	0	-4	1
x_2 =	1	-1	2
x_3 =	1	1	3

9.3. THE BRANCH-AND-BOUND METHOD FOR INTEGER AND MIXED INTEGER LINEAR FRACTIONAL PROGRAMMING

In this section we present the method branch-and-bound for solving problem (9.2.1)-(9.2.3) as defined in Section 9.2. For future developments, we recall the important assumptions 1) and 3) of Section 9.2. Additionally, we assume that degeneration is not present. The method branch-and-bound described for problem (9.2.1)-(9.2.3), or for the transformed problem, can be also utilized for solving the mixed integer linear fractional programming problem.

First, we present the method branch-and-bound applied to problem (9.2.1)-(9.2.3). The method is based on the work of Dakin[5] for the non-fractional problem. Solve first the problem (9.2.1)-(9.2.3), less the condition that x_j are integers, and obtain the solution $x \geq 0$. The maximum value of the objective function is denoted by z. If the solution x has all components integers, then it is the optimal solution of the integer linear fractional programming problem. Accept that not all the components of x are integers. Let $x_j = b_j$ be a component which is not integer. Solve two problems of fractional programming. The first problem has the objective function (9.2.1) and the constraints (9.2.2), $x_j \geq 0$ $x_j \leq [b_j]$. The second problem has the objective function (9.2.1) and the constraints (9.2.2), $x_j \geq 0$, $x_j \geq [b_j] + 1$, where $[b_j]$ represents the integer part of b_j. Let z' and z'' be the maximum values of the objective function. If, for example, the first problem is not feasible, then z' does not exist, and we shall explore only the branch for which $x_j \geq [b_j] + 1$. If z'' does not exist either, then x_j cannot be integer, and the initial problem does not have solution. Assume the first problem is solved, and the maximum value of the objective function is z'. Its optimal solution will have the x_j component an integer. Let $x_j = n$. Another component is selected which is not an integer. Let it be $x_k = c_k$. Solve two fractional programming problems with the additional constraints $x_j = n$, $x_k \leq [c_k]$ and $x_j = n$, $x_k \geq [c_k] + 1$. Proceed in the same way for the second problem too.
Continuing this argument one obtain a tree, with each of its vertices being characterized by a set of integer constraints. For example, the vertex associated with z' is characterized the constraint $x_j = n$. The branching stops if either we found an integer solution, or the current set of constraints does not have a solution. The solution of the integer linear fractional programming problem is obtained from the analysis of this tree and will be the solution which has all its components integers. The algorithm stops in a finite number of steps. Indeed, due to the assumptions made at the beginning, degeneration does not appear, the function $N(x)/D(x)$ can not be unbounded, and the algorithm for solving usual linear fractional programming problem is finite.

---

[5]Dakin, R. J., *A tree-search algorithm for mixed integer programming problem.* Computer J. 8(1965), 250-255.

Example 9.3.1. Maximize the function:

$$F(x) = (x_1 + 3x_2)/(4x_1 + 2x_2 + 1) \tag{9.3.1}$$

on the domain defined by the constraints:

$$21x_1 + 6x_2 \leq 14, \quad 4x_1 + 4x_2 \leq 7, \tag{9.3.2}$$

$$x_1, x_2 \geq 0, \quad x_1, x_2 \text{ integers.} \tag{9.3.3}$$

Solving this problem without the condition that x_1, x_2 are integers, yields the solution $\bar{x}_1 = 0$, $\bar{x}_2 = 7/4$ and $F(\bar{x}_1, \bar{x}_2) = 7/6$.
Then, solve two problems of linear fractional programming obtained from problem (9.3.1)-(9.3.2) by adding the constraints $x_2 \leq 1$ and $x_2 \geq 2$, respectively. The second problem, obtained by adding the constraint $x_2 \geq 2$, will have a empty domain, and needs not be considered. Hence, solve the problem defined by the function (9.3.1), the constraints (9.3.2), and the constraint $x_2 \leq 1$. Obtain the solution $x_1 = 0$, $x_2 = 1$, for which $F(x_1, x_2) = 1$. Since the components of the solution are integers, it follows that we have found the optimal solution of the integer fractional programming problem (9.3.1)-(9.3.3).

We present now a variant of the branch-and-bound method for solving the problem

$$\max \frac{N(x)}{D(x)} = \frac{\sum_{j=1}^{n} c_j x_j + c_0}{\sum_{j=1}^{n} d_j x_j + d_0} \tag{9.3.4}$$

subject to

$$\sum_{j=1}^{n} a_{ij} x_j \leq b_i, \quad i = 1, ..., m \tag{9.3.5}$$

$$x_j \geq 0, \quad j = 1, ..., n \tag{9.3.6}$$

$$x_j \text{ integers } j \in N_1 \subseteq N = \{1, ..., n\}. \tag{9.3.7}$$

The working hypotheses are the same as before.
In the case when $N_1 = N$, a method is obtained for solving the linear fractional programming problem with integer variables. Through the change of variable $y = tx$, $t \geq 0$, the problem (9.3.4)-(9.3.7) is transformed into the following the problem:

$$\max z(y, t) = \sum_{j=1}^{n} c_j y_j + c_0 t \tag{9.3.8}$$

subject to

$$\sum_{j=1}^{n} a_{ij} y_j - b_i t \leq 0, \quad i = 1, ..., m \tag{9.3.9}$$

$$\sum_{j=1}^{n} d_j x_j + d_0 t = 1 \tag{9.3.10}$$

$$y \geq 0, \quad t \geq 0 \tag{9.3.11}$$

$$y_j/t \text{ integers for } j \in N_1 \tag{9.3.12}$$

The connection between the problems (9.3.4)-(9.3.7) and (9.3.8)-(9.3.12) is given by the following theorem:

Theorem 9.3.1. a) *If there is an optimal solution of problem* (9.3.4)-(9.3.7), *then* (y^*, t^*), *where* $y^* = t^* x^*$ *and* $t^* = 1/(\sum_{j=1}^{n} d_j x_j^* + d_0)$ *is an optimal solution of problem* (9.3.8)-(9.3.12). b) *Conversely, if there is an optimal solution* (y^*, t^*) *of problem* (9.3.8)-(9.3.12), *then* $t^* > 0$ *and* $x^* = y^*/t^*$ *is an optimal solution of problem* (9.3.4)-(9.3.7).

The proof of this theorem is similar with that of Theorem 3.4.1, Chapter 3.

A variant of the branch-and-bound method for problem (9.3.4)-(9.3.7) is obtained by applying the method of Dakin for the transformed problem (9.3.8)-(9.3.12). Solve first the problem (9.3.8)-(9.3.11). Since, in the optimal solution, $t > 0$, then variable t is a base variable . Consider that the optimal Simplex table of problem (9.3.8)-(9.3.11) is

$$\max z = z^0 + \sum_{j \in J} c_j y_j,$$

$$y_{B_i} + \sum_{j \in J} a_{ij} y_j = \gamma_i, \quad i = 1, ..., m$$

$$t + \sum_{j \in J} d_j y_j = \delta$$

$$t \geq 0 \; ; \; y_i \geq 0 \; , \; j \in J \text{ and } j \in B_i (i = 1, ..., m)$$

in which we denoted by J the indices corresponding to the variables from outside the base whereas y_{B_i}, and t are base variables. From the above table it follows that the optimal solution of problem (9.3.8)-(9.3.11) is

$$y_{B_i} = \gamma_i, \quad i \in M = \{1, 2, ..., m\}, \quad t = \delta(> 0), \quad y_j = 0 \;\; j \in J \tag{9.3.13}$$

If this solution also verifies the constraint (9.3.12), then, according to Theorem 9.3.1, we can obtain an optimal solution for problem (9.3.4)-(9.3.7). Assume that the solution given by (9.3.13) does not verify constraint (9.3.12). Let γ_k/δ $(k \in N_1)$ be non-integer. Then, we have either a) $y_{B_k}/t \leq [\gamma_k/\delta]$, or b) $y_{B_k}/t \geq [\gamma_k/\delta] + 1$. Denote

$$\langle \gamma_k/t \rangle = [\gamma_k/\delta] + 1 \tag{9.3.14}$$

In Case a), since $t > 0$, we have

$$y_{B_k} \leq [\gamma_k/\delta] t \;\text{ or }\; \gamma_k - \sum_{j \in J} a_{kj} y_j \leq [\gamma_k/\delta](\delta - \sum_{j \in J} d_j y_j).$$

Rearranging yields

$$\sum_{j \in J} ([\gamma_k/\delta] d_j - a_{kj}) y_j \leq [\gamma_k/\delta]\delta - \gamma_k. \tag{9.3.15}$$

The solution given by (9.3.13) obviously does not verify constraint (9.3.15). Attach the constraint (9.3.15) to problem (9.3.10)-(9.3.11). Similarly, in Case b) we have the constraint

$$\sum_{j \in J} (a_{kj} - \langle \gamma_k/\delta \rangle \, d_j) y_j \leq \gamma_k - \langle \gamma_k/\delta \rangle \, \delta, \tag{9.3.16}$$

which we attach to problem (9.3.8)-(9.3.11). In this manner, another branch is obtained in the tree constructed for the branch-and-bound method. Then, one proceeds as in Dakin's method.

Chandra and Chandramohan [261] showed that, if one uses the penalties of Beale and Small[6], the efficiency of the method increases.
Accept that we have obtained the continuous optimal solution corresponding to a node. The objective function is

$$z = z^{+} + \sum_{j \in J^{+}} c_j^{+} y_j$$

where, J^{+} is the set of indices corresponding to the variables from outside the base and z^{+} is the value of the objective function in that node. If one creates a new branch by introducing the constraint

$$\sum_{j \in J^{+}} g_j y_j \leq -f \; (f > 0),$$

then one calculates a penalty:

$$p = f \frac{c_r^{+}}{g_r}, \text{ where } \frac{c_r^{+}}{g_r} = \min_{g_j < 0} \left\{ \frac{c_j^{+}}{g_j} \right\}$$

The new branch is consequently omitted if $z^{+} - p \leq z^{*}$, where z^{*} is the value of the objective function for a feasible solution of problem (9.3.4)-(9.3.7). In this way, a series of solutions which could not have been optimal, have been eliminated with a minimal computational effort.

The method branch-and-bound presented here is finite. Indeed, when a new branch is constructed, due to the correspondence between problems (9.3.4)-(9.3.7) and (9.3.8)-(9.3.12), when a new constraint added to problem (9.3.8)-(9.3.11), implies the constraint $x_j \leq \theta$ or $x_j \geq \theta$ (θ integer, $h \in N_1$) is added for problem (9.3.4)-(9.3.7). The finiteness of the method is assured by the fact that S (the set feasible solutions) is bounded. Note that, if we added the constraint $y_k - \theta t \leq 0$, and if later we add the constraint $y_k - \theta_1 t \leq 0 (\theta_1 < \theta)$, then, in order not to artificially increase the size of the Simplex table, the constraint $y_k - \theta t \leq 0$ is omitted.

Example 9.3.2. Reconsider the example of Section 3.1.
Problem 1. $\max F(x) = (3x_1 + 6x_2 + 2x_3 - 11)/(x_1 + 4x_2 + x_3 + 1)$,
subject to

$$3x_1 + 4x_2 + x_3 \leq 2,$$
$$x_1 + 4x_2 + x_3 \leq 1,$$
$$x_1, x_2, x_3 \geq 0, \quad x_1, x_2 \text{ integers.}$$

[6]E. M. L. Beale and R. E. Small, *Mixed integer programming by a branch and bound technique*. Proc. IFIP. Congr. 2 (W. A. Kalenich, Ed.), 450-451, 1965.

Through the change of variable $y_i = tx_i$ $(i = 1, 2, 3)$, Problem 1 becomes

Problem 2. $\max G(y, t) = 3y_1 + 6y_2 + 2y_3 - 11t$

subject to

$$3y_1 + 4y_2 + y_3 - 2t \le 0, \quad y_1 + 4y_2 + y_3 - t \le 0,$$
$$y_1 + 4y_2 + y_3 + t = 1,$$
$$y_1, y_2, y_3, t \ge 0, \quad y_1/t, \ y_2/t \text{ integers}$$

We solve the Problem 2 neglecting the constraint that y_1/t and y_2/t must be integers. The final Simplex table is given by Problem 3.

Problem 3. $\max z = -17/4 + 2y_2 + (1/2)y_5 + (23/4)y_6$

$$y_1 + (1/2)y_5 - (3/4)y_6 = 1/4$$
$$y_3 + 4y_2 - (1/2)y_5 + (5/4)y_6 = 1/4$$
$$t - (1/2)y_6 = 1/2$$
$$y_j \ge 0, \quad j = 1, ..., 6\ ; \quad t \ge 0$$

The optimal solution is $y_1 = 1/4$; $y_3 = 1/4$, $t = 1/2$, and the value of z is -17/4. Since $y_1/t = 1/2$ is not integer, we have not obtained the solution of Problem 1. Add the constraint $y_1 \le 0$ to Problem 3 i.e.,

$$-(1/2)y_4 + (3/4)y_6 + s_1 = -1/4.$$

where s_1 is a slack variable. The optimal solution is $y_1 = 0$, $y_2 = 0$, $y_3 = 1/2$, $t = 1/2$.

Table 9.3.1

			3	6	2	-11	0	0	0
			y_1	y_2	y_3	$y_4(t)$	y_5	y_6	s_2
3	y_1	1/4	1	0	0	0	1/2	-3/4	0
2	y_3	1/4	0	4	1	0	-1/2	5/4	0
-11	t	1/2	0	0	0	1	0	-1/2	0
0	s_2	-1/4	0	0	0	0	1/2	[-1/4]	1
		-17/4	0	2	0	0	1/2	23/4	0
3	y_1	1	1	0	0	0	-1	0	-3
2	y_3	-1	0	4	1	0	2	0	5
-4	t	1	0	0	0	1	-1	0	-2
0	y_6	1	0	0	0	0	-2	1	-4
		-10	0	2	0	0	12	0	23

Since $y_1/t = y_2/t = 0$, we have obtained a integer feasible solution for Problem 1. The solution is $x_1 = x_2 = 0$, $x_3 = 1$, with the value of F equal to 1/2. The other branch can be obtained by adding the constraint $y_1 \geq t$, i.e., $(1/2)y_5 - (1/4)y_6 + s_2 = -1/4$, where s_2 is a slack variable. The penalty p for adding this constraint is $p = (1/4)(23/4)(-4/1) = -23/4$. Since $-17/4 + 23/4 > 1/2$, we shall consider this branch too. Add the constraint $y_1 \geq t$ to Problem 3 i.e., $(1/2)y_5 - (1/4)y_6 + s_2 = -1/4$, where s_2 is a slack variable.

The new Simplex table is Table 9.3.1.

Since the line of y_3, corresponding to the negative component, has all the elements positive for the columns from outside the base, it follows that Problem 3 with constraint $y_1 \geq t$ does not have feasible solutions. Thus, we cannot proceed on branch $y_1 \geq t$. Since, on the previous branch, we found an integer solution, this solution also applies for Problem 1. This solution is $x_1 = 0$, $x_2 = 0$, $x_3 = 1$ and the value of the objective function is $F(x) = 1/2$.

9.4. HISTORICAL NOTES. COMMENTARIES

The algorithms presented in Section 9.1 for solving the problems (9.1.1)-(9.1.2) and (9.1.12)-(9.1.14) are due to Robillard [751], [752]. The parametric method of Section 9.1 has been developed by Florian and Robillard [394], [395]. If we apply the parametric method to problem (9.1.1)-(9.1.2) and if we consider $X^0 = 0$, then we obtain the algorithm presented by Hammer and Rudeanu in [465]. A study presented in [752] for the bivalent linear fractional programming problems with n = 1000, 2000, ..., 10000 variables and without constraints showed that the algorithm of [394] is better than the algorithm of [465]. The coefficients a_i and b_i were generated randomly and, for a given n, the procedure for generating the coefficients was repeated 100 times in order to obtain an average time.

Unconstrained 0-1 linear fractional programming was considered also by Hansen *et al.* [468]. They showed that this can be solved in linear time when the denominator is always positive. It is NP-hard if the denominator is unrestricted in sign. Finding an approximate solution with a value equal to a positive multiple of the optimal one is also NP-hard.

The literature also presents other methods for solving the bivalent linear fractional programming problem. Thus, Frieda and Daniel Granot [426] presented implicit enumeration algorithms for solving the problem :

$$\max \{(c'x + c_0)/(d'x + d_0) \mid Ax \leq b, x_j \in \{0, 1\}, j = 1, ..., n\}. \quad (9.4.1)$$

The method was applied directly to problem (9.4.1) and used the strongest surrogate constraints in order to strengthen the binary infeasibility and fathoming tests performed on the various partial solutions. Williams [1048] transformed a constrained (0,1) fractional programming problem into an equivalent mixed integer linear problem with $3n$ (n is the number of original variables) additional constraints and $n+1$ additional continuous variables and solved it by a branch-and-bound algorithm. Ţigan [994] presented a cascade algorithm for solving the pseudo-boolean linear fractional programming problem.

$$\max \{ \sum_{i=1}^{m}\sum_{j=1}^{n} a_{ij}x_{ij} \Big/ \sum_{i=1}^{m}\sum_{j=1}^{n} b_{ij}x_{ij} \mid \sum_{j=1}^{n} x_{ij} = 1, \ x_{ij} = 0 \text{ or } 1\} \quad (9.4.2)$$

where $i = 1, ..., m$, $j = 1, ..., n$, and $a_{ij} \geq 0$, $b_{ij} > 0$ ($\forall$ i, j) are given numbers. Solution of problem (9.4.2) can be achieved utilizing the parametric method presented in Section 9.1, i.e., through reduction to the solution of a finite number of auxiliary problems of pseudo-boolean linear programming, modification of the parameter being done after the the solution of each auxiliary problems. Due to the special form of the constraints (each variable intervenes only in one constraint), the solution of the auxiliary linear problems can be achieved by solving a sequence of linear subproblems with a single constraint. The proposed cascade algorithm follows from this particularity. Modification of the parameter is not done after the complete solution of a each auxiliary linear problem, but after the solution of each subproblem. In this way, the algorithm is considerably accelerated.

Saipe [765] presented a branch-and-bound method for solving the bivalent fractional programming problem

$$\max \{ \sum_{i=1}^{s}[(\sum_{j=1}^{T} a_{ij}x_{ij})/(\sum_{j=1}^{T} b_{ij}x_{ij})] \mid \sum_{j=1}^{T} x_j = n, \ x_j = 0 \ \text{or} \ 1, j = 1, ..., T \}$$

Three heuristic methods for quickly finding good feasible solutions to this problem were discussed.

The branch-and-bound method was also used by Molnár and Trčka [669] for solving the bivalent linear fractional programming problem. First, an algorithm for finding the local minimum without any conditions was given. Second, an algorithm for finding the absolute minimum in the case of a system of limiting linear conditions with non-negative coefficients was presented. The absolute minimum was found by the branch-and-bound method in the case where the local minimum was not included in the set of limiting conditions. The problem was branched in sequence of variables arranged increasingly according to the value of a particular fractional objective. For each node, an estimate concerning the possibility of achieving a feasible solution by means of further branching from this node, and the possibility of achieving a lower value of the fractional function by further branching were given.

Swarup and Sharma [964] considered the problem of form (9.4.1), in which the constraint $Ax \le b$ reduces to one inequality (linear or non-linear) of form $a'x \le b$. An equivalent fractional programming problem is considered

$$\max \{(c'x + c_0)/(d'x + d_0) \mid a'x \le b,$$
$$x \text{ extreme point for } I_n x \le 1, \quad x \ge 0\} \tag{9.4.3}$$

which he solved using the Primal Simplex algorithm and a special type of cuts (called angular cuts) to reach the optimal solution of the initial problem. In problem (9.4.3), I_n is an identity matrix and **1** is a vector for which each component is one.

Ţigan [987] considered the problem of fractional allocation

$$\max\{\sum_{i=1}^{n}\sum_{j=1}^{n} c_{ij}x_{ij} \Big/ \sum_{i=1}^{n}\sum_{j=1}^{n} d_{ij}x_{ij} \mid \sum_{i=1}^{n} x_{ij} = 1, \sum_{j=1}^{n} x_{ij} = 1, \ x_{ij} \ge 0, \ x_{ij} = 0 \ \text{or} \ 1\} \tag{9.4.4}$$

which he solved using an algorithm for linearization of the objective function. In a different work [988], Ţigan considered the problem

$$\min \{ \max \{a_{ij}x_{ij}\}/\max \{b_{ij}x_{ij}\} \mid \sum_{j=1}^{n} x_{ij} = 1,$$
$$\sum_{i=1}^{n} x_{ij} = 1, \ x_{ij} = 0 \ \text{or} \ 1, \ a_{ij}, b_{ij} \ge 0 \} \tag{9.4.5}$$

which was solved using the parametric method presented in Section 9.1.

The set covering problem with a fractional objective function

$$\min \{ \sum_{j=1}^{n} c_j x_j \Big/ (\sum_{j=1}^{n} d_j x_j + \beta)$$
$$\mid \sum_{j=1}^{n} a_{ij}x_j \ge 1, \ i = 1, ..., m, \ x_j = 0 \ \text{or} \ 1, j = 1, ..., n\} \tag{9.4.6}$$

was tackled by Arora, Puri and Swarup [76], Arora and Puri [75], Arora, Swarup and Puri [77] and Ishii, Nishida and Daino [507]. In references [75] and [76], the authors presented an enumerative technique which involves the evaluation of extreme points of a linear fractional functional programming problem derived from the given problem. The technique converges in a finite number of steps because it moves from one extreme point to the other extreme point solution of linear fractional problem, and these extreme points are always finite in number and none of them is repeated. In reference [77], the authors developed an iterative cutting plane technique. The method determines a prime cover solution from the optimal solution of the linear fractional problem. If this prime cover solution is not optimal, a linear cut is introduced and the process is continued until the optimal prime cover solution is reached. In reference [507], the authors presented an enumerative algorithm which combines the parametric procedure based on Dinkelbach's method [360] for fractional programming problems with the linear search technique of branch-and-bound method.

Daniel and Frieda Granot [427] proposed a Primal algorithm for the solution in a finite number of iterations of mixed and integer fractional programming problems. The algorithm constitutes an extension of the Primal algorithm for usual integer programming, and uses two types of cutting planes. Grunspan and Thomas [431] also used the method of cutting planes for solving the integer fractional problem.

The cutting plane method of Gomory was adapted by Hartmann [471], [472], [473], [474] for solving integer and mixed integer linear fractional programs.

Saxena, Patkar and Parkash [783] presented a Primal algorithm using Gomory cuts for solving the problem:

$$\max \{c'x + d'x/e'x \mid Ax = b, \ x_j \text{ integers } j = 1, ..., n\}$$

The first of the branch-and-bound methods presented in Section 9.3 was elaborated by Agrawal [37] (see, also [36]) and the second (see also [263] for the non-linear case) by Chandra and Chandramohan [261].

Patkar, Saxena and Parkash [716] used the first method of Section 9.3 for solving the problem

$$\max \{ F(x) = \sum_{i=1}^{s} ((c'x + c_0)/(d'x + d_0))^i \mid Ax \leq b, x \in I^+ = \{0, 1, 2, ...\} \}$$

Christov *et al.* [318] also presented a branch-and-bound type method for solving a linear-fractional integer optimization problem (see also [1035], [1036]). For mixed integer linear fractional programming the branch-and-bound method was applied by Zsigmond, Ildicó [1088].

The parametric method presented in Section 3.5, Chapter 3, for the continuous linear fractional programming problem, can be applied to solving fractional programming problems with integer variables [64], [432]. Such a method was used by Ishii, Ibaraki and Mine [509] for the solution of the knapsack problem with linear fractional objective function. Hashizume *et al.* [477] presented a fully polynomial time approximation scheme for the fractional 0 - 1 knapsack problem (see also reference [876]).

The change of variables method was used by Beale [111] for the solution of a linear fractional problem with zero-one variables.

Nikitin and Nuriev [693] presented a heuristic algorithm for solving a boolean linear fractional programming problem. In some cases this algorithm gives the optimal solution.

Seshan and Tikekar [833] presented two algorithms for solving integer linear fractional programming problem. The first algorithm is an extension of Bitran and Novaes' method [192], and the second algorithm is a refinement of the first.

Duality for mixed integer nonlinear fractional programming problem was presented by Craven and Mond [336].

Bector and Jolly [151] considered the pseudo-monotonic integer programming problem

$$\max \{ f(x) \mid Ax \leq b, \quad x \text{ an integer or mixed integer} \}$$

where f is a pseudo-monotonic function. Using the linearization technique of Kortanek and Evans, the solution of this problem is reduced to solving a finite set of linear integer programs.

Verma *et al.* [1023] used a special cut, and developed an algorithm which ranks the feasible solutions of an integer fractional programming problem in decreasing order of the values of the objective function.

Agrawal and Chand [38], [40], [42], [44] considered integer linear fractional programming with the additional requirement that the two specified sets of variables, u and v are complementary to each other and describe a branch-and-bound method for solving it. They also considered fractional interval programming [39], [43].

Verma [1018] considered an integer linear fractional programming problem for which the integer solution is required to satisfy any h out of n given sets of constraints.

$$\text{Maximize } Z = \frac{c'X + \alpha}{d'X + \beta}$$

subject to $X \in S$ and

$$X \in F = \bigcup_{(j_1, \dots, j_h) \in T} (F_{j_1} \cap F_{j_2} \cap \dots \cap F_{j_h})$$

where

$$S = \{ X \mid AX = b, X \geq 0 \}, A \in R^{m \times n}, b \in R^m, d, c, X \in R^n, \alpha, \beta \in R$$

$$F_i = \{ X \mid g_i(X) \leq (=, \geq) 0, X \geq 0 \text{ is an integer point} \}$$

for all $i \in I = \{1, \dots, n\}$, $g_i(X) \leq (=, \geq) \mathbf{0}$ is the set of secondary constraints in F_i, $T = [\{ j_1, \dots, j_h \} | \{ j_1, \dots, j_h \} \subset I]$ the set of all subsets of I taking h of its elements at a time, where h is a fixed integer less than or equal to n. A method for ranking and scanning all integer points was developed.

CHAPTER 10

FRACTIONAL TRANSPORTATION PROBLEM

10.1 TRANSPORTATION PROBLEM WITH LINEAR FRACTIONAL OBJECTIVE FUNCTION

In this section we deal with the following transportation problem with linear fractional objective function: Given two cost matrices $C = (c_{ij})$ and $D = (d_{ij}), i = 1, ..., m$; $j = 1, ..., n$, determine a matrix $X = (x_{ij})$ which minimizes the function:

$$Z(X) = \frac{N(X)}{D(X)} = \left[\sum_{i=1}^{m}\sum_{j=1}^{n} c_{ij}x_{ij} + c_0\right] \Big/ \left[\sum_{i=1}^{m}\sum_{j=1}^{n} d_{ij}x_{ij} + d_0\right] \tag{10.1.1}$$

with the constraints

$$\sum_{j=1}^{n} x_{ij} = a_i \,, \quad i = 1, ..., m, \tag{10.1.2}$$

$$\sum_{i=1}^{m} x_{ij} = b_j \,, \quad j = 1, ..., n, \tag{10.1.3}$$

$$x_{ij} \geq 0 \text{ for all } i \text{ and } j \tag{10.1.4}$$

where c_0 and d_0 are constants.

Denote by S the domain of the feasible solutions of the problem (10.1.1)-(10.1.4). Assume

$$\sum_{i=1}^{m}\sum_{j=1}^{n} d_{ij}x_{ij} + d_0 > 0 \text{ for all } X \in S, \tag{10.1.5}$$

$$\sum_{i=1}^{m} a_i = \sum_{j=1}^{n} b_j \tag{10.1.5'}$$

This last condition expresses the fact that the problem is balanced and constitutes the necessary and sufficient condition for the problem to have a solution. The optimal solution appears in a vertex of S and the local optimum is global.

As in the linear case, the solution of the fractional transportation problem takes place in two steps:

1. The determination of a initial solution,
2. The determination of the optimal solution.

The determination of the initial solution can be done using either the method of the North-West corner, or the method of the minimum element on the row (column or of the table), or Vogel's method (the procedure of maximum differences).
The second step consists in the verification of the solution optimality (starting with the initial solution) and, if it is not optimal, in its improvement.
Let $I = \{1, ..., m\} \times \{1, ..., n\}$ and $X = \{(x_{ij}) \mid (i, j) \in I\}$ be a basic feasible solution.
Denote by $I_X = \{(i, j) \in I \mid x_{ij} > 0, \; x_{ij} \in X\}$ the pairs of indices of the base variables.
Due to the condition (10.1.5') each non-degenerate basic solution will contain m+n-1 positive components.
Consider u_i^1, v_j^1 (u_i^2, v_j^2, respectively) the simplex multipliers (or dual variables) associated with the function N (D, respectively) defined such that to have

$$u_i^1 + v_j^1 = c_{ij} \qquad (10.1.6)$$

$$(i, j) \in I_X$$

$$u_i^2 + v_j^2 = d_{ij} \qquad (10.1.7)$$

For the variables outside the base, consider the comparative costs

$$\bar{c}_{ij} = c_{ij} - u_i^1 - v_j^1, \quad \bar{d}_{ij} = d_{ij} - u_i^2 - v_j^2 \quad (i, j) \in I - I_X$$

The systems (10.1.6) and (10.1.7) can be solved independently, each having $m + n - 1$ equations and $m + n$ unknowns. Their solution is obtained giving the value zero to one unknown. Being given the feasible solution X, consider another solution X^* which differs from X by only one component. More precisely, we assume that X^* is obtained from X by bringing into the base the variable x_{pq}. We have the following relations $N(X^*) = N(X) + \theta\bar{c}_{pq}$ and $D(X^*) = D(X) + \theta\bar{d}_{pq}$, where θ is a value attributed to the variable x_{pq}.
Since,

$$Z(X^*) - Z(X) = (N(X) + \theta\bar{c}_{pq})/(D(X) + \theta\bar{d}_{pq}) - N(X)/D(X) =$$
$$= \theta[\bar{c}_{pq}\, D(X) - \bar{d}_{pq}\, N(X)]/[D(X)(D(X) + \theta\bar{d}_{pq})]$$

it follows that X^* is a better solution than X if

$$\Delta_{pq} = \bar{c}_{pq}\, D(X) - \bar{d}_{pq}\, N(X) < 0.$$

This yields the optimality criterion for a solution.
Theorem 10.1.1. *A feasible solution* $X = (x_{ij})$ *is a local optimum of the problem (10.1.1)-(10.1.4) if*

$$\Delta_{pq} = \bar{c}_{pq}\, D(X) - \bar{d}_{pq}\, N(X) \geq 0, \quad (p, q) \in I - I_X.$$

For the variables which belong to the base we have $\Delta_{pq} = 0$. If for the variables from outside the base there are negative values Δ_{pq}, then $\Delta_{ij} = \min\{\Delta_{pq} \mid \Delta_{pq} < 0\}$ is determined and the variable x_{ij} is introduced in the base. Otherwise, one proceeds as in the usual transportation problem.

Numerical example. Solve the fractional transportation problem

$$Z(X) = \left[\sum_{i=1}^{3}\sum_{j=1}^{4} c_{ij}x_{ij}\right] \Big/ \left[\sum_{i=1}^{3}\sum_{j=1}^{4} d_{ij}x_{ij}\right],$$

where the coefficients c_{ij} and d_{ij} (together with the necessary and available quantities) are given in Table 10.1.1.

Table 10.1.1

						u^1	u^2
	10 2(d_{11}) 1(c_{11})	4 3	1 6	2 5	10(a_1)	0	0
	5 1 2	10 6 4	5 3 7	5 1	20	1	-1
	6 2	1 3	15 5 4	15 3 3	30	-2	1
	15(b_1)	10	20	15			
v^1	1	3	6	5			
v^2	2	7	4	2			

The initial solution (X^0), determined by the North-West corner method is given in Table 10.1.1.

The Simplex multipliers are determined from the systems

$$u_1^1 + v_1^1 = 1, \quad u_2^1 + v_3^1 = 7, \qquad u_1^2 + v_1^2 = 2, \quad u_2^2 + v_3^2 = 3,$$

$$u_2^1 + v_1^1 = 2, \quad u_3^1 + v_3^1 = 4, \qquad u_2^2 + v_1^2 = 1, \quad u_3^2 + v_3^2 = 5,$$

$$u_2^1 + v_2^1 = 4 \quad u_3^1 + v_4^1 = 3, \qquad u_2^2 + v_2^2 = 6, \quad u_3^2 + v_4^2 = 3.$$

These multipliers have the values given in Table 10.1.1, i.e.,

$$u_1^1 = 0, \quad v_1^1 = 1, \qquad u_1^2 = 0, \quad v_1^2 = 2$$

$$u_2^1 = 1, \quad v_2^1 = 3, \qquad u_2^2 = -1, \quad v_2^2 = 7$$

$$u_3^1 = -2, \quad v_3^1 = 6, \qquad u_3^2 = 1, \quad v_3^2 = 4$$

$$v_4^1 = 5, \qquad v_4^2 = 2$$

The comparative costs corresponding to the variables which do not belong to the base are given next in the parentheses $x_{12}(0, -3)$; $x_{13}(0, -3)$; $x_{14}(0, 0)$; $x_{24}(-5, 4)$; $x_{31}(3, 3)$; $x_{32}(2, -7)$.

The solution X^0 can be improved if there is

$$\Delta_{ij} = \bar{c}_{ij}\, D(X^0) - \bar{d}_{ij}\, N(X^0) < 0$$

Since $N(X^0)$ and $D(X^0)$ are positive numbers, we have $\Delta_{ij} \geq 0$ for $\bar{c}_{ij} \geq 0$ and $\bar{d}_{ij} \leq 0$. Consequently we calculate Δ_{ij} for the other cases. In our case we have $N(X^0) = 200$; $D(X^0) = 220$; $Z(X^0) = 200/220 = 0.909$; $\Delta_{24} = 220(-5) - 200(4) = -1900$, $\Delta_{31} = 220(3) - 200(3) = 60$. Since (Δ_{24}) is the only negative difference, x_{24} enters the base. The maximum value which can be attributed to this variable is $\theta = \min(5, 15) = 5$. The new solution is given in Table 10.1.2.

Table 10.1.2

					a_i	u_i^1	u_i^2
	10 2 1	4 3	1 6	2 5	10	0	0
	5 1 2	10 6 4	3 7	5 5 1	20	1	-1
	6 2	1 2	20 5 4	10 3 3	30	3	-3
b_j	15	10	20	15			
v_j^1	1	3	1	0			
v_j^2	2	7	8	6			

We have

$$N(X^1) = N(X^0) - 5\times 5 = 175; \quad D(X^1) = D(X^0) + 20 = 240$$

and

$$Z(X^1) = N(X^1)/(D(X^1)) = 175/240 = 0.729.$$

We test the optimality of X^1 and hence we construct the systems

$$u_1^1 + v_1^1 = 1, \quad u_2^1 + v_4^1 = 1, \qquad u_1^2 + v_1^2 = 2, \quad u_2^2 + v_4^2 = 5,$$

$$u_2^1 + v_1^1 = 2, \quad u_3^1 + v_3^1 = 4, \qquad u_2^2 + v_1^2 = 1, \quad u_3^2 + v_3^2 = 5,$$

$$u_2^1 + v_2^1 = 4, \quad u_3^1 + v_4^1 = 3, \qquad u_2^2 + v_2^2 = 6, \quad u_3^2 + v_4^2 = 3.$$

Their solutions are

$$u_1^1 = 0, \quad v_1^1 = 1, \qquad u_1^2 = 0, \quad v_1^2 = 2,$$

$$u_2^1 = 1, \quad v_2^1 = 3, \qquad u_2^2 = -1, \quad v_2^2 = 7,$$

$$u_3^1 = 3, \quad v_3^1 = 1, \qquad u_3^2 = -3, \quad v_3^2 = 8,$$

$$v_4^1 = 0, \qquad v_4^2 = 6.$$

We calculate the comparative costs corresponding to the variables which not belong to the base and get $x_{12}(0,\ -3)$, $x_{13}(5,\ -7)$, $x_{14}(5,\ -4)$, $x_{23}(5,\ -4)$, $x_{31}(-2,\ 7)$, $x_{32}(-3,\ -3)$. The differences Δ_{ij} are positive for the variables x_{12}, x_{13}, x_{14}, x_{23}. The other differences have the values $\Delta_{31} = 240(-2) - 175(7) = -1705$, $\Delta_{32} = 240(-3) - 175(-3) = -195$. Since $\Delta_{31} = \min\{\Delta_{31}, \Delta_{32}\}$, x_{31} enter the base. The maximum value which can be attributed of this variable is $\theta = \min(5, 10) = 5$. This quantity is transferred on the cycle given in Table 10.1.2 and the solution X^2 (Table 10.1.3) is obtained.

Table 10.1.3

					a_i	u_i^1	u_i^2
	10; 2 1	4 3	1 6	2 5	10	0	0
	1 2	10; 6 4	3 7	10; 5 1	20	−1	6
	5; 6 2	1 3	20; 5 4	5; 3 3	30	1	4
b_j	15	10	20	15			
v_j^1	1	5	3	2			
v_j^2	2	0	1	−1			

We have $N(X^2) = N(X^1) - 5 \times 2 = 165$; $D(X^2) = D(X^1) + 5 \times 7 = 275$ whereas $Z(X^2) = 165/275 = 0.6$.

Again, we test the optimality of X^2 and we construct the systems

$$\begin{array}{llll} u_1^1 + v_1^1 = 1, & u_3^1 + v_1^1 = 2, & u_1^2 + v_1^2 = 2, & u_3^2 + v_1^2 = 6, \\ u_2^1 + v_2^1 = 4, & u_3^1 + v_3^1 = 4, & u_2^2 + v_2^2 = 6, & u_3^2 + v_3^2 = 5, \\ u_2^1 + v_4^1 = 1, & u_3^1 + v_4^1 = 3, & u_2^2 + v_4^2 = 5, & u_3^2 + v_4^2 = 3, \end{array}$$

The solutions to these systems are given in Table 10.1.3.

The comparative costs are the following: $x_{12}(-2, 4)$; $x_{13}(3, 0)$, $x_{14}(3, 3)$, $x_{21}(2, -7)$, $x_{23}(5, -4)$, $x_{32}(-3, -3)$. The differences Δ_{ij} are positive for the variables x_{13}, x_{21}, x_{23}. The other differences have the values $\Delta_{12} = 275(-2) - 165 \times 4 = -1210$, $\Delta_{14} = 275 \times 3 - 165 \times 3 = 330$, $\Delta_{32} = 275(-3) - 165(-3) = -330$.

Since $\Delta_{12} = \min\{\Delta_{12}, \Delta_{32}\}$, x_{12} will enter the base. The maximum value which can be attributed to this variable is min (10, 5, 10) = 5. The new solution X^3 given in Table 10.1.4 is obtained.

For the solution X^3 we have $N(X^3) = N(X^2) - 5\times2 = 165 - 10 = 155$, $D(X^3) = D(X^2) + 5\times4 = 275 + 20 = 295$; $Z(X^3) = 155/295 = 0.525$.

The Simplex multipliers calculated for the solution X^3 are given in Table 10.1.4. With the help of these multipliers, we calculate the comparative costs $x_{13}(3, 0)$; $x_{14}(5, -1)$; $x_{21}(0,-3)$; $x_{23}(4, 0)$; $x_{32}(-1, -7)$; $x_{34}(2, -4)$;

$\Delta_{13} = 295 \times 3 - 155 \times 0 = 885$; $\Delta_{14} = 295 \times 5 + 155 = 1630$; $\Delta_{21} = 295 \times 0 + 155 \times 3 = 465$; $\Delta_{23} = 295 \times 4 - 155 \times 0 = 1180$; $\Delta_{32} = 295 \times (-1) - 155 \times (-7) = -295 + 1085 = 790$; $\Delta_{34} = 295 \times 2 - 155 \times (-4) = 590 + 620 = 1210$. Note that $\Delta_{ij} > 0$, i.e., X^3 represents the optimal solution of the problem we considered.

Table 10.1.4

					a_i	u_i^1	u_i^2
	5 2 1	5 4 3	 1 6	 2 5	10	0	0
	 1 2	5 6 4	 3 7	15 5 1	20	1	2
	10 6 2	 1 3	20 5 4	 3 3	30	1	4
b_j	15	10	20	15			
v_j^1	1	3	3	0			
v_j^2	2	4	1	3			

10.2. FRACTIONAL TRANSPORTATION PROBLEM WITH LIMITED CAPACITIES

In this section we consider again the fractional transportation problem (10.1.1)-(10.1.4) adding the new condition that, on each route, the quantity that is transported is limited. Hence, the problem is:

$$\min Z(X) = N(X)/D(X) = \left[\sum_{i=1}^{m}\sum_{j=1}^{n} c_{ij}x_{ij} + c_0\right] \Big/ \left[\sum_{i=1}^{m}\sum_{j=1}^{n} d_{ij}x_{ij} + d_0\right] \quad ,(10.2.1)$$

subject to

$$\sum_{j=1}^{n} x_{ij} = a_i , \quad i = 1, ..., m, \qquad (10.2.2)$$

$$\sum_{i=1}^{m} x_{ij} = b_j , \quad j = 1, ..., n, \qquad (10.2.3)$$

$$0 \le x_{ij} \le g_{ij} \text{ for all } i \text{ and } j \qquad (10.2.4)$$

From now on, we assume that the conditions (10.1.5) and (10.1.5') are satisfied. Due to the condition (10.2.4), equation (10.1.5') is a necessary but not a sufficient condition for the existence of a solution. Therefore, for the problem (10.2.1)-(10.2.4), we assume that the domain of the feasible solutions is non-empty. The solution of the problem (10.2.1)-(10.2.4) is obtained by transforming it into a three-dimensional transportation problem (without capacity conditions). To do this, we write the constraint (10.2.4) in the form $\sum_{k=1}^{2} y_{ijk} = E_{ij}$, where $E_{ij} = g_{ij}$, $y_{ij1} = x_{ij}$ and y_{ij2} is a slack variable corresponding to this constraint. Also, we introduce the notations

$$l_{ij1} = c_{ij}; \quad l_{ij2} = 0; \quad m_{ij1} = d_{ij}; \quad m_{ij2} = 0;$$

$$A_{i1} = a_i; \quad A_{i2} = \sum_{j=1}^{n} g_{ij} - a_i; \quad B_{1j} = b_j; \quad B_{2j} = \sum_{i=1}^{m} g_{ij} - b_j$$

With these notations, the problem (10.2.1)-(10.2.4) is written as the three-dimensional transportation problem

$$\min Z(X) = N(X)/D(X) = \left[\sum_{i=1}^{m}\sum_{j=1}^{n}\sum_{k=1}^{2} l_{ijk}x_{ijk} + c_0\right] \Big/ \left[\sum_{i=1}^{m}\sum_{j=1}^{n}\sum_{k=1}^{2} m_{ijk}x_{ijk} + d_0\right] \quad (10.2.5)$$

subject to

$$\sum_{j=1}^{n} x_{ijk} = A_{ik} \quad i = 1, ..., m; \ k = 1, 2 \qquad (10.2.6)$$

$$\sum_{i=1}^{m} x_{ijk} = B_{kj} \quad j = 1, ..., n; \ k = 1, 2 \qquad (10.2.7)$$

$$x_{ij1} + x_{ij2} = E_{ij} \quad i = 1, ..., m; \ j = 1, ..., n \qquad (10.2.8)$$

$$x_{ij1}, x_{ij2} \ge 0 \text{ for all } i \text{ and } j \qquad (10.2.9)$$

Additionally, the following conditions are fulfilled

$$\sum_{i=1}^{m} A_{ik} = \sum_{j=1}^{n} B_{kj} \; ; \; B_{1j} + B_{2j} = \sum_{i=1}^{m} E_{ij} \; ; \; \sum_{j=1}^{n} E_{ij} = A_{i1} + A_{i2} \; ;$$

$$\sum_{i=1}^{m} (A_{i1} + A_{i2}) = \sum_{j=1}^{n} (B_{1j} + B_{2j}) = \sum_{i=1}^{m} \sum_{j=1}^{n} E_{ij} \; . \qquad (10.2.10)$$

Let $I = \{1, ..., m\} \times \{1, ..., n\} \times \{1, 2\}$ and $I_X = \{(i, j, k) \in I \,|x_{ijk} > 0\}$

The Simplex multipliers u_{ik}^1, u_{ik}^2, v_{jk}^1, v_{jk}^2, w_{ij}^1, w_{ij}^2 $(i = 1, ..., m; \; j = 1, ..., n; \; k = 1, 2)$ are defined such that

$$u_{ik}^1 + v_{jk}^1 + w_{ij}^1 = l_{ijk} \qquad (10.2.11)$$

$$(i, j, k) \in I_X.$$

$$u_{ik}^2 + v_{jk}^2 + w_{ij}^2 = m_{ijk} \qquad (10.2.12)$$

These two systems can be solved independently. Each system has $mn + 2m + 2n$ unknowns and $mn + m + n - 1$ equations. For solution values are given to$m + n + 1$ variables such that the system can be solved.

Knowing the Simplex multipliers, one can evaluate the comparative costs for the variables from outside the base: $l'_{ijk} = l_{ijk} - u_{ik}^1 - v_{jk}^1 - w_{ij}^1$, $m'_{ijk} = m_{ijk} - u_{ik}^2 - v_{jk}^2 - w_{ij}^2$.

To establish an optimum criterion for the problem (10.2.5)-(10.2.9), we express the function Z in terms of the variables from outside the base

$$N(X) = \sum_{i=1}^{m} \sum_{j=1}^{n} \sum_{k=1}^{2} l_{ijk} x_{ijk} =$$

$$= \sum_{i=1}^{m} \sum_{j=1}^{n} \sum_{k=1}^{2} l_{ijk} x_{ijk} + \sum_{i=1}^{m} \sum_{k=1}^{2} \left[A_{ik} - \sum_{j=1}^{n} x_{ijk} \right] u_{ik}^1 +$$

$$+ \sum_{j=1}^{n} \sum_{k=1}^{2} \left[B_{kj} - \sum_{i=1}^{m} x_{ijk} \right] v_{jk}^1 + \sum_{i=1}^{m} \sum_{j=1}^{n} \left[E_{ij} - \sum_{k=1}^{2} x_{ijk} \right] w_{ij}^1 =$$

$$= \sum_{i=1}^{m} \sum_{j=1}^{n} \sum_{k=1}^{2} (l_{ijk} - u_{ik}^1 - v_{jk}^1 - w_{ij}^1) \, x_{ijk} + \sum_{i=1}^{m} \sum_{k=1}^{2} A_{ik} u_{ik}^1 + \sum_{j=1}^{n} \sum_{k=1}^{2} B_{kj} v_{kj}^1 +$$

$$+ \sum_{i=1}^{m} \sum_{j=1}^{n} E_{ij} w_{ij}^1 = \sum_{(i,j,k) \in I - I_X} l'_{ijk} x_{ijk} + V_1,$$

where

$$V_1 = \sum_{i=1}^{m} \sum_{k=1}^{2} A_{ik} u_{ik}^1 + \sum_{j=1}^{n} \sum_{k=1}^{2} B_{kj} v_{jk}^1 + \sum_{i=1}^{m} \sum_{j=1}^{n} E_{ij} w_{ij}^1$$

Similarly,

$$D(X) = \sum_{i=1}^{m} \sum_{j=1}^{n} \sum_{k=1}^{2} m_{ijk} x_{ijk} = \sum_{(i,j,k) \in I - I_X} m'_{ijk} x_{ijk} + V_2,$$

where

$$V_2 = \sum_{i=1}^{m} \sum_{k=1}^{2} A_{ik} u_{ik}^2 + \sum_{j=1}^{n} \sum_{k=1}^{2} B_{kj} v_{kj}^2 + \sum_{i=1}^{m} \sum_{j=1}^{n} E_{ij} w_{ij}^2$$

Hence, $Z(X) = \left[\sum_{(i,j,k) \in I - I_x} l'_{ijk} x_{ijk} + V_1 \right] \Big/ \left[\sum_{(i,j,k) \in I - I_x} m'_{ijk} x_{ijk} + V_2 \right]$.

For $(i, j, k) \in I - I_X$

$$\frac{\partial Z}{\partial x_{ijk}} = \frac{l'_{ijk} \left[\sum_{(i,j,k) \in I - I_x} m'_{ijk} x_{ijk} + V_2 \right] - m'_{ijk} \left[\sum_{(i,j,k) \in I - I_x} l'_{ijk} x_{ijk} + V_1 \right]}{\left[\sum_{(i,j,k) \in I - I_x} m'_{ijk} x_{ijk} + V_2 \right]^2}$$

Let $X^* = (x^*_{ijk})$ be a feasible solution of the problem (10.2.5)-(10.2.9). Since, in X^*, the variables from outside the base have zero value, the partial derivatives of Z calculated at X^* are

$$\frac{\partial Z}{\partial x_{ijk}} \Big|_{x_{ijk} = x^*_{ijk}} = (l'_{ijk} V_2 - m'_{ijk} V_1)/V_2^2$$

Denote $\Delta_{ijk} = l'_{ijk} V_2 - m'_{ijk} V_1$.

For the variables from the base, $\Delta_{ijk} = 0$, whereas for the variables from outside the base it is easy to calculate Δ_{ijk} knowing the Simplex multipliers. The solution $X^* = (x^*_{ijk})$ can be improved if there is at least a value $\Delta_{ijk} < 0$. Thus, we have the following optimum criterion.

Theorem 10.2.1. *A feasible solution* $X^* = (x^*_{ijk})$ *of the problem* (10.2.5)-(10.2.9) *is optimal solution if*

$$\Delta_{ijk} = l'_{ijk} V_2 - m'_{ijk} V_1 \geq 0$$

for all the variables from outside the base.

If we also have the values $\Delta_{ijk} < 0$, then one proceeds as in the usual transportation problem. Determine $\Delta_{i_0 j_0 k_0} = \min \{\Delta_{ijk} \mid \Delta_{ijk} < 0\}$ and then introduce in the base the variable $x_{i_0 j_0 k_0}$.

10.3. SPECIAL CASE OF FRACTIONAL TRANSPORTATION PROBLEM

In this section we analyze a fractional transportation problem with the objective function of the form

$$f(x) = (c'x + c_0)(d'x + d_0)/(e'x + e_0)^2 \tag{10.3.1}$$

where $c'x + c_0 \geq 0$, $d'x + d_0 > 0$, $e'x + e_0 > 0$.

Aggarwal [14] established that the function f defined by (10.3.1) is explicit quasiconcave. Such functions attain their minimum value in a vertex of the feasible solutions domain, and a local minimum is not necessarily a global minimum. The two-dimensional case was analyzed by Sharma [837], the three-dimensional case by Sharma [839], whereas the three- and multi-dimensional case was considered by Stancu-Minasian [895], [899], [901].

Consider the following three-dimensional transportation problem

$$\min Z = \frac{\left[\sum_{i=1}^{m}\sum_{j=1}^{n}\sum_{k=1}^{p} l_{ijk}x_{ijk}\right]\left[\sum_{i=1}^{m}\sum_{j=1}^{n}\sum_{k=1}^{p} m_{ijk}x_{ijk}\right]}{\left[\sum_{i=1}^{m}\sum_{j=1}^{n}\sum_{k=1}^{p} n_{ijk}x_{ijk}\right]^2} \tag{10.3.2}$$

subject to

$$\sum_{j=1}^{n}\sum_{k=1}^{p} x_{ijk} = a_i\,, \quad i = 1, ..., m \tag{10.3.3}$$

$$\sum_{i=1}^{m}\sum_{k=1}^{p} x_{ijk} = b_j\,, \quad j = 1, ..., n \tag{10.3.4}$$

$$\sum_{i=1}^{m}\sum_{j=1}^{n} x_{ijk} = c_k\,, \quad k = 1, ..., p \tag{10.3.5}$$

$$x_{ijk} \geq 0, \quad i = 1, ..., m; \; j = 1, ..., n; \; k = 1, ..., p \tag{10.3.6}$$

On the domain of the feasible solutions S, determined by the constraints (10.3.3)-(10.3.6), we assume

$$\sum_{i=1}^{m} a_i = \sum_{j=1}^{n} b_j = \sum_{k=1}^{p} c_k, \qquad a_i\,,\, b_j\,,\, c_k > 0, \tag{10.3.7}$$

$$\sum_{i=1}^{m}\sum_{j=1}^{n}\sum_{k=1}^{p} l_{ijk}x_{ijk} \geq 0, \quad \sum_{i=1}^{m}\sum_{j=1}^{n}\sum_{k=1}^{p} m_{ijk}x_{ijk} > 0,$$

$$\sum_{i=1}^{m}\sum_{j=1}^{n}\sum_{k=1}^{p} n_{ijk}x_{ijk} > 0\,. \tag{10.3.8}$$

The relations (10.3.7) represent the necessary and sufficient conditions for the existence of a solution of the problem (10.3.2)-(10.3.6).

To find a local minimum criterion, we shall proceed as in Section 10.2. Consider the Simplex multipliers u_i^1, u_i^2, u_i^3, $(i = 1, ..., m)$, v_j^1, v_j^2, v_j^3 $(j = 1, ..., n)$ and w_k^1, w_k^2, w_k^3 $(k = 1, ..., p)$ defined such that

$$u_i^1 + v_j^1 + w_k^1 = l_{ijk} \tag{10.3.9}$$

$$u_i^2 + v_j^2 + w_k^2 = m_{ijk} \qquad (i, j, k) \in I_X \tag{10.3.10}$$

$$u_i^3 + v_j^3 + w_k^3 = n_{ijk} \tag{10.3.11}$$

The systems (10.3.9), (10.3.10) and (10.3.11) can be solved independently. For example, the system (10.3.9) has $m + n + p - 2$ equations and $m + n + p$ variables. For solution, fix the values of two variables and find the other. Having determined the Simplex multipliers, one calculates the comparative costs for the variables from outside the base:

$$l'_{ijk} = l_{ijk} - u_i^1 - v_j^1 - w_k^1, \quad m'_{ijk} = m_{ijk} - u_i^2 - v_j^2 - w_k^2,$$

$$n'_{ijk} = n_{ijk} - u_i^3 - v_j^3 - w_k^3, \quad (i, j, k) \in I - I_X$$

The variable Z is expressed in terms of the variables from outside the base

$$\sum_{i=1}^{m}\sum_{j=1}^{n}\sum_{k=1}^{p} l_{ijk}x_{ijk} =$$

$$= \sum_{i=1}^{m}\sum_{j=1}^{n}\sum_{k=1}^{p} l_{ijk}x_{ijk} + \sum_{i=1}^{m}\left[a_i - \sum_{j=1}^{n}\sum_{k=1}^{p} x_{ijk}\right] u_i^1 +$$

$$+ \sum_{j=1}^{n}\left[b_j - \sum_{i=1}^{m}\sum_{k=1}^{p} x_{ijk}\right] v_j^1 + \sum_{k=1}^{p}\left[c_k - \sum_{i=1}^{m}\sum_{j=1}^{n} x_{ijk}\right] w_k^1 =$$

$$= \sum_{i=1}^{m}\sum_{j=1}^{n}\sum_{k=1}^{p} (l_{ijk} - u_i^1 - v_j^1 - w_k^1)\, x_{ijk} +$$

$$+ \sum_{i=1}^{m} a_i u_i^1 + \sum_{j=1}^{n} b_j v_j^1 + \sum_{k=1}^{p} c_k w_k^1 = \sum_{(i,j,k)\,\in\, I - I_x} l'_{ijk}x_{ijk} + V_1 ,$$

where

$$V_1 = \sum_{i=1}^{m} a_i u_i^1 + \sum_{j=1}^{n} b_i v_j^1 + \sum_{k=1}^{p} c_k w_k^1 .$$

Similarly, one obtains

$$\sum_{i=1}^{m}\sum_{j=1}^{n}\sum_{k=1}^{p} m_{ijk}x_{ijk} = \sum_{(i,j,k)\,\in\, I - I_x} m'_{ijk}x_{ijk} + V_2$$

$$\sum_{i=1}^{m}\sum_{j=1}^{n}\sum_{k=1}^{p} n_{ijk}x_{ijk} = \sum_{(i,j,k)\,\in\, I - I_x} n'_{ijk}x_{ijk} + V_3 .$$

where:

$$V_2 = \sum_{i=1}^{m} a_i u_i^2 + \sum_{j=1}^{n} b_j v_j^2 + \sum_{k=1}^{p} c_k w_k^2 ,$$

$$V_3 = \sum_{i=1}^{m} a_i u_i^3 + \sum_{j=1}^{n} b_j v_j^3 + \sum_{k=1}^{p} c_k w_k^3 .$$

Hence,

$$Z = \frac{\left[\sum_{(i,j,k)\,\in\, I - I_x} l'_{ijk}x_{ijk} + V_1\right]\left[\sum_{(i,j,k)\,\in\, I - I_x} m'_{ijk}x_{ijk} + V_2\right]}{\left[\sum_{(i,j,k)\,\in\, I - I_x} n'_{ijk}x_{ijk} + V_3\right]^2}$$

For $(i, j, k) \in I - I_X$, we have

$$\frac{\partial Z}{\partial x_{ijk}} = \frac{l'_{ijk}\left[\sum_{(i,j,k)\,\in\, I - I_x} m'_{ijk}x_{ijk} + V_2\right]\left[\sum_{(i,j,k)\,\in\, I - I_x} n'_{ijk}x_{ijk} + V_3\right]^2}{\left[\sum_{(i,j,k)\,\in\, I - I_x} n'_{ijk}x_{ijk} + V_3\right]^4} +$$

$$+\frac{m'_{ijk}\left[\sum_{(i,j,k)\in I-I_x} l'_{ijk}x_{ijk}+V_1\right]\left[\sum_{(i,j,k)\in I-I_x} n'_{ijk}x_{ijk}+V_3\right]^2}{\left[\sum_{(i,j,k)\in I-I_x} n'_{ijk}x_{ijk}+V_3\right]^4}-$$

$$-\frac{2n'_{ijk}\left[\sum_{(i,j,k)\in I-I_x} l'_{ijk}x_{ijk}+V_1\right]\left[\sum_{(i,j,k)\in I-I_x} m'_{ijk}x_{ijk}+V_2\right]\left[\sum_{(i,j,k)\in I-I_x} n_{ijk}x_{ijk}+V_3\right]}{\left[\sum_{(i,j,k)\in I_x} n'_{ijk}x_{ijk}+V_3\right]^4}$$

Evaluate these derivatives in the point $X^* = (x^*_{ijk})$,

$$\frac{\partial Z}{\partial x_{ijk}}\bigg|_{x_{ijk}=x^*_{ijk}} = \frac{l'_{ijk}V_2V_3^2 + m'_{ijk}V_1V_3^2 - 2n'_{ijk}V_1V_2V_3}{V_3^4} =$$

$$= \frac{l'_{ijk}V_2V_3 + m'_{ijk}V_1V_3 - 2n'_{ijk}V_1V_2}{V_3^3}.$$

If we denote by

$$\Delta_{ijk} = l'_{ijk}V_2V_3 + m'_{ijk}V_1V_3 - 2n'_{ijk}V_1V_2,$$

we note that the solution X^* can be improved if there is at least a value $\Delta_{ijk} < 0$. Thus, we arrive at the following local minimum criterion:

Theorem 10.3.1. *A feasible solution $X^* = (x^*_{ijk})$ of the problem (10.3.2)-(10.3.6) is a local optimum if*

$$\Delta_{ijk} = l'_{ijk}V_2V_3 + m'_{ijk}V_1V_3 - 2n'_{ijk}V_1V_2 \geq 0,$$

for all the variables from outside the base.

If a certain solution is not optimal, the variable which enters the base, and the variable which exits the base are determined, and then a new solution is considered with the same argumentation.

Numerical example. Consider the problem with $a_1 = 20$; $a_2 = 7$; $a_3 = 15$; $a_4 = 6$; $b_1 = 10$; $b_2 = 16$; $b_3 = 17$; $b_4 = 5$; $c_1 = 14$; $c_2 = 25$; $c_3 = 9$ and with the three-dimensional cost matrix as in Figure 10.3.1.

An initial solution X_0 obtained via the North-West corner method is given in Figure 10.3.2.

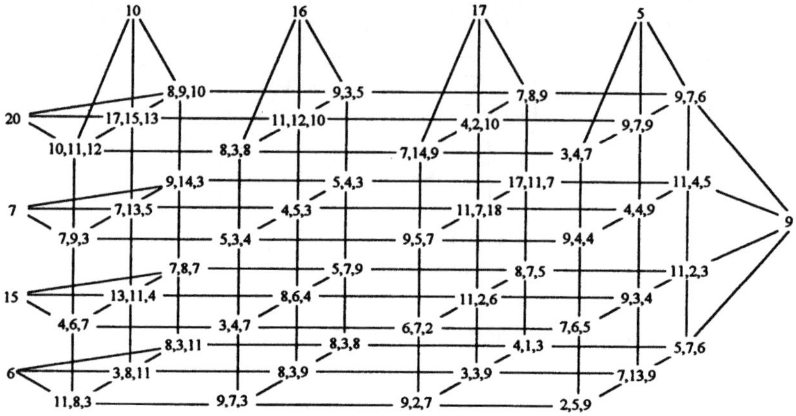

Figure 10.3.1

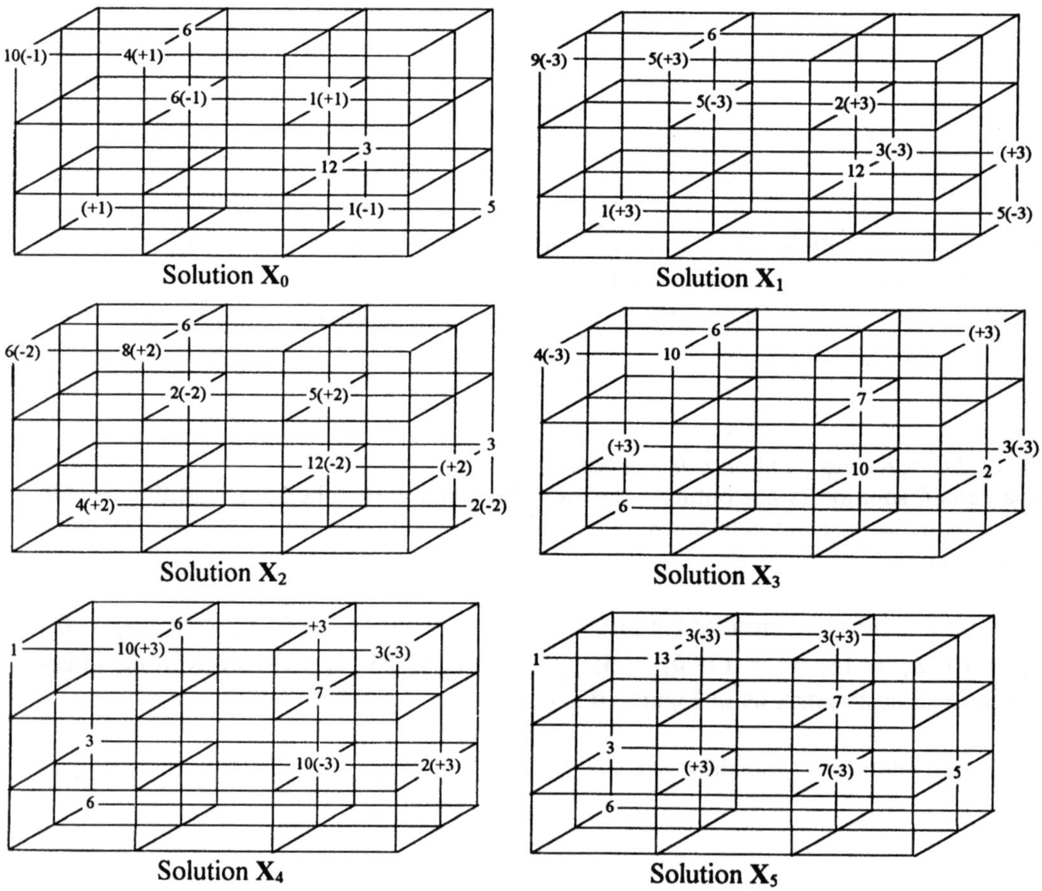

Figure 10.3.2

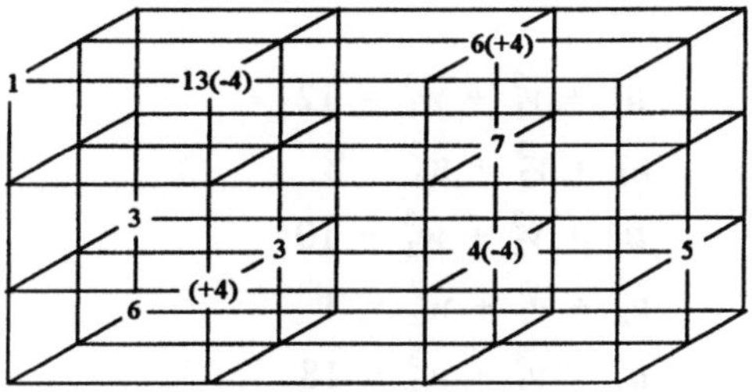

Solution $\mathbf{X}_6$

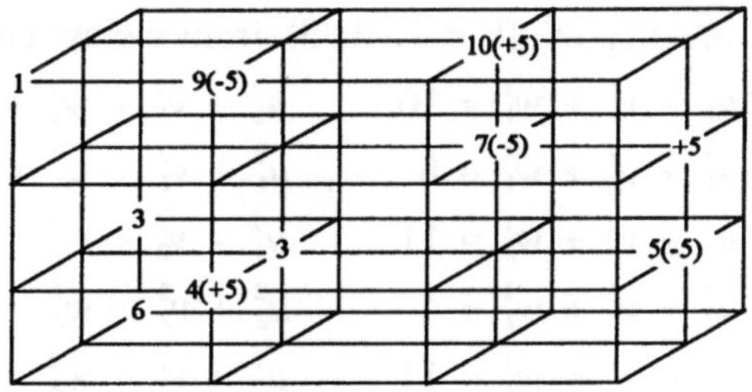

Solution $\mathbf{X}_7$

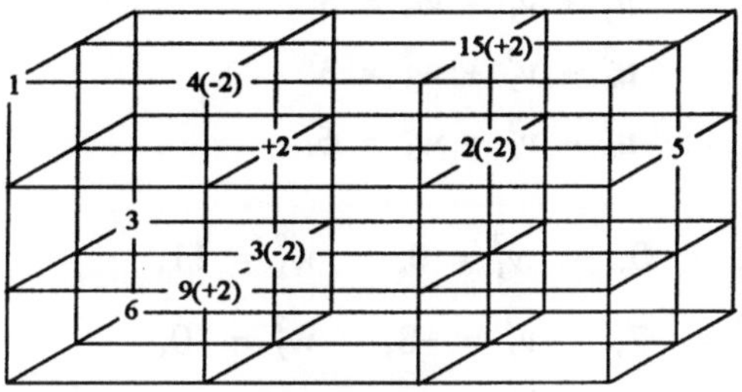

Solution $\mathbf{X}_8$

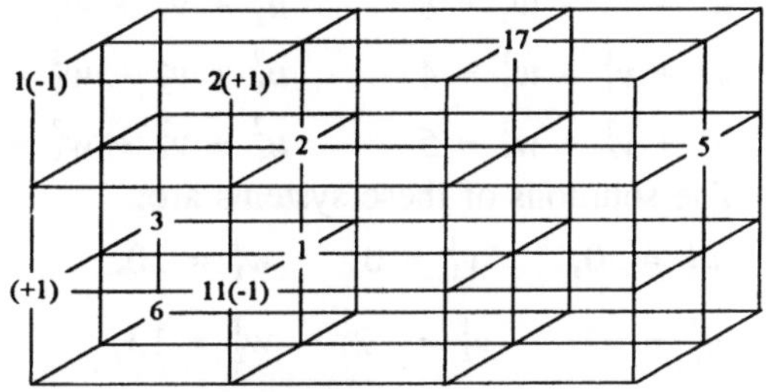

Solution $\mathbf{X}_9$

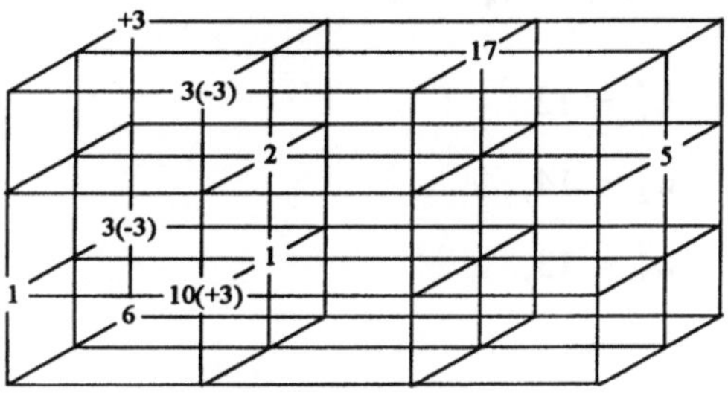

Solution $\mathbf{X}_{10}$

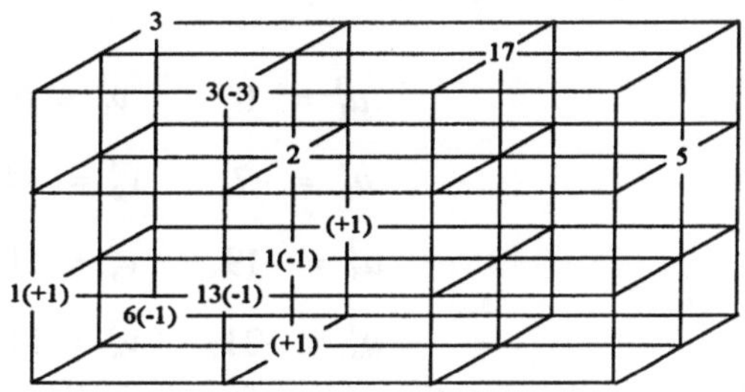

Solution $\mathbf{X}_{11}$

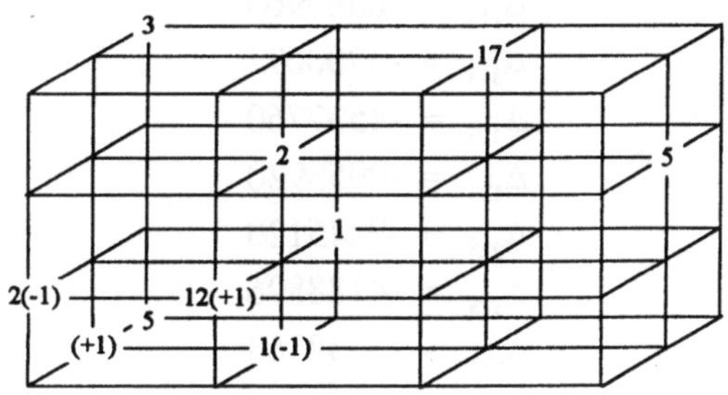

Solution $\mathbf{X}_{12}$

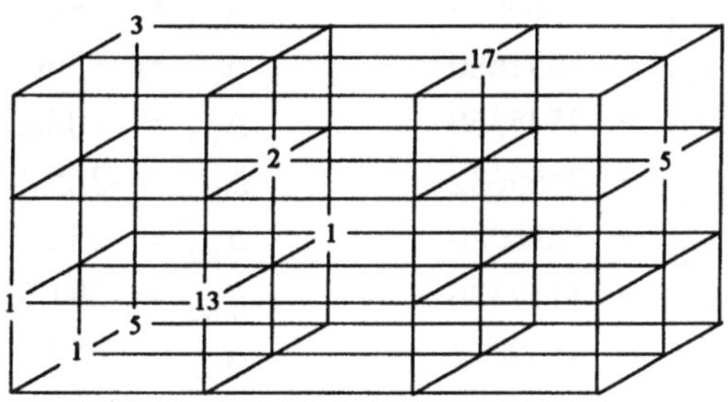

Solution $\mathbf{X}_{13}$

Figure 10.3.2 (continued)

Determine the quantities u_i^1, u_i^2, u_i^3 (i = 1, 2, 3, 4), v_j^1, v_j^2, v_j^3 (j = 1, 2, 3, 4) and w_k^1, w_k^2, w_k^3 (k = 1, 2, 3) from the systems

$$
\begin{array}{lll}
u_1^1 + v_1^1 + w_1^1 = 10 & u_1^2 + v_1^2 + w_1^2 = 11 & u_1^3 + v_1^3 + w_1^3 = 12 \\
u_1^1 + v_2^1 + w_1^1 = 8 & u_1^2 + v_2^2 + w_1^2 = 3 & u_1^3 + v_2^3 + w_1^3 = 8 \\
u_1^1 + v_2^1 + w_2^1 = 11 & u_1^2 + v_2^2 + w_2^2 = 12 & u_1^3 + v_2^3 + w_2^3 = 10 \\
u_2^1 + v_2^1 + w_2^1 = 4 & u_2^2 + v_2^2 + w_2^2 = 5 & u_2^3 + v_2^3 + w_2^3 = 3 \\
u_2^1 + v_3^1 + w_2^1 = 11 & u_2^2 + v_3^2 + w_2^2 = 7 & u_2^3 + v_3^3 + w_2^3 = 18 \\
u_3^1 + v_3^1 + w_2^1 = 11 & u_3^2 + v_3^2 + w_2^2 = 2 & u_3^3 + v_3^3 + w_2^3 = 6 \\
u_3^1 + v_3^1 + w_3^1 = 8 & u_3^2 + v_3^2 + w_3^2 = 7 & u_3^3 + v_3^3 + w_3^3 = 5 \\
u_4^1 + v_3^1 + w_3^1 = 4 & u_4^2 + v_3^2 + w_3^2 = 1 & u_4^3 + v_3^3 + w_3^3 = 3 \\
u_4^1 + v_4^1 + w_3^1 = 5 & u_4^2 + v_4^2 + w_3^2 = 7 & u_4^3 + v_4^3 + w_3^3 = 6.
\end{array}
$$

The solutions of these systems are:

$$
\begin{array}{llllll}
u_1^1 = 0, & v_1^1 = 0, & w_1^1 = 10, & u_1^2 = 0, & v_1^2 = 0, & w_1^2 = 11, \\
u_2^1 = -7, & v_2^1 = -2, & w_2^1 = 13, & u_2^2 = -7, & v_2^2 = -8, & w_2^2 = 0, \\
u_3^1 = -7, & v_3^1 = 5, & w_3^1 = 10, & u_3^2 = -12, & v_3^2 = -6, & w_3^2 = 5, \\
u_4^1 = -11, & v_4^1 = 6, & & u_4^2 = -18, & v_4^2 = 0 &
\end{array}
$$

$$
\begin{array}{lll}
u_1^3 = 0, & v_1^3 = 0, & w_1^3 = 12, \\
u_2^3 = -7, & v_2^3 = -4, & w_2^3 = 14, \\
u_3^3 = -19, & v_3^3 = 11, & w_3^3 = 13, \\
u_4^3 = -21, & v_4^3 = 14, &
\end{array}
$$

Calculate the quantities $\Delta_{i,j,k}$ for $(i, j, k) \in I - I_X$

Δ_{211} = 1750048	Δ_{441} = 4581120	Δ_{442} = 1055584
Δ_{311} = -2460064	Δ_{112} = -4902400	Δ_{113} = 8560960
Δ_{411} = 5551680	Δ_{212} = 6364800	Δ_{313} = -3700672
Δ_{221} = 7535360	Δ_{312} = -1082304	Δ_{413} = -4537760
Δ_{321} = -2465632	Δ_{412} = -3657216	Δ_{123} = -9953920
Δ_{421} = 5886080	Δ_{322} = -2582688	Δ_{223} = -7245120
Δ_{131} = 4117536	Δ_{422} = -2913696	Δ_{323} = -4188896
Δ_{231} = 3539072	Δ_{132} = 4591680	Δ_{423} = -3338368
Δ_{331} = 2445568	Δ_{432} = -6866560	Δ_{233} = 3487840
Δ_{431} = 1577280	Δ_{142} = 1807936	Δ_{143} = 1904928
Δ_{141} = 2386432	Δ_{242} = 8270400	Δ_{243} = 1988576

$\Delta_{241} = 3912480$ $\Delta_{342} = 1905920$ $\Delta_{343} = -1582720$

$\Delta_{341} = 1368800$

Since not all the quantities Δ_{ijk} are non-negative, we calculate $\Delta_{413} = \min \{\Delta_{ijk} | \Delta_{ijk} < 0\} = -4\,537\,760$. The variable x_{413} is brought into the base, and the solution X_1 is obtained (Figure 10.3.2). In Figure 10.3.2 one can follow the successive solutions until the optimal solution X_{13} is obtained, whereas in Table 10.3.1 one can follow the values of Z corresponding to this solution.

Table 10.3.1

Sol.	X_0	X_1	X_2	X_3	X_4	X_5	X_6
V_1	418	427	452	472	439	427	397
V_2	312	318	263	245	212	212	194
V_3	368	387	429	457	454	459	445
$Z = \frac{V_1 V_3}{V_3^2}$	0.96301	0.87812	0.64592	0.55370	0.51542	0.44500	0.38890

Sol.	X_7	X_8	X_9	X_{10}	X_{11}	X_{12}	X_{13}
V_1	349	264	232	223	219	217	211
V_2	199	188	173	167	173	176	179
V_3	457	437	417	413	419	422	424
$Z = \frac{V_1 V_3}{V_3^2}$	0.33087	0.25298	0.23081	0.21833	0.21580	0.21446	0.21008

10.4. DUALITY FOR FRACTIONAL TRANSPORTATION PROBLEM

We introduce two dual problems for fractional transportation problem. The first dual problem was considered by Swarup [954], and the second by Stancu-Minasian and Teghem, Jr. [918].

First, we construct the dual problem for fractional transportation problem of the form

$$\min Z = N(X)/D(X) = \sum_{i=1}^{m}\sum_{j=1}^{n} c_{ij}x_{ij} \Big/ \sum_{i=1}^{m}\sum_{j=1}^{n} d_{ij}x_{ij} \tag{10.4.1}$$

subject to

$$\sum_{j=1}^{n} x_{ij} = a_i \,, \quad i = 1, ..., m \tag{10.4.2}$$

$$\sum_{i=1}^{m} x_{ij} = b_j \,, \quad j = 1, ..., n, \tag{10.4.3}$$

$$x_{ij} \geq 0 \text{ for all } i \text{ and } j. \tag{10.4.4}$$

We assume that the problem (10.4.1)-(10.4.4) is balanced, that the domain of the feasible solutions (S) is a regular set, and that the denominator of Z is strict positive on S. The problem (10.1.1)-(10.1.4) can be brought to the form (10.4.1)-(10.4.4). We will show that the dual of the problem (10.4.1)-(10.4.4) is the problem

$$\max W = \left[\sum_{i=1}^{m} a_i u_i + \sum_{j=1}^{n} b_j v_j\right] \Big/ \left[\sum_{i=1}^{m} a_i p_i + \sum_{j=1}^{n} b_j q_j\right], \tag{10.4.5}$$

subject to

$$(d_{ij}-p_i-q_j)\left[\sum_{i=1}^{m} a_i u_i + \sum_{j=1}^{n} b_j v_j\right] - (c_{ij}-u_i-v_j)\left[\sum_{i=1}^{m} a_i p_i + \sum_{j=1}^{n} b_j q_j\right] \leq 0 \tag{10.4.6}$$

for all i and j,

$$\sum_{i=1}^{m} a_i p_i + \sum_{j=1}^{n} b_j q_j \geq 0, \tag{10.4.7}$$

where u_i, v_j, p_i and q_j for all indices i and j does not have sign restrictions, whereas

$\sum_{i=1}^{m} a_i u_i + \sum_{j=1}^{n} b_j v_j$ and $\sum_{i=1}^{m} a_i p_i + \sum_{j=1}^{n} b_j q_j$ are not simultaneously zero.

The duality theorems are given next.

Theorem 10.4.1. *If* $x_{ij}(i = 1, ..., m; j = 1, ..., n)$ *is a feasible solution for the problem* (10.4.1)-(10.4.4) *whereas* u_i, v_j, p_i, q_j $(i = 1, ..., m; j = 1, ..., n)$ *is a feasible solution for the problem* (10.4.5)-(10.4.7), *then*

$$\frac{\sum_{i=1}^{m}\sum_{j=1}^{n} c_{ij}x_{ij}}{\sum_{i=1}^{m}\sum_{j=1}^{n} d_{ij}x_{ij}} \geq \frac{\sum_{i=1}^{m} a_i u_i + \sum_{j=1}^{n} b_j v_j}{\sum_{i=1}^{m} a_i p_i + \sum_{j=1}^{n} b_j q_j} \tag{10.4.8}$$

Proof. Multiplying relations (10.4.2) and (10.4.3) by u_i, p_i, v_j and q_j, and summing up after i or j, yields

$$\sum_{i=1}^{m}\sum_{j=1}^{n} u_i x_{ij} = \sum_{i=1}^{m} a_i u_i, \quad \sum_{i=1}^{m}\sum_{j=1}^{n} p_i x_{ij} = \sum_{i=1}^{m} a_i p_i \tag{10.4.9}$$

$$\sum_{j=1}^{n}\sum_{i=1}^{m} v_j x_{ij} = \sum_{j=1}^{n} b_j v_j, \quad \sum_{j=1}^{n}\sum_{i=1}^{m} q_j x_{ij} = \sum_{j=1}^{n} b_j q_j. \tag{10.4.10}$$

Multiplying (10.4.6) by $x_{ij} \geq 0$ and summing after all i and j, yields

$$\left\{\sum_{i=1}^{m}\sum_{j=1}^{n} d_{ij}x_{ij} - \sum_{i=1}^{m}\sum_{j=1}^{n} p_i x_{ij} - \sum_{j=1}^{n}\sum_{i=1}^{m} q_j x_{ij}\right\}\left[\sum_{i=1}^{m} a_i u_i + \sum_{j=1}^{n} b_j v_j\right] - \tag{10.4.11}$$

$$- \left\{\sum_{i=1}^{m}\sum_{j=1}^{n} c_{ij}x_{ij} - \sum_{i=1}^{m}\sum_{j=1}^{n} u_i x_{ij} - \sum_{j=1}^{n}\sum_{i=1}^{m} v_j x_{ij}\right\}\left[\sum_{i=1}^{m} a_i p_i + \sum_{j=1}^{n} b_j q_j\right] \leq 0$$

or, using relations (10.4.9)-(10.4.10),

$$\left[\sum_{i=1}^{m}\sum_{j=1}^{n} d_{ij}x_{ij} - \sum_{i=1}^{m} p_i a_i - \sum_{j=1}^{n} q_j b_j\right]\left[\sum_{i=1}^{m} a_i u_i + \sum_{j=1}^{n} b_j v_j\right] -$$

$$- \left[\sum_{i=1}^{m}\sum_{j=1}^{n} c_{ij}x_{ij} - \sum_{i=1}^{m} a_i u_i - \sum_{j=1}^{n} b_j v_j\right]\left[\sum_{i=1}^{m} a_i p_i + \sum_{j=1}^{n} b_j q_j\right] \leq 0.$$

A simple calculation yields the inequality

$$\left[\sum_{i=1}^{m}\sum_{j=1}^{n} d_{ij}x_{ij}\right]\left[\sum_{i=1}^{m} a_i u_i + \sum_{j=1}^{n} b_j v_j\right] - \left[\sum_{i=1}^{m}\sum_{j=1}^{n} c_{ij}x_{ij}\right]\left[\sum_{i=1}^{m} a_i p_i + \sum_{j=1}^{n} b_j q_j\right] \leq 0. \tag{10.4.12}$$

Two cases are possible:

Case 1. If $\sum_{i=1}^{m} a_i p_i + \sum_{j=1}^{n} b_j q_j > 0$, then (10.4.12) yields the inequality (10.4.8).

Case 2. If $\sum_{i=1}^{m} a_i p_i + \sum_{j=1}^{n} b_j q_j = 0$ then $\sum_{i=1}^{m} a_i u_i + \sum_{j=1}^{n} b_j v_j \neq 0$, and from (10.4.12) it follows

$$\left[\sum_{i=1}^{m} a_i u_i + \sum_{j=1}^{n} b_j v_j\right]\left[\sum_{i=1}^{m}\sum_{j=1}^{n} d_{ij}x_{ij}\right] \leq 0, \text{ i.e., } \sum_{i=1}^{m} a_i u_i + \sum_{j=1}^{n} b_j v_j < 0$$

since we assumed that the denominator of Z is strict positive pe S.

Hence, for the feasible solution u_i, v_j, p_i and q_j of the dual problem, the right hand side of (10.4.8) approaches $-\infty$. It follows that the inequality (10.4.8) is true in this case too.

Theorem 10.4.2. *If* $\hat{x}_{ij}(i = 1, ..., m; j = 1, ..., n)$ *is a feasible solution for the problem* (10.4.1)-(10.4.4) *whereas* $\hat{u}_i$, $\hat{v}_j$, $\hat{p}_i$, $\hat{q}_j$ $(i = 1, ..., m; j = 1, ..., n)$ *is a feasible solution for the problem* (10.4.5)-(10.4.7) *such that*

$$\frac{\sum_{i=1}^{m}\sum_{j=1}^{n} c_{ij}\hat{x}_{ij}}{\sum_{i=1}^{m}\sum_{j=1}^{n} d_{ij}\hat{x}_{ij}} = \frac{\sum_{i=1}^{m} a_i\hat{u}_i + \sum_{j=1}^{n} b_j\hat{v}_j}{\sum_{i=1}^{m} a_i\hat{p}_i + \sum_{j=1}^{n} b_j\hat{q}_j}, \tag{10.4.13}$$

then $\hat{x}_{ij}$ *is optimal solution for the problem* (10.4.1)-(10.4.4) *whereas* $\hat{u}_i$, $\hat{v}_j$, $\hat{p}_i$, $\hat{q}_j$ *is optimal solution for the problem* (10.4.5)-(10.4.7).

Proof. If $x_{ij}(i = 1, ..., m; j = 1, ..., n)$ is another feasible solution of the problem (10.4.1)-(10.4.4), whereas u_i, v_j, p_i, q_j $(i = 1, ..., m; j = 1, ..., n)$ is another feasible solution for the problem (10.4.5)-(10.4.7), then, according to Theorem 10.4.1, we have

$$\frac{\sum_{i=1}^{m}\sum_{j=1}^{n} c_{ij}x_{ij}}{\sum_{i=1}^{m}\sum_{j=1}^{n} d_{ij}x_{ij}} \geq \frac{\sum_{i=1}^{m} a_i\hat{u}_i + \sum_{j=1}^{n} b_j\hat{v}_j}{\sum_{i=1}^{m} a_i\hat{p}_i + \sum_{j=1}^{n} b_j\hat{q}_j} = \frac{\sum_{i=1}^{m}\sum_{j=1}^{n} c_{ij}\hat{x}_{ij}}{\sum_{i=1}^{m}\sum_{j=1}^{n} d_{ij}\hat{x}_{ij}} \geq \frac{\sum_{i=1}^{m} a_iu_i + \sum_{j=1}^{n} b_jv_j}{\sum_{i=1}^{m} a_ip_i + \sum_{j=1}^{n} b_jq_j}.$$

Hence, it follows that $\hat{x}_{ij}$ is an optimal solution of the problem (10.4.1)-(10.4.4), whereas $\hat{u}_i, \hat{v}_j, \hat{p}_i, \hat{q}_j$ is an optimal solution for the problem (10.4.5)-(10.4.7).

Theorem 10.4.3. *If the primal problem* (10.4.1)-(10.4.4) *has an optimal basic solution* $\hat{X} = (\hat{x}_{ij})$, *then the problem* (10.4.5)-(10.4.7) *also has an optimal solution, and the optimal values of the objective functions are equal.*

Proof. Since $\hat{X} = (\hat{x}_{ij})$ is an optimal solution, then

$$\Delta_{ij} = \bar{c}_{ij}D(\hat{X}) - \bar{d}_{ij}N(\hat{X}) = (c_{ij} - u_i^{(1)} - v_j^{(1)})\, D(\hat{X}) - \\ - (d_{ij} - u_i^{(2)} - v_j^{(2)})\, N(\hat{X}) \geq 0, \qquad (10.4.14)$$

where

$$N(\hat{X}) = \sum_{i=1}^{m}\sum_{j=1}^{n} c_{ij}\hat{x}_{ij} = \sum_{i=1}^{m} u_i^{(1)}a_i + \sum_{j=1}^{n} v_j^{(1)}b_j,$$

$$D(\hat{X}) = \sum_{i=1}^{m}\sum_{j=1}^{n} d_{ij}\hat{x}_{ij} = \sum_{i=1}^{m} u_i^{(2)}a_i + \sum_{j=1}^{n} v_j^{(2)}b_j$$

whereas $u_i^{(1)}, v_j^{(1)}, u_i^{(2)}, v_j^{(2)}$ are the Simplex multipliers.
The conditions (10.4.14) become

$$(c_{ij} - u_i^{(1)} - v_j^{(1)})\left[\sum_{i=1}^{m} u_i^{(1)}a_i + \sum_{j=1}^{n} v_j^{(1)}b_j\right] - \\ - (d_{ij} - u_i^{(2)} - v_j^{(2)})\left[\sum_{i=1}^{m} u_i^{(2)}a_i + \sum_{j=1}^{n} v_j^{(2)}b_j\right] \geq 0. \qquad (10.4.15)$$

Choose $u_i^{(1)} = u_i$, $v_j^{(1)} = v_j$; $u_i^{(2)} = p_i$, $v_j^{(2)} = q_j$, and hence the inequality (10.4.15) becomes

$$(c_{ij} - u_i - v_j)\left[\sum_{i=1}^{m} u_ia_i + \sum_{j=1}^{n} v_jb_j\right] - \\ - (d_{ij} - p_i - q_j)\left[\sum_{i=1}^{m} u_ia_i + \sum_{j=1}^{n} v_jb_j\right] \geq 0$$

which is the relation (10.4.6). We also have $\sum_{i=1}^{m} a_i p_i + \sum_{j=1}^{n} b_j q_j > 0$ i.e., (10.4.7) i.e., u_i, v_j, p_i, q_i constitutes a feasible solution for the problem (10.4.5)-(10.4.7).
We have:

$$W = \frac{\sum_{i=1}^{m} a_i u_i + \sum_{j=1}^{n} b_j v_j}{\sum_{i=1}^{m} a_i p_i + \sum_{j=1}^{n} b_j q_j} = \frac{\sum_{i=1}^{m} a_i u_i^{(1)} + \sum_{j=1}^{n} b_j v_j^{(1)}}{\sum_{i=1}^{m} a_i u_i^{(2)} + \sum_{j=1}^{n} b_j v_j^{(2)}} = \frac{\sum_{i=1}^{m}\sum_{j=1}^{n} c_{ij}\hat{x}_{ij}}{\sum_{i=1}^{m}\sum_{j=1}^{n} d_{ij}\hat{x}_{ij}} = \min Z.$$

According to Theorem 10.4.2, it follows that u_i, v_j, p_i, q_j is an optimal solution for the problem (10.4.5)-(10.4.7).
We introduce now another dual problem for fractional transportation problem, proposed by Stancu-Minasian and Teghem, Jr. [918]. Consider the following fractional transportation problem as a primal problem (PP)

$$\min f(X) = \frac{\sum_{i=1}^{m}\sum_{j=1}^{n} c_{ij}x_{ij}}{\sum_{i=1}^{m}\sum_{j=1}^{n} d_{ij}x_{ij}} \tag{10.4.16}$$

subject to

$$\sum_{j=1}^{n} x_{ij} \le a_i \qquad i = 1, ..., m \tag{10.4.17}$$

$$\sum_{i=1}^{m} x_{ij} \ge b_j \qquad j = 1, ..., n \tag{10.4.18}$$

$$x_{ij} \ge 0 \qquad i = 1, ..., m; \quad j = 1, ..., n \tag{10.4.19}$$

Assume that

(i) The set of feasible solutions is regular, i.e. non-empty and bounded.

(ii) $\sum_{i=1}^{m}\sum_{j=1}^{n} d_{ij}x_{ij} > 0$ for all feasible solutions.

Obviously, a necessary condition for problem (PP) to have a solution is

$$\sum_{i=1}^{m} a_i \ge \sum_{j=1}^{n} b_j$$

The dual problem (DP) is

$$\max g(U) = \frac{\sum_{i=1}^{m}\sum_{j=1}^{n} c_{ij}u_{ij}}{\sum_{i=1}^{m}\sum_{j=1}^{n} d_{ij}u_{ij}} \tag{10.4.20}$$

subject to

$$(-p_i + q_j) - c_{ij}\left[\sum_{i=1}^{m}\sum_{j=1}^{n} d_{ij}u_{ij}\right] + d_{ij}\left[\sum_{i=1}^{m}\sum_{j=1}^{n} c_{ij}u_{ij}\right] \le 0 \quad \forall\, i, j \tag{10.4.21}$$

$$-\sum_{i=1}^{m} a_i p_i + \sum_{j=1}^{n} b_j q_j \geq 0 \tag{10.4.22}$$

$$u_{ij}\ ,\ p_i\ ,\ q_j \geq 0 \quad \forall\ i, j \tag{10.4.23}$$

Likewise, for excluding the indeterminate form, we assume that

(iii) $\sum_{i=1}^{m}\sum_{j=1}^{n} c_{ij}u_{ij}$ and $\sum_{i=1}^{m}\sum_{j=1}^{n} d_{ij}u_{ij}$ are not simultaneously zero

To simplify the formulation, let us note

- $f_1(X) = \sum_{i=1}^{m}\sum_{j=1}^{n} c_{ij}x_{ij}$
- $f_2(X) = \sum_{i=1}^{m}\sum_{j=1}^{n} d_{ij}x_{ij}$
- $g_1(U) = \sum_{i=1}^{m}\sum_{j=1}^{n} c_{ij}u_{ij}$
- $g_2(U) = \sum_{i=1}^{m}\sum_{j=1}^{n} d_{ij}u_{ij}$

Theorem 10.4.4 (Weak duality theorem). *If $X = (x_{ij})$ $(i = 1, ..., m;\ j = 1, ..., n)$ is a feasible solution of problem (PP) and $U = (u_{ij})$, $P = (p_i), Q = (q_j)$ $(i = 1, ..., m;\ j = 1, ..., n)$ is a feasible solution of problem (DP), then*

$$f(X) \geq g(U) \tag{10.4.24}$$

Proof. From relations (10.4.17) and (10.4.18), multiplying by $(-p_i)$ and q_j respectively, and making sum for $i = 1, ..., m$ and $j = 1, ..., n$ we have

$$-\sum_{i=1}^{m}\sum_{j=1}^{n} p_i x_{ij} \geq -\sum_{i=1}^{m} a_i p_i$$

$$\sum_{i=1}^{m}\sum_{j=1}^{n} q_j x_{ij} \geq \sum_{j=1}^{n} b_j q_j$$

Or, summing these two relations, we get

$$\sum_{i=1}^{m}\sum_{j=1}^{n} (-p_i + q_j)x_{ij} \geq -\sum_{i=1}^{m} a_i p_i + \sum_{j=1}^{n} b_j q_j \tag{10.4.25}$$

Due to (10.4.19), from condition (10.4.21) of the dual problem, multiplying by x_{ij} and making sum for $i = 1, ..., m$ and $j = 1, ..., n$ we get

$$\sum_{i=1}^{m}\sum_{j=1}^{n} (-p_i + q_j)x_{ij} - f_1(X)\ g_2(U) + f_2(X)\ g_1(U) \leq 0 \tag{10.4.26}$$

From (10.4.22), (10.4.25) and (10.4.26), we obtain

$$f_1(X)\ g_2(U) - f_2(X)\ g_1(U) \geq 0 \tag{10.4.27}$$

<u>Case I</u> : If $g_2(U) > 0$

then relation (10.4.24) is obtained immediately.

Case II : If $g_2(U) = 0$
then $g_1(U) < 0$ (because of (10.4.27) and using the fact that $g_1(U)$ and $g_2(U)$ do not vanish simultaneously) so that $g(U) \to -\infty \leq f(X)$, i.e. relation (10.4.24).
The Theorem is thus completely proven.

Theorem 10.4.5. *If $\tilde{X}$ is a feasible solution of problem (PP) and $\tilde{U}$, $\tilde{P}$, $\tilde{Q}$ is a feasible solution of problem (DP) so that*

$$f(\tilde{X}) = g(\tilde{U})$$

then $\tilde{X}$ is an optimal solution for problem (PP) and $\tilde{U}$, $\tilde{P}$, $\tilde{Q}$ is an optimal solution for problem (DP).

Proof. According to Theorem 10.4.4., for each X, feasible solution of problem (PP), we have

$$f(X) \geq g(\tilde{U}) = f(\tilde{X})$$

and hence $\tilde{X}$ is an optimal solution of problem (PP).
Also, if U, P, Q is a feasible solution of problem (DP), then according to Theorem 10.4.4., we have

$$g(U) \leq f(\tilde{X}) = g(\tilde{U})$$

and hence $\tilde{U}$, $\tilde{P}$, $\tilde{Q}$ is an optimal solution of problem (DP).

Theorem 10.4.6. (Direct duality theorem). *If X^0 is an optimal solution for problem (PP), then there exist P^0 and Q^0 such that $U^0 = X^0$, P^0, Q^0 is an optimal solution for problem (DP) and the optimal values of the objective functions in both cases are equal.*

Proof. Let X^0 be feasible optimal solution for problem (PP). Let us consider a linearization of the objective function of problem (PP). Thus, we consider the linear transportation problem (LPP).

$$\min \bar{f}(X) = \sum_{i=1}^{m}\sum_{j=1}^{n} f_{ij}x_{ij}$$

subject to (10.4.17), (10.4.18) and (10.4.19) where

$$f_{ij} = c_{ij}\, f_2(X^0) - d_{ij}\, f_1(X^0) \tag{10.4.28}$$

Obviously, X^0 is a feasible solution of problem (LPP). We shall show that X^0 is an optimal solution to problem (LPP). Assume by *reductio ad absurdum* that $X^*(X^* \neq X^0)$ is the optimal solution to problem (LPP), i.e.

$$\bar{f}(X^*) < \bar{f}(X^0) \tag{10.4.29}$$

As $\bar{f}(X^0) = 0$, from 10.4.28 it follows that

$$f_1(X^*)f_2(X^0) - f_1(X^0)f_2(X^*) < 0$$

and due to assumption (ii) for X^* and X^0, it results that

$$f(X^*) < f(X^0) \tag{10.4.30}$$

But relation (10.4.30) contradicts the hypothesis that X^0 is an optimal solution for (PP). Hence, X^0 is an optimal solution to problem (LPP).
Consider the dual (DLPP) of the linear transportation problem (LPP), which is the linear problem

$$\max\ G(P, Q) = -\sum_{i=1}^{m} a_i p_i + \sum_{j=1}^{n} b_j q_j$$

subject to

$$-p_i + q_j \leq f_{ij} \quad \forall\ i, j \tag{10.4.31}$$
$$p_i,\ q_j \geq 0 \quad \forall\ i, j \tag{10.4.32}$$

Let P^0, Q^0 be an optimal solution of (DLPP). By the duality Theorem of linear programming

$$\max\ G(P^0, Q^0) = \min \bar{f}(X^0), \text{ i.e.}$$

$$-\sum_{i=1}^{m} a_i p_i^0 + \sum_{j=1}^{n} b_j q_j^0 = 0 \tag{10.4.33}$$

From (10.4.31), (10.4.32) and (10.4.33), it results that ($U^0 = X^0$, P^0, Q^0) is a feasible solution to (DP) such that

$$f(X^0) = g(U^0)$$

According to Theorem 10.4.5, it results that ($U^0 = X^0$, P^0, Q^0) is an optimal solution for problem (DP).

Theorem 10.4.7. (Converse duality theorem). *If (U^0, P^0, Q^0) is an optimal solution for problem (DP), then there exists X^0 an optimal solution for problem (PP) and the optimal values of the objective functions in both cases are equal.*

Proof : Let (U^0, P^0, Q^0) be an optimal solution for problem (DP). This is also an optimal solution for the problem (LDP).

$$Max\ h(U) = g(U^0) + \bar{g}(U)$$

subject to (10.4.21), (10.4.22) and (10.4.23)
where

$$\bar{g}(U) = \sum_{i=1}^{m}\sum_{j=1}^{n} g_{ij} u_{ij}$$

and

$$g_{ij} = c_{ij}\left[\sum_{i=1}^{m}\sum_{j=1}^{n} d_{ij} u_{ij}^0\right] - d_{ij}\left[\sum_{i=1}^{m}\sum_{j=1}^{n} c_{ij} u_{ij}^0\right]$$

Obviously, (U^0, P^0, Q^0) is a feasible solution of problem (LDP). We shall show that (U^0, P^0, Q^0) is also optimal for problem (LDP).
We assume that (U^*, P^*, Q^*) is the optimal solution to problem (LDP), i.e.

$$\bar{g}(U^0) < \bar{g}(U^*)$$

or

$$g_2(U^*)g_1(U^0) < g_1(U^*)g_2(U^0) \tag{10.4.34}$$

Case I : If $g_2(U^*) > 0$ and $g_2(U^0) > 0$, then from (10.4.34) it results that

$$g(U^0) < g(U^*)$$

Case II : If $g_2(U^*) = 0$ and $g_2(U^0) > 0$, then $g_1(U^*) > 0$ and $g(U^*) \to \infty > g(U^0)$.

Case III : If $g_2(U^*) > 0$ and $g_2(U^0) = 0$, then $g_1(U^0) < 0$ and $g(U^0) \to -\infty < g(U^*)$.

In all these three possible cases, we obtained a contradiction to (U^0, P^0, Q^0) is an optimal solution to problem (DP). Thus, (U^0, P^0, Q^0) is an optimal solution to problem (LDP). The dual problem of problem (LDP) is

(DLDP) $\quad Min\ H(Y, \theta) = g(U^0)$

subject to

$$\sum_{j=1}^{n} y_{ij} \le \theta a_i, \quad i = 1, ..., m$$

$$\sum_{i=1}^{m} y_{ij} \ge \theta b_j, \quad j = 1, ..., n$$

$$g_{ij} \le -d_{ij} f_1(Y) + c_{ij} f_2(Y) \quad \forall\ i, j$$

$$y_{ij},\ \theta \ge 0, \quad i = 1, ..., m; \ j = 1, ..., n$$

where y_{ij} and θ are the dual variables associated to constraints (10.4.21) and (10.4.22) respectively. The duality Theorem (in linear programming) ensures the existence of an optimal solution ($Y^0 = (y^0_{ij})$, θ) for problem (DLDP) with $\theta > 0$ due to assumption (i). Using the variable transformation $y_{ij} = \theta x_{ij}$ ($\theta > 0$), problem (DLDP) becomes

$$\min H(X) = g(U^0)$$

subject to (10.4.17), (10.4.18) and (10.4.19)

$$g_{ij} \le -\theta d_{ij} f_1(X) + \theta c_{ij} f_2(X) \quad \forall\ i, j \tag{10.4.35}$$

$$\theta > 0$$

so that $x^0_{ij} = y^0_{ij}/\theta$ is a feasible solution to problem (PP). We shall show that $x^0_{ij} = y^0_{ij}/\theta$ is an optimal solution to problem (PP).

From (10.4.35) multiplying by $U^0 > 0$ and making sum we have

$$-g_2(U^0)f_1(X^0) + g_1(U^0)f_2(X^0) \ge 0 \tag{10.4.36}$$

Case I : If $g_2(U^0) > 0$, because of $f_2(X^0) > 0$, (10.4.36) becomes

$$g(U^0) \ge f(X^0) \tag{10.4.37}$$

Case II : If $g_2(U^0) = 0$ then $g_1(U^0) > 0$ (because of (10.4.36) and using the fact that $g_1(U)$ and $g_2(U)$ do not vanish simultaneously) so that

$$g(U^0) \to +\infty \ge f(X^0) \tag{10.4.38}$$

From (10.4.37) and (10.4.38) and using Theorem 10.4.4 it results

$$f(X^0) = g(U^0)$$

Using weak duality Theorem (Theorem 10.4.5) it results that X^0 is an optimal solution to problem (PP). Obviously, the optimal values of the objective functions are equal.

Numerical example.
Primal problem

$$Minimize\ f(X) = \frac{x_{11} + 3x_{12} + 2x_{21} + x_{22}}{2x_{11} + 4x_{12} + x_{21} + 5x_{22}} \qquad \text{(PP)}$$

subject to

$$\begin{aligned} x_{11} + x_{12} &\le 10 \\ x_{21} + x_{22} &\le 20 \\ x_{11} + x_{21} &\ge 15 \\ x_{12} + x_{22} &\ge 15 \\ x_{ij} \ge 0 \quad i, j &= 1, 2 \end{aligned}$$

Solving, we get that $X^0(x_{11} = 10; x_{21} = 5; x_{22} = 15)$ is an optimal solution and the optimal value of the objective function is $f(X^0) = 35/100 = 0.35$
Dual problem

$$Maximize\ \ g(U) = \frac{u_{11} + 3u_{12} + 2u_{21} + u_{22}}{2u_{11} + 4u_{12} + u_{21} + 5u_{22}} \qquad \text{(DP)}$$

subject to

$$\begin{aligned} & -p_1 + q_1 + 2u_{12} + 3u_{21} - 3u_{22} \le 0 \\ & -p_1 + q_2 - 2u_{11} + 5u_{21} - 11u_{22} \le 0 \\ & -p_2 + q_1 - 3u_{11} - 5u_{12} - 9u_{22} \le 0 \\ & -p_2 + q_2 + 3u_{11} + 11u_{12} + 9u_{21} \le 0 \\ & -10p_1 - 20p_2 + 15q_1 + 15q_2 \ge 0 \\ & u_{ij}, p_i, q_j \ge 0, \quad i, j = 1, 2 \end{aligned}$$

The problem (DLPP) has the form

$$Maximize\ \ (-10p_1 - 20p_2 + 15q_1 + 15q_2)$$

subject to

$$\begin{aligned} -p_1 + q_1 &\le 30 \\ -p_1 + q_2 &\le 160 \\ -p_2 + q_1 &\le 165 \\ -p_2 + q_2 &\le -75 \\ p_1, p_2, q_1, q_2 &\ge 0 \end{aligned}$$

The optimal solution of the problem (DLPP) is

$$(p_i^0, q_j^0) = (p_1^0 = 210;\ p_2^0 = 75;\ q_1^0 = 240;\ q_2^0 = 0)$$

According to Theorem 10.4.6. $(U^0 = X^0, p_i^0, q_j^0) =$
$= (u_{11}^0 = 10;\ u_{12}^0 = 0;\ u_{21}^0 = 5;\ u_{22}^0 = 15;\ p_1^0 = 210;\ p_2^0 = 75;\ q_1^0 = 240;\ q_2^0 = 0)$ is also optimal solution of the dual problem and the optimal value of the objective function is $g(U^0) = f(X^0) = 0.35$.
This is not a basic feasible solution (extreme point solution) and therefore, when we solve the dual problem we may not get the same solution as (X^0, p_i^0, q_j^0).

Solving the dual problem, we get that$(U^*, p_i^*, q_i^*) = (u_{11}^* = 0;\ u_{12}^* = 0.0682;\ u_{21}^* = 0;\ u_{22}^* = 0.1455;\ p_1^* = 2{,}1;\ p_2^* = 0.75;\ q_1^* = 2.4;\ q_2^* = 0)$ is an optimal solution and the optimal value of the objective function is $g(U^*) = 0.35$.
Generally, if A is the matrix associated with primal problem (PP), its rank is m + n - 1. The optimal solution (X^0, p_i^0, q_j^0) will have at most 2(m + n -1) nonzero components, while the optimal solution of dual problem will have at most m + n +1 nonzero components.

Now we consider that (U^0, p_i^0, q_j^0), the optimal solution of the dual problem (DP) is known.
The problem (LDP) has the form

$$Maximize\ \ h(U) = 0.35 + 0.3u_{11} + 1.6u_{12} + 1.65u_{21} - 0.75u_{22}$$

subject to

$$\begin{aligned}
&- p_1 + q_1 + 2u_{12} + 3u_{21} - 3u_{22} \le 0\\
&- p_1 + q_2 - 2u_{11} + 5u_{21} - 11u_{22} \le 0\\
&- p_2 + q_1 - 3u_{11} - 5u_{12} - 9u_{22} \le 0\\
&- p_2 + q_2 + 3u_{11} + 11u_{12} + 9u_{21} \le 0\\
&- 10p_1 - 20p_2 + 15q_1 + 15q_2 \ge 0\\
&p_1, p_2, q_1, q_2, u_{11}, u_{12}, u_{21}, u_{22} \ge 0
\end{aligned}$$

The dual problem of the problem (LDP) is

$$Minimize\ \ 0.35 \left[= \frac{\sum_{i=1}^{m}\sum_{j=1}^{n} c_{ij}u_{ij}^0}{\sum_{i=1}^{m}\sum_{j=1}^{n} d_{ij}u_{ij}^0} \right] \qquad \text{(DLDP)}$$

subject to

$$\begin{array}{rrrrrcr}
 & -2y_2 & -3y_3 & +3y_4 & & \ge & 0.3\\
2y_1 & & -5y_3 & +11y_4 & & \ge & 1.6\\
-3y_1 & +5y_2 & & +9y_4 & & \ge & 1.65\\
-3y_1 & -11y_2 & -9y_3 & & & \ge & -0.75\\
-y_1 & -y_2 & & & +10y_5 & \ge & 0\\
 & & -y_3 & -y_4 & +20y_5 & \ge & 0\\
y_1 & & +y_3 & & -15y_5 & \ge & 0\\
 & y_2 & & +y_4 & -15y_5 & \ge & 0
\end{array}$$

$$y_i \ge 0, \quad i = 1, ..., 5$$

An optimal solution is

$$y_1^0 = 0.1;\ y_2^0 = 0;\ y_3^0 = 0.05;\ y_4^0 = 0.15;\ y_5^0 = 0.01$$

and therefore

$$X^0 = \left(y_1^0/y_5^0 = 10;\ \ 0\ ;\ \ y_3^0/y_5^0 = 5;\ \ y_4^0/y_5^0 = 15\right)$$

is the optimal solution of the primal problem (PP).

10.5. HISTORICAL NOTES. COMMENTARIES.

The fractional transportation problem of Section 10.1 can be considered as particular case of the linear fractional programming problem. Consequently, for its solution, one can use any of the methods presented in Chapter 3. In particular, one can utilize the change of variable method presented in Section 3.4 of Chapter 3 (see Hirche [482]), or the method of the reduction to a sequence of non-fractional transportation problems (see Hirche [482] and Hirche *et al.* [492]). In Section 10.1 we used the method of double potentials (see Swarup [946]) for the solution of the fractional transportation problem. The method is based on the fact that the optimal solution is obtained in a vertex of the solutions polyhedron and hence, it can be obtained through a Simplex-type technique. Aggarwal [15] utilized the same method for a problem of transport with linear constraints and objective function (of maximum) of the form

$$Z = \left[\sum_{i=1}^{m}\sum_{j=1}^{n} c_{ij}x_{ij} + c_0\right]^2 \Bigg/ \left[\sum_{i=1}^{m}\sum_{j=1}^{n} d_{ij}x_{ij} + d_0\right]^2$$

Objective function Z is pseudomonotonic (i.e. pseudoconvex and pseudoconcave both) [17] and has the property that maximum occurs at the vertex of the feasible solution set and local maximum is global maximum.

Kanchan *et al.* [529] also applies the same Simplex-type technique for a problem of transport with objective function of the form sum of a linear and linear fractional function i.e.

$$Z = \sum_{i=1}^{m}\sum_{j=1}^{n} c_{ij}x_{ij} + \frac{\sum_{i=1}^{m}\sum_{j=1}^{n} d_{ij}x_{ij}}{\sum_{i=1}^{m}\sum_{j=1}^{n} q_{ij}x_{ij}}$$

subject to constraints (10.1.2)-(10.1.4). Misra and Das [654] generalize these results for a three dimensional transportation problem i.e.

$$Z = \sum_{i=1}^{m}\sum_{j=1}^{n}\sum_{k=1}^{p} l_{ijk}x_{ijk} + \frac{\sum_{i=1}^{m}\sum_{j=1}^{n}\sum_{k=1}^{p} m_{ijk}x_{ijk}}{\sum_{i=1}^{m}\sum_{j=1}^{n}\sum_{k=1}^{p} n_{ijk}x_{ijk}}$$

subject to constraints (10.3.3)-(10.3.6) and (10.3.7)-(10.3.8).

Solomon [880] proposes an algorithm based on a combination of the ideas of Dantzig-Wolfe decomposition and linear fractional programming.

Kaltinska [525] also presents an algorithm for the solution of the fractional transportation problem. Ruiz-Canales *et al.* [761] present a finite method for finding an optimal solution of linear fractional programming problem, subject to constraints (10.1.2)-(10.1.4).
Sharma and Swarup [843] present a transportation technique for time minimization in fractional programming problem with an objective function of the form

$$\text{Minimize} \; \frac{\sum_{i=1}^{m}\sum_{j=1}^{n} c_{ij}x_{ij}}{\underset{(i,j)}{Max}[t_{ij} \mid x_{ij} > 0]}$$

Kabe [521] presents direct solutions to a fractional transportation problem by using the representation theory of the optimal estimators of linearly restricted normal multivariate regression models. The work of Černov and Lange [306] is also mentioned since it gives the approximate method for the solution of a separable fractional transportation problem. Extensions to the three-dimensional case of solution methods for fractional transportation problems have been obtained by Bhatia [177] for simple sums and by Corban [324], [327] for simple and double sums.
Ryang So Yon considers [1067] the fractional n - index transportation problem, proves optimality theorems and gives a solution method on the basis of the potential method. In another work [1068], Ryang So Yon considers the fractional n - index transportation problem with pass capacity, proves sufficient optimality condition and gives a solution method on the basis of the generalized potential method.
Sharma *et al.* [844] consider a particular three-dimensional convex transportation problem

$$\text{Maximize} \quad Z = \frac{\left[\sum_{i=1}^{m}\sum_{j=1}^{n}\sum_{k=1}^{p} c_{ijk}x_{ijk}\right]^2}{\sum_{i=1}^{m}\sum_{j=1}^{n}\sum_{k=1}^{p} d_{ijk}x_{ijk}}$$

subject to constraints (10.3.3)-(10.3.6) and they give an algorithm for obtaining a local maximum.
Stancu-Minasian presents in [895] a simplex-type technique for solution and computational tests for local minimum for three-dimensional transportation problem.

$$\text{Minimize} \quad Z = \frac{\prod_{l=1}^{q}\left[\sum_{i=1}^{m}\sum_{j=1}^{n}\sum_{k=1}^{p} c_{ijk}^{l}x_{ijk}\right]^{a_l}}{\left[\sum_{i=1}^{m}\sum_{j=1}^{n}\sum_{k=1}^{p} c_{ijk}^{0}x_{ijk}\right]^{a_0}}$$

subject to constraints (10.3.3)-(10.3.6) and $a_0 \geq \sum_{l=1}^{q} a_l$.
Reeta Gupta [321] applies the decomposition technique for the solution of some fractional transportation problems and states that the procedure is useful for problems with a large number of destinations relative to the number of depots. Though the given problem is transformed in a new problem which may however results in a substantial increase in the

number of columns, however the Simplex procedure used generates only columns of interest for entry into the successive basis.

Chandra and Saxena present fractional transportation problems with impurities [269] and with penalty costs and impurities [270].

The method from Section 10.2. for the solution of the fractional transportation problem with limited capacities was developed by Suresh Chandra [255]. The same problem is treated by Körth [583], [584], whereas the method which he proposes represents a modification of the solution for the fractional transportation problem without bounded variables. The transportation problem with limited capacities was considered also by Manju Lata [596] who makes a sensitivity analysis of the optimal solution when the problem data depend linearly on a parameter.

The first form of the dual problem for fractional transportation problem presented in Section 10.4 was given by Swarup [954] based on his earlier paper [951]. He proved Weak and Direct duality theorems but he could not demonstrate the Converse duality theorem. Corban [325] - [327] extended the Swarup's results to three- and multi-index transportation problems having simple or double sums with linear fractional objective functions. Corban also demonstrated only Weak and Direct duality theorems and did not prove the Converse duality theorem.

The second dual problem for the fractional transportation problem presented in Section 10.4, was elaborated by Stancu-Minasian and Teghem, Jr. [918]. As different from previous works, in [918] they proved Weak, Direct and Converse duality theorems. The duality results of Swarup for linear fractional transportation problem have been extended by Neelam Datta and Davinder Bhatia [346] to a linear fractional multiple objective transportation problem. Using the duality results, [346] gives the method to determine an initial efficient basic solution for a linear fractional multiple objective transportation problem.

Chandra and Saxena [271] present a technique for shipment completion date-total shipping cost tradeoffs in quadratic fractional transportation problem

$$\text{Minimize } Z = \frac{\left[\sum_{i=1}^{m}\sum_{j=1}^{n} c_{ij}x_{ij} + \alpha\right]^2}{\left[\sum_{i=1}^{m}\sum_{j=1}^{n} d_{ij}x_{ij} + \beta\right]^2}$$

$$\text{Minimize } [\max_{(i,j)} t_{ij} \mid x_{ij} > 0]$$

subject to constraints (10.1.2) - (10.1.4).

The algorithm generates all solutions that are Pareto-optimal with respect to cost and completion date.

BIBLIOGRAPHY

1. **Abadie, J. M; Williams, A.C.:** *Dual and parametric methods in decomposition*, In : *Recent Advances in Mathematical Programming*, R.L. Graves and P. Wolfe (eds.) McGraw-Hill, 1963, 149-158.
2. **Abramov, Yu S.:** *Quasiconvexity of the values of some extremal problems generated by selfadjoint operators*, (Russian). Optimizatsiya No. **40**(1987)57, 106-110, 160.
3. **Abrham J.; Luthra, S.:** *Comparison of duality models in fractional linear programming*, Z. Oper. Res. Ser. A-B, **21**(1977)3, 125-130.
4. **Abrham, J.; Buie, R.N.:** *Duality in continuous fractional programming*, Utilitas Math., **17**(1980), 35-44.
5. **Aggarwal, S.P.:** *Stability of the solution to linear fractional functional programming problem*, Z. Angew. Math. Mech., **46**(1966)6, 343-349.
6. **Aggarwal, S.P.:** *A simplex technique for a particular convex programming problem*, CORS Journal (Journal of the Canadian Operational Research Society), **4**(1966)2, 82-88.
7. **Aggarwal, S.P.:** *Contributions to the Theory of Fractional Programming*, Ph. D. Thesis, Delhi University, Delhi, 1967.
8. **Aggarwal, S.P.:** *Standard error fractional functional programming*, Istanbul Üniv. Fen Fak. Mecm. Ser. A, **30** (1965), 45-51, (1968).
9. **Aggarwal, S.P.:** *A note on quasi convex programming*, Metrika, **12**(1968)2-3, 97-105.
10. **Aggarwal, S.P.:** *Parametric linear fractional functionals programming*, Metrika, **12**(1968)2-3, 106-114.
11. **Aggarwal, S.P.:** *Analyses of the solution to a linear fractional functionals programming*, Metrika, **16**(1968), 9-26.
12. **Aggarwal, S.P.:** *Variation in parameters of a quadratic fractional functionals programming*, Rev. Belge Statist. Informat. Rech. Operat., **11**(1972)4, 3-12.
13. **Aggarwal, S.P.:** *Quadratic fractional functionals programming with nonlinear constraints* (Italian), Ricerca Operativa, **2**(1972), 51-53.
14. **Aggarwal, S.P.:** *Indefinite quadratic fractional programming*, Ekonom.- Mat. Obzor, **8**(1972)2, 191-199.
15. **Aggarwal, S.P.:** *Transportation technique for quadratic fractional programming*, Rev. Belge Statist. Informat. Rech. Operat., **12**(1972)2, 3-7.
16. **Aggarwal, S.P.:** *Upper bounds and quadratic fractional functional programming*, Rev. Belge Statist. Informat. Rech. Operat., **12**(1973)4, 17-21.
17. **Aggarwal, S.P.:** *Quadratic fractional functionals programming*, Cahiers Centre Études Rech. Opér., **15**(1973)2, 157-165.
18. **Aggarwal. S.P.:** *Indefinite quadratic fractional programming with a quadratic constraint*, Cahiers Centre Études Rech. Opér., **15**(1973)4, 405-410.
19. **Aggarwal, S.P.; Arora, S.:** *A special class of non-linear fractional functional programming problems*, SCIMA, J. Management Sci. Appl. Cybernetics, **3**(1974)2, 30-39.
20. **Aggarwal, S.P.; Parkash Om:** *Duality in general linear fractional functional programming*, Cahiers Centre Études Rech. Opér., **20**(1978)1, 75-81.
21. **Aggarwal, S.P.; Patkar, V.N.:** *Dual of a linear fractional program through geometric programming*, Portugal. Math., **37**(1978)1-2, 81-86.
22. **Aggarwal, S.P.; Saxena, P.C.:** *Duality theorems for fractional functional programming with quadratic constraint*, Ekonom.- Mat. Obzor, **10**(1974)1, 86-92.
23. **Aggarwal, S.P.; Saxena, P.C.:** *Duality theorems for non-linear fractional programs*, Z. Angew. Math. Mech., **55**(1975), 523-524.

24. **Aggarwal, S.P.; Saxena, P.C.:** *Decomposition method for linear and piecewise linear program*, (Polish), Przeglad Statyst., 23(1976)2, 211-216.
25. **Aggarwal, S.P.; Saxena, P.C.:** *A class of fractional functional programming problems*, New Zealand Oper. Res., **7**(1979)1, 79-90.
26. **Aggarwal, S.P.; Sharma, I.C.:** *Maximization of the transmission rate of a discrete, constant channel*, Unternehmensforschung, **14**(1970), 152-155.
27. **Aggarwal, S.P.; Swarup, K.:** *Fractional functional programming with a quadratic constraint*, Operations Res., **14**(1966), 950-956.
28. **Aggarwal, Shashi; Davinder, Bhatia; Neelam, Lau:** *Duality in multiple right-hand choice linear fractional problems*, J. Inform. Optim. Sci. **12**(1991)1, 13-24.
29. **Aggarwal, V.; Aneja, Y.P.; Nair, K.P.K.:** *Ratio rewards in networks*, RAIRO Rech. Opér. **15**(1981)2, 129-138.
30. **Aggarwal, V.; Chandrasekaran. R.; Nair, K.P.K.:** *Markov ratio decision processes*, J. Optim. Theory Appl., **21**(1977)1, 27-37.
31. **Aggarwal, V.; Chandrasekaran, R.; Nair, K.P.K.:** *Discounted stochastic ratio games*, SIAM J. Algebraic Discrete Methods, **1**(1980)2, 201-210.
32. **Aggarwal, V.; Nair, K.P.K.; Chandrasekaran, R.:** *Non terminating stochastic ratio game*, RAIRO, Rech. Opér., **14**(1980)1, 21-30.
33. **Aggarwal, R.C.:** *A new approach to planning and programming in agriculture-linear fractional functionals programming*, Indian J. of Agriculture Economics, **24**(1969), 24-25.
34. **Agrawal, S.C.:** *A primal integer programming algorithm with parabolic constraints*, Opsearch, **11**(1974)2-3, 59-80.
35. **Agrawal, S.C.:** *On integer solutions to linear fractional functional programming problems*, Acta Cienc. Indica, **1**(1975)3, 203-208.
36. **Agrawal, S.C.:** *On integer solution to linear fractional functional by a branch and bound technique*, Acta Cienc. Indica, **2**(1976)1, 75-78.
37. **Agrawal, S.C.:** *An alternate method of integer solutions to linear fractional functionals by a branch and bound technique*, Z. Angew. Math. Mech., **57**(1977), 52-53.
38. **Agrawal, S.C.; Chand, M.:** *On integer solutions to complementary programming problems with linear fractional objective function by a branch and bound technique*, Acta Cienc. Indica, **4**(1978)3, 283-289.
39. **Agrawal, S.C.; Chand, M.:** *On intersection cuts in fractional interval integer programming*, Acta Cienc. Indica, **5**(1979)3, 140-142.
40. **Agrawal, S.C.; Chand, M.:** *On mixed integer solutions to complementary programming problems with linear fractional objective functions*, Aligarh Bull. Math., **9**(1979/80)10, 21-30.
41. **Agrawal, S.C.; Chand, M.:** *A note on the sum of linear and fractional interval programming*, Revista de Informatica e Investigacion Operativa, **20**(1980), 33-36.
42. **Agrawal, S.C.; Chand, M.:** *On integer solutions to complementary programming problems with linear fractional objective functions*, Ricerca Operativa, **10**(1980)13, 19-30.
43. **Agrawal, S.C.; Chand, M.:** *A note on integer solutions to linear fractional interval programming problems by a branch & bound technique*, Naval Res. Logist. Quart., **28**(1981)4, 671-677.
44. **Agrawal, S.C.; Chand, M.:** *An alternative method on integer solutions to complementary programming problems with linear fractional objective function by branch and bound technique*, Acta Cienc. Indica Math., **11**(1985)3, 213-221.
45. **Agrawal. S.C; Verma, R.K.:** *p-Variables replacement in linear fractional programming*, Acta Cienc. Indica, **6**(1980)2 95-103.
46. **Agrawal, S.C.; Verma, R.K.:** *A suboptimization method for the sum of linear and linear fractional interval programming problems*, Acta Cienc. Indica Math., **7**(1981)1-4, 14-24.
47. **Agrawal, S.C.; Verma, R.K.:** *On the solutions of the sum of linear and linear fractional interval programming problems*, Pure Appl. Math. Sci. **XVII** (1983)1-2, 73-81.
48. **Agrawal, U.; Swarup, K.; Garg, K.C.:** *Goal programming problem with linear fractional objective function*, Cahiers Centre Études Rech. Opér., **26**(1984)1-2, 33-41.
49. **Ahuja, R.K.:** *Minimum cost-reliability ratio path problem*, Comput. Oper. Res. **15**(1988)1, 83-89.
50. **Ahuja, R.K.; Batra, J.L.; Gupta, S.K.:** *Combinatorial optimization with rational objective functions : A communication*, Math. Oper. Res., **8**(1983)2, 314.
51. **Almogy, Y.; Levin, O.:** *Optimal cargo shipping problem with hiperbolic objective function*, Departament of Industrial and Management Engineering. Mimeograph Series No.24, Technion Institute of Technology, Haifa, Israel, 1968.

52. **Almogy, Y.; Levin, O.:** *Parametric analysis of a multi-stage stochastic shipping problem*, In: Lawrence, J. (Ed.), *Operational Research'69*, Tavistock Publications, London, 359-370, 1969.
53. **Almogy, Y.; Levin, O.:** *A class of fractional programming problems*, Operations Res., **19**(1971), 57-67.
54. **Almogy, Y.; Levin, O.:** *The fractional fixed-charge problem*, Naval Res. Logist. Quart., **18**(1971), 307-315.
55. **Anand, P.:** *Dual and parametric methods in decomposition for linear fractional program*, Studia Sci. Math. Hungar., **6**(1971), 267-275.
56. **Anand, P.:** *Decomposition principle for indefinite quadratic programe*, Trabajos Estadistica, **23**(1972)1-2, 61-71.
57. **Anand, P.:** *Decomposition procedure for linear fractional programs with upper bounds*, Z. Angew. Math. Mech., **53**(1973)10, 635-636.
58. **Anand, P.; Swarup, K.:** *The procedure for local separable programming*, Z. Angew. Math. Mech., **50**(1970)3, 320-321.
59. **Aneja, Y.P.; Nair, K.P.K.:** *Ratio dynamic programs*, Oper. Res. Lett., **3**(1984)3, 167-172.
60. **de Angelis, V.:** *Linear programming with uncertain objective function : minimax solution for relative loss*, Calcolo **XVI** (1979), fasc. II, 125-141.
61. **de Angelis, V.:** *Stochastic linear programming in the objective function : Minimax solution for relative loss*, Proceedings of the First World Conference on Mathematics at the Service of Man (Barcelona, 1977), Vol.I, 248-271, Univ. Politec., Barcelona, 1980.
62. **Anstreicher, K.M.:** *A monotonic projective algorithm for fractional linear programming*, Algorithmica, **1**(1986)4, 483-498.
63. **Anstreicher, K.M.:** *A combined Phase I-Phase II projective algorithm for linear programming*, Math. Programming, Ser. A **43**(1989), 209-223.
64. **Anzai, Y.:** *On integer fractional programming*, J. Operations Res. Soc. Japan, **17**(1974)1, 49-66.
65. **Arbuzova, N.I.:** *Interrelation of the stochastic* ε *-stability of linear and fractional-linear programming problems of a special form*, (Russian), Èkonom. i Mat. Metody, **4**(1968)1, 108-110.
66. **Arisawa, S.:** *Two extensions of GERT*, M.Sc.Thesis, North Carolina State University, Raleigh, N.C., June 1970.
67. **Arisawa, S.; Elmaghraby, S.E.:** *Optimal time-cost trade offs in GERT-Networks*, Management Sci., **18**(1972), 589-599.
68. **Armstrong, R.; Charnes, A.; Cooper, W.W.; Haksever, C.:** *Effective solution of non-convex multi-objective ratio goals problems*, Research Report CCS 390, Center for Cybernetic Studies, University of Texas, Austin, 1980. Also in : J.P. Brans (ed.) : *Operational Research'81*. Amsterdam, North Holland, Preproceedings 1981, 251-259.
69. **Armstrong, R.; Charnes, A.; Haksever, C.:** *Successive linear programming for ratio goal problems*, European J. Oper. Res. **32**(1987)3, 426-434.
70. **Armstrong, R.; Charnes, A.; Haksever, C.:** *Implementation of successive linear programming algorithms for non-convex goal programming*, Comput. Oper. Res. **15**(1988), 37-49.
71. **Arora, S.R.:** *A set partitioning problem with linear fractional objective functions*, Indian J. Pure Appl. Math., **8**(1977), 961-968.
72. **Arora, S.R.:** *A note on fractional fixed charge problems*, New Zealand Oper. Res., **5**(1977), 66-71.
73. **Arora, S.; Aggarwal, S.P.:** *Linear fractional functional programming with a parameter in an activity vector*, Econom. Comp. Econom. Cybernet. Stud. Res., **3**(1977), 37-55.
74. **Arora, S.; Aggarwal, S.P.:** *Dynamic programming approach to linear fractional functional programming*, Rev. Belge Statist. Informat. Rech. Operat., **17**(1977)3, 10-23.
75. **Arora, S.R.; Puri, M.C.:** *Enumeration technique for the set covering problem with linear fractional functional as its objective function*, Z. Angew. Math. Mech., **56**(1977), 181-186.
76. **Arora, S.R.; Puri, M.C.; Swarup, K.:** *The set covering problem with linear fractional functional*, Indian J. Pure Appl. Math., **8**(1977)5, 578-588.
77. **Arora, S.R.; Swarup, K.; Puri, M.C.:** *Cutting plane technique for the set covering problem with linear fractional functional*, Z. Angew. Math. Mech., **57**(1977)10, 597-602.
78. **Arsham, H.; Kahn, A.B.:** *A fast algorithm to identify any "bad-point" in linear fractional programs*, The Joint National Meeting of TIMS/ORSA, New York, 1989.
79. **Arsham, H.; Kahn, A.B.:** *A complete algorithm for linear fractional programs*, Comput. Math. Appl. **20**(1990)7, 11-23.
80. **Artjuhin, A.V.:** *An algorithm for the solution of the distribution problem of parametric fractional linear programming*, In: Some Mathematical Optimization Methods and Their Application in the Economy of Kirghizia. Izdat. "Ilim" Frunze, 30-36, 1973.

81. **Artjuhin, A.V.:** *Some applications of parametric fractional linear programming*, In: Some Mathematical Optimization Methods and Their Application in the Economy of Kirghizia. Izdat. "Ilim" Frunze, 37-54, 1973.

82. **Ashton, D.J.; Atkins, D.R.:** *Multicriteria programming for financial planning*, J. Oper. Res.Soc. **30**(1979), 259-270.

83. **Auslender, A.:** *Minimization de functions localement lifschitziennes : applications à la programmation mi-convexe, mi-différentiable*, Nonlinear programming, 3 (Proc.Sympos., Special Interest Group Math. Programming, Univ. Wisconsin, Madison, Wis, 1977), pp. 429-460. Academic Press, New York, 1978.

84. **Avriel, M.; Diewert, W.E.; Schaible, S.; Ziemba, W.T.:** *Introduction to concave and generalized concave functions*, In: Schaible, S. and Ziemba, W.T., (eds.), *Generalized Concavity in Optimization and Economics*, Academic Press, New York, 1981, 21-50.

85. **Avriel, M.; Diewert, W.E.; Schaible, S.; Zang, I.:** *Generalized Concavity*, Mathematical Concepts and Methods in Science and Engineering, 36. New York etc. Plenum Press 1988.

86. **Avriel, M.; Williams, A.C.:** *Complementary geometric programming*, SIAM J. Appl. Math., **19**(1970)1, 125-143.

87. **Awerbuch, S.; Ecker, J.G.; Wallace, W.A.:** A note: *Hidden nonlinearities in the application of goal programming*, Management Sci., **22**(1976), 918-920.

88. **Aylawadi, D.R.:** *Duality for homogeneous fractional programming with non-linear constraints*, J. Math. Sci., **12-13**(1977-1978), 29-32.

89. **Babaev, D.A.:** *Methode der losung einer klasse nichtlinearer Programmierungsprobleme*, Akad. Nauk Azerbaidžan SSR Dokl., **30**(1974)9, 3-6.

90. **Babaev, D.A.:** *A general fractional programming problem*, (Russian), *Mathematical programming and related questions* (Proc. Seventh Winter Scool, Drogobych, 1974), Numerical Methods (Russian), 98-105. Central. Ekonom. - Math. Inst. Akad. Nauk SSSR, Moscow, 1976.

91. **Babaev, D.A.; Karaeva, È.M.:** *The solution of certain nonlinear problems of mathematical programming by the introduction of a parameter* (Russian), Questions of mathematical cybernetics and applied mathematics, No.2 (Russian), pp. 54-64, Izdat, "Élm", Baku, 1976.

92. **Bajalinov, È.B.:** *On a question of duality in fractional-linear programming* (Russian), Izv. Akad. Nauk. Kirghiz. SSR **2**(1981), 10-20, 101.

93. **Bajalinov, È.B.:** *The effect of changes in the conditions of a linear-fractional programming problem on the optimum of the target function* (Russian), *Application of mathematical-economics methods in improving control of the economy*, 144-145, "Ilim", Frunze, 1982.

94. **Bayalinov, È.B.:** *A method of successive reduction of residue for solving a problem of fractional-linear programming* (Russian), *Application of mathematics in economics* (Russian), 153-162, 194 "Ilim", Frunze, 1986.

95. **Bayalinov, È.B.:** *Economic interpretation of dual estimates in linear-fractional programming* (Russian), Izv. Akad. Nauk. Kirgiz. SSR 1987, no.3, 8-11, 81.

96. **Bayalinov, È.B.:** *On the economical sense of dual variables of linear-fractional programming* (Russian), Èkonom. i Mat. Metody **24**(1988)3, 558-561.

97. **Bayalinov, È.B.:** *On a system of correlated dual estimates*, (Russian), Izv. Akad. Nauk Kirgiz. SSR, 1988, no.2, 8-12, 91.

98. **Bayalinov, È.B.:** *Coincidence of optimal plans for problems of linear and linear-fractional programming*, (Russian), Izv. Akad. Nauk Kirgiz. SSR 1988, no.3, 9-15, 91.

99. **Bayalinov, È.B.:** *On a system of three interrelated mathematical programming problems*, (Russian), Kibernetika (Kiev) 1989, no.6, 120-122.

100. **Bayalinov, È.B.:** *On decomposition of dual variables in linear programming and its economic interpretation*, IVth International Workshop on Generalized Convexity, Pécs/ Ungary, August 31 - September 2, 1992.

101. **Bayalinov, È.B.; Pannell, D.J.:** *GULF A general, user-friendly linear and linear-fractional programming package*, IVth International Workshop on Generalized Convexity, Pécs/Ungary, August 31 - September 2, 1992.

102. **Bakhshi, H.C.:** *Sensitivity analysis in linear fractional functionals programming problems with extreme point restriction*, SCIMA (J. Management Sci. Appl. Cybern.), **8**(1979), 6-15.

103. **Bakhshi, H.C.:** *A study of sensitivity in extreme point linear fractional functional programming problems*, J. Math. Sci., **14-15**(1979-1980), 12-20(1982).

104. **Bakhshi, H.C.; Puri, M.C.:** *An efficient technique for extreme point mathematical programming problems*, Cahiers Centre Études Rech. Opér., **21**(1979)3, 257-268.

105. **Bansal, Sushma:** *Absolute value linear fractional programming*, Cahiers Centre Études Rech. Opér., **23**(1981)1, 43-52.

106. **Bansal, Sushma:** *On extreme point programming problem*, J. Inform. Optim. Sci., **3**(1982)2, 173-184.

107. **Bansal, Sushma; Puri, M.C.:** *Absolute value programming with indefinite quadratic function under linear constraints*, Cahiers Centre Ètudes Rech. Opér., **24**(1982)1, 49-54.

108. **Barrodale, I.:** *Best rational approximation and strict quasi-convexity*, SIAM J. Numer. Anal. **10**(1973), 8-12.

109. **Barrodale, I.; Powell, J.D.; Roberts, F.D.K.:** *The differential correction algorithm for rational l_∞ approximation*, SIAM J. Numer. Anal. **9**(1972), 493-504.

110. **Băluţă, M.; Dobrescu, V.; Păun, G.:** *Algorithm for solving a combinatorial problem of optimal selection*, Econom. Comput. Econom. Cybernet. Stud. Res., **13** (1979)1, 87-95.

111. **Beale, E.M.L.:** *Fractional programming with zero-one variables*, Extremal Methods and Systems Analysis (Internat. Sympos. Univ. Texas, Austin, Texas, 1977) Lecture Notes in Econom. and Math. Systems, 174, Springer Berlin, 430-432, 1980.

112. **Beato-Moreno, Antonio; Pascual Ruiz Canales:** *A comparison of the efficiency of two methods for integral fractional linear programming* (Spanish), Proceedings of the XVth Portuguese-Spanish Conference on Mathematics, Vol. IV (Évora, 1990), 355-360, Univ. Évora, Évora, 1991.

113. **Bector, C.R.:** *General non-linear fractional functional programming*, Presented at the Conference of Bhart Ganita Prashid Lucknow University, Lucknow (India), April 1966.

114. **Bector, C.R.:** *Indefinite quadratic programming with standard errors in objective*, Cahiers Centre Études Rech. Opér., **10**(1968)4, 247-253.

115. **Bector, C.R.:** *Indefinite cubic programming with standard errors in objective function*, Unternehmensforschung, **12**(1968)2, 113-120.

116. **Bector, C.R.:** *Certain aspects of duality in non-linear indefinite functional programming*, Ph. D. Thesis, Dept. of Maths. I.I.T. Kanpur (India), 1968.

117. **Bector, C.R.:** *Duality in fractional and indefinite programming*, Z. Angew. Math. Mech., **48**(1968)6, 418-420.

118. **Bector, C.R.:** *Programming problems with convex fractional functions*, Operations Res., **16**(1968), 383-390.

119. **Bector, C.R.:** *Non-linear fractional functional programming with non-linear constraints*, Z. Angew. Math. Mech., **48**(1968)4, 284-286.

120. **Bector, C.R.:** *Some aspects of non-linear indefinite fractional functional programming*, Cahiers Centre Études Rech. Opér., **12**(1970)1, 22-34.

121. **Bector, C.R.:** *Some aspects of quasi-convex programming*, Z. Angew. Math. Mech., **50**(1970), 495-502.

122. **Bector, C.R.:** *Indefinite quadratic fractional functional programming*, Metrika, **18**(1971)1, 21-30.

123. **Bector, C.R.:** *Duality in non-linear fractional programming*, In: Proceedings of the Second Manitoba Conference on Numerical Mathematics held at the University of Manitoba, Winnipeg, Man, October 5-7, R.S.D. Thomas and H.C. Williams (eds.), Congressus Numerantium, No. VII, Utilitas Mathematica Publishing Inc., Winnipeg, Man., vi + 468, 81-93, 81-93, 1973.

124. **Bector, C.R.:** *Duality in nonlinear fractional programming*, Z. Oper. Res. Ser. A-B, **17**(1973), 183-193.

125. **Bector, C.R.:** *Duality in linear fractional programming*, Utilitas Math., **4**(1973), 155-168.

126. **Bector, C.R.:** *On convexity, pseudo-convexity and quasi-convexity of composite functions*, Cahiers Centre Études Rech. Opér., **15**(1973)4, 411-428.

127. **Bector, C.R.:** *A note on a dual fractional program*, Cahiers Centre Études Rech. Opér., **10**(1974)2, 107-115.

128. **Bector, C.R.; Bector K. Meena; Klassen E. John:** *Duality for a nonlinear programming problem*, Utilitas Math., **11**(1977), 87-99.

129. **Bector, C.R.; Meena K. Bector:** *FRITZ JOHN sufficient optimality conditions and duality for a generalized minmax program*, J. Inform. Optim. Sci. **10**(1989)1, 193-205.

130. **Bector, C.R.; Bhatia, B.L.:** *An optimization theorem with applications in some mathematical programming problems*, Utilitas Math., **26**(1984), 249-258.

131. **Bector, C.R.; Bhatia, B.L.:** *Generalized duality for nonlinear programming in complex space*, Econom. Comput. Econom. Cybernet. Stud. Res., **20**(1985)2, 75-80.

132. **Bector, C.R.; Bhatia, B.L.:** *Nature of RENYI's entropy and associated divergence function*, Naval Res. Logist. **34**(1986)4, 741-746.

133. **Bector, C.R.; Bhatt, S.K.:** *A linearization technique for solving interval linear fractional programs*, In: Proceedings of the Fifth Manitoba Conference on Numerical Mathematics, Congressus Numerantium, No. XVI, Utilitas Math. Publ., Winnipeg, Man. 1976, 221-229.

134. **Bector, C.R.; Davinder Bhatia; Shashi Aggarwal:** *Multiobjective fractional programming duality: a nondifferentiable case*, Proceedings of the Twenty-first Manitoba Conference on Numerical Mathematics and Computing (Winnipeg, MB, 1991). Congr. Numer. 87 (1992), 77-85.

135. **Bector, C.R.; Bhatt, S.K.:** *Duality for a non-convex program in a real Banach space*, Z. Angew. Math. Mech., **57**(1977)3, 193-194.
136. **Bector, C.R.; Bhatt, S.K.:** *Pseudo-monotonic interval programming*, Naval Res.Logist. Quart., **25**(1978)2, 309-314.
137. **Bector, C.R.; Bhatt, S.K.:** *Nonlinear programming in complex space: Necessary and sufficient conditions*. Rev. Roumaine Math. Pures Appl. **30**(1985)7, 497-503.
138. **Bector, C.R.; Cambini, A.:** *Fractional programming-some recent results*. In: [229], pag. 86-98.
139. **Bector, C.R.; Chandra, S.:** *First order duality in minmax multiobjective non-linear fractional programming*, Research Report 85-1, Faculty of Administrative Studies, University of Manitoba, 1985.
140. **Bector, C.R.; Chandra, S.:** *Second order duality for generalized fractional programming*, Methods Oper. Res., **56**(1986), 11-28.
141. **Bector, C.R.; Chandra, S.:** *Duality for pseudolinear min- max programs*, Asia-Pacific J. Oper. Res. **3**(1986), 86-94.
142. **Bector, C.R.; Chandra, S.:** *First and second order duality for a class of nondifferential fractional programming problem*, J. Inform. Optim. Sci. **7**(1986)3, 335-348.
143. **Bector, C.R.; Chandra, S.:** *(Generalized)- bonvexity and higher order duality for fractional programming*, Opsearch, **24**(1987)3, 143-154.
144. **Bector, C.R.; Chandra, S.; Bector, M.K.:** *Duality for minmax programs*, Proceedings of the Twentieth Manitoba Conference on Numerical Mathematics and Computing (Winnipeg, MB, 1990). Congr. Numer. **80**(1991), 33-48.
145. **Bector, C.R.; Chandra, S.; Gulati, T.R.:** *Duality for complex nonlinear fractional programming over cones*, In: Proceedings of the Third Manitoba Conference on Numerical Mathematics (Winnipeg, Man. 1973), 87-103, Utilitas Math., Winnipeg, Man. 1974.
146. **Bector, C.R.; Chandra, S.; Gulati, T.R.:** *Duality for fractional control problem*, In: Proceedings of the Fifth Manitoba Conference on Numerical Mathematics (Univ. Manitoba, Winnipeg, Man), 231-241, 1975, Congressus Numerantium, No. XVI, Utilitas Math. Publ., Winnipeg., 1976.
147. **Bector, C.R.; Chandra, S.; Gulati, T.R.:** *A Langrangian approach to duality for complex nonlinear fractional programming over cones*, Math. Operations Forsch. Statist. Optimiz., **8**(1977)1, 17-25.
148. **Bector, C.R.; Chandra, S.; Husain, I.:** *Generalized continuous fractional programming duality: a parametric approach*, Utilitas Math. **42**(1992), 39-60.
149. **Bector, C.R.; Dahl, M.:** *Simplex type finite iteration technique and duality for a special type of pseudo-concave quadratic program*, Cahiers Centre Études Rech. Opér., **16**(1974)3, 207-222.
150. **Bector, C.R.; Grover, T.R.:** *Minimizing certain non-convex quadratic fractional programs by ranking the extreme points*, In: Proceedings of the Second Manitoba Conference on Numerical Mathematics, held at the University of Manitoba, Winnipeg, Man., October 5-7, 1972, R.S.D. Thomas and H.C. Williams (eds.), Congressus Numerantium, No. VII. Utilitas Mathematica Publishing Inc., Winnipeg, Man., vi + 468, 95-100, 1973.
151. **Bector, C.R.; Jolly, P.L.:** *Pseudo-monotonic integer programming*, In: D. Mac Carthy and H.C. Williams (eds.): Proceedings of the Seventh Manitoba Conference on Numerical Mathematics and Computing, 211-218, Congressus Numerantium XX, Utilitas Math., Winnipeg, 1978.
152. **Bector, C.R.; Jolly, P.L.:** *Programming problems with pseudomonotonic objectives*, Math. Operationsforsch. Statist. Ser. Optim., **15**(1984)2, 217-229.
153. **Bector, C.R.; Klassen, J.E.:** *A non-parametric self-dual algorithm for linear fractional functionals programming*, TIMS/ORSA Bull (USA), No.1, p.102, 1976 ORSA/TIMS, 1976 Joint National Meeting (abstract only). Philadelphia, Pa. USA, 31 March-2 April, 1976.
154. **Bector, C.R.; Kumar, U.:** *Duality for multiple-objective fractional programs*, Cahiers Centre Études Rech. Opér., **26**(1984)3-4, 201-207.
155. **Bector, C.R.; Suneja, S.K.:** *Duality in nondifferentiable generalized fractional programming*, Asia-Pacific J. Oper. Res. **5**(1988)2, 134-139.
156. **Bector, C.R.; Suneja, S.K.:** *Duality in generalized fractional programming involving nondiferentiable functions*, Seventeenth Manitoba Conference on Numerical Mathematics and Computing, D.S. MEEK, R.G. Stanton and G.H.J. Van Rees (eds.) Congr. Numer. 62 (1988), Utilitas Mathematica Publishing, Inc., Winnipeg, MB, 1988, 159-164.
157. **Bector, C.R.; Chandra, S.; Bector, M.K.:** *Generalized fractional programming duality: a parametric approach*, J. Optim. Theory Appl. **60**(1989)2, 243-260.

158. **Bector, C.R.; Chandra, S.; Durga Prasad, M.V.:** *Nonlinear fractional programming duality for HANSON-MOND class of functions with nondifferentiable terms*, Seventeenth Manitoba Conference on Numerical Mathematics and Computing. D.S. Meek, R.G. Stanton and G.H.J. Van Rees (eds.) Congr. Numer. **62**(1988), Utilitas Mathematica Publishing, Inc., Winnipeg, 1988, 165-170.

159. **Bector, C.R.; Chandra, S.; Durga Prasad, M.V.:** *Duality in pseudolinear multiobjective programming*, Asia-Pacific J. Oper. Res. **5**(1988)2, 150-159.

160. **Bector, C.R.; Chandra, S.; Singh, C.:** *Duality in multiobjective fractional programming*, In: [229], pag. 232-241.

161. **Bector, Meena; Husain, I.; Chandra, S.; Bector, C.R.:** *A duality model for a generalized minmax program*, Nav.Res. Logist. **35**(1988)5, 493-501.

162. **Bedi, Mohinder Kaur:** *Duality for a special class of quasi-convex programming problem*, Z. Angew. Math. Mech., **58**(1978), 165-166.

163. **Belen'kii, A.S.:** *Minmax problem with linear constraints*, Avtomat. i Telemeh. 1980, no.4, 151-158 (Russian); Translated as Automat Remote Control **41**(4), part.2(1980), 562-568.

164. **Bell, E.J.:** *Primal-Dual decomposition programming*, Ph.D. Thesis, Operations Research Center, University of California at Berkeley, Report ORG 65-92, 1965.

165. **Belykh, V.M.; Gavurin, M.K.:** *An algorithm for minimizing a linear-fractional function*, Vestnik Leningrad Univ. Math. **13**, 271-276 (1981). Translation from Vestnik Leningrad. Univ. Mat. Meh. Astronom. (Russian), **19**(1980)4, 10-15, 115.

166. **Benadada, Y.:** *Approaches de résolution du problème de programmation fractionnaire généralisée*, Thése de doctorat, Université de Montreal (Montreal, Canada, 1989).

167. **Benadada, Y.; Ferland, J.A.:** *Partial linearization for generalized fractional programming*, Z. Oper. Res. **32**(1988)2, 101-106.

168. **Benadada, Y.; Crouzeix, J.P.; Ferland, J.A.:** *An interval-type algorithm for generalized fractional programming*, In : [229], pag. 106-120.

169. **Benson, H.P.:** *Finding certain weakly-efficient vertices in multiple objective linear fractional programming*, Management Sci., **31**(1985)2, 240-245.

170. **Beoni, C.A.:** *A generalization of Fenchel duality theory*, J. Optim. Theory Appl., **49**(1986)3, 375-386.

171. **Bereanu, B.:** *Programme de risque minimal en programmation linéaire stochastique*, C.R. Acad. Sci. (Paris), **259**(1964)5, 981-983.

172. **Bereanu, B.:** *Soluţii cu risc minim în programarea liniară*, An. Univ. Bucureşti, Mat.-Mec., **13**(1964), 121-140.

173. **Bereanu, B.:** *On the use of computers in planning under conditions of uncertainty*, Computing, **15**(1975), 11-32.

174. **Bergthaller, C.:** *A quadratic equivalent for the minimum risk problem*, Rev. Roumaine Math. Pures Appl., **15**(1970), 17-23.

175. **Bernard, J.C.:** *Quelques aspects théoriques en programmation fractionnaire généralisée et algorithmes*, Thése de doctorat, Université de Montréal (Montréal, Canada, 1986).

176. **Bernard, J.C.; Ferland, J.A.:** *Convergence of interval-type algorithms for generalized fractional programming*, Math. Programming Ser. A **43**(1989)3, 349-363.

177. **Bhatia, H.L.:** *Solid transportation problem in linear fractional programming*, Rev. Belge Statist. Informat. Rech. Operat., **18**(1978)4, 35-49.

178. **Bhatia, D.; Budhraya, R.K.:** *On a class of fractional functional programming problems*, Opsearch **27**(1990)4, 225-238.

179. **Bhatia, D.; Datta, N.:** *Necessary conditions and subgradient duality for non-diferentiable and non-convex multi-objective programming problem*, Cahiers Centre Études Rech. Opér., **27**(1985)1-2, 131-139.

180. **Bhatia, D.; Gupta, B.:** *Efficiency in certain nonlinear fractional vector maximization problem*, Indian J. Pure Appl. Math. **11**(1980)5, 669-672.

181. **Bhatia, D.; Pandey, S.:** *A note on multiobjective fractional programming*, Cahiers Centre Études Rech. Opér. **33**(1991)1-2, 3-11.

182. **Bhatia, Davinder; Pandey, Shashi:** *Subgradient duality and duality for multiobjective fractional programming involving invex functions*, Cahiers Centre Études Rech. Opér. **33**(1991)3-4, 167-181.

183. **Bhatia, Davinder; Jain, Pushp:** *Nondiferentiable multiobjective fractional programming with Hanson-Mond classes of functions*, J. Inform. Optim. Sci. **12**(1991), no.1, 35-47.

184. **Bhatt. S.K.:** *An existence theorem for a fractional control problem*, J. Optim. Theory Appl., **11**(1973)4, 379-385.

185. **Bhatt, S.K.:** *Sequential unconstrained minimization technique for a non-convex program*, Cahiers Centre Études Rech. Opér. **15**(1973)4, 429-435.

186. **Bhatt, S.K.:** *Linearization technique for linear fractional and pseudomonotonic programs revisited*, Cahiers Centre Études Rech. Opér., **23**(1981)1, 53-56.

187. **Bhatt, S.K.; Rosenbloom, E.S.:** *A dynamic programming approach to generalized linear fractional programs*, Cahiers Centre Études Rech. Opér., **27**(1985)3-4, 207-212.

188. **Bhatt, S.K.:** *Equivalence of various linearization algorithms for linear fractional programming*, Methods Models Oper. Res. **33**(1989)1, 39-43.

189. **Bitran, G.R.:** *Experiments with linear fractional problems*, Naval Res. Logist. Quart., **26**(1979)4, 689-693.

190. **Bitran, G.R.; Magnanti, T.L.:** *Fractional programming: Duality, algorithms, sensitivity analysis and applications*, Technical report No.92, Operations Research Center, Massachusetts Institute of Technology, June 1974.

191. **Bitran, G.R.; Magnanti, T.L.:** *Duality and sensitivity analysis for fractional programs*, Operations Res., **24**(1976)4, 675-699.

192. **Bitran, G.R.; Novaes, A.G.:** *Linear programming with a fractional objective function*, Operations Res., **21**(1973)1, 22-29.

193. **Blau, R.A.:** *Descomposition techniques for the Chebyshev problem*, Operations Res., **21**(1973)5, 1157-1163.

194. **Boncompte, M.:** *Programacion fraccional generalizada*; Tesina de licenciatura, Universidad de Barcelona, Spain, 1985.

195. **Boncompte, M.; Martinez-Legaz, J.E.:** *Fractional programming by lower subdifferentiability techniques*, J.Optim. Theory Appl. **68**(1991)1, 95-116.

196. **Borde, J.; Crouzeix, J.P.:** *Convergence of a Dinkelbach-type algorithm in generalized fractional programming*, Z.Oper. Res. Ser. A-B, **31**(1987)1, A31-A54.

197. **Borwein, J.M.:** *Fractional programming without differentiability*, Math. Programming, **11**(1976)3, 283-290.

198. **Boyadzhiev, A.V.:** *On a nonlinear programming problem* (Bulgarian), God. Vissh. Uchebn. Zaved., Prilozhna Mat. **17**(1981)1, 107-112.

199. **Boyd, G.; Färe, R.:** *Measuring the efficiency of decizion making units: A comment*, European J. Oper. Res. **15**(1984)3, 331-332.

200. **Bradley, G.H.:** *Transformation of integer programs to knapsack problems*, Discrete Math. **1**(1971)1, 29-45.

201. **Bradley, S.P.; Frey, C. Sherwood, Jr.:** *Fractional programming with homogeneous functions*, Operations Res., **22**(1974)2, 350-357.

202. **Brosh, Israel:** *Optimal cargo allocation on board a plane: A sequential linear programming approach*, European J.Oper. Res., **8**(1981)1, 40-46.

203. **Brosowski, Bruno; Guerreiro, Claudia:** *On the characterization of a set of optimal points and some applications*, Approximation and optimization in mathematical physics (Oberwolfach, 1982), 141-174, Methoden Verfahren Math. Phys., 27, Lang, Frankfurt am Main, 1983.

204. **Burley, S.P.:** *Dynamic generalization of the von-Neumann model*, Math. Mod. Econ., Amsterdam, 1974, 27-33.

205. **Bühler, W.:** *A note on fractional interval programming*, Z. Oper. Res. A-B, **19**(1975)1, 29-36.

206. **Bühler, W.; Newinger, N.:** *On the convergence of Martos'hyperbolic programming algorithm*, Arbeitsbericht, Lehrstuhl für unternehmensforschung, Rheinish-Westfälische Technische Hochschule, Aachen, 1973.

207. **Bykadorov, I.A.:** *Finite systems of linear-fractional inequalities* (Russian), Optimizatsiya **35**(1982)52, 43-50.

208. **Bykadorov, I.A.:** *A problem of linear-fractional programming* (Russian), Optimizatsiya **35**(1985)52, 51-55, 159.

209. **Bykadorov, I.A.:** *On conditions of quasi-convexity of sums of linear-fractional functions* (Russian), Optimizatsiya **39**(1986)56, 25-41, 158.

210. **Bykadorov, I.A.:** *On quasiconvexity in fractional programming*, IVth International Workshop on Generalized Convexity, Pécs/Ungary, August 31 - September 2, 1992.

211. **Cabot, V.:** *Maximizing the sum of certain quasi-concave functions using generalized Benders descomposition*, Naval Res. Logist. Quart., **25**(1978)3, 473-481.

212. **Callahan, J.R.; Bector, C.R.:** *Optimization with general stochastic objective functions*, In: Proceedings of the Third Manitoba Conference on Numerical Mathematics 1973, 127-137; also in: Z. Angew. Math. Mech., **55**(1975), 528-530.

213. **Cambini, A.:** *Un algoritmo per il massimo del quoziente di due forme affini con vincoli lineari*, Publication No.42, series A. Dept. di Ricerca Operativa e Scienze Statistiche, Univ. of Pisa, 1977.

214. **Cambini A.:** *Sulla programmazione lineare frazionaria stocastica*, Publication No.50, series A, Dept. di Ricerca Operativa e Scienze Statistiche, Università di Pisa, Italy, 1978.

215. **Cambini, A.**: *Sulla esistenza di moltiplicatori esponenziali per i problemi di programmazione lineare frazionaria*, Publication No.67, serie A. Dept. di Ricerca Operativa e Scienze Statistiche. Università di Pisa, Italy, 1979.

216. **Cambini, A.**: *An algorithm for special class of fractional programs*, In: Schaible, S., Ziemba W.T. (eds.), Generalized Concavity in Optimization and Economics. Academic Press, New York, 1981, 491-508.

217. **Cambini, A.; Martein, L.**: *A modified version of Martos algorithm for a class of fractional programs*, Meth. Oper. Res., **53**(1985), 33-41.

218. **Cambini, A.; Martein, L.**: *On the Fenchel-like and Langrangian duality in fractional programming*, Publication No. 115, Universita di Pisa, Dipartimento di matematica, Italy, 1984. Also in : Beeckmann, M.J. et al. (eds.): X Symp. on Operations Research, München 1985. Methods of Operations Research **53**(1986), 21-32.

219. **Cambini, A.; Martein, L.**: *Linear fractional and bicriteria linear fractional programs*, In: [229], pag. 155-166.

220. **Cambini, A.; Martein, L.**: *Equivalence in linear fractional programming*, Department of Statistics and Applied Mathematics, University of Pisa, report n.28, 1989. Also in: Optimization, **23**(1992)2, 41-51.

221. **Cambini, A.; L., Martein; Pellegrini, L.**: *A decomposition algorithm for a particular class of nonlinear programs*, Mathematiques appliques, 1er Colloq. AFCET-SMF, Palaiseau 1978, II, 179-189 (1978).

222. **Cambini, A.; Martein, L.; Pellegrini, L.**: *Decomposition methods for a class of non-linear fractional programs*, 10th International Symposium on Mathematical Programming, Montreal, August 27-31, 1979.

223. **Cambini, A.; Martein, L.; Schaible, S.**: *On maximizing a sum of ratios*, Report no.2, Universita di Pisa, Italy. Dipartimento Di Statistica e Matematica Applicata All'Economia, 1987. Also in: J. Inform. Optim. Sci. **10**(1989)1, 65-79.

224. **Cambini, A.; Martein, L.; Sodini, C.**: *An algorithm for two particular non linear fractional programs*, Publication No. 104, Università di Pisa, Dipartimento di matematica, Italy 1983. Also in: Methods of Operations Research **45**(1983), 61-70.

225. **Cambini, A.; Schaible, S.; Sodini, C.**: *Parametric linear fractional programming for an unbounded feasible region*, Journal of Global Optimization **2**(1992).

226. **Cambini, A.; Sodini, C.**: *Un algorithmo per un problema di programmazione frazionaria non lineare derivante da un problema di selezione del portafoglio*, Atti del Quinto Convegno. A.M.A.S.E.S., Perugia, 22-24 Ottobre 1981, 49-61, 1981.

227. **Cambini, A.; Sodini, C.**: *Sulla equivalenza di alcuni algorithmi di programmazione lineare frazionaria*, Publication No.105, Universita di Pisa, Dipartimento di matematica, Italy, 1983.

228. **Cambini, A.; Sodini, C.**: *On parametric linear fractional programs*, Department of Statistics and Applied Mathematics, University of Pisa, Report no.6, 1987.

229. **Cambini, A.; Castagnoli, E.; Martein, L.; Mazzoleni, P.; Schaible, S.** (Eds.): *Generalized Convexity and Fractional Programming with Economic Applications*, Proceedings of the International Workshop on "Generalized Concavity, Fractional Programming and Economic Applications" held at the University of Pisa, Italy, May 30-June 1, 1988. Lecture Notes in Economics and Mathematical Systems, 345, Springer Verlag, 1990.

230. **Chadha, S.S.**: *A decomposition principle for fractional programming*, Opsearch, **4**(1967)3, 123-132.

231. **Chadha, S.S.**: *An extension of upper bounded technique for a linear fractional program*, Istanbul Üniv. Fen Fak. Mecm. Ser. A, **34**(1969), 91-96, Istanbul 1971.

232. **Chadha, S.S.**: *A linear fractional functional program with a two parameter objective function*, Z. Angew. Math. Mech., **51**(1971), 479-481.

233. **Chadha, S.S.**: *A dual fractional program*, Z. Angew. Math. Mech., **51**(1971), 560-561.

234. **Chadha, S.S.**: *A linear fractional functional program with variable coefficients*, Istanbul Üniv. Fen Fak. Mecm. Ser. A., **36**(1971), 7-13, Istanbul 1974, Also in : J. Math. Sci. **7**(1972), 15-20 (with R.N. Kaul).

235. **Chadha, S.S.**: *Duality theorems for a generalized linear and linear-fractional program*, Cahiers Centre Études Rech. Opér., **15**(1973)2, 167-173, Also in : Ekonom.-Mat. Obzor, **8**(1972)4, 410-415.

236. **Chadha, S.S.**: *A generalized upper bounded technique for a linear fractional program*, Metrika, **20**(1973)1, 25-35.

237. **Chadha, S.S.**: *Quadratic parametric linear programming*, Cahiers Centre Études Rech. Opér., **25**(1983)1-2, 23-28.

238. **Chadha, S.S.**: *A dual non-linear program*, Acta Math. Appl. Sinica Engl. Ser. **1**(1984), 163-167.

239. **Chadha, S.S.**: *Hyperbolic programming-new criteria*, Econom. Comput. Econom. Cybernet. Stud. Res., **22**(1987)4, 83-88.

240. **Chadha, S.S.:** *Duality theorems for a class of non-linear programming problems*, Trab. Invest. Oper. **3**(1988)1, 141-148.

241. **Chadha, S.S.:** *A parametric linear fractional program*, J. Combin. Inform. System Sci. **13**(1988)3-4, 106-113.

242. **Chadha, S.S.:** *Multiparametric linear fractional functionals programming*, Trab. Invest. Oper. **4**(1989)1, 3-10.

243. **Chadha, S.S.:** *Duality in nonlinear fractional functional programming*, Econom. Comput. Econom. Cybernet. Stud. Res. **26**(1992)1-4, 87-91.

244. **Chadha, S.S.:** *Optimization of a quadratic fractional function*, Cahiers Centre Études Rech. Opér. **34**(1992)1, 3-6.

245. **Chadha, S.S.:** *Dual of the sum of a linear and linear fractional program*, European J. Oper. Res. **67**(1993)1, 136-139.

246. **Chadha, S.S.; Gupta, J.M.:** *Sensitivity analysis of the solution on a generalized linear and piece-wise linear program*, Cahiers Centre Études Rech. Opér., **18**(1976), 309-321.

247. **Chadha, S.S.; Gupta, J.M.:** *A decomposition principle for a generalized linear and piece-wise linear program*, Trabajos Estadist. Investigación Oper., **28**(1977)1, 85-92.

248. **Chadha, S.S.; Kaul, R.N.:** *A dual non-linear program*, Metrika, **19**(1972)1, 18-22.

249. **Chadha, S.S.; Shivpuri, Saroj:** *A simple class of parametric linear fractional functionals programming*, Z. Angew. Math. Mech., **53**(1973)10, 644-646.

250. **Chadha, S.S.; Shivpuri, Saroj:** *Parametrization of a generalized linear and piece-wise linear programme*, Trabajos Estadist. Investigación Oper., **28**(1977)2-3, 151-160.

251. **Chadha, S.S.; Shivpuri, Saroj:** *Multiparametric linear and piecewise linear programming*, Pure Appl. Math. Sci., **8**(1978)1-2, 31-37.

252. **Chadha, S.S.; Shivpuri, Saroj:** *Enumerative technique for an extreme point fractional program*, European J. Oper. Res., **4**(1980)1, 54-59.

253. **Chambers, D.:** *Programming the allocation of funds subject to restrictions on reported results*, Oper. Res. Quart. **10**(1967), 407-431.

254. **Chandra, S.:** *Assignment problem with linear fractional objective function*, Presented at the 33rd Conference of Indian Mathematical Society, Lucknow 27-29, Dec. 1967.

255. **Chandra, S.:** *The capaciated transportation problem in linear fractional functionals programming*, J. Oper. Res. Soc. Japan, **10**(1967)1-2, 18-26.

256. **Chandra, S.:** *Decomposition principle for linear fractional functional programs*, Rev. Francaise Informat. Recherche Opérationnelle **10**-R2(1968), 65-72.

257. **Chandra, S.:** *Linear fractional functional programming*, J. Oper. Res. Soc. Japan, **10**(1967)1-2.

258. **Chandra, S.:** *Duality in Banach spaces, Transformation, Assignment and Large-Structured Fractional Programs*, Ph. D. Thesis, Dept. of Maths., Indian Institute of Technology, Kanpur (India), 1969.

259. **Chandra, S.; Chandramohan, M.:** *Duality and algorithmic aspects of integer linear fractional programming*, Research Report, Department of Mathematics, Indian Institute of Technology. New Delhi, 1976.

260. **Chandra, S.; Chandramohan, M.:** *Solving mixed integer fractional programms*, Presented at the 10th Annual Convention of O.R. Society of India, New Delhi, Dec. 1977.

261. **Chandra, S.; Chandramohan, M.:** *An improved branch and bound method for mixed integer-linear fractional program*, Z. Angew. Math. Mech., **59**(1979)10, 575-577.

262. **Chandra, S.; Chandramohan, M.:** *Duality in mixed integer non-convex and non-differentiable programming*, Z. Angew. Math. Mech., **59**(1979)4, 205-209.

263. **Chandra, S.; Chandramohan, M.:** *A branch and bound method for integer non-linear fractional programs*, Z. Angew. Math. Mech., **60**(1980)12, 735-737.

264. **Chandra, S.; Chandramohan, M.:** *A note on integer linear fractional programming*, Naval Res. Logist. Quart., **27**(1980)1, 171-174.

265. **Chandra, S.; Craven, B.D.; Mond, B.:** *Vector-valued Lagrangian and multi-objective fractional programming duality*, Numer. Funct. Anal. Optim. **11**(1990)3-4, 239-254.

266. **Chandra, S.; Gulati, T.R.:** *A duality theorem for a non-differentiable fractional programming problem*, Management Sci., **23**(1976), 32-37.

267. **Chandra, S.; Husain, I.:** *Symmetric dual continuous fractional programming*, J. Inform. Optim. Sci., **10**(1989)1, 241-255.

268. **Chandra, S.; Craven, B.D.; Mond, B.:** *Multiobjective fractional programming duality, A Lagrangian approach*, Optimization **22**(1991)4, 549-556.

269. **Chandra, S.; Saxena, P.K.:** *Fractional transportation problem with impurities*, Advances in Management Studies, **2**(1983), 335-349.

270. **Chandra, S.; Saxena, P.K.:** *Fractional transportation problem with penalty costs and impurities*, Advances in Management Studies, **3**(1984), 57-65.

271. **Chandra, S.; Saxena. P.K.:** *Cost/completion-data tradeoffs in quadratic fractional transportation problem*, Econom. Comput. Econom. Cybernet. Stud. Res., **22**(1987)3, 67-72.

272. **Chandra, S.; Craven, B.D.; Husain. I.:** *Continuous programming containing arbitrary norms*, J. Austral. Math. Soc. Ser. **A39**(1985), 28-38.

273. **Chandra, S.; Craven, B.D.; Mond, B.:** *Nonlinear programming duality and matrix game equivalence*, J. Austral. Math. Soc. Ser. **B26**(1989), 422-429.

274. **Chandra, S.; Craven, B.D.; Mond, B.:** *Symmetric dual fractional programming*, Z. Oper. Res. Ser. A-B, **29**(1985)1, A59-A64.

275. **Chandra, S.; Craven, B.D.; Mond, B.:** *Generalized fractional programming duality : a ratio game approach*, J. Austral. Math. Soc. Ser. **B28**(1986)2, 170-180.

276. **Chandra, S.; Mond, B.; Durga Prasad, M.V.:** *Constrained ratio games and generalized fractional programming*, ZOR, Methods Models Oper. Res. **32**(1988)5, 307-314.

277. **Chandrasekaran, R.:** *Minimal ratio spanning trees*, Networks, **7**(1977), 335-342.

278. **Chandrasekaran, R.; Tamir, A.:** *Optimization problems with algebraic solutions : quadratic fractional programs and ratio games*, Math. Programming **30**(1984)3, 326-339.

279. **Chandrasekaran, R.; Tamir, A.:** *Polynomial testing of the query "Is $a^b \geq c^d$?" with application to finding a minimal cost reliability ratio spanning tree*, Discrete Appl. Math. **9**(1984), 117-123.

280. **Chandrasekaran, R.; Aneja, Y.P.; Nair, K.P.K.:** *Minimal cost-reliability ratio spanning tree*, In: Studies on graphs and discrete programming. Proceedings of the Workshop on Applications of Graph Theory to Management, held in Brussels, March 20-21, 1979. Pierre Hansen (ed.). North-Holland Mathematics Studies, 59. Ann. Discrete Math. 11. North-Holland Publishing Co., Amsterdam-New York, 1981, 53-60.

281. **Charnes, A.; Cooper, W.W.:** *Programming with linear fractional functionals*, Naval Res. Logist. Quart., **9**(1962)3-4, 181-186.

282. **Charnes, A.; Cooper, W.W.:** *Programming with fractional functionals: I, Linear fractional programming*, Systems Research Group, The Technological Institute, North-Western Univ. ONR Research Memorandum No.50, February 1962.

283. **Charnes, A.; Cooper, W.W.:** *Systems evaluation and repricing theorems*, Management Sci., **9**(1962), 33-49.

284. **Charnes, A.; Cooper, W.W.:** *Programming with linear fractional functionals*, Naval Res. Logist. Quart., **10**(1963)3, 273-274.

285. **Charnes, A.; Cooper, W.W.:** *An explicit general solution in linear fractional programming*, Naval Res. Logist. Quart., **20**(1973), 449-467.

286. **Charnes, A.; Cooper, W.W.:** *Goal programming and multiple objective optimization*, European J. Oper. Res., **1**(1977), 39-54.

287. **Charnes, A.; Cooper, W.W.:** *The effective solution of the class of nonconvex multi-objective problems with ratio goals*, 9th Triennial Conference on Operational Research, Hamburg Germany, July 20-24, 1981.

288. **Charnes, A.; Cooper, W.W.:** *The non-Archimedian CCR ratio for efficiency analysis: A rejoinder to BOYD and FÄRE*, European J. Oper. Res. **15**(1984)3, 333-334.

289. **Charnes, A.; Granot, D.:** *Constrained non-cooperative von Neumann ratio games*, Working paper No. 368, Faculty of Commerce and Business Administration, University of British Columbia, Vancouver, 1976.

290. **Charnes, A.; Neralic, L.:** *Sensitivity analysis in data envelopment analysis*, Glas. Mat. Ser. III **24**(44)(1989)1, 211-226.

291. **Charnes, A.; Cooper, W.W.; Rhodes, E.:** *Measuring the efficiency of decision making units*, European J. Oper. Res. **2** (1978), 429-444.

292. **Charnes, A.; Cooper, W.W.; Rhodes, E.:** *Measuring the efficiency of decision making units*, European J. Oper. Res. **3**(1979), 339.

293. **Charnes, A.; Cox, L.; Lane, M.:** *A note on the redesigning of a rate structure for allocation of state funds to educational institutions*, Working paper 70-49 of Project GUM, University of Texas at Austin, 1970.

294. **Charnes, A.; Granot, D.; Granot, Frieda:** *On improving bounds for variables in linear integer programs by surrogate constraints*. INFOR-Canad. J. Oper. Res. Inform. Process., **13**(1975)3, 260-269.

295. **Charnes, A.; Granot, D.; Granot, Frieda:** *A note on explicit solution in linear fractional programming problems*, Naval Res. Logist. Quart. **23**(1976)1, 161-167.

296. **Charnes, A.; Granot, D.; Granot, Frieda:** *An algorithm for solving general fractional interval programming problems*, Naval Res. Logist. Quart., **23**(1976)1, 57-65.

297. **Charnes, A.; Granot, D.; Granot, Frieda:** *On solving linear fractional interval programming problems*, Cahiers Centre Études Rech. Opér., **20**(1978)2, 45-47.

298. **Chatterjee, S.; Sen, R.:** *On a certain type of bicriterion programming problems*, Bull. Math. Soc. Sci. Math. R.S. Roumanie (N.S.), **29(77)**(1985), 297-306.

299. **Černov, Ju. P.:** *Several problems of parametric fractional linear programming*, (Russian), Optimal planirovanie vyp, **16**(1970), 98-111.

300. **Černov, Ju.P.:** *A certain problem of parametric linear-fractional programming*, (Russian), Izd. Akad. Nauk. Kirgiz SSR **3**(1970), 20-27.

301. **Černov, Ju. P.:** *On fractional programming problems with linear separable and quadratic functions*, (Russian), Èkonom. i Mat. Metody, **7**(1971), 721-732.

302. **Černov, Ju. P.:** *An application of the δ - method to the solution of fractional programming problems with separable functions*, (Russian), In: Mathematical Methods for the Solution of Economic Problems. (Supplement to Èkonom. i Mat. Metody, Collection no-3), 68-73, Izdat. "Nauka", Moscow 1972.

303. **Černov, Ju. P.; Bajalinov, È.B.:** *Linear analogue of a fractional linear programming problem* (Russian), *Mathematical modeling of economic processes*. 109-115, 236, "Ilim", Frunze, 1981.

304. **Černov, Ju. P.; Bajalinov, È.B.:** *A dual fractional-linear programming problem* (Russian), *Mathematical modeling of economic processes*, 115-122, 236, "Ilim", Frunze, 1981.

305. **Černov, Ju. P.; Lange, È.G.:** *Non-linear programming problems with specific economic indices, Methods and applications* (Russian), "Ilim" Frunze, 1978, 291 p.

306. **Černov, Ju. P.; Lange, È.G.:** *A transport problem of fractional programming*, (Russian), Optimal planirovanie Vyp., **16**(1970), 112-132.

307. **Černov, Ju.P.; Lange, È.G.:** *An application of the method of succesive computations to the solution of a certain class of fractional concave programming problems*, (Russian), In: Mathematical Methods for the Solution of Economic Problems. (Supplement to Èkonom. i Mat. Metody, Collection No.3), 37-49, Izdat. "Nauka", Moscow, 1974.

308. **Černov, Ju. P.; Lange, È.G.; Zhusupbaev, A.:** *Eine Awendung der methode der sukzessiven berechungen für die lösung eines problems der produktionsanordung mit einem gebrochenen-konvexenfractional*, Izv. Akad. Nauk Kirgiz. SSR **2**(1979), 27-34 (Russisch).

309. **Kim Lin Chew; Choo, E.-U.:** *Pseudolinearity and efficiency*, Math. Programming, **28**(1984)2, 226-239.

310. **Chong, Ji Huan; Kim, Jae Ok:** *A solution of linear and fractional linear programming with both restrictions* (Korean) Su-hak kwa Mul-li, **19**(1975)4, 9-15.

311. **Choo, E.-U.:** *Multicriteria linear fractional programming*, Unpublished Ph. D. Dissertation, University of British Columbia 1980.

312. **Choo, E.-U.:** *Proper efficiency and the linear fractional vector maximum problem*, Oper. Res., **32**(1984)1, 216-220.

313. **Choo, E.-U.; Atkins, D.R.:** *Bicriteria linear fractional programming*, J. Optim. Theory Appl., **36**(1982)2, 203-220.

314. **Choo, E.-U.; Atkins, D.R.:** *Connectedness in multiple linear fractional programming*, Management Sci., **29**(1983)2, 250-255.

315. **Chowdhuri, S.; Breuer, M.A.:** *Optimization algorithms for a class of nonlinear programming problems*, Comput. Math. Appl. **15**(1988)3, 175-184.

316. **Christov, G.:** *Hyperbolic optimization problem*, Paper presented at the IX International Symposium on Mathematical Programming, Budapest, 23-27 August 1976.

317. **Christov, G.:** *Properties and method for solving fractional-linear optimization problems*, C.R. Acad. Bulgare Sci., **36**(1983)1, 61-64.

318. **Christov, G.; Karamiteva, Z.; Stoyanov, T.E.:** *A numerical method and program for solving a linear-fractional integer optimization problem* (Bulgarian), Mathematics and mathematical education (Sunny Beach (Slunchev Bryag, 1985), 566-570, Bulgar. Akad. Nauk, Sofia, 1985.

319. **Chung, Kun Jen:** *A note on maximal mean/standard deviation ratio in an undiscounted MPD*, Oper. Res. Lett. **8**(1989)4, 201-203.

320. **Cimoca, Gh.; Ţigan, Şt.:** *O problemă polinomială de max-min cu restricţii liniare*, Seminarul "Th. Angheluţă". Inst. Politehnic Cluj-Napoca (România), 10-12 iunie 1983, 35-38.

321. **Climaco, J.C.N.; Cardoso, D.M.:** *Linear fractional programming: a new bicriteria approach*, Belgian J. Oper. Res. Stat. Comput. Sci. **29**(1989)3, 3-24.

322. **Conde, E.; Ruiz Canales, P.:** *A branch-and-bound algorithm for discrete fractional programming*, (Spanish) Proceedings of the XVth Portuguese-Spanish Conference on Mathematics. Vol. IV (Évora, 1990), 361-366. Univ. Évora, Évora, 1991.

323. **Cook, M.D.; Kirby M.J.L.; Mehndiratta, S.L.:** *A linear fractional max-min problem*, Operations Res., **23**(1975)3, 511-521.

324. **Corban, A.:** *Programming with fractional linear objective function*, Rev. Roumaine Math. Pures Appl., **18**(1973)5, 633-637.

325. **Corban, A.:** *Dualitate pentru probleme de transport în programarea fracționară*, Stud. Cerc. Mat., **25**(1973)3, 347-357.

326. **Corban, A.:** *Dualitate în programarea neliniară*, Stud. Cerc. Mat., **26**(1974)3, 375-399.

327. **Corban, A.:** *Non-linear three-dimensional programming*, Rev. Roumaine Math. Pures Appl., **20**(1975)9, 1043-1059.

328. **Craven, B.D.:** *Duality for generalized convex fractional programs*, In: Schaible, S., and W.T. Ziemba, (eds.) Generalized Concavity in Optimization and Economics. Academic Press, New York, 1981, 473-490.

329. **Craven, B.D.:** *Fractional programming - a survey*, Opsearch **25**(1988)3, 165-176.

330. **Craven, B.D.:** *Fractional programming*, Heldermann Verlag Berlin 1988, Sigma series in applied mathematics, Vol.4.

331. **Craven, B.D.; Mond, B.:** *On fractional programming and equivalence*, Naval Res. Logist. Quart., **22**(1975)2, 405-410.

332. **Craven, B.D.; Mond, B.:** *Duality for homogeneous fractional programming*, Cahiers Centre Études Rech. Opér., **18**(1976)4, 413-417.

333. **Craven, B.D.; Mond, B.:** *The dual of a fractional linear program*, J. Math. Anal. Appl., **42**(1973)3, 507-512, erratum in J. Math. Anal. Appl., **55**(1976)3, 807.

334. **Craven, B.D.; Mond, B.:** *On duality for fractional programming*, Z. Angew. Math. Mech., **59**(1979)6, 278-279.

335. **Craven, B.D.; Mond, B.:** *A note on duality in homogeneous fractional programming*, Naval Res. Logist. Quart., **26**(1979)1, 153-155.

336. **Craven, B.D.; Mond, B.:** *Modified duality for mixed-integer programs*, Utilitas Math. **21C**(1982), 235-248.

337. **Craven, B.D.; Mond, B.:** *On maximizing a ratio of optimization problems*, Cahiers Centre Études Rech. Opér., **25**(1983)1-2, 29-34.

338. **Crouzeix, J.-P.:** *A duality framework in quasiconvex programming*, In: S. Schaible and W.T. Ziemba (eds.). Generalized Concavity in Optimization and Economics. Academic Press, New York, 1981, 207-225.

339. **Crouzeix, J.-P.; Ferland, J.A.:** *Algorithms for generalized fractional programming*, Math. Programming 52(1991)2, Ser. B, 191-207.

340. **Crouzeix, J.-P.; Ferland, J.A.; Schaible, S.:** *Duality in generalized linear fractional programming*, Technical Report No. 399, Départment d'Informatique et de Recherche Opérationelle, Université de Montréal, March, 1981. Also in: Math. Programming **27**(1983)3, 342-354. Erratum in: Math. Programming **29**(1984)2, 243.

341. **Crouzeix, J.-P.; Ferland, A.; Schaible, S.:** *An algorithm for generalized fractional programs*, J. Optim. Theory Appl., **47**(1985)1, 35-49.

342. **Crouzeix, J.-P.; Ferland, J.A.; Schaible, S.:** *A note on algorithm for generalized fractional programs*, J. Optim. Theory Appl., **50**(1986)1, 183-187.

343. **Dantzig, G.B.; Blattner, W.; Rao, M.R.:** *Finding a cycle in a graph with minimum cost to time ratio with applications to a ship routing problem*, Theory of Graphs International Symposium, Dunod, Paris and Gordon and Breach, New York, 77-83, 1966.

344. **Das, C.; Swarup, K.:** *Complex fractional functionals programming with nonlinear constraints*, Z. Angew. Math. Mech., **55**(1975), 441-442.

345. **Datta, N.:** *Efficiency in multiobjective fractional functional programming*, J. Inform. Optim. Sci. **3**(1982)3, 262-268.

346. **Datta, N.; Bhatia, D.:** *Algorithm to determine an initial efficient basic solution for a linear fractional multiple objective transportation problem*, Cahiers Centre Études Rech. Opér., **26**(1984)1-2, 127-136.

347. **Datta, N.; Bhatia, D.:** *A note on duality theory for concave convex fractional programming problem in complex space*, Indian J. Pure Appl. Math., **15**(1984)12, 1289-1295.

348. **Derman, C.:** *An sequential decisions and Markov chains*, Management Sci., **9**(1962)1, 16-24.

349. **Deumlich, R.; Elster, K.-H.:** *Duality theorems for nonconvex optimization problems*, Math. Operationsforsch. Statist., Ser. Optim., **9**(1978), 335-347.

350. **Deumlich, R.; Elster, K.-H.:** *Duality theorems and optimality conditions for nonconvex optimization problems*, Math. Operationsforsch Statist. Ser. Optimization, **11**(1980)2, 181-219.

351. **Deumlich, R.; Elster, K.-H.:** *A contribution to duality theory of nonlinear programming*, Methods of mathematical programming (S. Walukiewicz and A.P. Wierzbicki eds.) PWN- Warszawa, 31-40, 1981.

352. **Deumlich, R.; Elster, K.-H.:** *Fractional programming in view of generalized conjugation*, Methods Oper. Res., **49**(1985), 3-16.

353. **Deumlich, R.; Elster, K.-H.:** *On a class of nonconvex optimization problems*, In: Selected topics in operations research and mathematical economics. G. Hammer and D. Pallaschke (eds.) Proceedings of the eighth symposium on operations research held at the University of Karlsruhe, August 22-25, 1983. Lecture Notes in Economics and Mathematical Systems, 226, Springer-Verlag, Berlin-New York, 1984, 13-29.

354. **Deumlich, R.; Elster, K.-H.:** *Φ-conjugation and nonconvex optimization, A survey* I-III. Optimization **14**(1983)1, 125-140, ibid. **15**(1984)4, 499-515, ibid. **16**(1985)6, 789-803.

355. **Deumlich, R.; Elster, K.-H.:** *Contributions to duality of certain nonconvex optimization problems*, In: Contributions to operations research. Proceedings of the conference on operations research held in Oberwolfach, February 26 - march 3, 1984. Edited by K. Neumann and D. Pallaschke. Lecture Notes in Economics and Mathematical Systems, 240, 10-40, Springer-Verlag, 1985.

356. **Deumlich, R.; Elster, K.-H.:** *On perturbations of certain nonconvex optimization problems*, J. Optim. Theory Appl. **48**(1986)1, 81-93.

357. **Dezhurko, L.F.; Fam Tkhe Long:** *On methods for solving the general fractional-linear programming problem*, (Russian), Dokl. Akad. Nauk BSSR **27**(1983)7, 595-598.

358. **Ding, Baoyan; Wang, Chung-Lie:** *Stability analysis for linear fractional programs*, Proceedings of the Twentieth Manitoba Conference on Numerical Mathematics and Computing (Winnipeg, MB, 1990) Congr. Numer. 80(1991), 107-115.

359. **Dinkelbach, W.:** *Die maximierung eines quotienten zweier linearer functionen unter linearen nebenbedingungen*, Z. Wahrscheinlichkeistheorie und Verw. Gebiete, **1**(1962), 141-145.

360. **Dinkelbach, W.:** *On nonlinear fractional programming*, Management Sci., **13**(1967)7, 492-498.

361. **Dormány, Mihály:** *Specialis alaku parametrikus linearis es hiperbolikus programozasi feladatok megoldasa revide alt szimplex modszerral*, Nehez. musz. Egyetem Közl., **4**, **25**(1979)1, 53-62.

362. **Dormány, Mihály:** *On solving a bicriterion program*, (Hungarian), Alkalmaz. Mat. Lapok **9** (1983) 3-4, 393-404.

363. **Dorn, W.S.:** *Linear fractional programming*, I.B.M. Research Report, RC-830, Nov. 27, 1962.

364. **Dragomirescu, M.:** *An algorithm for the minimum-risk problem of stochastic programming*, Operations Res., **20**(1972), 154-164.

365. **Duca, D.:** *Asupra unei probleme de programare fracţionară*, Stud. Cerc. Mat., **29**(1977)5, 487-497.

366. **Duca, D.:** *Ordonarea soluţiilor extremale într-o problemă de programare fracţionară*, Stud. Cerc. Mat., **32**(1980)2, 179-185.

367. **Duca, Eugenia:** *Asupra ordonării punctelor eficiente extremale într-o problemă de programare vectorială*, Studia Univ. Babeş-Bolyai, Math., **1**(1979), 57-63.

368. **Duca, Eugenia:** *An algorithm for the calculation of efficient extreme points in problems of vector programming*, Operations research, Proc. 3-rd Colloq. Cluj-Napoca/Rom. 1978, 90-97, 1979.

369. **Dumitru, V.; Luban, Florica:** *Mathematical programming real variable models for integer and mixed-integer problems*, Econom. Comput. Econom. Cybernet. Stud. Res., **14**(1980)3, 69-79.

370. **Dumitru, V.; Luban, Florica:** *Mathematical programming real variable models for integer and mixed-integer problems and some applications to production scheduling*, Math., Rev. Anal. Numér. Théor. Approx., Math., **23** (**46**)(1981), 11-23.

371. **Dutta, D.; Rao, J.R.; Tiwari, R.N.:** *Sensitivity analysis in fuzzy linear fractional programming problem*, Fuzzy Sets and Systems 48(1992)2, 211-216.

372. **Dutta, D.; Rao, J.R.; Tiwari, R.N.:** *Sensitivity analysis in fractional programming-the tolerance approach*, Internat. J. Systems Sci. **23**(1992)5, 823-832.

373. **Dutta, D.; Tiwari, R.N.; Rao, J.R.:** *Multiple objective linear fractional programming - A fuzzy set theoretic approach*, Fuzzy Sets and Systems **52**(1992)1, 39-45.

374. **Dutta, D.; Rao, J.R.; Tiwari, R.N.:** *Effect of tolerance in fuzzy linear fractional programming*, Fuzzy Sets and Systems **55**(1993)2, 133-142.

375. **Dutta, D.; J.R. Rao; Tiwari, R.N.:** *A restricted class of multiobjective linear fractional programming problems*, European J. Oper. Res. **68**(1993)3, 352-355.

376. **Egorova, E.N.:** *One approach to solving special problems of cube-fractional programming*, (Russian), Optimizatsiya **39** (**56**)(1986)1, 81-98.

377. **Egudo, R.R.:** *Multiobjective fractional duality*, Bull. Austral. Math. Soc. **37**(1988)3, 367-378.

378. **Ellero, A.; Moretti Tomasin, E.:** *Equivalence between two algorithms for linear bicriteria programming problems*, Generalized Concavity for Economic Applications, Piera Mazzoleni (ed.). Proceedings of the Workshop held in Pisa, April 2, 1992, 41-52.

379. **Elmaghraby, S.E.; Arisawa, S.:** *On hyperbolic programming with a single constraint and upper-bounded variables*, Management Sci., **19**(1972), 42-45.

380. **Elster, K.-H.:** *On duality results in nonconvex optimization, In: Functional analysis, optimization, and mathematical economics*, 235-247, Oxford Univ. Press New York, 1990.

381. **Elster, K.-H.; Deumlich, R.:** *Lagrange duality for nonconvex optimization problems*, Operations Research Verfahren, **43**(1981), 53-61.

382. **Elster, K.-H.; Wolf, A.:** *Generalized convexity and fractional programming*, In: [229], pag. 219-231.

383. **Ermolev, L.G.:** *A problem of fractional nonlinear programming*, (Russian), Kibernetika (Kiev) **8**(1972)2, 45-47, also in Cybernetics, **8**(1972)2, 224-227.

384. **Faarland, Bruce H.; Nancy Jacob, L.:** *The linear fractional portfolio selection problem*, Management Sci., **27**(1981)12, 1383-1389.

385. **Faiziev, N.G.; Šerman, I.M.; Nuritdinov, Š.:** *Application of a certain class of target functions in the optimization of engineering processes*, (Russian), Dokl. Akad. Nauk UzSSR, **6**(1976), 16-18.

386. **Falk, J.E.:** *Maximization of signal-to-noise ratio in an optical filter*, SIAM J. Appl. Math., **17**(1969), 582-592.

387. **Falk, J.E.; Susan W. Palocsay:** *Optimizing the sum of linear fractional functions*, Recent advances in global optimization (Princeton, NJ, 1991), 221-258, Princeton ser. Comput. Sci., Princeton Univ. Press, Princeton, NJ, 1992.

388. **Ferland, J.A.; Potvin, J.-Y.:** *Generalized fractional programming: algorithms and numerical experimentation*, European J. Oper. Res., **20**(1985)1, 92-101.

389. **Ferrero, O.:** *Conditions for convexity for a function defined on a polyhedra and applications*, (Italian), Rend. Semin. Mat., Torino **45**(1987)2, 25-39.

390. **Filipovich, E.I.:** *On solving fractional programming problem*, (Russian), Sbornic "Trudi po voprosam primeneniia electronih vicislitelnih masin narodnom hoziaistve", Izd. Gor'skogo isledovatel'scogo fiziko-matematiceshogo Institute Gorki, 1964.

391. **Flachs, J.:** *Global saddle-point duality for quasi-concave programs*, I, II, Math. Programming, **20**(1981), 327-347, ibid., **24**(1982), 326-345.

392. **Flachs, J.; Pollatschek, M.A.:** *Equivalence between a generalized Fenchel duality theorem and a saddle-point theorem for fractional programs*, J. Optim. Theory Appl., **37**(1982), 23-32.

393. **Flachs, J.:** *Generalized Cheney-Loeb-Dinkelbach-type algorithms*, Math. Oper. Res., **10**(1985)4, 674-687.

394. **Florian, M.; Robillard, P.:** *Programmation hyperbolique en variables bivalents*, Rev. Francaise Informat. Recherche Opérationnelle V-1(1971), 3-9.

395. **Florian, M.; Robillard. P.:** *Programmation hyperbolique pseudo-booleenne en variables bivalentes*, Department d'Informatique, Université de Montreal, Publication No.36, mai 1970.

396. **Florian, M.; Robillard, P.:** *A note on hyperbolic programming*, Department d'Informatique, Université de Montreal, Publication No.31, mars 1976.

397. **Fox, B.:** *Markov renewal programming by linear fractional programming*, SIAM J. Appl. Math., **14**(1966)6, 1418-1432.

398. **Fox, B.:** *Finding minimal cost-time ratio circuits*, Operations Res., **17**(1969)3, 546-551.

399. **Frankel, E.; Novaes, A.; Pollack, E.:** *Optimization and integration of shipping ventures* (*A parametric linear programming algorithm*). International Shipbuilding Progress, July 1965.

400. **Gabasov, R.; Dezhurko, L.F.:** *The adaptive method for solving the general problem of fractional programming* (Russian), Dokl. Akad.Nauk BSSR **29**(1985)8, 685-687.

401. **Galambos, Gábor; Imreh, Balazs:** *Solution of one-dimensional cutting stock problems by column-generation*, (Hungarian), Alkalmaz. Mat. Lapok **10**(1984)1-2, 73-85.

402. **Gamidov, S.I.; Dem'yanov, V.F.:** *Minimization of the ratio of maximum and minimum functions* (Russian), Vestn. Leningr. Univ. 1984, No. 13, Mat. Mech. Astron. No.3, 14-18, 1984.

403. **Garg, K.C.; Swarup, K.:** *Complementary programming with linear fractional objective function*, Cahiers Centre Études Rech. Opér., **20**(1978)1, 83-94.

404. **Garg, K.C.; Swarup, K.:** *Linear fractional functional complementary programming with extreme point optimization*, Indian J. Pure Appl. Math., **9**(1978)6, 559-566.

405. **Garg, K.C.; Swarup, K.:** *The use of cuts in linear fractional functional complementary programming*, Z. Angew. Math. Mech., **60**(1980)1, 53-54.

406. **Gasparotto, G.:** *On the Charnes-Cooper transformation in linear fractional programming*, Report n.3, University of Pisa, Department of Statistics and Applied Mathematics, 1987.

407. **Gavurin, M.K.:** *On minimization of a function expressed in terms of linear fractional functions*, Vestn. Leningr. Univ. Math., **14**(1982), 255-259. Translation from Vest. Leningr. Vestn. Univ. Mat. Meh. Astron., **19**(1981)4, 5-8, (1981).

408. **Gavurin, M.K.:** *Linear fractional programming on an unbounded set*, (Russian), Vestnik Leningrad. Univ. Mat. Meh. Astronom. (Russian), **4**(1982), 12-16, 110.

409. **Gilmore, P.G.; Gomory, R.E.:** *A linear programming approach to the cutting stock problem*, part II. Operations Res., **11**(1963), 863-888.

410. **Glassey, C.R.:** *Explicit duality for convex homogeneous programs*, Math. Programming, **10**(1976), 176-191.

411. **Glower, F.; Woosley, R.E.:** *Aggregating diophantine equations*, University of Colorado, Report 70-4, oct 1970.

412. **Gogia, N.K.:** *The multiplex method for linear fractional functional programming*, Cahiers Centre Études Rech. Opér., **9**(1967)3, 123-133.

413. **Gogia, N.K.:** *Revised simplex algorithm for the linear fractional functional programming problem*, Math. Student, **36**, 1968, 55-57, 1969.

414. **Gogia, N.K.:** *The non-linear fractional functional programming problem with separable functions*, J. Math. Sci., **4**(1969)2, 77-84.

415. **Goldfard, D.; Mehrotra, Sanjay:** *A self-correcting version of Karmarkar's algorithm*, SIAM J. Numer. Anal. **26**(1989)4, 1006-1015.

416. **Golitschek, M.V.:** *The cost-to-time ratio problem for large or infinite graphs*, Discrete Appl. Math. **16**(1987)1, 1-9.

417. **Golovanov, L.V.; Sarkisyan, S.A.:** *Forecasting the development of large technical systems*, Tech. Forecasting & Soc. Change, **12**(1978), 175-184.

418. **Gol'shtein, E.G.:** *Dual problems of convex and fractionally-convex programming in functional spaces*, (Russian), Dokl. Akad. Nauk. SSSR, **172**(1967)5, 1007-1010.

419. **Gol'shtein, E.G.; Judin, D.B.:** *New directions in the linear programming*, Izdat. "Sovet Radio", Moscow, 1966.

420. **Gol'shtein, E.G.; Borisova, Eh.P.; Dudson, M.S.:** *A dialog system of vector optimization with fractional-linear criteria and linear constraints*, (Russian), Wiss. Z. Tech. Hochsch. Ilmenau **30**(1984)5, 17-31.

421. **Golub, G.H.; Underwood, R.:** *Stationary values of the ratio of quadratic forms subject to linear constraints*, Z. Angew. Math. Phys., **21**(1970), 318-326.

422. **Gondran, M.:** *Les problémes de ratio minimum en optimization combinatoire*, Note EDF HI/3433-02, 1980.

423. **Gondran, M.:** *Optimization with rational objective functions*, Bull. Direction Études Rech. Sér. C. Math. Inform., **1**(1982), 49-54.

424. **Goswami, M.K.; Sharma, J.K.:** *Cutting plane technique for linear fractional functional complementary programming problem*, Acta Cienc. Indica **14**(1988)2, 75-83.

425. **Goswami, M.K.; Sharma, J.K.:** *Fractional fixed charge complementary programming problem*, Methods Oper. Res. **61**(1990), 21-35.

426. **Granot, D.; Granot, Frieda:** *On solving fractional (0-1) programs by implicit enumeration*, INFOR-Canad. J. Oper. Res. Inform. Process, **14**(1976)3, 247-249.

427. **Granot, D.; Granot, Frieda:** *On integer and mixed integer fractional programming problems*, Ann. Discrete Math., **1**(1977), 221-231.

428. **Grinold, R.C.; Stanford, R.E.:** *Limiting distributions in a linear fractional flow model*, SIAM J. Appl. Math., **30**(1976)3, 402-406.

429. **Grover, T.R.:** *A note maximization of the transmission rate of discrete, constant channel*, Z. Oper. Res. Ser. A-B, **17**(1973)1, A13-A14.

430. **Grunspan, M.:** *Fractional programming : A survey*, Technical Report No.50. Project THEMIS, Systems Research Center, Industrial and Systems Engineering Department, University of Florida, Jan. 1971.

431. **Grunspan, M.; Thomas, M.E.:** *Hyperbolic integer programming*, Naval Res. Logist. Quart., **20**(1973)2, 341-356.

432. **Grunspan, M.; Thomas, M.E.:** *On integer solution of the hyperbolic programming problem*, Technical report TR-61, Project THEMIS, Systems Research Center, Industrial and Systems Engineering Department, Univ. Florida, Gainesvielle pp. 91, Febr. 1974.

433. **Guerra, Francisco; Rolando Verdaguer:** *Un algorithmo para la programacion lineal fraccionaria bicriterio*, Revista de Investigación Operational **9**(1988)3, 3-13.

434. **Gulati, T.R.:** *Duality for a nondifferentiable fractional program*, Cahiers Centre Études Rech. Opér., **21**(1971)4, 325-330.

435. **Gulati, T.R.:** *Optimality criterion and duality in complex fractional and indefinite programming*, Ph. D. Thesis, I.I.T. New Delhi, 1975.

436. **Gulati, T.R.; Chandra, S.:** *A duality theorem for complex fractional programming*, Z. Angew. Math. Mech., **55**(1975)6, 348-349.

437. **Gulati, T.R.; Islam, M.A.:** *Proper efficiency in a linear fractional vector maximum problem with generalized convex constraints*, European J. Oper. Res. **36**(1988)3, 339-345.

438. **Gulati, T.R.; Islam, M.A.:** *Efficiency in fractional vector maximization problems with nonlinear constraints*, Optimization **20**(1989)4, 477-482.

439. **Gulati, T.R.; Nadia Talaat:** *Duality in nonconvex vector minimum problems*, Bull. Austral. Math. Soc. **44**(1991)3, 501-509.

440. **Gulati, T.R.; Nadia Talaat:** *Duality in nonconvex multiobjective programming*, Asia-Pacific J. Oper. Res. **8**(1991)1, 62-69.

441. **Gupta, B.:** *Finding the set of all efficient solutions for the linear fractional multi-objective program with zero-one variables*, Opsearch, **18**(1981)4, 204-214.

442. **Gupta, B.:** *Existence and duality relations for multi-objective programs in complex space*, Opsearch, **19**(1982)3, 178-182.

443. **Gupta, B.:** *Linear fractional vector maximization problem, Existence and duality*, Cahiers Centre Études Rech. Opér., **25**(1983)1-2, 35-40.

444. **Gupta, B.:** *Programming with multi-objective linear fractionals*, Acta Cienc. Indica Math., **9**(1983)1-4, 195-201.

445. **Gupta, R.:** *Decomposition method and tranportation type problems with a fractional objective function*, Z.Angew. Math. Mech., **57**(1977), 81-88.

446. **Gupta, R.K.:** *Basic feasible solutions and decomposition principle for linear fractional functionals programming problem*, Trabajos Estadist., **2**(1971)1-2, 185-193.

447. **Gupta, R.K.:** *A simple class of parametric linear fractional functionals programming problem*, Cahiers Centre Études Rech. Oper., **15**(1973)2, 185-196, erratum in ibid. **16**(1974), 179.

448. **Gupta, S.K.; Bector, C.R.:** *Nature of quotients, products and rational powers of convex (concave)-like functions*, Math. Student, **36**(1968), 63-67.

449. **Gupta, S.N.; Jain, R.K.:** *Stochastic fractional programming under chance constraints with random technology matrix*, Acta Cienc. Indica, **12**(1986)3, 191-198.

450. **Gupta, S.N.; Jain, A.K.:** *Optimization with the ratio of independent normal variates*, Acta Cienc. Indica **12**(1986)3, 209-212.

451. **Gupta, S.N.; Jain, A.K.; Swarup, K.:** *Stochastic linear fractional programming with ratio of independent Chauchy variates*, Nav. Res. Logist., **34**(1987), 293-305.

452. **Gupta, A.K.; Sharma, J.K.:** *Fractional functional programming: A brief survey*, Saharanpur, 1981.

453. **Gupta, R.K.; Swarup, K.:** *Approximate method of solution for non-linear fractional programming problem*, Z. Angew. Math. Mech., **49**(1969)12, 753-756.

454. **Gupta, R.K.; Swarup, K.:** *A cutting plane algorithm for extreme point linear fractional functionals programming*, Cahiers Centre Études Rech. Opér., **16**(1974)2, 161-177.

455. **Gupta,R.K.; Swarup, K.:** *On extreme point fractional programming problem*, Portugal. Math., **37**(1978)1-2, 13-29, 1981.

456. **Gupta, S.N.; Swarup, K.:** *Stochastic fractional functional programming*, Ric. Oper., **9**(1979)10, 65-79.

457. **Gupta, S.N.; Swarup, K.:** *Duality in stochastic fractional programming under chance constraints*, Ric. Oper., **10**(1980)3, 53-63.

458. **Gupta, S.N.; Swarup, K.; Banwarilal:** *Stochastic fractional programming under chance constraints with random technology matrix*, Gujarat Statist. Rev., **8**(1981)1, 23-34.

459. **Gwinner, J.:** *A general FARKAS lemma and applications in duality*, Methods Oper. Res. **57**(1987), 25-48.

460. **Gwinner, J.:** *Duality in minmax fractional programming*, International Conference on Operations Research, Vienna, Austria, August 28-31, 1990.

461. **Gwinner, J.; Jeyakumar, V.:** *A solvability theorem and minmax fractional programming*, ZOR - Methods and Models of Operations Research **37**(1993)1, 1-12.

462. **Hagopian, J.D.; Frisch, I.T.:** *Capacitance, inductance, and resistance minimization in RLC networks*, IEEE Trans. Circuit Theory CT-19(1972), 383-385.

463. **Haimovici, A.; Rimer, Şarlota:** *On a problem of generalized linear programming* (romanian), Bul. Inst. Politehn. Iaşi (N.S.), **8(12)**(1962), fasc. 1-2, 9-14.

464. **Halpern, J.:** *The sequential covering problem under uncertainty*, Working Paper no. 76-003. Department of Industrial Engineering, University of Toronto, Canada, January 1976.

465. **Hammer, P.L.; Rudeanu, S.:** *Boolean Methods in Operations Research and Related Areas*, Springer-Verlag Berlin, Heidelberg, New York, 151-159, 1968.

466. **Hannan, E.L.:** *Effects of substituting a linear goal for a fractional goal in the goal programming problem*, Management Sci., **24**(1977)1, 105-107.

467. **Hannan, E.L.:** *An interpretation of fractional objectives in goal programming as related to papers by Awerbuch et al., and Hannan*, Management Sci., **27**(1981)7 847-848.

468. **Hansen, P.; Marcus V. Poggi de Aragao; Ribeiro, C.C.:** *Hyperbolic 0-1 programming and query optimization in information retrieval*, Math. Programming **52**(1991)2, 255-263.

469. **Hansen, P.; Marcus V. Poggi de Aragao; Ribeiro, C.C.:** *Boolean query optimization and the 0-1 hyperbolic sum problem*, Annals of Mathematics and Artificial Intelligence. Vol.1, 1990.

470. **Hanssman, Fred:** *Probability of survival as in investment criterion*, Management Sci. **15**(1968)1, 33-48.

471. **Hartmann, K.:** *Einige aspekte der Ganzzahligen linearen quotientenoptimierung*, Wiss Z. Tech. Hochsch. Chem. Leuna-Merseburg., **15**(1973)4, 413-418.

472. **Hartmann, K.:** *Zur anwendung des schnittverfahrens von Gomory auf gemischt ganzzahlige lineare Quotientenotimierungsprobleme*, III. Internat. Tagung "Mathematik and Kybernetik in der Ökonomie" vom 14, bis May 16, 1973.

473. **Hartmann, K.:** *Rein ganzzahlige lineare quotientenoptimierung nach den schmittverfahren von Gomory*, Math. Operationsforsch. Statist. **6**(1975), 33-53.

474. **Hartmann, K.:** *Gemischt Ganzzahlige Lineare quotientenoptimierung nach den schmittverfahren von Gomory*, Math. Operationsforsch. Statist. **6**(1975)6, 845-854.

475. **Hartwig, H.:** *Ein simplexartigen Lösungsalgorithmus für pseudolineare Optimierungsprobleme*, Studia Sci. Math. Hungar. **10**(1975)1-2, 213-236.

476. **Harvey, Charles M.:** *Stochastic programming models for decreasing risk aversion*, J. Oper. Res. Soc. **32**(1981)10, 885-889.

477. **Hashizume, S.; Fukushima, M.; Katoh, N.; Ibaraki, T.:** *Approximation algorithms for combinatorial fractional programming problems*, Math. Programming **37**(1987)3, 255-267.

478. **Hassin, R.; Tamir, A.:** *Maximizing classes of two parameter objectives over matroids*, Meth. Oper. Res. **14**(1989)2, 362-375.

479. **Havrda, Jan:** *Mathematiké programovani (Czech.) Mathematical Programming*, Statni Nakladatelatki Technicke Literatury, Prague, 1972.

480. **Helbig, S.:** *Optimality criteria in disjunctive optimization and some applications*, Methods of Operations Research **62**(1990), 67-78.

481. **Hiriart-Urruty, J.-B.:** *Gradients généralises de functions composés, Applications*, C.R. Acad. Sci. Paris, **285**(1977), 781-784.

482. **Hirche, J.:** *Lösung hyperbolischer transportproblememittels linearer ersatzprobleme*, Berichte der jahrestagung numerische mathematik, Postdam-Cecilienhof. Nov. 25 bis Dec. 1, 1973, 59-45.

483. **Hirche, J.:** *Zur extremwertannahme und dualität bei optimierungsproblem mit linearem und gebrochen linearem zielfungktionsanteil*, Z. Angew. Math. Mech. **55**(1975), 184-185.

484. **Hirche, J.:** *Birichtigung : Über eine klasse nichtkonvexer optimierungsproblems*, (Z. Angew. Math. Mech. **57**(1977)4, 247-253), by Hirche,J. and Ho Khac Tan, Z. Angew. Math. Mech. **58**(1978)8, 367.

485. **Hirche, J.:** *Zur Lösung von optimierungsproblemen mit monoton-linear zusammengesetzten zielfunktionen*, Beiträge Numer. Math., **9**(1981), 87-94.

486. **Hirche, J.:** *A note on minimizing a polynomial of a linear fractional function*, Rev. Roumaine Math. Pures Appl. **26**(1981), 1193-1195.

487. **Hirche, J.:** *Verallgemeinerte konvexität bei summen und producten linearer und gebrochen linearer funcktionen*, Wiss. Z. Martin-Luther-Univ. Halle-Wittenberg Math.-Natur. Reihe **32**(1983)5, 91-99.

488. **Hirche, J.:** *On programming problems with a linear-plus-linear-fractional objective function*, Cahiers Centre Études Rech. Opér. **26**(1984)1-2, 59-64.

489. **Hirche, J.:** *Some remarks on generalized convexity of sums and products*, Z. Angew. Math. Mech. **65**(1985)1, 62-63.

490. **Hirche, J.:** *Lösung einer Klasse ganzzahliger Optimierungs-probleme mit zusammengesetzten Zielfunktionen*, Wiss. Z. Univ. Halle Math. -Naturwiss **38**(1989)4, 99-107.

491. **Hirche, J.; Ho Khac Tan:** *Über eine Klasse nichtkonvexer optimierung probleme*, Z. Angew. Math. Mech. **57**(1977), 247-253.

492. **Hirche, J.; Köhler, J.; Volkman Stiebitz:** *Ein effecktives lösungaverfahren für das hyperbolische transportproblem*, Beiträge Anal. **7**(1975), 151-156.

493. **Ho, James K.:** *A parametric subproblem for dual methods in decomposition*, Math. Oper. Res. **11**(1986), 644-650.

494. **Hodgson, T.J.; Lowe, T.J.:** *Production lot sizing with material handling cost considerations*, AIIE Trans. **14**(1982), 44-51.

495. **Horst, R.:** *On the convexification on nonlinear programming problems : An applications-oriented survey*, European J. Oper. Res. **15**(1983)3, 382-392.

496. **Horwáth, I.:** *Programare fracționară liniară cu restricții suplimentare*, Revista de Analiză Numerică și Teoria Aproximației, 3(1974), 71-77.

497. **Horváth, I.:** *Asupra programării fracționare lineare cu restricții suplimentare*, Informatica pentru Conducere. Orizont'81. Realizări și aplicații, Cluj-Napoca, 101-102. 1981.

498. **Hughes, J.B.:** *Interior efficient solutions in bicriterion linear fractional programming - A geometric approach*, Math. Comput. Modelling, **17**(1993)6, 23-28.

499. **Ibaraki, T.:** *Exploring recent advances in mathematical programming V. Fractional programming*, (Japanese), Syst. & Control (Japan), **24**(1980)12, 787-797.

500. **Ibaraki, T.:** *Solving mathematical programming problems with fractional objective functions*, In: Schaible, S. and W.T. Ziemba, (eds.) Generalized Concavity in Optimization and Economics. Academic Press, New York, 1981, 441-472.

501. **Ibaraki, T.:** *Parametric approaches to fractional programs*, Math. Programming **26**(1983)3, 345-362.

502. **Ibaraki, T.; Ishii, H.; Iwase, J.; Hasegawa, T.; Mine, H.:** *Algorithms for quadratic fractional programming problems*, J. Oper. Res. Soc. Japan, **19**(1976)2, 174-191.

503. **Iida, Koji:** *Optimal stopping of a contact investigation in two-stage search*, Math. Japon., **34**(1989)2, 169-190.

504. **Isbell, J.R.; Marlow, W.H.:** *Attrition Games*, Naval Res. Logist. Quart. **3**(1956), 71-94.

505. **Ishii, H.; Han, S.S.; Fujii, S.:** *MINMAX-MINRATIO bicriterion path problem.* Asia-Pacific Operations Research'88 (ed. B.H. Ahn), North-Holland, Amsterdam, Netherlands, 1990, 163-170.

506. **Ishii, H.; Nishida, T.:** *Stochastic linear knapsack problem : Probability maximization model*, Math. Japon. **29**(1984)2, 273-281.

507. **Ishii, H.; Nishida, T.; Daino, Akira:** *Fractional set covering problems*, Tech. Rep. Osaka Univ. **29**(1979) no.1492, 319-326.

508. **Ishii, H.; Ibaraki, T.; Mine, H.:** *A primal cutting plane algorithm for integer fractional programming problems*, J. Oper. Res. Soc. Japan **19**(1976)3, 228-244.

509. **Ishii, H.; Ibaraki, T.; Mine, H.:** *Fractional knapsack problems*, Math. Programmming **13**(1977)3, 255-271.

510. **Jagannathan, R.:** *The programming approach in multiple character studies*, Econometrica **33**(1965)1, 236-237.

511. **Jagannathan, R.:** *A method for solving a nonlinear programming problem in sample surveys*, Econometrica **33**(1965)4, 841-846.

512. **Jagannathan, R.:** *On some properties of programming problems in parametric form pertaining to fractional programming*, Management Sci. **12**(1966)7, 609-615.

513. **Jagannathan, R.:** *Duality for nonlinear fractional programs*, Z. Oper. Res. A-B **17**(1973)1, 1-3.

514. **Jagannathan, R.:** *An algorithm for a class of nonconvex programming problems with nonlinear fractional objectives*, Management Sci. **31**(1985)7, 847-851.

515. **Jagannathan, R.; Schaible, S.:** *Duality in generalized fractional programming via Farkas'lemma*, J. Optim. Theory Appl. **41**(1983)3, 417-424.

516. **Jagannathan, R.; Schaible, S.:** *An application of Farkas'lemma to a nonconvex minimization problem*, General inequalities 4, Mem. E.F. Beckenbach, 4-th. Int. Conf., Oberwolfach/Ger. 1983, ISNM71, 365-367, 1984.

517. **Jain, O.P.:** *Duality for fractional functional programming*, Cahiers Centre Études Rech. Opér. **21**(1979)1, 81-86.

518. **Jeyakumar, V.:** *First and second order fractional programming duality*, Opsearch **22**(1985)1, 24-41.

519. **Joksch, H.C.:** *Programming with fractional linear objective functions*, Naval Res. Logist. Quart. **11**(1964)2-3, 197-204.

520. **Jüttler, H.:** *Die lineare quotientenoptimierung als hilfsmittel für die entscheidungsfindung*, Rechentechnik - Datenverarbeitung, Heft 11(1967).

521. **Kabe, D.G.:** *Direct solutions to the m-median and fractional transportation problems*, Indust. Math. **30**(1980)1, 1-27.

522. **Kabe, D.G.:** *On a certain linear fractional programming problem*, Indian J. Pure Appl. Math. **11**(1980)1, 1411-1413.

523. **Kabe, D.G.:** *Direct solutions to some linear programming problems*, Indust. Math. **34**(1984)1, 1-20.

524. **Kacnelson, L.Z.; Neizvestnyi, M.M.:** *Isoextremal Cebysev fractions with a denominator of second degree*, (Russian), Latvian mathematical yearbook, **17** (Russian), 24-29, 276 Izdat, "Zinatne" Riga, 1976.

525. **Kaltinska, R.:** *An algorithm for the solution of the hyperbolic tranportation problem*, Seminarber. Humboldt-Univ. Berlin, Sekt. Math. **39**, 1981, (Russian), 104-113.

526. **Kalberg, J.G.; Ziemba, W.T.:** *Generalized concave functions in stochastic programming and portfolio theory*, In: Schaible, S. and Ziemba W.T. (eds.) Generalized Concavity in Optimization and Economics. Academic Press. New York, 1981, 719-767.

527. **Kanchan, P.K.:** *Linear fractional functional programming*, Acta Cienc. Indica **2**(1976)4, 401-405.

528. **Kanchan, P.K.:** *Upper bounds in linear and piecewise linear programming*, Acta Cienc. Indica **3**(1977), 357-360.

529. **Kanchan, P.K.; Holland, A.S.B.; Sahney, B.N.:** *Transportation techniques in linear plus-fractional programming*, Cahiers Centre Étude Rech. Opér. **23**(1981)2, 153-157.

530. **Kang, Ri Sob; Chong, Yun Chol:** *On the minimization of the sum of two linear fractional functions*, (Korean), Cho-son In-min Kong-hwa-kuk Kwa-hak-won T'ong-bo 1989, no.4, 2-4.

531. **Karp, R.M.:** *A characterization of the minimum cycle mean in a digraph*, Memorandum No. UCB/ERL/47, Electronic Research Laboratory. College of Engineering, University of California at Berkeley, 1977.

532. **Kant Uma; Gupta, G.C.:** *Second order duality in nonlinear fractional programming*, Math. Student **XLIII** (3), 165 (1975), Abstracts of paper presented at the 40th Conference of the Indian Mathematical Society, held at Indian Institute of Technology, Powai Bombay, Dec. 27-29, 1974.

533. **Kant Uma; Gupta, G.C.:** *Symmetric dual nonlinear fractional programs*, Paper presented at the 41-th Conference of the Indian Mathematical Society, Kanpur, Dec. 24-26, 1975.

534. **Kaska, J.:** *Duality in linear fractional programming*, Econom.-Mat. Obzor **5**(1969)4, 442-453.

535. **Kaska, J.; Pisek, M.:** *Linearni lomene programovani a jeho uplat neni v planovani stavebni viroby*, Pozemni Stavby 2(1964).

536. **Kaska, J.; Pisek, M.:** *Linearni lomene programovani*, Statistika a demografie V(1963), 191-207.

537. **Kaska, J.; Pisek, M.:** *Kvadraticko linearni lomene programovani*, Ekonom.-Mat. Obzor **2**(1966)2, 169-173.

538. **Kaska, J.; Pisek, M.:** *Konvexne konkavne lomene programovani*, Ekonom. -Mat. Obzor **3**(1967)4, 457-464.

539. **Kasyanyuk, S.A.:** *A certain problem in infinite linear fractional programming*, (Russian), Ž. Vyčisl. Mat. i Mat. Fiz. **9**(1969), 413-417.

540. **Kasyanyuk, S.A.; Kucker, B.N.:** *On monotonous programming*, Dopovidi Akad. Nauk Ukrain. SSR, Ser. A **1**(1967), 18-21.

541. **Kaul, R.N.; Bhatia, D.:** *Generalized linear fractional programming*, Ekonom.- Math. Obzor **10**(1974)3, 322-330.

542. **Kaul, R.N.; Chadha, S.S.:** *Duality in non-linear fractional programming problems*, Ekonom.- Mat. Obzor **7**(1971)2, 141-148.

543. **Kaul, R.N.; Datta, N.:** *Efficiency in multi-objective fractional functional programming*, Presented at the annual meeting of the society of mathematical sciences (Delhi, India) held on May 7, 1980. Abstract in J. Math. Sci. **14-15**(1979-1980) 95.

544. **Kaul, R.N.; Datta, N.:** *On the solution of separable programming problem with a fractional objective function*, Cahiers Centre Études Rech. Opér. **23**(1981)2, 150-169.

545. Kaul, R.N.; Gupta, B.: *Efficiency and linear fractional vector maximum value problem*, Z. Angew. Math. Mech. **60**(1980)2, 112-113.

546. **Kaul, R.N.; Gupta, B.**: *Multi-objective programming in complex space*, Z. Angew. Math. Mech. **61**(1981)11, 599-601.

547. **Kaul, R.N.; Kaur, S.; Lyall, V.**: *Duality in inexact fractional programming with set-inclusive constraints*, J. Optim. Theory Appl. **50**(1986)2, 279-288.

548. **Kaul, R.N.; Lata, M.**: *A method of decomposition for linear fractional programming*, Opsearch **11**(1974)4, 183-192.

549. **Kaul, R.N., Lyall, V.**: *A note on nonlinear fractional vector maximization*, Opsearch **26**(1989)2, 108-121.

550. **Kaur, S.**: *Inexact fractional programming with set-inclusive constraints*, Cahiers Centre Études Rech. Opér. **23**(1981)2, 171-181.

551. **Kaur, S.**: *Subgradient duality in fractional programming*, Indian J. Pure Appl. Math. **13**(1982)3, 287-298.

552. **Kaur, S.; Bhatia, D.**: *Duality theory for generalized fractional programmes*, Indian J. Pure Appl. Math. **14**(1983)2, 257-264.

553. **Kaur, S.; Gupta, S.**: *Duality in multiple objective fractional programming problems involving non-convex functions*, Opsearch **27**(1990)4, 239-253.

554. **Khalitov, N.T.**: *A linear-fractional programming problem in singular case*, J. Sov. Math. 40, No. 6, 725-727, 1988, translation from Issled. Prikl. Mat. 1, 38-40, 1973.

555. **Khan, Zulfiqar Ali**: *Converse duality in nonlinear fractional programming*, Asia-Pacific J. Oper. Res. **7**(1990)1, 9-15.

556. **Klein, M.**: *Inspection-maintenance-replacement schedules under markovian deterioration*, Management Sci. 9(1962)1, 25-32.

557. **Kleinemann, P.**: *Quantitative sensitivitätsanalyse bei parametrischen optimierungsaufgaben*, Seminarberichte, 9, Sektion Mathematik, Humboldt Universität, Berlin, 1978, iv + 105 pp (Cap.2) 62-76.

558. **Klevachev, V.I.**: *Solving the problem of linear fractional programming*, (Russian), Kibernetika (Kiev) 4(1968)6, 27-31. Also in: Cybernetics 4(1968)6, 31-36.

559. **Konno, H.; Inori, M.**: *Bond portfolio optimization by bilinear fractional programming*, J. Oper. Res. Soc. Japan 32(1989)2, 143-158.

560. **Konno, H.; Yajima, Yasutoshi**: *Minimizing and maximizing the product of linear fractional functions*, Recent advances in global optimization (Princeton, N J, 1991), 259-273, Princeton Ser. Comput. Sci., Princeton Univ. Press, Princeton, N J, 1992.

561. **Konno, Hiroshi; Kuno, Takahito**: *Generalized linear multiplicative and fractional programming*, Computational methods in global optimization. Ann. Oper. Res. 25(1990)1-4, 147-161.

562. **Kornbluth, J.S.H.**: *A survey of goal programming*, Omega 1(1973)2, 193-205.

563. **Kornbluth, J.S.H.**: *Indifference regions and marginal utility weights in multiple objective linear fractional programming*, Working paper 79-02-03, Department of decision sciences, The Warton School, University of Pennsylvania, 1979.

564. **Kornbluth, J.S.H.**: *Multiple objective linear fractional programming algorithms, Some computational experience*, Lecture Notes in Econom. and Math. Systems, 190, 1981, 173-198.

565. **Kornbluth, J.S.H.**: *Ratio goals in manpower planning models*, INFOR-Canad. J. Oper. Res. Inform. Process, 21(1983)2, 151-154.

566. **Kornbluth, J.S.H.**: *Max-min programming with linear fractional functionals : algorithms and examples*, Essays and surveys on multiple criteria decision making (Mons, 1982), 204-213, Lecture Notes in Econom. and Math. Systems, 209, Springer, Berlin-New York, 1983.

567. **Kornbluth, J.S.H.**: *On the use of multiple objective linear programming algorithms to solve problems with fractional objectives*, European J. Oper. Res. 23(1986)1, 78-81.

568. **Kornbluth, J.S.H.; Salkin, G.R.**: *A note on the economic interpretation of the dual variables in linear fractional programming*, Z. Angew. Math. Mech. 52(1972), 175-178.

569. **Kornbluth, J.S.H.; Salkin, G.R.**: *The optimal dual solution in linear fractional decomposition problems*, Operations Res. 22(1974)1, 183-189.

570. **Kornbluth, J.S.H.; Salkin, G.R.**: *A note on returns to scale in linear fractional programming*, Z. Angew. Math. Mech. 55(1975)12, 757-758.

571. **Kornbluth, J.S.H.; Steuer R.E.**: *On computing the set of all weakly efficient vertices in multiple objective linear fractional programming*, In: *Multiple criteria decision making theory and applications*, (Proc. Third Conf. Hagen/Koenigswinter, 1979), 189-202, Lecture Notes in Econom. and Math. Systems, 177, Springer, Berlin, 1980.

572. **Kornbluth, J.S.H.; Steuer, R.E.**: *Multiple objective linear fractional programming*, Management Sci. 27(1981)9, 1024-1039.

573. **Kornbluth, J.S.H.; Steuer, R.E.**: *Goal programming with linear fractional criteria*, European J. Oper. Res. 8(1981), 58-65.

574. **Kovacs, A.; Stahl, J.**: *Decomposition procedure in case of maximizing the index of enterprise interest*, (Hungarian) Szigma 6(1973), 105-114.

575. **Kovacs, A.; Stahl, J.**: *On large scale linear fractional programs*, Lecture Notes Computer Sci. Germ. 41, 353-361, 1976, Optimization Tech. Modeling Optimization Serv. Man. 7th (IFIP Conf. Proc. II. Nice, 1975).

576. **Kovalëv, M.M.; Do Zui Chin**:' *A problem of parametric convex discrete optimization*, (Russian); Vestnik Beloruss. Gos. Univ. Ser. I Fiz. Mat. Mekh. 1990, no.3, 35-38, 79.

577. **Köhler, J.**: *An algorithm for the solution of integer hyperbolic programming problems*, Mathematische optimierungtheorie und anwendungen, Internationale tagung/1981 Eisenach, DDR, 1981, 89-92.

578. **Köhler, J.; Hirche, J.**: *Mehrfache zerlegung hyperbolische optimierungsausfgaben*, Beiträge Anal. 7(1975), 143-149.

579. **Körth, H.**: *Quotientenoptimierung* (hyperbolische optimierung), XII internationales wissenschaftliches kolloquium wirschaftsmatematik (1), T.H. Ilmenau 11, p. 95-99, Sept. 15, 1967.

580. **Körth, H.**: *Ein zerlegungsprinzip für die hyperbolische optimierung*, Wiss. Z. Humboldt-Univ. Berlin, Ges., Sprachw, Reihe 18(1969), 827-829.

581. **Körth, H.**: *Untersuchungen zur nichtlinearen optimierung ökonomischer erscheinungen und prozesse unter besonderer berucksichtigung der quotientenoptimierung sovie der Lösung Okonomishchmathematischer Modelle bei Existenz mehrerer Zielfunktionen*, Habilitation, eingereicht an der Humbold-Universität, Berlin, Sektion Wirtschaftwissenschaften, 1969.

582. **Körth, H.**: *Transportoptimierung mit hyperbolischer Zielfunktion*, Wiss. Z. Humboldt-Univ. Berlin, Gesellschafts-und sprachwissenschaftiliche Reihe **XIX**(1970)6, 737-740.

583. **Körth, H.**: *Hyperbolische transportoptimierung mit beschränkung der variablen*; Wiss. Z. Humboldt-Univ. Berlin, Ges.- Sprachw R. **XX**(1971), 521-524.

584. **Körth, H.**: *Hyperbolische transportoptimierung mit beschränkung der variablen*, Wissenschaftliche zeitschrift der hochsschule für architektur und Bauwesen Weimar **19**(1972)2, 184-188.

585. **Körth, H.**: *Verallgemeinerte hyperbolische optimierungsaufgobe*, 24. Internationales Wissenschaftliches Kolloquium, Technische Hochschule Ilmeanu, Heft **4**(1979), 41-43.

586. **Kreutzberger, O.**: *Bemerkungen Zur Quotientenund produktoptimierung mit anwendung beim risikoproblem der stochastischen optimierung*, Wiss. Z. Martin-Luther-Univ. Halle-Wittenberg. Math. Natur. Reihe. Heft. **3**(1978), 75-79.

587. **Krupitskij, A.E.**: *Minimization of the sum of two linear-fractional functions on a convex polyhedral set*, Vestnik Leningrad Univ. Math. **16**(1984), 155-161. Translation from Vestnik Leningrad. Univ. 1983, No.13, Math. Mekh. Astron. No. 3, 15-21 (Russian).

588. **Kucher, B.N.**: *On one algorithm for solving a problem of monotone programming*, (Russian) Dopovidi Akad. Nauk. Ukrain. RSR, Ser. A **10**(1987), 869-872.

589. **Kuznecov, V.F.; Baranovscaja, G.C.**: *Problem of the optimal distribution of auxiliary loading and unloading media with the goal of minimizing the maximal time of handling the incoming loads*, (Russian), *Theory of optimal solutions*; (Proc. Seminar, Kiev, 1969), No.2 (Russian) Akad. Nauk. Ukrain. SSR, Kiev, 61-74, 1979.

590. **Kydland, F.**: *Simulation of liner operations*, Institute for Shipping Research, Norwegian School of Economics and Business Administration, Bergen, translated and reprinted from Sosialekonomen **23**(1969)9.

591. **Kydland, F.**: *Duality in fractional programming*, Naval Res. Logist. Quart. **19**(1972)4, 691-697.

592. **Lal, S.N.; Mukherjee, R.N.; Singh, A.K.**: *Sufficiency of exact penalty minimization and fractional programming*, Opsearch. **27**(1990)3, 165-170.

593. Lal, S.N.; Nath, Baij; Kumar, Ajay: *duality for some nondifferentiable static multiobjective programming problems*, J. Math. Rerch. Appl. 186(1993)3, 862-867.

594. **Lange, E.G.; Artjuhin, A.V.**: *Some problems of linear-fractional parametric programming*, (Russian), *Mathematical Methods for the Solution of Mathematical-Economic Problems*, Izdat. "Ilim", Frunze, 1974, 5-19.

595. **Lara, P.**: *Multiple objective fractional programming and Livestock ration formulation : a case study for dairy cow diets in Spain*, Agricultural Systems **41**(1992)3, 321-334.

596. **Lata, M.**: *An operator theory for a class of linear fractional programming problems*, **I**, **II**, Z. Angew. Math. Mech., **55**(1975)3, 133-140; **56**(1976), 75-88.

597. **Lata, M.; Mittal, B.S.**: *Decomposition method for interval linear fractional programming*, Z. Angew. Math. Mech. **56**(1976)4, 153-159.

598. **Lau, N.; Lata, M.**: *Fractional transportation problem with quantity discounts*, Presented at the Anual meeting of the Society of Mathematical Sciences (Delhi, India) held on february, 1, 1978. Abstract in J. Math. Sci. Vol. **14-15**(1979-1980).

599. **Lawler, E.L.**: *Optimal cycles in doubly weighted directed linear graphs*, In: Théorie des graphes, P. Rosenthiel ed., Dunod 1966, 209-213.

600. **Lawler, E.L.**: *Optimal cycles in graphs and the minimal cost-to-time ratio problem*, Technical report ERL-M343, Department of electrical engineering. University of California, Berkley, May 1972.

601. **Lawler, E.L.**: *Combinatorial optimization: Networks and matroids* (Holt, Rinehart and Winston, New York, 1976), 94-97.

602. **Lawrence, J.A.**: *A simple proof for a vertex following procedure for hyperbolic programming*, Proceedings of the American Institute for Decision Sciences Eleventh Anual Meeting, Western Regional Conference, San Diego, CA, USA, 17-19 March 1982 (Atlanta, GA, USA: AIDS 1982), 158-159.

603. **Lee, B.; Dong-Wan Tcha**: *An iteractive procedure for fuzzy programming problems with linear fractional objectives*, Comput. Ind. Eng. (UK), **16**(1989)2, 269-275.

604. **Lee, B.; Chung, N.; Tcha, D.**: *A parallel algorithm and duality for a fuzzy multiobjective linear fractional programming problem*, Computers & Industrial Engineering **20**(1991)3, 367-372.

605. **Lee, G.M.**: *On efficiency in nonlinear fractional vector maximization problem*, Optimization **25**(1992)1, 47-52.

606. **Lee, Haijune; Pulat, Simin:** *Geometric interpretation of linear fractional programming*, ORSA/TIMS Joint

National Meeting, Denver, 23-26 Oct. 1988.
607. **Leleno, J.**: *Remarks on the algorithm of B. Martos for the solution of hyperbolic programming problem*, (Polish), Przeglad Statyst. **24**(1977)3, 399-408.
608. **Lin, Cuo-Yun**: *Comparison of duality models in nonlinear fractional programming*, (Chinese), Numer. Math. J. Chinese Univ. **3**(1981)3, 270-272.
609. **Lin, Cuo-Yun**: *Duality theory for multiobjective fractional programming*, (Chinese), Numer. Math. J. Chinese Univ. **4**(1982)4, 289-299.
610. **Lin, Cuo-Yun**: *The fundamental theorem of multiobjective fractional programming*, (Chinese), Acta Math. Appl. Sinica **6**(1983)2, 247-250.
611. **Lintner, J.**: *The valution of risk assets and the selection of risky investment in stock portfolios and capital budgets*, Rev. Econom. Statist., **47**(1965), 13-37.
612. **Liu, San Yang**: *Symmetric duality of fractional programming*, (Chinese), Xi'an Jiaotong Daxue Xuebao **24**(1990)6, 135-138.
613. **Lommis, L.H.**: *On a theorem of von-Neumann*, Proc. Nat. Acad. Sci. USA **32**(1946), 213-215.
614. **Luhandjula, M.K.**: *Fuzzy approaches for multiple objective linear fractional optimization*, Fuzzy Sets and Systems **13**(1984)1, 11-23.
615. **Lupşa, L.**: *Asupra alurii unor funcţii hiperbolice*, Studia Univ. Babeş-Bolyai Math. **23**(1978)2, 66-72.
616. **Lupşa, L.**: *Remarques concernant le rapport entre les problèmes de programmation quadratique indéfinie et les problèmes de programmation hyperbolique*, Studia Univ. Babeş-Bolyai Math. **23**(1978)1, 50-54.
617. **Lyall, Vinod; Suneja Surjeet Kaur**: *Decomposition in generalized fractional programming and its optimal dual solution*, Indian J. Pure Appl. Math. **18**(1987)11, 973-978.
618. **Lysenko, A.I.; Dabagijian, A.V.; Gorelyi, A.V.**: *Analytic solution of a nonlinear programming problem*, (Russian), Avtomat. Sistemy Upravlenija i Pribory Avtomat. No. **42**(1977)3, 53-62.
619. **Mahajan, D.G.; Vartak, M.N.**: *Generalization of some duality theorems in nonlinear programming*, Math. Programming **12**(1977)3, 293-317.
620. **Major, D.C.**: *Benefit-cost ratios for projects in multiple objective investment programs*, Water Resources Research **6**(1969)5, 1174-1178.
621. **Makhmudov, R.A.**: *On an algorithm for solving a covering problem with linear fractional functional*, (Russian), Izv. Akad. Nauk Az. SSR, Ser. Fiz.-Tekh. Mat. Nauk 1988, No.3, 128-130, 1988.
622. **Manas, M.**: *On transformations of quasi-convex programming problems*, Ekonom.-Mat. Obzor **4**(1968)1, 93-99.
623. **Mangasarian, O.L.**: *Nonlinear fractional programming*, J. Oper. Res. Soc. Japan, **12**(1969)1, 1-10.
624. **Mangasarian, O.L.**: *Convexity, pseudo-convexity and quasi-convexity of composite functions*, Cahiers Centre Études Rech. Opér. **12**(1970)2, 114-122.
625. **Manjurov, D.M.**: *Investigation of a linear-fractional programming problems when certain parameters are varied*, (Russian), Izd Akad. Nauk Azerbaidzan SSR. Ser. Fiz.-Tehn. Mat. Nauk **4**(1968), 89-95.
626. **Manjurov, D.M.**: *A certain generalized problem of linear-fractional programming*, (Russian), Numerical Mathematics (Russian), 88-103. Izdat. "Élm", Baku, 1973.
627. **Manjurov, D.M.**: *Block methods for fractional linear programming*, (Russian), *Questions of Mathematical Cybernetics and Applied Mathematics* No.1 (Russian) edited by Dž. É. Allahverdiev, K.F. Sirinov, D.A. Babev and S.R. Musaev. Izdat. "Élm", Baku, 51-71, 1975.
628. **Marchi, Ezio**: *Equilibrium points of rational n-person games*, J. Math. Anal. Appl. **54**(1976)1, 1-4.
629. **Marchi, A.**: *Sulla relazione tra un problema bicriteria ed un problema frazionario*, Department of Statistics and Applied Mathematics; University of Pisa, report no. 33, 1990.
630. **Marchi, A.**: *Solving a quadratic fractional program by means of a complementarity approach*, Research Report n.32, Dept. of Statistics and Applied Mathematics, University of Pisa, Italy, 1990.
631. **Martein, Laura**: *Maximum of the sum of linear function and a linear-fractional function*, (Italian), Riv. Mat. Sci. Econom. Social. **8**(1985)1, 13-20.
632. **Martein, Laura**: *On generating the set of all efficient points of a bicriteria linear fractional problem*, Report n.13. Universita di Pisa, Italy. Dipartimento Di Statistica e Matematica Applicata All'Economia, 1988.
633. **Martein, Laura**: *Applicazioni della programazione frazionaria nel campo economico-finanziario*, Report n.14. Universita di Pisa, Italy. Dipartimento Di Statistica e Matematica Applicata All'Economia, 1988.
634. **Martein, Laura; Pellegrini, L.**: *Un algoritmo per la determinazione del massimo di una particulare funzione razionale fratta soggetta a vincoli lineari*, Paper No. A-45, Dept. of Operations Research, University of Pisa, Italy, 1977.

635. **Martein, Laura; Pellegrini, L.:** *On an extension of a particular class of fractional programming problems*, (Italian), Proceedings of the first AMASES Meeting. Held in Pisa, Nov. 4-5, (1977), G. Giappichelli, Editore, Turin viii + 393, 213-228, 1979.

636. **Martein, Laura; Schaible, S.:** *On solving a linear program with one quadratic constraint*, University of Pisa. Department of Statistics and Applied Mathematics, Report n.8, 1987.

637. **Martein, Laura; Sodini, C.:** *Un algoritmo per un problema di programmazione frazionaria non lineare e non convessa*, Publication No.93, Serie A, Dept. di Ricerca Operativa e Scienze Statistiche, Universita di Pisa, Italy, 1982.

638. **Martinez-Legaz, J.E.:** *Some new results on exact quasionvex duality*, Methods Oper. Res. **49**(1985), 47-62.

639. **Martins, E.Q.V.:** *An algorithm to determine a path with minimal cost/capacity ratio*, Discrete Appl. Math. **8**(1984), 189-194.

640. **Martos, B.:** *Hiperbolikus programozas*, Publications Math. Inst. Hungarian Academy Sci. **5**, 383-406, 1960, Serie B. A translation by Andrew and Veronika Whinston appeared under the title "Hiperbolic programming" in Naval Res. Logist. Quart. **11**(1964), 135-155.

641. **Martos, B.:** *Hyperbolic programming by simplex method*, Deuxième Congress Mathématique Hongrois, 1960, Vol. **II**, 44-48, Budapest, 1961.

642. **Martos, B.:** *The direct power of adjacent vertex programming methods*, Management Sci. **12**(1965), 241-252.

643. **Maruşciac, I.:** *Asupra unei programări hiperbolice*, Stud. Cerc. Mat. **26**(1974)3, 419-430.

644. **Mazzoleni, P.:** *Stationary values of the ratio of quadratic polynomials*, Tech. Rep. The Hartfield Polytechnic, 45, 1973.

645. **Mazzoleni, P.:** *Teoria della dualita per una classe di problemi con funzione obiectivo frata*, Boll. Un. Mat. Ital. **IV**, Ser. **11**. 571-577, 1975.

646. **Mazzoleni, P.:** *The dual for a particular fractional program*, Technical report No. 59, Numerical Optimization Centre, The Hartfield Polytechnic. September, 1974.

647. **Mazzoleni, P.:** *Some experience on a moving-truncation method applied to a nonlinear programming problem with fractional objective function*, In: Towards Global Optimization (L.C.W. Dixon, G. P. Szegö eds.), Proceedings of a Workshop at the University of Cagliary 1974, North-Holland Publishing Company. Amsterdam and New York, 350-360, 1975.

648. **Megiddo, N.:** *Combinatorial optimization with rational objective functions*, Conference Record of the 10-th Anual ACM Symposium on Theory of Computing (papers presented at the Symposium held in San Diego Calif., May 1-3, 1978) Association of Computing Machinery, New York, 1978, 1-12.

649. **Megiddo, N.:** *Combinatorial optimization with rational objective functions*, Math. Oper. Res. **4**(1979)4, 414-424.

650. **Megiddo, N.:** *Applying parallel computation algorithms in the design of serial algorithms*, Proc. 22nd IEEE Symposium on Foundation of Computer Science, 1981, 399-408.

651. **Meister, B.; Oettli, W.:** *On the capacity of a discrete, constant channel*, Inform. and Control **11**(1967), 341-351.

652. **Meister, B.; Oettli, W.:** *Two classes of algorithms for concave optimization, and the calculation of the capacity of discrete memoryless channels*, Elektron. Informationsverarb. Kybernet, **9**(1973), 189-195.

653. **Mine, H.; Fukushima, M.; Ryang, Y.J.:** *Parametric nonlinear programming for general cases and its application to some problems*, Mem. Fac. Engrg. Kyoto Univ. **40**(1978)3, 198-211.

654. **Misra, S.; Das, C.:** *The sum of a linear and linear fractional function and a three dimensional transportation problem*, Opsearch **18**(1981), 139-157.

655. **Miščenko, I.I.:** *A possible method of optimization on infinite matrices of limiting conditions*, (Russian), Issled. Operacii i ASU No. 15, 76-80, 135, 1980.

656. **Mititelu, Şt.; Stancu-Minasian, I.M.:** *Invexity at a point: generalisations and classification*, Bull. Austral. Math. Soc. **48**(1993)1, 117-126.

657. **Mjelde, K.M.:** *Sufficiency of Kuhn-Tucher optimality conditions for a fractional programming problem*, BIT **18**(1978)4, 454-456.

658. **Mjelde, K.M.:** *Convex-concave fractional programming with each variable occuring in a single constraint*, BIT **18**(1978)2, 202-210.

659. **Mjelde, K.M.:** *Allocation of resources according to a fractional objective*, European J. Oper. Res., **2**(1978), 116-124.

660. **Mjelde, K.M.:** *Convex-concave fractional programming-evaluation of solutions and optimality conditions*, BIT **19**(1979)2, 270.

661. **Mjelde, K.M.:** *Location of a discrete resource and its allocation according to a fractional objective*, European J. Oper. Res. **4**(1980)4, 49-53.

662. **Mjelde, K.M.:** *Properties of optimal allocations of resources according to a fractional objective*, J. Oper. Res. Soc. **32**(1981)5, 405-408.

663. **Mjelde, K.M.:** *Improvement of a method of the evaluation of allocation of resource to activities*, European J. Oper. Res. **8**(1981)1, 86-87.

664. **Mjelde, K.M.:** *Cost-effective allocations of bounded and binary resources in polynomial time*, European J. Oper. Res. **11** (1982), 176-180.

665. **Mjelde, K.M.:** *Fractional resource allocation with S-shaped return functions*, J. Oper. Res. Soc. **34**(1983)7, 627-632.

666. **Mjelde, K.M.:** *Methods of Allocation of Limited Resources*, John Wiley and Sons Ltd., 1983.

667. **Mjelde, K.M.:** *Componentwise fractional programming with application to resource allocation*, MIC-Modeling Identification Control **4**(1983)2, 117-123.

668. **Mjelde, K.M.:** *An incremental and parametrical algorithm for convex-concave fractional programming with a single constraint*, European J. Oper. Res. **23**(1986)3, 391-395.

669. **Molnár, Z.; Trčka, V.:** *Solution of a hyperbolic zero-one programming problem by the branch and bound method*, (Czech), Ekonom.-Mat. Obzor **12**(1976)4, 428-437.

670. **Mond, B.:** *Fractional programming*, In: Optimization Proc. Sem. Austral. Nat. Univ. Canberra, 172-182, 1971. Univ. Queensland Press, Sf. Lucia, 1972.

671. **Mond, B.:** *A class of nondiferentiable fractional programming problems*, Z. Angew. Math. Mech. **58**(1978), 337-341.

672. **Mond, B.:** *On the solution of a linear fractional programming problem*, 10th International Sympposium on Mathematical Programming Montreal, August 27-31, 1979.

673. **Mond, B.:** *On algorithmic equivalence in linear fractional programming*, Math. Comp. **37**(1981)155, 185-187.

674. **Mond, B.; Craven, B.D.:** *A note on mathematical programming with fractional objective functions*, Naval Res. Logist. Quart, **20**(1973), 577-581.

675. **Mond, B.; Craven, B.D.:** *Nonlinear fractional programming*, Bull. Austral. Math. Soc. **12**(1975)3, 391-397.

676. **Mond, B.; Craven, B.D.:** *A duality theorem for a nondifferentiable nonlinear fractional programming problem*, Bull. Austral. Math. Soc. **20**(1979)3, 397-406.

677. **Mond, B.; Schechter, M.:** *A duality theorem for a homogeneous fractional programming problem*, J. Optim. Theory Appl. **25**(1978)3, 349-352.

678. **Mond, B.; Schechter, M.:** *Converse duality in homogeneous fractional programming*, Preprint, 1979.

679. **Mond, B.; Schechter, M.:** *Duality in homogeneous fractional programming*, J. Inform. Optim. Sci., **1**(1980)3, 271-280.

680. **Mond, B.; Weir, T.:** *Generalized concavity and duality*, In: Schaible, S. and W.T. Ziemba (eds.), Generalized concavity in optimization and economics. Academic Press, New York, 1981, 263-279.

681. **Mond, B.; Weir, T.:** *Duality for fractional programming with generalized convexity conditions*, J. Inform. Optim. Sci. **3**(1982), 105-124.

682. **Morita, H.; Ishii, H.; Nishida, T.:** *Stochastic linear knapsack programming problem and its application to a portfolio selection problem*, European J. Oper. Res. **40**(1989)3, 329-336.

683. **Morris, A.J.:** *Generalization of dual structural optimization problems in terms of fractional programming*, Quart. Appl. Math. **36**(1978)2, 115-119.

684. **Motanov, V.G.:** *Optimization problems with a linear-fractional functional*, (Russian), Èkonom. i Mat. Metody **7**(1971)4, 586-592.

685. **Mukherjee, R.N.:** *Generalized convex duality for multiobjective fractional programs*, J. Math. Anal. Appl. 162(1991)2, 309-316.

686. **Munteanu, E., Radó, F.:** *Calculul şarjelor celor mai economice la cuptoarele de topit fontă*, Studii Cerc. Mat. (Cluj), **XI**, fascicola anexă, 149-158, 1960.

687. **Nabeya, Seiji:** *On linear fractional programming*, Hitotsubashi J. Arts Sci. **5**(1965)1, 58-64.

688. **Nakayama, Hirotaka:** *Statisficing trade-off method for problems with multiple linear fractional objectives and its applications*, In : Multiobjective problems of mathematical programming. A. Lewandowski and V. Volkovich (eds.). Proceedings of the International Conference held in Ialta, October 26 - November 2, 1988. Lecture Notes in Economics and Mathematical Systems, 351, Springer - Verlag, Berlin, 1991, 42-50.

689. **Narihisa, H.:** *An algorithm for solving the stochastic programming problem*, Mem. Defense Acad. **18**(1978)3, 151-160.

690. **Nath, Baij; Raghawendra; Mukherjee, R.N.:** *Necessary conditions for nondifferentiable fractional multi-objective programming*, Internat. J. Management Systems **4**(1988)3, 153-159.

691. **Nauss, R.M.:** *On the use of internal rate of return in linear and integer programming*, Oper. Res. Lett. 7(1988)6, 258-289.

692. **Neumann, von John:** *A model of general economic equilibrium*, Rev. Econom. Stud. **13**(1945-1946), 1-9.

693. **Nikitin, A.I.; Nuriev, U.G.:** *A heuristic algorithm for solving a linear-fractional boolean programming problem*, (Russian), Izv. Akad. Nauk. Azerbaidzhan SSR, Ser. Fiz.-Tekh. Mat. Nauk **3**(1982)5, 112-117, 1983.

694. **Nykowski, I.:** *Problem pogodźenia kilku kryteriow w jednym programie liniowym*, Przeglad Statyst. **13**(1966)4, 367-375.

695. **Nykowski, I.; Zolkiewski, Z.:** *On some connections between bicriteria and fractional programming problems*, Fifth International Conference on Multiple Criteria Decision Making. Mons. 1982, Essays and Surveys on Multiple Criteria Decision Making. Springer-Verlag, Berlin-New York, 1983, 300-309.

696. **Nykowski, I.; Zolkiewski, Z.:** *A linear model with a linear fractional objective function and multi-objective programming*, (Polish), Przeglad Statyst. **28**(1982)3-4, 181-196.

697. **Nykowski, I.; Zolkiewski, Z.:** *A compromise procedure for the multiple objective linear fractional programming problem*, European J. Oper. Res. **19**(1985)1, 91-97.

698. **Ohlson, J.A.; Ziemba, W.T.:** *Portfolio selection in a log-normal market when the investor has a power utility function*, Journal of Financial and Quantitative Analysis. **11**(1976), 57-71.

699. **Okabe, A.:** *Spatial distributions maximizing or minimizing Geary's contiguity ratio*, J. Fac. Engrg. Univ. Tokio Ser. B, **36**(1982)3, 525-528.

700. **Pandey, Shashi:** *Duality for multiobjective fractional programming involving generalized η - bonvex functions*, Opsearch **28**(1991)1, 36-43.

701. **Pang, Jong-Shi:** *A parametric linear complementarity technique for optimal portfolio selection with a risk-free assets*, Operations Res. **28**(1980)4, 927-941.

702. **Pardalos, P.M.:** *An algorithm for a class of nonlinear fractional problems using ranking of the vertices*, BIT **26**(1986)3, 392-395.

703. **Pardalos, P.M.; Phillips, A.T.:** *Global optimization of fractional programs*, J. of Global Optimization **1**(1991)2, 173-182.

704. **Parkash, Om; Saxena, P.C.; Patkar, V.N.:** *Duality in a class of nonlinear fractional programming problems*, Nat. Acad. Sci. Letters (India), **2**(1979)7, 267-268.

705. **Parkash, Om; Saxena, P.C.; Patkar, V.N.:** *Nondiferentiable fractional programming in complex space*, Z. Angew. Math. Mech. **64**(1984)1, 59-62.

706. **Passy, U.:** *Fractional programming using pseudo duality*, Operations Research Verfahren vol. **31**(1979), 481-493.

707. **Passy, U.:** *Duality and pseudo duality in mathematical programs with quotients and products of finitely many functionals*, Operations Research, Statistics & Economics. Mimeograph Series No.**218**, 1978, Technion, Haifa, Israel.

708. **Passy, U.:** *Pseudo-duality in mathematical programs with quotients and products*, J. Optim. Theory Appl. **33**(1981)3, 349-374.

709. **Passy, U.:** *Pseudo duality and non-convex programming*, In: Schaible, S. and W.T. Ziemba (eds.), Generalized concavity in optimization and economics. Academic Press, New York, 1981, 239-261.

710. **Passy, U.; Keslassy, Annie:** *Pseudo duality and duality for explicitly quasiconvex functions*, Operations Research, Statistics & Economics. Mimeograph Series No.249, Technion, Haifa, Israel.

711. **Passy, U.; Keslassy, A.:** *Duality for a class of quasiconvex programs*, J. Optim. Theory Appl. **40**(1983)4, 515-536.

712. **Passy, U.; Prisman, E.Z.:** *A convex-like duality scheme for quasi-convex programs*, Math. Programming **32**(1985)3, 278-300.

713. **Patkar, V.N.:** *Fractional programming models for sharing of urban development responsabilities*, Nagarlok **22**(1990)4, 88-94.

714. **Patkar, V.N.; Saxena, P.C.; Parkash, Om:** *Linear piecewise linear programs with variable coefficients*, Pure Appl. Math. Sci. **10**(1979)1-2, 51-56.

715. **Patkar, V.N.; Saxena, P.C.; Parkash, Om:** *On a class of quadratic fractional programming problems*, Nat. Acad. Sci. Letters (India) **2**(1979)1, 29-30.

716. **Patkar, V.N.; Saxena, P.C.; Parkash, Om:** *On a discrete non-linear fractional programming problem*, Nat. Acad. Sci. Letters (India) **3**(1980)7, 204-205.

717. **Patkar, V.N.; Saxena, P.C.; Parkash, Om:** *Dual program for a convex fractional function*, Econom. Comput. Econom. Cybernet. Stud. Res. **15**(1981)1, 77-80.

718. **Patkar, V.N.; Stancu-Minasian, I.M.:** *Approaches for solving a class of nondifferentiable fractional programming problems*, Nat. Acad. Sci. Letters (India). **4**(1981)12, 477-480.

719. **Patkar, V.N.; Stancu-Minasian, I.M.:** *On disjunctive linear fractional programming*, Econom. Comp. Econom. Cybernet. Stud. Res. **17**(1982)2, 87-96.

720. **Patkar, V.N.; Stancu-Minasian, I.M.:** *Parametric algorithm for a class of disjunctive linear fractional programs*, Bull. Math. Soc. Sci. Math. R.S. Roumaine (N.S.), **29**(77)(1985), nr.3, 279-284.

721. **Patkar, V.N.; Stancu-Minasian, I.M.:** *Duality in disjunctive linear fractional programming*, European J. Oper. Res. **21**(1985), 101-105.

722. **Patkar, V.N.; Stancu-Minasian, I.M.:** *Recent results in disjunctive linear fractional programming*, In: [229] pag. 99-105.

723. **Patkar, V.N.; Stancu-Minasian, I.M.:** *A disjunctive linear fractional max-min problem*, Portugal. Math. **48**(1991), fasc. 1, 67-73.

724. **Păun, Gh.:** *Two methods for multicriteria hierarchization of discrete alternatives*, RAIRO Recherche Opérationnelle/Operations Research, **18**(1984)3, 263-278.

725. **Peteanu, V.; Ţigan, Şt.:** *O nouă problemă de drum optim în grafe*, Al 2-lea Simpozion de Informatică şi Conducere, 20-23 aprilie, Cluj-Napoca, 1976, 169-173, 1977.

726. **Peteanu, V.; Ţigan, Şt.:** *Unele probleme de optimizare neliniară în grafe*, Seminarul de Teoria Grafurilor II, 8-9 oct. 1982, Iaşi, 1982.

727. **Peteanu, V.; Ţigan, Şt.:** *Asupra unor probleme de max-min în grafe*, Lucrările celui de al III-lea Seminar de Teoria grafurilor, Braşov 21-22 oct. 1983, 43-53.

728. **Peteanu, V.; Ţigan, Şt.:** *Interval goal programming with linear fractional criteria*, Preprint Babeş-Bolyai University of Cluj-Napoca, 1984, 1-22.

729. **Peteanu, V.; Ţigan, Şt.:** *On some discrete fractional max-min problems, Application to max-min problems in graphs*, Rev. Anal. Numér. Théor. Approx. **13**(1984)2, 167-173.

730. **Peteanu, V.; Ţigan, Şt.:** *The multiobjective linear-fractional programming and interval goal programming*, Itinerant Seminar on Functional Equations, Approximation and Convexity. Iaşi 26 Oct. 1986, Univ. "Al. I. Cuza", Facultatea de matematică, 40-45.

731. **Picard, Jean-Claude; Queyranne, Maurice:** *A network flow solution to some nonlinear 0-1 programming problems, with applications to graph theory*, Networks **12**(1982)2, 141-159.

732. **Plávka, J.:** *The $O(n^3)$ algorithm for a special case of the maximum cost-to-time ratio cycle problem and its coherence with an eigenproblem of a matrix*; Z. Oper. Res. **36**(1992)5, 417-422.

733. **Podkaminer, L.:** *Maksymalizacja ilorazu dwu form liniowych przy liniowych warunkach ograniczjacych*, Przeglad Statyst., **17**(1970), 93-103.

734. **Podkaminer, L.:** *Ceny dualne i inne parametry zadanis oprymalizacyjnego z ilorazowa funkcjia kryterium*, Przeglad Statyst. **18**(1971), 332-338.

735. **Pogodin, V.P.:** *Study of a certain parametric linear-fractional programming*, (Russian), In: Mathematical Methods in Economic Research. Suppliment to Èkonom. i Mat. Metody, Izdat. "Nauka", Moscow, 44-50, 1974.

736. **Prabha, S.:** *Parametrizing a column vector in a linear fractional programming problem*, J. Inform. Optim. Sci. **3**(1982)3, 290-304.

737. **Preda, V.:** *On minmax programming problems containing n - set functions*, Optimization **22**(1991)4, 527-537.

738. **Preda, V.:** *On duality for multiobjective fractional programs*, Stud. Cerc. Mat. **45**(1993)4, 319-324.

739. **Puri, M.C.:** *Enumerative technique for extreme point linear fractional programming problem*, SCIMA (J. Manag. Sci. Appl. Cybernetics) **2**(1973)1, 1-8.

740. **Puri, M.C.:** *Strong cut cutting plane method for extreme point linear fractional programming*, Cahiers Centre Études Rech. Opér. **17**(1975)1, 65-69.

741. **Puri, M.C.; Swarup, K.:** *Extreme point linear fractional programming*, Z. Oper. Res. Ser. A-B **18**(1974), 131-139.

742. **Ramos, Paulo Cesar F.:** *Generalized fractional programming*, Ph. D. Thesis. Florida State University, 1981-1982.

743. **Rani, O.; Kaul, R.N.:** *Duality theorems for a class of nonconvex programming problems*, J. Optim. Theory Appl., **11**(1973), 305-308.

744. **Rani, O.; Manju Lata:** *A method of decomposition for linear fractional programming*, Paper presented at the annual meeting of the Society of Mathematical Sciences (India), held on 22nd October 1977. Abstract in J. Math. Sci. Vol. **12-13**(1977-1978), 92-93.

745. **Rani, O.; Shivpuri, Saroj:** *An algorithm for linear fractional functions programming problems*, Z. Angew. Math. Mech. **57**(1977), 75-80.

746. **Rao, M.R.:** *Cluster analysis and mathematical programming*, J. Amer. Statist. Assoc. **66**(1971), 335, 622-626.

747. **Rapcsák, T.:** *On pseudolinear functions*; European J. Oper. Res. **50**(1991)3, 353-360.

748. **Reiss, S.P.:** *Rational search*, Inf. Proc. Let. **8**(1979), 89-90.

749. **Rhode, P.:** *Verfahren zur ganzzahligen lineren quotientenprogrammierung*, Diplomarbeit, Köln, 1978.

750. **Ritter, K.:** *A parametric method for solving certain nonconcave maximization problems*, J. Comput. System Sci., **1**(1967), 44-54.

751. **Robillard, P.:** *Hyperbolic programming with bivalent variables*, Publication No.41, Department d'informatique, Université de Montreal, Août 1970.

752. **Robillard, P.:** *(0,1) hyperbolic programming problems*, Naval Res. Logist. Quart. **18**(1971), 47-57.

753. **Robinson, S.M.:** *Computational solution of ratio games by iterative linear programming*, Transactions of the 17-th Conference of Army Mathematicians (U.S. Army Missile Command. Redistone Arsenal. Ala. 1971), p. 265-290 A.U.S. Army Research Office (Durham, N.C.), Report No. 72-1, U.S. Army Research Office Durham, N.C., 1972.

754. **Rodova, A.; Saska, J.:** *Program řešeni lomeneho linearniho programovani metodu nahradni funkce*, TOVUS, Praha, 1964.

755. **Rote, G.:** *The convergence rate of the Newton method in fractional programming*, IVth International Workshop on Generalized Convexity, Pécs/Ungary, August 31-September 2, 1992.

756. **Rothblum, Uriel G.:** *Ratios of affine functions*, Math. Programming, **32**(1985)3, 357-365.

757. **Rubal'skiĭ, G.B.:** *Integer minimization of fractional - separable functions*, (Russian), Zh. Vychisl. Mat. i Mat. Fiz. **30**(1990), no.10, 1454-1466.

758. **Ruiz, José Arzola:** *Modela matematico para la planificacion calendaria optima de los talleres de hornos Martin*, Investigación Oper., **14**(1975), 6-19.

759. **Ruiz-Canales, P.:** *Estructuras de dominancia definidas por un Cono : Programación Fraccionaria Multiple*, Tesis Doctoral 1986.

760. **Ruiz-Canales, P.; Osuna-Gomez, R.; Antonio Mayor-Gallego, J.:** *A cutting primal-dual algorithm in integer linear fractional programming*, IVth International Workshop on Generalized Convexity, Pécs / Ungary, August 31-September 2, 1992.

761. **Ruiz-Canales, P.; Osuna-Gomez, R.; Beato-Moreno, A.:** *On the transport problem with linear fractional objective*, IVth International Workshop on Generalized Convexity, Pécs/Ungary, August 31-September 2, 1992.

762. **Rutledge, R.W.:** *A simplex method for zero-one mixed integer linear programs*, J. Math. Anal. Appl. **18**(1967), 377-390.

763. **Rybasov, M.V.; Dudnikov, E.E.:** *A parametric method of solving fractional programming problems on analogue computers*, (Russian), Dokl. Akad. Nauk SSSR **161**(1965), 1289-1290.

764. **Rybin, S.V.:** *Analysis of a conical programming problem*, (Russian), Vestn. Leningr. Univ. 1985, No. 15, Mat. Mekh. Astron. No.3, 105-107, 1985.

765. **Saipe, A.I.:** *Solving a (0,1) hyperbolic program by branch and bound*, Naval Res. Logist. Quart. **22**(1975)3, 497-515.

766. **Sakaguchi, Minoru:** *A maximization for Markov chains related to trafic flow problems*, Rep. Statist. Appl. Res. Un. Japan. Sci. Engrs. **22**(1975)3, 20-23.

767. **Sakawa, M.; Yano, H.:** *Interactive decision making for multiobjective linear fractional programming problems with fuzzy parameters*, Cybernet. Systems **16**(1985)4, 377-394.

768. **Sakawa, M.; Yano, H.:** *An interactive fuzzy satisficing method for multiobjective linear fractional programming problems with fuzzy parameters*, Preprint of VII-th International Conference on Multiple Criteria Decision Making-Toward Interactive and Intelligent Decision Support Systems, Kyoto, Japan, 18-22 Aug. 1986 (Kyoto, Japan), Japan Inst. Syst. Res. 1986, 723-731.

769. **Sakawa, M.; Yano, H.:** *An interactive fuzzy satisfying method for multiobjective linear fractional programming problems*, Fuzzy Sets and Systems **28**(1988)2, 129-144.

770. **Sakawa, M.; Yano, H.:** *Interactive decision making for multiobjective linear fractional programming problems with fuzzy parameters based on solution concepts incorporating fuzzy goals*, Japanese J. Fuzzy Theory Systems **3**(1991)1, 45-62.

771. **Sakawa, M.; Yumine, T.:** *Interactive fuzzy decision making for multiobjective linear fractional programming problems*, (Japanese), Systems and Control **27**(1983)2, 138-146. Also in : Large Scale Syst. **5**(1983), 105-114.

772. **Sakawa, M.; Yano, H.; Yumine, T.:** *An interactive fuzzy satisfying method for multiobjective linear fractional programming problems*, (Japanese), Trans. Inst. Electron. & Commun. Eng. Jpn. Part A (Japan) Vol. J69A, no.1, 1986, 32-41.

773. **Sakawa, M.; Yano, H.; Takahashi, J.:** *PARETO optimality for multiobjective linear fractional programming problems with fuzzy parameters*, (Japanese), Trans. Inst. Electron. Inf. Commun. Eng. A (Japan), Vol. J72A, no.6, 1989, 931-937.

774. **Sakawa, M.; Yano, H.; Takahashi, J.:** *Pareto optimality for multiobjective linear fractional programming problems with fuzzy parameters*, Inform. Sci. **63**(1992)1-2, 33-53.

775. **Samyrkanov, S.:** *Solution of a fractional quadratic programming problem*, (Russian), In: Some Mathematical Optimization Methods and Their Application in the Economy of Kirghizia (Russian), p. 55-56, 104-105, Izdat. "Ilim", Frunze, 1973.

776. **Sannomiya, N.; Nishikawa, Y.; Inazu, M.:** *A method for solving multi-objective programming problems with linear fractional functions*, (Japanese), Systems and Control **23**(1979)1, 61-62.

777. **Saxena, P.C.:** *Duality theorem for fractional functional programming in complex space*, Portugal. Math. Vol. **37**(1978), Fasc. 1-2, 87-92.

778. **Saxena, P.C.:** *A duality theorem for fractional functions programming in complex space*, Z. Angew. Math. Mech., (to appear).

779. **Saxena, P.C.; Aggarwal, S.P.:** *Parametric linear fractional functional programming*, Econom. Comput. Econom. Cybernet. Stud. Res. **14**(1980)1, 87-97.

780. **Saxena, P.C.; Parkash, Om; Patkar, V.N.:** *A note on duality for a pseudoconvex functional programming problem*, Nat. Acad. Sci. Letters (India), **2**(1979)6, 231-232.

781. **Saxena, P.C.; Patkar, V.N.:** *Linear fractional functional programming in complex space*, Portugal. Math., **37**(1978)1-2, 73-84.

782. **Saxena, P.C.; Patkar, V.N.:** *Non-linear non-differentiable fraction programming in complex space*, Cahiers Centre Études Rech. Opér. **20**(1978)2, 183-193.

783. **Saxena, P.C.; Patkar, V.N.; Parkash, Om:** *A note on an algorithm for integer solution to linear and piecewise linear programs*, Pure Appl. Math. Sci. **9**(1979)1-2, 31-36.

784. **Saxena, P.C.; Patkar, V.N.; Parkash, Om:** *Linear fractional functional programming in complex space*, Z. Angew. Math. Mech. **59**(1979)6, 276-278.

785. **Saxena, P.C.; Patkar, V.N.; Parkash, Om:** *Generalized linear and piecewise linear program with absolute-value functionals*, J. Univ. Bombay, Nos. 75-76 Vols. XLVIII-XLIX(1979-1980), 24-31.

786. **Scaruppe, L.:** *Die quotientenoptimierung*; Diplomarbeit, eingereicht an der Humboldt-Universität Berlin II Mathematisches Institut, 1967.

787. **Schaible, S.:** *Beitrage zur quasikonvexen programmierung*, Dissertation, Universitat Köln, Juli, 1971.

788. **Schaible, S.:** *Quasi-convex optimization in general real-linear spaces*, Z. Oper. Res. Ser. A-B, **16**(1972)5, 205-213.

789. **Schaible, S.:** *Transformation nichtlinearer quotientenprogramme in Konvexe programme*, Proc. Operat. Res. Z. Würzburg Wien, 351-361, 1972-1973.

790. **Schaible, S.:** *Fractional programming: Transformations, duality and algorithmic aspects*, Stanford University, Department of Operations Research, Technical Report, No.73-79, 1973.

791. **Schaible, S.:** *Maximization of quasi-concave quotients and products of finitely many functionals*, Cahiers Centre Études Rech. Opér. **10**(1974)1, 47-53.

792. **Schaible, S.:** *Nonlinear fractional programming*, Operations Research Verfahren. Vol. **19**(1974), 109-115.

793. **Schaible, S.:** *Parameter-free convex equivalent and dual programs of fractional programming problems*, Z. Oper. Res. Ser. A-B, **18**(1974)5, 187-196.

794. **Schaible, S.:** *A note on "Quadratic fractional functionals programming" by S.P. Aggarwal*, Cahiers Centre Études Rech. Oper. **17**(1975)1, 95-96.

795. **Schaible, S.:** *Marginal werte in der quotientenoprogrammierung*, Zeitschrift für Betriebswirtschaft **45**(1975), 649-659.

796. **Schaible, S.:** *Dual fractional programs*, Report TR-75-5, Stanford Univ. California, p.21, Febr. 1975.

797. **Schaible, S.:** *Duality in fractional programming: A unified approach*, Operations Res. **24**(1976), 452-461.

798. **Schaible, S.:** *Minimization of ratios*, J. Optim. Theory Appl. **19**(1976)2, 347-352.

799. **Schaible, S.:** *Fractional programming. I, Duality*, Management Sci. **22**(1976)8, 858-867.

800. **Schaible, S.:** *Fractional programming II, On Dinkelbach's algorithm*, Management Sci. **22**(1976)8, 868-873.

801. **Schaible, S.:** *A note on the sum of a linear and linear-fractional function*, Naval Res. Logist. Quart. **24**(1977)4, 691-693.

802. **Schaible, S.:** *Duality in ratio optimization (fractional programming)*, Operations Research Verfahren-Methods Oper. Research, **25**(1977), 131-132.

803. **Schaible, S.:** *Recent results in fractional programming*, Operations Research Verfahren vol. **23**(1977), 271-272.

804. **Schaible, S.:** *Analyse und Anwendungen von Quotientenprogrammen-Ein Beitrag zur Planun mit Hilfe des nichtlinearen Programmierung*, Mathematical Systems in Economics No.42 (Hain, Meisenheim), Königstein, vii + 259, 1978.

805. **Schaible, S.:** *Fractional programming-state of the art*, 10-th International Symposium on Mathematical Programming. Montreal, August 27-31, 1979.
806. **Schaible, S.:** *Generalized convex quadratic functions, A unified approach*, 10-th International Symposium on Mathematical Programming, Montreal, August 27-31, 1979.
807. **Schaible, S.:** *Fractional programming - state of the art*, Operational research'81. Proceedings of the ninth IFORS International Conference on Operational Research held in Hamburg, Germany. 20-24 July 1981, Edited by Jean-Pierre Brans. North-Holland Publishing Co., Amsterdam - New York, 1981, 479-493.
808. **Schaible, S.:** *A survey of fractional programming*, In: Schaible, S. and W.T. Ziemba, (eds.). Generalized Convexity in Optimization and Economics. Academic Press, New York, 1981, 417-440.
809. **Schaible, S.:** *Fractional programming : applications and algorithms*, European J. Oper. Res. **7**(1981)2, 111-120.
810. **Schaible, S.:** *Bibliography in fractional programming*, Z. Oper. Res. A-B **26**(1982)7, 211-241.
811. **Schaible, S.:** *Simultaneous optimization of absolute and relative terms*, Faculty of business administration and commerce, University of Alberta, Edmonton, January 1982.
812. **Schaible, S.:** *Fractional programming*, Z. Oper. Res. Ser. A-B **27**(1983)1, 39-54.
813. **Schaible, S.:** *Bicriteria quasiconcave programs*, Cahiers Centre Études Rech. Opér. **25**(1983)1-2, 93-101.
814. **Schaible, S.:** *Fractional programming with several ratios*, Methods Oper. Res. **49**(1985), 77-83.
815. **Schaible, S.:** *Multi-ratio fractional programming - a survey*, In: *Optimization, Parallel Processing and Applications*, Proceedings of the Oberwolfach Conference on Operations Research, February 16-21, 1987 and the Workshop on Advanced Computation Techniques, Parallel Processing and Optimization, Held at Karlsruhe, West Germany, February 22-25, 1987, Lect. Notes Econ. Math. Syst. **304** : Springer-Verlag, Berlin, West Germany, 1988, 57-66.
816. **Schaible, S.:** *Fractional programming-some recent developments*, J. Inform. Optim. Sci. **10**(1989)1, 1-14.
817. **Schaible, S.:** *Some recent results in fractional programming*, In: *Generalized Concavity for Economic Applications*, Piera Mazzoleni (ed.). Proceedings of the Workshop held in Pisa, April 2, 1992, 7-14.
818. **Schaible, S.; Ziemba, W.T. (eds.):** *Generalized Concavity in Optimization and Economics*, Academic Press, New York, 1981.
819. **Schaible, S.; Ziemba, W.T.:** *On the concavity of the sum of lognormals is lognormal approximation in portfolio theory*, Working Paper No.317, University of California, Los Angeles (1982). Also in: Z. Oper. Res. **29**(1986), 161-186.
820. **Schaible, S.; Ibaraki, T.:** *Fractional programming*, European J. Oper. Res., **12**(1983), 325-338.
821. **Schaible, S.; Lowe, T.:** *A note on a material control problem*, IIE Transactions, **15**(1983), 177-179.
822. **Schechter, M.:** *An extension of the Charnes-Cooper method in linear fractional programming*, J. Inform. Optim. Sci. **10**(1989)1, 97-104.
823. **Schroeder, R.G.:** *Linear programming solution to ratio games*, Operations Res. **18**(1970), 300-305.
824. **Scott, C.H.; Jefferson, T.R.:** *Fractional programming duality via geometric programming duality*, J. Austral. Math. Soc. Ser., **21**(1980)4, 398-401.
825. **Scott, C.H.; Jefferson, T.R.:** *Conjugate duality for fractional programs*, J. Math. Anal. Appl. **84**(1981)2, 381-389.
826. **Scott, C.H.; Jefferson, T.R.:** *Non-standard posynomial geometric programs*, Internat. J. Systems Sci. **18**(1987), 1467-1474.
827. **Scott, C.H.; Jefferson, T.R.:** *Conjugate duality in generalized fractional programming*, J. Optim. Theory Appl. **60**(1989)3, 475-483.
828. **Sen, R.; Chatterjee, S.:** *Page cuts for mixed integer interval linear fractional programming*, 1983 Proceedings of the International Conference on Systems, Man and Cybernetics (Cat. No. 83 CH 1962-0), Bombay and New Delhi, India, 29 Dec. 1983-7 Jan. 1984 (New York, USA: IEEE 1983), 813-816, Vol. 2.
829. **Sen, R.; Chatterjee, S.:** *On an algorithm for solving absolute value linear fractional programming with integer interval linear constraints*, Proceedings of the 1984 IEEE International Conference on Systems, Man and Cybernetics (Cat. No. 84 CH 2063-6), Halifax, NS, Canada, 10-12 Oct. 1984 (New York, USA: IEEE 1984) 158-162.
830. **Seshan, C.R.:** *An algorithm for ranking the extreme points for a linear fractional objective function*, J. Indian Inst. Sci. **62**(1980)6, 119-121.
831. **Seshan, C.R.:** *On duality in linear fractional programming*, Proc. Indian Acad. Sci. Sect. A. Math. Sci. **89**(1980)1, 35-42.
832. **Seshan, C.R.; Achary, K.K.:** *A branch and bound algorithm for a tranportation type problem with piecewise linear convex objective function*, Z. Angew. Math. Mech. **60**(1980), 303-307.

833. **Seshan, C.R.; Tikekar, V.G.:** *Algorithms for integer fractional programming*, J. Indian Inst. Sci. **62**(1980)2, 9-16.

834. **Shapley, L.S.:** *Stochastic games*, Proc. Nat. Acad. Sci. U.S.A. **39**(1953), 1095-1100.

835. **Sharma, I.C.:** *Generalized transportation technique in linear fractional programming*, Presented at the International Conference on Operational Research at Delhi University, August 10-12, 1967.

836. **Sharma, I.C.:** *Feasible direction approach to fractional programming problems*, Opsearch **4**(1967)2, 61-72.

837. **Sharma, I.C.:** *Transportation technique in non-linear fractional programming*, Trabajos de Estadistica **24**(1973)1-2, 131-139.

838. **Sharma, I.C.; Swarup, K.:** *On duality in linear fractional functionals programming*, Z. Oper. Res. Ser. A-B **16**(1972), 91-100.

839. **Sharma, J.K.:** *Programming with fractional non-linear objective function and transportation technique*, Rev. Roumaine Math. Pures Appl. **23**(1978)8, 1227-1234.

840. **Sharma, J.K.:** *Extension and special cases of transportation problem : a survey*, Indian J. Pure Appl. Math. **9**(1978)9, 928-940.

841. **Sharma, J.K.; Goswami, M.K.:** *Enumerative technique for fractional complementary programming problem*, Methods Oper. Res. **56**(1986), 93-101.

842. **Sharma, J.K.; Gupta, A.K.; Gupta, M.P.:** *Extension of simplex technique for solving fractional programming problem*, Indian J. Pure Appl. Math. **11**(1980)8, 961-968.

843. **Sharma, J.K.; Swarup, K.:** *Transportation fractional programming with respect to time*, Ricerca Operativa **7**(1978), 49-58.

844. **Sharma, M.M.; Sharma, J.K.; Agrawal, S.C.:** *A particular convex programming problem and tranportation techique*, Acta Cienc. Indica **3**(1977)4, 353-356.

845. **Shepilov, M.A.:** *Methods for solving fractional mathematical programming problems*, Cybernetics (USA) **16**(1980)1, 104-111. Translation of Kibernetica (USSR) **16**(1980)1, 93-98.

846. **Shi, Ying-Guang:** *A minimization problem in the mean norm using generalized rational functions*, Math. Numer. Sinica **8**(1986)2, 205-208.

847. **Shivpuri, Saroj; Chadha, S.S.:** *Multiparametric linear fractional functionals programming*, Cahiers Centre Études Rech. Opér, **20**(1978)1, 103-108.

848. **Shor, N.Z.; Solomon, D.I.:** *Decomposition methods in linear fractional programming*, (Russian), Chişinău, "Ştiinţa", 1989.

849. **Shukla, D.P.; Kanchan, P.K.:** *Sum of linear and linear fractional programming*, Acta Cienc. Indica, **4**(1978)2, 199-201.

850. **Shvartzman, A.P.:** *On an algorithm of the fractional-linear programming*, (Russian), Èkonom. i Mat. Metody **1**(1965)4, 558-566.

851. **Sideri, E.A.:** *A cutting plane algorithm for min-max fractional programming*, J. Inform. Optim. Sci. **10**(1989)1, 177-192.

852. **Sideri, E.A.:** *A modified KELLEY's cutting plane algorithm for some special nonconvex problem*, In : [229] pag. 121-142.

853. **Singh, C.:** *Optimality conditions in fractional programming*, J. Optim. Theory Appl. **33**(1981)2, 287-294.

854. **Singh, C.:** *Convex programming with set-inclusive constraints and its applications to generalized linear and fractional programming*, J. Optim. Theory Appl. **38**(1982)1, 33-42.

855. **Singh, C.:** *Optimality conditions for fractional minmax programming*, J. Math. Anal. Appl. **100**(1984)2, 409-415.

856. **Singh. C.:** *A class of multiple-criteria fractional programming problems*, J. Math. Anal. Appl. **115**(1986), 202-213.

857. **Singh, C.:** *Nondifferentiable fractional programming with Hanson-Mond classes of functions*, J. Optim. Theory Appl. **49**(1986)3, 431-447.

858. **Singh, C.:** *Generalized fractional programming with HANSON-MOND classes of functions*, J. Inform. Optim. Sci. **9**(1988)2, 219-230.

859. **Singh, C.; Dass, B.K. (eds.):** *Continuous-Time, Fractional and Multiobjective Programming*, Proceedings of the Conference at St. Lawrence University, Canton, New York, July 29-Aug. 1, 1986. Analytic Publishing Company, 1989. Also in: J. Inform. Optim. Sci. **10**(1989), no.1.

860. **Singh, C.; Hanson, M.A.:** *Saddle-point theory for nondifferentiable multiobjective fractional programming*, J. Inform. Optim. Sci. **7**(1986)1, 41-48.

861. **Singh, C.; Hanson, M.A.:** *Multiobjective fractional programming duality theory*, Naval Res. Logist. **38**(1991)6, 925-933.

862. **Singh, C.; Rueda, N.:** *Generalized fractional programming: Optimality and duality theory*, J. Optim. Theory Appl. **66**(1990)1, 149-159.
863. **Singh, C.; Suneja, S.K.; Rueda, N.G.:** *Preinvexity in multiobjective fractional programming*, J. Inform. Optim. Sci. **13**(1992)2, 293-302.
864. **Sinha, S.M.; Aylawadi, D.R.:** *Optimality conditions for a class of nondifferential fractional programming problem*, Indian J. Pure Appl. Math. **14**(1983)2, 167-174.
865. **Sinha, S.M.; Swarup, K.:** *Mathematical Programming : A survey*, J. Math. Sci. **2**(1967)2, 125-146.
866. **Sinha, S.M.; Vijay, Wadhwa:** *Programming with a special class of nonlinear functionals*, Unternehmernforschung **14**(1970), 215-219.
867. **Slowinski, R.:** *A multicriteria fuzzy linear programming method for water supply system development planning*, Fuzzy Sets and Systems **19**(1986), 217-237.
868. **Slowinski, R.:** *An interactive method for multiobjective linear programming with fuzzy parameters and its application to water supply planning*, In: *Optimization models using fuzzy sets and possibility theory*, Theory Decis. Libr., Ser. **B4**, 396-414, 1987.
869. **Slusarczyk, C.:** *Modification of the BITRAN-NOVAES method for soving problems with a fractionally linear objective function in the case of an unbounded set of feasible solutions*, (Polish) Przeglad Statyst. **28**(1981)1-2, 63-73, 1982.
870. **Slusarczyk, C.:** *On a certain property of linear-fractional functions and its application*, (Polish) Przeglad Statyst. **33**(1986)4, 403-413.
871. **Smith, J.D.:** *A mathematical investigation into the optimum ratio of plough and conveyor speeds in multi-plough bidirectional cutting*, Int. J. Rock. Mech. & Min. Sci. (G.B.) **9**(1972)6, 767-781.
872. **Smyreva, N.V.:** *Ein algorithmus für das problem der parametrischen gebrochen linearen programmierung*, Optimizacija Vyp. **14**(1976)31, 83-102 (russisch), 187.
873. **Sniedovich, M.:** *A new lock at fractional programming*, J. Optim. Theory Appl. **54**(1987)1, 113-120.
874. **Sniedovich, M.:** *Fractional programming revisited*, European J. Oper. Res. **33**(1988)3, 334-341.
875. **Sniedovich, M.:** *Analysis of a class of fractional programming problems*, Math. Programming Ser. A **43**(1989)3, 329-347.
876. **Sniedovich, M.; Shamsedin Vazirinejad:** *A solution strategy for a class of nonlinear knapsack problems*, Amer. J. Math. Management Sci. **10**(1990)1-2, 51-71.
877. **Sobel, Matthew J.:** *Maximal mean/standard deviation ratio in an undiscounted MDP*, Oper. Res. Lett. **4**(1985)4, 157-159.
878. **Sodini, C.:** *Equivalence and parametric analysis in linear fractional programming*, In: [229] pag. 143-154.
879. **Solomon, D.I.:** *On a certain procedure to transform mathematical programming problems having connected constraints and variables*, (Russian), *Mathematical modeling of economic systems*; Mat. Issled. no. **52**(1979), 199-205, 225-226.
880. **Solomon, D.I.:** *Generalized linear fractional programming problems*, (Russian), *Mathematical modeling of economic systems*, Mat. Issled. No. **52**(1979), 206-215, 226.
881. **Solomon, D.I.:** *On a method in linear-fractional programming with a block-diagonal matrix* (Russian), Izv. Akad. Nauk Mold. SSR, Ser. Fiz. Tekh. Mat. Nauk, **3**(1979), 68-70.
882. **Solomon, D.I.:** *An application of generalized gradient descent method for solving linear-fractional programming problems*, (Russian), Izv. Akad. Nauk Moldav. SSR, Ser. Fiz. Tekh. Mat. Nauk 1983, No. 1, 7-13.
883. **Solomon, D.I.:** *A problem of integer fractional-linear programming*, (Russian), *Mathematical models and methods of optimization of economic systems*, Mat. Issled. No. **72**(1983), 122-131.
884. **Solomon, D.I.:** *Generalization of the linear and linear-fractional transport problem*, (Russian), Izv. Akad. Nauk Mold. SSR, Ser. Fiz.-Tech. Mat. Nauk 1984, No.1, **13-18**(1984).
885. **Solomon, D.I.:** *The principle of decomposition by variables with application of the method of generalized gradient descent in the solution of fractional-linear programming problems*, (Russian), Mat. Issled., No.82 Mat. Modeli Met. Optim. (1985), 121-134, 156.
886. **Solomon, D.I.:** *An iterative algorithm for the solution of a generalized linear-fractional programming problem* (Russian), Mat. Issled. **87**(1986), 161-164.
887. **Solomon, D.I.:** *A parametric method for solution of problems of linear-fractional programming*, (Russian), Mat. Issled. 96, 124-134, 1987.
888. **Solomon, D.I.:** *Decomposition methods in linear-fractional programming*, (Russian), Mat. Issled. No.100 Sistemy Optim. Obrab. Dannykh (1988), 115-132, 152.
889. **Solomon, D.I.:** *Decomposition algorithms for solving generalized linear-fractional programming problems*, (Russian), Mat. Issled. No.100 Sistemy Optim. Obrab. Dannykh (1988), 133-141, 153.

890. **Solomon, D.I.; Stancu-Minasian, I.M.:** *Metode de descompunere în programarea fracţionară separabilă*, Econom. Comput. Econom. Cybernet. Stud. Res. (to appear).

891. **Soyster, A.L.; Lev, B.:** *An interpretation of fractional objectives in goal programming as related to papers by Awerbuch et al., and Hannan*, Management Sci. **24**(1978)14, 1546-1549.

892. **Stahl, J.:** *Two new methods for solution of hyperbolic programming*, (Hungarian), Publications of the Mathematical Institute of the Hungarian Academy of Science **9**, Series B, fasc. 4, 743-754 (1964).

893. **Stahl, J.:** *On the decomposition of fractional programming problems*, Szigma **15**(1982)4, 289-292.

894. **Stancu-Minasian, I.M.:** *On stochastic programming with multiple objective functions*, Proceedings of the 5-th Conference on Probability Theory, Braşov-România, 1974, Edit. Acad. RSR 1977, 429-436.

895. **Stancu-Minasian, I.M.:** *A three-dimensional transportation problem with a special structured objective function*, Bull. Math. Soc. Sci. Math. R. S. Roumanie, **18(66)**(1974)3-4, 385-397.

896. **Stancu-Minasian, I.M.:** *Asupra problemei lui Kataoka*, Stud. Cerc. Mat., **28**(1976)1, 95-111.

897. **Stancu-Minasian, I.M.:** *Asupra problemei cu risc minim multiplu I: cazul a două funcţii-obiectiv*, Stud. Cerc. Mat. **28**(1976)5, 617-623.

898. **Stancu-Minasian, I.M.:** *Asupra problemei de risc minim multiplu II: cazul a* r(r>2) *funcţii-obiectiv*, Stud. Cerc. Mat. **28**(1976)6, 723-734.

899. **Stancu-Minasian, I.M.:** *Criterii multiple în programarea stochastică*; Teză de doctorat, Centrul de Statistică Matematică, Bucureşti, 1976.

900. **Stancu-Minasian, I.M.:** *On a class of nonlinear fractional programming problems*, Rev. Roumaine Math. Pures Appl. **23**(1978)2, 285-290.

901. **Stancu-Minasian, I.M.:** *On the transportation problem with multiple objective functions*, Bull. Math. Soc. Sci. Math. Roumanie (N.S.) Tome **22(70)**(1978)3, 315-328.

902. **Stancu-Minasian, I.M.:** *On the multiple minimum risk problem*, Bull. Math. Soc. Sci. Math. Roumanie (N.S.) **23(71)**(1979)4, 427-437.

903. **Stancu-Minasian, I.M.:** *Applications of the fractional programming*, Econom. Comput. Econom. Cybernet. Stud. Res. **14**(1980)1, 69-86.

904. **Stancu-Minasian, I.M.:** *Asupra unei probleme de programare fracţionară*, Buletin Ştiinţific. Seria Tehnică-Matematică, Volumul IV, Institutul de învăţământ superior Sibiu, 1981, 37-42.

905. **Stancu-Minasian, I.M.:** *A survey of methods used for solving the problems of fractional programming, The linear case I, II*, Bull. Math. Soc. Sci. Math. Roumanie (N.S) **25(73)**(1981)3, 313-320, ibid. **25(73)**(1981)4, 415-430.

906. **Stancu-Minasian, I.M.:** *A survey of methods used for solving the linear fractional programming problems with several objective functions*, Operation Research Verfahren Vol. **40**(1981), 159-162.

907. **Stancu-Minasian, I.M.:** *Bibliography of fractional programming: 1960-1976*, Pure Appl. Math. Sci. **13**(1981)1-2, 35-69.

908. **Stancu-Minasian, I.M.:** *Fractional programming in complex space. The state of the art*, Rev. Roumaine Math. Pures Appl. **26**(1981)3, 481-491.

909. **Stancu-Minasian, I.M.:** *A second bibliography of fractional programming : 1977-1981*, Pure Appl. Math. Sci. **17**(1983)1-2, 87-102.

910. **Stancu-Minasian, I.M.:** *Stochastic Programming with Multiple Objective Functions*, Editura Academiei Române, Bucureşti, România and D. Reidel Publishing Company, Dordrecht, 1984.

911. **Stancu-Minasian, I.M.:** *A third bibliography of fractional programming*, Pure Appl. Math. Sci. **22**(1985)1-2, 109-122.

912. **Stancu-Minasian, I.M.:** *An overview of separable fractional programming problem*, Anal. Numér. Théor. Approx. **14**(1985)1, 91-96.

913. **Stancu-Minasian, I.M.:** *A fourth bibliography of fractional programming*, Optimization **23**(1992)1, 53-71.

914. **Stancu-Minasian, I.M.:** *Stochastic programming with multiple fractile criteria*, Rev. Roumaine Math. Pures Appl. **37**(1992)10, 939-944.

915. **Stancu-Minasian, I.M.:** *Methods of solving the fractional programming problems* (Romanian); Editura Academiei Române, Bucureşti, 1992.

916. **Stancu-Minasian, I.M.; Duca, I.D.; Nishida T.:** *Multiple objective linear fractional optimization in complex space*, Math. Japonica **35**(1990)1, 195-203.

917. **Stancu-Minasian, I.M.; Patkar, V.N.:** *A note on nonlinear fractional max-min problem*, Nat. Acad. Sci. Letters (India) **8**(1985)2, 39-41.

918. **Stancu-Minasian, I.M.; Teghem, J.Jr.:** *Duality for fractional transportation problem*, Cahiers Centre Études Rech. Opér. (to appear).

919. **Stancu-Minasian, I.M.; Ţigan, Şt.:** *The minimum-risk approach to special problems of mathematical programming, The distribution function of the optimal value*, Rev. Anal. Numér. Théor. Approx. **13**(1984)2, 175-187.

920. **Stancu-Minasian, I.M.; Ţigan, Şt.:** *The minimum-risk approach to max-min bilinear programming*, An. Ştiinţ. Univ. "Al.I. Cuza" Iaşi, Secţ. I-a Mat. (N.S.) **31**(1985)2, 205-209.

921. **Stancu-Minasian, I.M.; Ţigan, Şt.:** *The minimum-risk approach to the bottleneck transportation problem*, Itinerant Seminar on Functional Equations, Approximation and Convexity (Cluj-Napoca, 1985), 203-208, Preprint 85-6, Univ. "Babeş-Bolyai", Cluj-Napoca, 1985.

922. **Stancu-Minasian, I.M.; Ţigan, Şt.:** *The vectorial minimum-risk problem, In: Aproximation and optimization*, Proc. Colloq., Cluj-Napoca/Rom. 1984, 321-328, 1985.

923. **Stancu-Minasian, I.M.; Ţigan, Şt.:** *Criteriul riscului minim în programarea stohastică*, Lucrările Sesiunii ştiinţifice a Centrului de Calcul al Universităţii Bucureşti, 20-21 februarie 1987, 392-397, 1987.

924. **Stancu-Minasian, I.M.; Ţigan, Şt.:** *The stochastic linear-fractional max-min problem*, Itinerant Seminar on Functional Equations, Approximation and Convexity (Cluj-Napoca, 1987), 275-280, Preprint 87-6, Univ. "Babeş-Bolyai", Cluj-Napoca, 1987.

925. **Stancu-Minasian, I.M.; Ţigan, Şt.:** *Inexact mathematical programming*, Seminar on Optimization Theory, 99-116, Preprint 87-8, Univ. "Babeş-Bolyai", Cluj-Napoca, 1987.

926. **Stancu-Minasian, I.M.; Ţigan, Şt.:** *A stochastic approach to some linear fractional goal programming problems*, Kybernetika, **24**(1988)2, 139-149.

927. **Stancu-Minasian, I.M.; Ţigan, Şt.:** *Generalized pseudofractional max-min problems*, Itinerant Seminar on Functional Equations, Approximation and Convexity (Cluj-Napoca, 1988), 295-302, Preprint 88-6, Univ. "Babeş-Bolyai", Cluj-Napoca, 1988.

928. **Stancu-Minasian, I.M.; Ţigan, Şt.:** *On some fractional programming models occurring in minimum-risk problems*, In:[229], pag. 295-324.

929. **Stancu-Minasian, I.M.; Ţigan, Şt.:** *Multiobjective mathematical programming with inexact data*, R. Slowinski and J. Teghem (eds.) Stochastic versus Fuzzy Approach to Multiobjective Mathematical Programming under Uncertainty. 395-418, 1990, Kluwer Academic Publishers.

930. **Stancu-Minasian, I.M.; Ţigan, S.:** *Fractional programming under uncertainty*, IVth International Workshop on Generalized Convexity, Pécs/Ungary, August 31-September 2, 1992.

931. **Stancu-Minasian, I.M.; Ţigan, Şt.:** *On some methods for solving fractional programming problems with inexact data*, Stud. Cerc. Mat. **45**(1993)6, 517-532.

932. **Stancu-Minasian, I.M.; Ţigan, Şt.; Lozovanu, D.D.; Solomon, D.I.:** *On a parametrical method for solving bicriterion max-min fractional problems*, Buletinul Academiei de Ştiinţe a Republicii Moldova **3**(9)(1992), 56-61.

933. **Steuer, Ralph E.:** *Multiple criteria optimization; Theory, computation and application*, Wiley Series in Probability and Mathematical Statistics : Applied Probability and Statistics. New York : John Wiley & Sons, Inc. XXII, 546p, 1986.

934. **Storey, C.:** *Optimization using rational functions*, Operations Research Verfahren. Vol. **31**(1979), 613-617.

935. **Storoy, S.:** *Ranking of vertices in the linear fractional programming problem*, BIT **23**(1983)3, 403-405.

936. **Subrahmanyan, M.B.:** *A note on best constants in discrete inequalities*, Aequationes Math. **30**(1986), 208-211.

937. **Suneja, S.K.; Singh, C.; Kaul, R.N.:** *Optimality and duality in continuous-time nonlinear fractional programming*, J. Austral. Math. Soc. Ser. B34 (1992) Part 2, 229-244.

938. **Suneja, S.K.; Gupta, S.:** *Duality in multiple objective fractional programming problems involving non-convex functions*, Opsearch **27**(1990)4, 239-253.

939. **Suneja, S.K.; Bector, C.R.; Singh, C.:** *Duality in multiobjective fractional programming involving strongly and weakly invex and related functions*, Opsearch **28**(1991)3, 153-164.

940. **Suppe, C.:** *Hyperbolische optimierungsprobleme mit homogenenfuncktionen*, Ekonom.-Mat. Obzor **12**(1976)4, 438-443.

941. **Swarup, K.:** *Programming with quadratic fractional functional*, Opsearch **2**(1965)3-4, 23-30.

942. **Swarup, K.:** *Linear fractional functionals programming*, Operations Res. **13**(1965)6, 1029-1036.

943. **Swarup, K.:** *Some aspects of linear fractional functionals programming*, Austral. J. Statist. **7**(1965)3, 90-104.

944. **Swarup, K.:** *Fractional programming with non-linear constraints*, Z. Angew. Math. Mech. **46**(1966)7, 468-469.

945. **Swarup, K.:** *Programming with indefinite quadratic function with linear constraints*, Cahiers Centre Études Rech. Opér. **8**(1966)2, 132-136.

946. **Swarup, K.:** *Transportation technique of linear fractional functionals programming*, Journal of Royal Naval Scientific Service **21**(1966)5, 256-260.

947. **Swarup, K.:** *Computational technique for linear fractional program*, J. Math. Sci. **2**(1967)1, 17-20.

948. **Swarup, K.:** *Some properties of fractional programming*, Cahiers Centre Études Rech. Opér. **9**(1967)2, 82-86.

949. **Swarup, K.:** *Mathematical programming*, Ph. D. Thesis, Delhi University, Delhi, 1967.

950. **Swarup, K.:** *Some aspects of duality for linear fractional functionals programming*, Z. Angew. Math. Mech. **47**(1967)3, 204-205.

951. **Swarup, K.:** *Duality in fractional programming*, Unternehmensforshung **12**(1968)2, 106-112.

952. **Swarup, K.:** *Fractional programming*, Math. Student **36**(1968-1969), 58-62.

953. **Swarup, K.:** *Note on linear fractional functionals programming*, Metrika, **13**(1968)1, 72-77.

954. **Swarup, K.:** *Duality for transportation problem in fractional programming*, Cahiers Centre Études Rech. Opér. **10**(1968)1, 46-54.

955. **Swarup, K.:** *On varying all the parameters in a linear fractional functionals programming problem*, Metrika **13**(1968)2-3, 196-205.

956. **Swarup, K.:** *Upper bounds in linear fractional functionals programming*, Metrika **15**(1970), 81-85.

957. **Swarup, K.:** *Some aspects of fractional programming*, Mat. Vesnik, **9**(1972)24, 97-100.

958. **Swarup, K.:** *On duality in non-linear fractional programming problems*, Z. Angew. Math. Mech. **54**(1974), 734.

959. **Swarup, K.; Aggarwal, S.P.; Gupta, R.K.:** *Stochastic indefinite quadratic programming*, Z. Angew. Math. Mech. **52**(1972), 371-373.

960. **Swarup, K.; Mohinder Kaur Bedi:** *Convex simplex method and non-linear programming problems*, Indian J. Pure Appl. Math. **6**(1975)2, 190-209.

961. **Swarup, K.; Mohinder Kaur Bedi:** *Linear fractional functionals programming with absolute-value fractional functionals*, Cahiers Centre Études Rech. Opér., **18**(1976), 367-375.

962. **Swarup, K.; Das, C.:** *Various kinds of convexity and their relations with harmonic convexity*, Orissa Mathematical Society, 1973.

963. **Swarup, K.; Sharma, I.C.:** *Programming with linear fractional functionals in complex space*, Cahiers Centre Études Rech. Opér. **12**(1970)2, 103-109.

964. **Swarup, K.; Sharma, I.C.:** *A primal-like algorithm for zero-one integer fractional programming problem*, Trabajos de Estadistica **24**(1973)3, 123-136.

965. **Taha, H.A.:** *Hyperbolic programming with bivalent variables*, Tech. Rep. no. 71-7. Dept. of Ind. Eng. Univ. of Arkansas, Fayetteville, 1971.

966. **Taha, A.H.:** *An algorithm for zero-one fractional programming*, AIIE Trans. **7**(1975)1, 29-34.

967. **Tammer, Elisabeth-Charlotte:** *Dualitäts-und Aquivalenzbeziehungenzwischen stückweise Linearen Konvexen*, Hyperbolischen und Linearen Optimierungs Problemen, XVIII Intern. Wiss. Koll, der TH Illmenau, Heft 1, Vortragareihe Al. 43-47, 1973.

968. **Tammer, Elisabeth-Charlotte:** *Ein problem der parametrischen Hyperbolisschen Optimierung*, Ekonomico-matematika Laborator, II Celostatni Konference o Mathematickych Metodick ekonomii Zadov, 285-301, 1973.

969. **Tammer, Elisabeth-Charlotte:** *Dualitätstheorie für hyperbolische und strückweise lineare konvexe optimierungsprobleme*, Math. Operationsforschung u Statist. **5**(1974), 93-108.

970. **Tammer, Elisabeth-Charlotte:** *Parametrische Hyperboliche Optimierungsprobleme mit Pararmetern in der Zielfunktion*, Math. Operationsforsch. Statist. Ser. Optimization **8**(1977)2, 207-225.

971. **Tang, Huan Wen; Li, Guo Bin:** *KARMARKAR's algorithm and fractional linear programming*, (Chinese) J. Dalian Inst. Tech. 25 (1986), Special Issue on Mathematics, suppl. 79-83.

972. **Teterev, A.G.:** *A certain generalization of linear and fractional linear programming*, (Russian), Èkonom. i Mat. Metody 5(1969)3, 440-447.

973. **Teterev, A.G.:** *On generalization of linear and piecewise-linear programming*, Matekon **6**(1970)3, 246-259.

974. **Thanassoulis, E.:** *An adaptation of PASSEB for the solution of multi-objective linear fractional programming problems*, J. Oper. Res. Soc. **36**(1985)2, 155-161.

975. **Thang, Nguen Ngok:** *Generalization of Beale's method to a pseudoconvex function*, (Russian), Bull. Math. Soc. Sci. Math. Roumanie (N.S.) **21(69)**(1977), no.1-2, 67-81.

976. **Tobin, J.:** *Liquidity preference as behavior toward risk*, Rev. Econom. Stud. **26**(1958), 65-86.

977. **Tomlinson, R.E.:** *Numerical realization of attrition games*, Washington D.C. The George Washington University Logistics Research Project, January, 1963.

978. **Tran, Quoc Chien:** *Duality in vector optimization; Part. III. Vector partially quosiconcave programming and vector fractional programming*, Kybernetika **20**(1984)6, 458-472.

979. **Tran, Quoc Chien:** *Nondifferentiable and quasidifferentiable duality in vector optimization theory*, Kybernetika 21(1985)4, 298-312.

980. **Tran, Quoc Chien:** *Fenchel-Lagrange duality in vector fractional programming via abstract duality scheme*, Kybernetika 22(1986)4, 299-319.

981. **Trattner, E.; Zidăroiu, C.:** *Eficienţa economico-socială a investiţiilor în funcţie de costul creării noilor locuri de muncă*, Studii şi Cercetări de Calcul Economic şi Cibernetică Economică; nr.1, 81-85, 1969.

982. **Tung Chi Tung; Chan Gin Hor; Chew, Kim Li:** *Finding a minimal-ratio elementary path in a network*, Asia-Pacific J. Oper. Res. 4(1987)2, 151-157.

983. **Tzeng, L.Y.-R.; G.J.Y. Hsu:** *A new algorithm for solving multiobjective linear fractional programming: The CONNISE Method*, Proceedings of the tenth International Conference on Multiple Criteria Decision Making (in Taipei'92) Vol.3, July 19-24, 1992. 459-470.

984. **Trzeciak, J.M.; Sposito, V.A.:** *Some duality results in goal programming with fractional objectives*, Bull. Oper. Res. Soc. Am. (USA), 23, suppl.3, p.B 34 (fall 1975) (ORSA) TIMS National Meeting Las Vegas, USA, Nov. 17-19, 1975.

985. **Ţigan, Eugenia:** *Algorithms to minimize the cost capacity ratio*, Econom. Comput. Econom. Cybernet. Stud. Res. 24(1989)4, 53-58.

986. **Ţigan, Şt.:** *Sur une méthode de décomposition pour le problème de programmation monotone*, Mathematica (Cluj), 13(36)(1971)2, 347-354.

987. **Ţigan, Şt.:** *Sur quelques problèmes d'affection*, Direction Scientifique. Note de Travail no.157, SEMA, Paris, Août 1971.

988. **Ţigan, Şt.:** *Sur quelques problèmes d'affection a applications économiques*, Die Referate der 13. Internationale Hochschulwocke 1972. Der Südosteuropa - Gesellschaft October 1972. Elektrönische Datenverarbeitung in der wissenschft, Wirtschaft und verwaltung 107-123, München, 1972.

989. **Ţigan, Şt.:** *Asupra unei probleme de programare neliniară fracţionară*, Rev. Anal. Numér. Théor. Aprox. 1(1972)2, 215-226.

990. **Ţigan, Şt.:** *On a method for fractional optimization problems, Application to stochastic optimization problems*, Proc. of the Computer Science Conference, Székesfehérvar, Hungary, 351-355, 1973.

991. **Ţigan, Şt.:** *Sur une méthode pour la résolution d'un probléme d'optimization fractionaire par segments*, Rev. Anal. Numér. Théor. Approx. 4(1975)1, 87-97.

992. **Ţigan, Şt.:** *Sur le probleme de la programmation vectorielle fractionaire*, Rev. Anal. Numér. Théor. Approx. 4(1975), 99-103.

993. **Ţigan, Şt.:** *Un algoritm pentru o problemă de programare dinamică fracţionară*, Al 2-lea Simpozion de Informatică şi Conducere, Cluj-Napoca, 20-23 aprilie 1976, 12-14, 1977.

994. **Ţigan, Şt.:** *Algoritm in cascadă pentru rezolvarea unei probleme speciale de programare fracţionară*, In: *Probleme actuale de informatică şi conducere*, Comunicările selective ale celui de al III-lea simpozion de informatică, 11-18 mai 1977. Cluj-Napoca şi Satu Mare, Editura Dacia, Cluj-Napoca, 310-317, 1977.

995. **Ţigan, Şt.:** *On the max-min nonlinear fractional problem*, Rev. Anal. Numér. Théor. Approx. 9(1980)2, 283-288.

996. **Ţigan, Şt.:** *Asupra problemei de maximin fracţionare*, Seminarul itinerant de ecuaţii funcţionale, aproximare şi convexitate, Cluj-Napoca, mai 1980, 139-143.

997. **Ţigan, Şt.:** *Remarque sur certain problems de programation pseudolineare par morceaux*, Rev. Anal. Numér. Théor. Appox. 9(1980)1, 129-132.

998. **Ţigan, St.:** *Asupra unor probleme de programare pseudofracţionară*, Al VII-lea Simpozion Informatică şi Conducere, 20-23 mai, 1981, Cluj-Napoca.

999. **Ţigan, Şt.:** *Asupra unor metode de rezolvare a unor probleme particulare de programare fracţionară*, In: Informatica pentru conducere, Orizont'81, Realizări şi aplicaţii, Cluj-Napoca, 92-93, 1981.

1000 **Ţigan, St.:** *Eficienţă şi propriu-eficienţă pentru programarea fracţionară vectorială*, Seminarul itinerant de ecuaţii funcţionale, aproximare şi convexitate, Cluj-Napoca, 1982, 349-358.

1001 **Ţigan, Şt.:** *Sur quelques problémes de programmation pseudo-fractionnaire*, Rev. Anal. Numér. Théor. Approx. 11(1982)1-2, 167-174.

1002 **Ţigan, Şt.:** *Problema fracţionară a arborelui minim*, Buletinul Român de Informatică, 1(1982), 9-15.

1003 **Ţigan, St.:** *A parametrical method for max-min nonlinear fractional problems*, Seminarul itinerant de ecuaţii funcţionale, aproximare şi convexitate, Cluj-Napoca, 19-21 mai 1983, 175-184.

1004 **Ţigan, Şt.:** *Numerical methods for solving some max-min pseudofractional problems*, Proceedings of the Second Symposium of Mathematics and its Applications (Timişoara 30-31 October, 1987), 93-97, Res. Centre, Acad. S.R. Roumanie, Timişoara, 1988.

1005 **Ţigan, Şt.**: *On some procedures for solving fractional max-min problems*, Math., Rev. Anal. Numér. Théor. Approximation, Anal. Numér. Théor. Approximation. **17**(1988)1, 73-91.

1006 **Ţigan, Şt.**: *Asupra unor probleme fracţionare de maximin*, Studii şi Cercetări de Calcul Economic şi Cibernetică Economică. **27**(1992)1-2, 53-57.

1007 **Ţigan, St.; Stancu-Minasian, I.M.**: *Criteriul riscului minim pentru problema Cebîşev*, Lucrările celui de al IV-lea Simpozion "Modelarea cibernetică a proceselor de producţie" 26-28 mai 1983, ASE-Bucureşti, Vol.I, 338-342.

1008 **Ţigan, St.; Stancu-Minasian, I.M.**: *The stochastic max-min problem*, Cahiers Centre Études Rech. Opér. **27**(1985)3-4, 247-254.

1009 **Ţigan, Şt.; Stancu-Minasian, I.M.**: *Fractional goal programming with inexact data*, Itinerant Seminar on Functional Equations, Approximation and Convexity. (Cluj-Napoca 1989) 311-318, Preprint 89-6, Univ. "Babeş-Bolyai", Cluj-Napoca, 1989.

1010 **Ţigan, Şt.; Stancu-Minasian, I.M.**: *An application of Warburton procedure to a max-min fractional programming problem*, Itinerant Seminar on Functional Equations Approximation and Convexity. 83-92, Preprint 90-6, Univ. "Babes-Bolyai", Cluj-Napoca, 1990.

1011 **Ţigan, Şt.; Stancu-Minasian, I.M.**: *On a bicriterion max-min fractional problem*, Anal. Numér. Théor. Approx. **20**(1991)1-2, 117-125.

1012 **Ţigan, Şt.; Stancu-Minasian, I.M.**: *On a variant of bicriterion max-min fractional problem*, Multicriteria Decision Making. Methods, Algorithms, Applications. Proceedings of the International Workshop held at Liblice Czechoslovakia, March 18-22, 1991, Martin Černy, Dagman Glückaufová and Dimitrij Loula (eds.). The Institute of Economics, Czechoslovak Academy of Sciences, Prague, 1992, 152-161.

1013 **Varma, G.K.**: *Some aspects of parametric linear fractional programming*, J. Indian Statist. Assoc. **7**(1969), 162-167.

1014 **Varma, G.K.**: *General parametric linear fractional programming*, Metrika **19**(1972)1, 11-17.

1015 **Varma, G.K.**: *On parametric linear fractionals functional programming*, Trabajos Estadist. Investigación Oper. **23**(1972)1-2, 149-157.

1016 **Vartak, M.N.; Gupta, I.**: *Duality theory for fractional programming problems under* η *- convexity*, Opsearch **24**(1987)3, 163-174.

1017 **Verdaguer, Rolando; Iglesias, Leonardo**: *A modified fractional simplex method*, (Spanish), Investigación Oper. **11**(1990)1, 3-10.

1018 **Verma, V.**: *Constrained integer linear fractional programming problem*, Optimization **21**(1990)5, 749-757.

1019 **Verma, V.; Khanna, S.; Puri, M.C.**: *"Bad-points" in linear fractional program : A comparative study*, Cahiers Centre Études Rech. Opér. **29**(1987)1-2, 123-131.

1020 **Verma, V.; Khanna, S.; Puri, M.C.**: *On Martos' and Charnes-Cooper's approach vis-à-vis "singular points"*, Optimization **20**(1989)4, 415-420.

1021 **Verma, V.; Puri, M.C.; Arora, S.R.**: *Some special situations in linear fractional programming problem: an algorithmic comparison*, Opsearch **26**(1989)2, 96-107.

1022 **Verma, V.; Puri, M.C.**: *On Wolf's method for solving linear fractional programming problem*, Opsearch **27**(1990)3, 176-179.

1023 **Verma, V.; Bakhshi, H.C.; Puri, M.C.**: *Ranking in integer linear fractional programming problems*, Z. Oper. Res. **34**(1990)5, 325-334.

1024 **Vial, Jean-Philippe**: *A unified approach to projective algorithms for linear programming*, Optimization, Proc. 5th French-German Conf., Varetz/Fr. 1988, Lecture Notes Math. 1405, 191-220, 1989.

1025 **Vijay, Wadhwa**: *Programming with separable fractional functionals*, J. Math. Sci. **4**(1969)2, 51-60.

1026 **Vijay, Wadhwa**: *Linear fractional programs with variable coefficients*, Cahiers Centre Études Rech. Opér. **14**(1972)4, 223-232.

1027 **Vijay, Wadhwa**: *Parametric linear fractional programming*, SCIMA (J. Management Sci. Appl. Cybernetics) **3**(1974)2, 21-29.

1028 **Vlach, M.**: *Rubinstein duality scheme for vector optimization*, In: [229], pag. 252-264.

1029 **Wagner, H.; Yuan. J.S.C.**: *Algorithmic equivalence in linear fractional programming*, Tech. Rep. 17, Graduate School of Business, Stanford Univ., Stanford 1966. Also in: Management Sci. **14**(1968), 301-306.

1030 **Wang, Chung-Lie**: *The principle and models of dynamic programming, I-III*. J. Math. Anal. Appl. **118**(1986)2, 287-308, ibid. **135**(1988)1, 268-283, 284-296.

1031 **Wang, Chung-Lie; Yanan Wu**: *Optimal control problems with nonstandard cost functionals*, Proceedings of the Twentieth Manitoba Conference on Numerical Mathematics and Computing (Winnipeg, MB, 1990). Congr. Numer. **80**(1991), 129-137.

1032 **Warburton, A.R.:** *Topics in multicriteria optimization*, Ph. D. Thesis, University of British Columbia, Vancouver, 1981.

1033 **Warburton, A.R.:** *Parametric solution of bicriterion linear fractional programs*, Oper. Res. **33**(1985)1, 74-84.

1034 **Warren, D.; Thomas, J.B.:** *Signal-to-noise ratio and central limit theorem considerations in non-Gaussian defection*, Proceedings of the Twenty-third Annual Allerton Conference on Communication, Control and Computing, Monticello, IL, USA, 2-4 Oct. 1985 (Urbana-Champaign, IL, USA; Univ. Illinois 1985) 35-44.

1035 **Wdowiak, J.:** *Discrete hiperbolic programming problem I*, (Polish), Przeglad Statyst. **24**(1977)4, 483-497.

1036 **Wdowiak, J.:** *A discrete linear fractional programming problem II*, (Polish), Przeglad Statyst. **25**(1978), 133-140.

1037 **Weber, R.:** *Pseudomonotonic multiobjective programming*, Cahiers Centre Études Rech. Opér. **25**(1982)1-2, 115-128, 1982.

1038 **Weir, T.:** *A note on duality for fractional programming problems*, Opsearch **22**(1985)4, 241-247.

1039 **Weir, T.:** *A dual for a multiple objective fractional programming problem*, J. Inform. Optim. Sci. **7**(1986)3, 261-269.

1040 **Weir, T.:** *A duality theorem for a multiple objective fractional optimization problem*, Bull. Austral. Math. Soc. **34**(1986), 415-425.

1041 **Weir, T.:** *A note on invex functions and duality in generalized fractional programming*, Report No.4, Department of Mathematics, Australian Defence Force Academy, 1986.

1042 **Weir, T.:** *On duality in multiobjective fractional programming*, Opsearch **26**(1989)3, 151-158.

1043 **Weir, T.:** *Duality for nondifferentiable multiple objective fractional programming problems*, Utilitas Math. **36**(1989), 53-64.

1044 **Weir, T.:** *On strong pseudoconvexity in nonlinear programming duality*, Opsearch **27**(1990)2, 117-121.

1045 **Weir, T.:** *Symmetric dual multiobjective fractional programming*, J. Austral. Math. Soc. Ser. **A50**(1991)1, 67-74.

1046 **Weir, T.; Mond, B.:** *Duality for fractional programming without a constraint qualification*, Utilitas Math. **38**(1990), 193-197.

1047 **Werner, Jochen:** *Duality in generalized fractional programming*, Trends in mathematical optimization, 4th French-German Conf., Irsee/FRG 1986, 341-351, Internat. Schriftenreihe Numer. Math., 84, Birkhäuser, Basel-Boston, MA, 1988.

1048 **Williams, H.P.:** *Experiments in the formulation of integer programming problems*, Math. Programming Stud., vol. **II**(1974), 180-197.

1049 **Wolf, H.:** *A new solution approach to the linear fractional programming problem*, Diskussionsbeiträge des Fachbereichs Wirtschaftswissenschaft der Fernuniversität Hagen, Nr.67, 1983.

1050 **Wolf, H.:** *Die parametrische Analyse eines linearen Quotientenprogramms mit einem Skalarparameter in der rechten Seite*, Diskussionsbeiträge des Fachbeireichs Wirtschaftswissenschaft der Fernuniversität Hagen, No.74, 1983.

1051 **Wolf, H.:** *A parametric method for solving the linear fractional programming problem*, Oper. Res. **33**(1985)4, 835-841.

1052 **Wolf, H.:** *Solving special nonlinear fractional programming problems via parametric linear programming*, European J. Oper. Res. **23**(1986)3, 396-400.

1053 **Wolf, H.:** *Parametric analysis in linear fractional programming*, Oper. Res. **34**(1986)6, 930-937.

1054 **Wolf, H.:** *Parametric analysis for fractional programs*, Methods Oper. Res. **53**(1986), 215-222.

1055 **Wolfe, O.B.; Hawaleshka, O.; Mohamed, A.M.:** *User friendly micro computer program for solving fractional and linear programming problems*, Comput. & Ind. Eng. (G.B.), Vol. 11, no. 1-4, 225-231, 1986 (Proceedings of the 8th Annual Conference on Computers and Industrial Engineering, Orlando, FL, USA, 19-221 March, 1986).

1056 **Xiaolin Wu; Kaizhou Chen:** *Optimality conditions and duality for a class of minmax nondifferentiable fractional programming problems*, ICIAM 91 - Second International Conference on Industrial and Applied Mathematics, July 8-12, 1991. Sheraton Washington Hotel, Washington, DC, USA.

1057 **Xu, Z.K.:** *A method for a fractional programming problem*, Journal of Shanghai University of Technology, **1**(1983), 66-72.

1058 **Xu, Z.K.:** *Saddle-point type optimality criteria for generalized fractional programming*, J. Optim. Theory Appl. **57**(1988)1, 189-196.

1059 **Xu, Z.K.:** *On inexact fractional programming*, (Chinese), J. Systems Sci. Math. Sci. **10**(1990)4, 377-382.

1060 **Xu, Z.K.:** *Duality in generalized nonlinear fractional programming*, J. Math. Anal. Appl. **169**(1992)1, 1-9.

1061 **Yadav, Shri Ram; Mukherjee, R.N.:** *A generalized Big-M method for fractional programming and some problems on parametric fractional programming*, Bull. Tech. Univ. Istanbul **37**(1984)4, 465-475.

1062 **Yadav, Shri Ram; Mukherjee, R.N.:** *On parametric fractional programming*, Bull. Tech. Univ. Istanbul **37**(1984)3, 273-282.

1063 **Yadav, Shri Ram; Mukherjee, R.N.:** *Duality for fractional minmax programming problems*, J. Austral. Math. Soc., Ser. **B31**(1990)4, 484-492.

1064 **Yadav, Shri Ram; Shiv Prasad; Mukherjee, R.N.:** *A dual differentiable exact penalty function in fractional programming*, Indian J. Pure Appl. Math. **19**(1988)6, 513-515.

1065 **Yamada, K.:** *On linear fractional programming*, (Japanese), The Hitotsubashi Review **49**(1963).

1066 **Yano, H.; Sakawa, M.:** *Interactive fuzzy decision making for generalized multiobjective linear fractional programming problems with fuzzy parameters*, Fuzzy Sets and Systems **32**(1989)3, 245-262.

1067 **Yon, Ryang So:** *A solution method of the fractional n-index transportation problem* (Korean), Suhak **3**(1986), 1-8.

1068 **Yon, Ryang So:** *A solution method of the fractional n-index transportation problem with pass capacity (Korean)*, Suhak, **1**(1987), 21-28.

1069 **Yon, Ryang So:** *A solution method for linear fractional functional programming with a parameter objective function (Chinese)*, Suhak 1987, No.2, 20-24, 1987.

1070 **Zajiček, F.; Daucha T.; Foltijn, I.:** *Linearni a linearni a lomene programovani*, În : Kundrat, J. Matematicke metody a programy provypocy systemîi modelu strednedoheho a provadeciho planu v zemedelstvi, VUEZVz, Praha, 1974, p. 2-1-2-95.

1071 **Zhak, S.V.; Melkadze, A.G.:** *On inter-repairable covers optimization*, Èkonom. i Mat. Metody **7**(1971)5, 765-769.

1072 **Zalmai, G.J.:** *Duality for a class of continuous-time homogeneous fractional programming problems*, Z. Oper. Res. Ser. A-B, **30**(1986)1, 43-48.

1073 **Zalmai, G.J.:** *Optimality conditions for a class of nondifferentiable minmax programming problems*, Optimization **17**(1986)4, 453-465.

1074 **Zalmai, G.:** *Duality for a class of continuous-time fractional programming problems*, Utilitas Math. **31**(1987), 209-218.

1075 **Zalmai, G.J.:** *Optimality conditions and subgradient duality for minmax programming problems with applications*, Utilitas Math. **34**(1988), 193-222.

1076 **Zalmai, G.J.:** *Optimality conditions and duality for constrained measurable subset selection problems with minmax objective functions*, Optimization **20**(1989)4, 377-395.

1077 **Zalmai, G.J.:** *Duality for generalized fractional programs involving n-set functions*, J. Math. Anal. Appl. **149**(1990)2, 339-350.

1078 **Zalmai, G.J.:** *Optimality conditions and duality for a class of continuous-time generalized fractional programming problems*, J. Math. Anal. Appl. **153**(1990)2, 356-371.

1079 **Zemel, E.:** *On search over rationals*, Oper. Res. Lett. **1**(1981/1982)1, 34-38.

1080 **Zheng, Quan:** *Theory and methods for global optimization-an integral approach, In: Advances in optimization and control*, Proc. Conf., Optimization Days, Montreal/Can. 1986. Lect. Notes Econ. Math. Syst. 302, 15-37, 1988.

1081 **Ziemba, W.T.; Brooks-Hill, F.J.; Parkan, C.:** *Calculating investment porfolios when the returns are normally distributed*, University of British Columbia, Working Paper, Vancouver, British Columbia, Canada, 1971.

1082 **Ziemba, W.T.; Parkan, C.; Brooks-Hill, R.:** *Calculation of investment portfolios with risk free borrowing and lending*, Management Sci. **21**(1974)2, 209-222.

1083 **Zietak, Krystyna:** *On the optimum rational connected with the ADI-method*, Zastos Mat. **11**(1969-1970), 337-352.

1084 **Zionts, S.:** *Programming with linear fractional functionals*; Naval Res. Logist. Quart. **15**(1968)3, 449-451.

1085 **Żolkiewski, Z.:** *Multiobjective linear fractional programming problem* (Polish), Przeglad Statyst. **31**(1984), 359-373.

1086 **Żolkiewski, Z.:** *On computing L_∞ - compromise solutions of the multiple objective linear fractional programming (MOLFP) problem*, Ekonom.-Mat. Obzor **20**(1984)2, 197-202.

1087 **Żolkiewski, Z.:** *A multicriteria linear programming model with linear fractional objective functions*, (Polish), Ph. D. Thesis. The Central School of Planning and Statistics, Warsaw, 1983.

1088 **Zsigmond, Ildikó:** *Mixed integer linear fractional programming by a branch and bound technique*, Ann. Univ. Sci. Budapest, Rolando Eötvös, Sect. Comput. 7 (1987), 117-130 (1988).

1089 **Žusupbaev, A.:** ***The production allocation problem with a fractional linear of fractional convex functional and with unknown quantitites of production and consumption*****,** **(Russian), In: Some Mathematical Optimization Methods and Their Application in the Economy of Kirghizia (Russian) 19-29, 103, Izd. "Ilim", Frunze, 1973.**

AUTHOR INDEX

SUBJECT INDEX

Other *Mathematics and Its Applications* titles of interest:

B.S. Razumikhin: *Physical Models and Equilibrium Methods in Programming and Economics.* 1984, 368 pp. ISBN 90-277-1644-7

N.K. Bose (ed.): *Multidimensional Systems Theory. Progress, Directions and Open Problems in Multidimensional Systems.* 1985, 280 pp. ISBN 90-277-1764-8

J. Szep and F. Forgo: *Introduction to the Theory of Games.* 1985, 412 pp. ISBN 90-277-1404-5

V. Komkov: *Variational Principles of Continuum Mechanics with Engineering Applications.* Volume 1: Critical Points Theory. 1986, 398 pp. ISBN 90-277-2157-2

V. Barbu and Th. Precupanu: *Convexity and Optimization in Banach Spaces.* 1986, 416 pp. ISBN 90-277-1761-3

M. Fliess and M. Hazewinkel (eds.): *Algebraic and Geometric Methods in Nonlinear Control Theory.* 1986, 658 pp. ISBN 90-277-2286-2

P.J.M. van Laarhoven and E.H.L. Aarts: *Simulated Annealing: Theory and Applications.* 1987, 198 pp. ISBN 90-277-2513-6

B.S. Razumikhin: *Classical Principles and Optimization Problems.* 1987, 528 pp. ISBN 90-277-2605-1

S. Rolewicz: *Functional Analysis and Control Theory. Linear Systems.* 1987, 544 pp. ISBN 90-277-2186-6

V. Komkov: *Variational Principles of Continuum Mechanics with Engineering Applications.* Volume 2: Introduction to Optimal Design Theory. 1988, 288 pp. ISBN 90-277-2639-6

A.A. Pervozvanskii and V.G. Gaitsgori: *Theory of Suboptimal Decisions.* Decomposition and Aggregation. 1988, 404 pp. *out of print,* ISBN 90-277-2401-6

J. Mockus: *Bayesian Approach to Global Optimization.* Theory and Applications. 1989, 272 pp. ISBN 0-7923-0115-3

Du Dingzhu and Hu Guoding (eds.): *Combinatorics, Computing and Complexity.* 1989, 248 pp. ISBN 0-7923-0308-3

M. Iri and K. Tanabe: *Mathematical Programming.* Recent Developments and Applications. 1989, 392 pp. ISBN 0-7923-0490-X

A.T. Fomenko: *Variational Principles in Topology.* Multidimensional Minimal Surface Theory. 1990, 388 pp. ISBN 0-7923-0230-3

A.G. Butkovskiy and Yu.I. Samoilenko: *Control of Quantum-Mechanical Processes and Systems.* 1990, 246 pp. ISBN 0-7923-0689-9

A.V. Gheorghe: *Decision Processes in Dynamic Probabilistic Systems.* 1990, 372 pp. ISBN 0-7923-0544-2

Other *Mathematics and Its Applications* titles of interest:

A.G. Butkovskiy: *Phase Portraits of Control Dynamical Systems.* 1991, 180 pp.
ISBN 0-7923-1057-8

A.A. Zhigljavsky: *Theory of the Global Random Search.* 1991, 360 pp.
ISBN 0-7923-1122-1

G. Ruhe: *Algorithmic Aspects of Flows in Networks.* 1991, 220 pp.
ISBN 0-7923-1151-5

S. Walukuwiecz: *Integer Programming.* 1991, 196 pp. ISBN 0-7923-0726-7

M. Kisielewicz: *Differential Inclusions and Optimal Control.* 1991, 320 pp.
ISBN 0-7923-0675-9

J. Klamka: *Controllability of Dynamical Systems.* 1991, 260 pp.
ISBN 0-7923-0822-0

V.N. Fomin: *Discrete Linear Control Systems.* 1991, 302 pp. ISBN 0-7923-1248-1

L. Xiao-Xin: *Absolute Stability of Nonlinear Control Systems.* 1992, 180 pp.
ISBN 0-7923-1988-5

A. Halanay and V. Rasvan: *Applications of Liapunov Methods in Stability.* 1993, 238 pp. ISBN 0-7923-2120-0

D. den Hertog: *Interior Point Approach to Linear, Quadratic and Convex Programming.* 1994, 208 pp. ISBN 0-7923-2734-9

V.S. Tanaev, V.S. Gordon and Y.M. Shafranksy: *Scheduling Theory.* Single-Stage Systems. 1994, 380 pp. ISBN 0-7923-2853-1

V.S. Tanaev, Y.N. Sotskov and V.A. Strusevich: *Scheduling Theory.* Multi-Stage Systems. 1994, 412 pp. ISBN 0-7923-2854-X

L.D. Akulenko: *Problems and Methods of Optimal Control.* 1994, 356 pp.
ISBN 0-7923-2855-8

C. Udrişte: *Convex Functions and Optimization Methods on Riemannian Manifolds.* 1994, 348 pp. ISBN 0-7923-3002-1

B.S. Jensen: *The Dynamic Systems of Basic Economic Growth Models.* 1994, 355 pp. ISBN 0-7923-3091-9

V. Barbu: *Mathematical Methods in Optimization of Differential Systems.* 1994, 259 pp. ISBN 0-7923-3176-1

A. Prékopa: *Stochastic Programming.* 1995, 556 pp. ISBN 0-7923-3482-5

R. Lucchetti and J. Revalski (eds.): *Recent Developments in Well-Posed Variational Problems.* 1995, 266 pp. ISBN 0-7923-3576-7

A. Cheremensky and V. Fomin: *Operator Approach to Linear Control Systems.* 1995, 410 pp. ISBN 0-7923-3765-4

Other *Mathematics and Its Applications* titles of interest:

V.N. Afanas'ev, V.B. Kolmanovskii and V.R. Nosov: *Mathematical Theory of Control Systems Design*. 1996, 681 pp. ISBN 0-7923-3724-7

D. Pallaschke and S. Rolewicz: *Foundations of Mathematical Optimization.* Convex Analysis without Linearity. 1997, 594 pp. ISBN 0-7923-4424-3

A. Chikrii: *Conflict-Controlled Processes*. 1997, 424 pp. ISBN 0-7923-4522-3

I.M. Stancu-Minasian: *Fractional Programming*. Theory, Methods and Applications. 1997, 426 pp. ISBN 0-7923-4580-0

MIX
Papier aus verantwortungsvollen Quellen
Paper from responsible sources
FSC® C105338

If you have any concerns about our products, you can contact us on
ProductSafety@springernature.com

In case Publisher is established outside the EU, the EU authorized representative is:
Springer Nature Customer Service Center GmbH
Europaplatz 3, 69115 Heidelberg, Germany

Printed by Libri Plureos GmbH
in Hamburg, Germany